UNIVERSITY PHYSICS

Volume I

HARCOURT BRACE JOVANOVICH COLLEGE OUTLINE SERIES

UNIVERSITY PHYSICS

Volume I

Kenneth E. Jesse

Department of Physics
Illinois State University
Normal, Illinois

Books for Professionals
Harcourt Brace Jovanovich, Publishers
San Diego   New York   London

Requests for permission to make copies of any part of the work should be mailed to:

>   Permissions
>   Harcourt Brace Jovanovich, Publishers
>   Orlando, Florida 32887

Printed in the United States of America

Library of Congress Cataloging-in-Publication Data

Jesse, Kenneth E.
  University physics.

  (Harcourt Brace Jovanovich college outline series)
  (Books for professionals)
  Includes index.
  1. Physics. I. Title II. Series. III. Series:
Books for professionals.
QC23.J47 1987   530   87-19633
ISBN 0-15-601668-0 (v. 1)

First edition

A   B   C   D   E

PREFACE

Do not *read* **this Outline** — *use* **it.** You can't learn physics simply by reading about it: You have to *do* it. Solving specific, practical problems is the best way to master — and to demonstrate your mastery of — the theories, laws, and definitions upon which the science of physics is based. Outside the laboratory, you need three tools to do physics: a pencil, paper, and a calculator. Add a fourth tool, this Outline, and you're all set.

This HBJ College Outline has been designed as a tool to help you sharpen your problem-solving skills in physics. Each chapter covers a single topic, whose fundamental principles are broken down in outline form for easy reference. The outline text is filled with worked-out examples, so you can see immediately how each new idea is applied in problem form. Each chapter also contains a Summary of all the principal equations presented and a Raise Your Grades section. Taken together, these two features give you an opportunity to review the primary principles of a topic and the problem-solving techniques implicit in those principles.

Most important, this Outline gives you plenty of problems to practice on. Work the Solved Problems, and check yourself against the step-by-step solutions provided. Test your mastery of the material in each chapter by doing the Supplementary Exercises. (In the Supplementary Exercises, you're given answers only — the details of the solutions are up to you.) Finally, you can review all the topics covered in the Outline by working the problems in the Midsemester and Semester Exams.

The level of difficulty is graded in the Examples, Solved Problems, and Exams. Most Examples involve the straightforward use of a single concept or equation. You should strive to understand the connection between the question asked and the solution presented. The Solved Problems are somewhat more complex, but again you should try to discover the reasoning behind the sequence of steps shown. (The reasoning is stated clearly as the parts of each problem are worked out: Try to anticipate where the solution should go next.) Everything is fair game in the Exams, but don't panic. Just apply the skills you learned in the Examples and Solved Problems.

Having the tools is one thing; knowing how to use them is another. The solution to any problem in physics requires six procedures: (1) UNDERSTANDING, (2) ANALYZING, (3) PLANNING, (4) EXECUTING, (5) CHECKING, (6) REPORTING. Let's look at each of these procedures in more detail.

1. **UNDERSTANDING:** Read the problem carefully and be sure you understand every part of it. If you have difficulty with any of the terms or ideas in the problem, reread the text material on which the problem is based. (In this Outline, important ideas, principles, laws, and terms are printed in boldface type, so they will be easy to find.) Make certain that you understand what kind of answer will be required. If the problem is quantitative, make an estimate of the magnitude of the answer.

2. **ANALYZING:** Break the problem down into its components. Ask yourself

    - What are the data?
    - What is (are) the unknown(s)?
    - What law or definition connects the data to the unknowns? What is the equation that encompasses the law?

3. **PLANNING:** Trace a connection between the data and the unknowns as a series of discrete operations (steps). This often involves manipulating one or more mathematical or physical expressions to isolate unknown quantities. Once you have a clear, stepwise path between data and solution, take note of any steps that require ancillary operations, such as substituting in equations or converting units. (Keep a sharp watch on units — they are often useful clues.)

4. EXECUTION: Follow your plan and execute any mathematical operations. You should work with symbols whenever possible: Substituting for variables should be the *last* thing you do. Make sure you've used the correct signs, exponents, and units.

5. CHECKING: Never consider a problem solved unitl you have checked your work. Does your answer

- make sense?
- have the right units?
- answer the question?

Is your math correct?

6. REPORTING: Make sure you have shown your reasoning and method clearly, and that your answer is readable. (It can't hurt to write the word "Answer" in front of your answer. That way, you — and your instructor — can find it at a glance, saving time and trouble all 'round.)

We have assumed that you own a calculator. Many hand-held calculators enable you to perform mathematical calculations that once required tables of logarithms and trigonometric functions. Calculators also permit you to eliminate the tedious methods of computation that arise from the properties of logarithms. Therefore, we provide neither trigonometric tables nor logarithm tables in this Outline. A hand-held calculator is essential for solving many of the problems. Even solutions that don't require the use of a calculator should be verified with a calculator.

The problem of precision and significant figures arises whenever something is measured. When you're applying your knowledge of physics to problems in the real world, you must be sure that your results do not imply more accuracy than your measurements warrant. In this Outline, which emphasizes principles and methods rather than measurements, you may assume that all numbers given in the Examples and Problems are precise — so you may express your answers in as many figures as your calculator allows. You may find it convenient, however, to give the final answers to the Examples and Problems to three or four significant figures, as we have done frequently in the Outline. But be careful of rounding errors! Many Problems contain several parts, and the answer to one part may be the starting point for calculations in the next part. Repeated approximations can introduce errors that accumulate over several steps. The problem of rounding errors arises often in physics applications, so in any problem based on measurements, round your answer to the proper number of significant figures only as the last step before reporting your result.

I wish to thank all the people in the Books for Professionals division of HBJ for their support during the preparation of this volume, Emily Thompson for guiding the project and especially Philip Unitt for his exceptionally helpful work editing the manuscript. Florence Kawahara has done an excellent job producing the fine drawings in this volume and special thanks to Monique Waters who typed most of the manuscript. Finally recognition must go to my wife, Sandra, for her understanding and support during the preparation of this volume.

K. E. J.

CONTENTS

CHAPTER 1   **Physical Units and Vector Techniques**   **1**
    1-1: Quantities and Units   1
    1-2: Vector and Scalar Quantities   3
    1-3: Vector Addition   3
    1-4: Vector Multiplication   7
    Solved Problems   10

CHAPTER 2   **Motion in One Dimension**   **14**
    2-1: Displacement, Speed, and Average Velocity   14
    2-2: Instantaneous Velocity   15
    2-3: Acceleration   15
    2-4: Constant Accelerated Motion: Kinematic Equations   16
    2-5: Bodies in Free Fall   18
    2-6: Applications of the Calculus for the Kinematic Equations   19
    Solved Problems   21

CHAPTER 3   **Motion in Two Dimensions**   **26**
    3-1: Displacement, Velocity, and Acceleration   26
    3-2: Motion with Constant Acceleration: Kinematic Equations   29
    3-3: Projectile Motion   29
    3-4: Uniform Circular Motion   31
    3-5: Accelerated Circular Motion   31
    3-6: Relative Velocity   33
    3-7: Rotational Motion   33
    Solved Problems   35

CHAPTER 4   **Newton's Laws of Motion**   **41**
    4-1: Newton's Three Laws of Motion   41
    4-2: Force, Mass, and Weight   41
    4-3: Frictional Forces   42
    4-4: Statics—The First Condition of Equilibrium   43
    4-5: Dynamics—Motion of a Particle   44
    4-6: Motion Along a Curve   45
    4-7: Velocity-Dependent Forces   46
    4-8: Accelerated Frames of Reference   47
    Solved Problems   50

CHAPTER 5   **Work and Kinetic Energy**   **58**

     5-1: Work—The Dot Product   58

     5-2: Work Done by a Changing
          Force   59

     5-3: The Work–Energy Principle   59

     5-4: Power   60

     Solved Problems   62

CHAPTER 6   **Conservative Forces and Conservation
of Energy**   **67**

     6-1: Conservative Forces   67

     6-2: Gravitational Potential
          Energy   68

     6-3: Elastic Potential Energy   69

     6-4: Conservation of Energy   70

     Solved Problems   72

CHAPTER 7   **Harmonic Motion**   **79**

     7-1: Simple Harmonic Motion
          (SHM) and Hooke's Law   79

     7-2: Equations of SHM—The
          Mass–Spring Oscillator   79

     7-3: The Simple Pendulum   83

     7-4: Damped Harmonic Motion   85

     7-5: Forced Harmonic Motion   87

     Solved Problems   89

CHAPTER 8   **Impulse and Conservation of
Momentum**   **94**

     8-1: Impulsive Forces and
          Momentum   94

     8-2: Conservation of Linear
          Momentum   95

     8-3: Collisions: Elastic and
          Inelastic   97

     8-4: Center of Mass (CM)   98

     8-5: Collision of Two Bodies in
          CM Frame of Reference   99

     8-6: Rocket Propulsion   102

     Solved Problems   104

EXAM 1   **110**

CHAPTER 9   **Rigid Bodies I: Statics**   **115**

     9-1: Center of Mass   115

     9-2: Torque—The Vector Cross
          Product   116

     9-3: Equilibrium Conditions for
          Rigid Bodies   117

     Solved Problems   121

CHAPTER 10   Rigid Bodies II: Dynamics   **129**
         10-1: Moment of Inertia   129
         10-2: Kinetic Energy of Rotation   132
         10-3: Angular Momentum   134
         10-4: Torque and Angular
                Acceleration   135
         10-5: Conservation of Angular
                Momentum   138
                Solved Problems   140

CHAPTER 11   Gravitation   **147**
         11-1: Newton's Law of Universal
                Gravitation   147
         11-2: Gravitational Force Due to
                Nonspherical Bodies   148
         11-3: The Gravitational Field   149
         11-4: Gravitational Energy   150
         11-5: Satellite Motion   151
                Solved Problems   153

CHAPTER 12   Special Relativity   **158**
         12-1: Einstein's Postulates   158
         12-2: Length Contraction and
                Time Dilation   159
         12-3: Addition of Velocities   160
         12-4: Relativistic Momentum and
                Energy   161
                Solved Problems   164

CHAPTER 13   Elasticity of Matter   **169**
         13-1: Stress and Strain   169
         13-2: Young's Modulus   170
         13-3: Poisson's Ratio   171
         13-4: Shear Modulus   171
         13-5: Bulk Modulus   172
                Solved Problems   174

CHAPTER 14   Fluid Statics   **178**
         14-1: Density and Specific Gravity   178
         14-2: Static Pressure and Pascal's
                Principle   180
         14-3: Archimedes' Principle   181
         14-4: Surface Tension   182
                Solved Problems   184

CHAPTER 15   Fluid Dynamics   **188**
         15-1: Equation of Continuity   188
         15-2: Bernoulli's Equation   188
         15-3: Viscosity and Poiseuille's
                Equation   189
         15-4: Stokes' Laws   190
         15-5: Reynolds Number   191
                Solved Problems   192

EXAM 2   **197**

CHAPTER 16   Temperature and Thermal Expansion   202
                16-1: The Celsius and Fahrenheit
                       Temperature Scales   202
                16-2: The Kelvin and Rankine Scales   203
                16-3: Thermal Expansion   203
                16-4: Thermal Stress   205
                       Solved Problems   206

CHAPTER 17   Heat and Heat Transfer   209
                17-1: Quantity of Heat   209
                17-2: Heat Capacity   209
                17-3: Change of Phase   211
                17-4: Heat Transfer   213
                       Solved Problems   217

CHAPTER 18   Ideal Gas and Kinetic Theory   222
                18-1: Equation of State   222
                18-2: Pressure of an Ideal Gas   224
                18-3: Density of an Ideal Gas   226
                18-4: Specific Heats of Gases   227
                18-5: Equipartition of Energy   229
                       Solved Problems   231

CHAPTER 19   The First Law of Thermodynamics   236
                19-1: Work and Internal Energy   236
                19-2: The First Law of
                       Thermodynamics   237
                19-3: Isothermal Processes   237
                19-4: Isovolumic Processes   238
                19-5: Isobaric Processes   239
                19-6: Adiabatic Processes   239
                       Solved Problems   241

**CHAPTER 20   The Second Law of Thermodynamics
                and Entropy   246**
                20-1: Heat Engines   246
                20-2: Efficiency of Heat Engines   247
                20-3: Refrigerators   248
                20-4: The Carnot Cycle   249
                20-5: The Second Law of
                       Thermodynamics   250
                20-6: Entropy   251
                       Solved Problems   254

CHAPTER 21   Mechanical Waves   259
                21-1: Types of Waves   259
                21-2: Periodic Waves   259
                21-3: The Wave Equation   262
                21-4: Velocities of Waves   263
                21-5: Energy and Power in Waves   264
                21-6: Interference of Waves   266
                       Solved Problems   271

CHAPTER 22 **Sound**   **278**
              22-1: The Speed of Sound   278
              22-2: Intensity Level   279
              22-3: Musical Intervals   280
              22-4: Resonance of Air Columns   280
              22-5: Beats   283
              22-6: The Doppler Effect   283
                     Solved Problems   287

FINAL EXAM   **291**

APPENDIX A: **The International System of Units**   **302**
APPENDIX B: **Commonly Used SI Prefixes**   **303**
APPENDIX C: **Constants**   **304**
APPENDIX D: **Greek Alphabet**   **304**

INDEX   **305**

PHYSICAL UNITS AND VECTOR TECHNIQUES

THIS CHAPTER IS ABOUT

☑ **Quantities and Units**
☑ **Vector and Scalar Quantities**
☑ **Vector Addition**
☑ **Vector Multiplication**

1-1. Quantities and Units

A. Base quantities and units

We need only seven fundamental quantities to describe phenomena in physics: *length, time, mass, electric current, temperature, luminous intensity,* and *amount of substance*. These quantities, called **base quantities**, can't be defined in terms of other quantities—they have to be defined *operationally*, in terms of rules of measurement. Base quantities are measured in **base units**. Two systems of base units are currently used:

(1) The *metric system*, which features the most modern and important **SI** (*Système International* [International System]) units (meter-kilogram-second) and the older, but still used, **cgs** (centimeter-gram-second) units.

(2) The *British engineering system*, which features the unwieldy British units (foot-pound-second) and crops up just often enough that we have to be able to convert British units into metric units.

Table 1-1 lists the seven base quantities in physics, their standard SI units, and their equivalents in cgs and British units. Note that these systems differ only in the ways that length, mass, and temperature are measured.

TABLE 1-1: Base Quantities

SI base quantity	SI base unit	cgs unit and conversion relationship	British unit and conversion relationship
Length	meter (m)	centimeter $1 \text{ cm} = 1 \times 10^{-2} \text{ m}$	foot $1 \text{ ft} = 0.3048 \text{ m}$
Time	second (s)	second	second
Mass	kilogram (kg)	gram $1 \text{ g} = 1 \times 10^{-3} \text{ kg}$	[Force*] pound $1 \text{ lb} = 0.453\,592\,37 \text{ kg}$, where $g = 9.806\,65 \text{ m s}^{-2}$
Electric current	ampere (A)	ampere	ampere
Temperature	kelvin (K)	degree Celsius $T(K) = T(C) + 273.15$	degree Fahrenheit $T(F) = 1.8\ T(C) + 32$
Luminous intensity	candela (cd)	candela	candela
Amount of substance	mole (mol)	mole	mole

* Force, not mass, is considered fundamental in the British system. The British unit of mass is the slug: 1 slug = 14.6 kg.

B. Derived quantities and units

Quantities that are defined in terms of other quantities, such as force, pressure, and energy, are called **derived quantities**. Derived quantities are measured in **derived units**, which are combinations of base units. For example,

Quantity	Unit	SI equivalent
Force	newton (N)	$1\,N = 1\,\text{kg}\,\text{m}\,\text{s}^{-2}$
Pressure	pascal (Pa)	$1\,Pa = 1\,\text{kg}\,\text{m}^{-1}\text{s}^{-2}$
Energy	joule (J)	$1\,J = 1\,\text{kg}\,\text{m}^2\text{s}^{-2}$

Definitions of derived quantities and their corresponding units will be given as needed in later chapters. (Appendix A lists most of the important quantities and units in physics.)

C. Conversion of units

TABLE 1-2: Conversion of Metric Units

Prefix	Symbol	Multiplying factor
giga	G	10^9
mega	M	10^6
kilo	k	10^3
deci	d	10^{-1}
centi	c	10^{-2}
milli	m	10^{-3}
micro	μ	10^{-6}
nano	n	10^{-9}

Conversion of standard metric units into smaller and larger metric units is simple: We just affix a prefix indicating a power of 10 (see Table 1-2) to the unit, and then multiply the number by that power of 10 or its reciprocal. For example, $1\,m = 1 \times 10^{-3}\,km = 1 \times 10^2\,cm = 1 \times 10^3\,mm = 1 \times 10^6\,\mu m$; and $1\,km = 1 \times 10^3\,m$, $1\,cm = 1 \times 10^{-2}\,m$, $1\,mm = 1 \times 10^{-3}\,m$, etc.

We also have to be able to convert from one system of units to another. To do this, we use **conversion relationships** and **unity conversion factors**, which are derived from the conversion relationships. Examples of conversion relationships are $1\,ft = 0.3048\,m$, $1\,mi = 5280\,ft$, and $1\,day = 24\,h$.

- Two unity conversion factors—reciprocals—can be obtained from a conversion relationship.

Consider, for example, the conversion relationship $1\,ft = 0.3048\,m$.

(a) Divide this relationship by 1 ft to obtain

$$1 = \frac{0.3048\,m}{1\,ft} = 0.3048\,m/ft$$

which converts feet to meters.

(b) Divide the relationship by 0.3048 m to obtain

$$1 = \frac{1\,ft}{0.3048\,m} \cong 3.281\,ft/m$$

which converts meters to feet.

Notice that units are treated like algebraic quantities.

EXAMPLE 1-1: Use the conversion relationships $1\,mi = 5280\,ft$, $1\,ft = 0.3048\,m$, and $1\,km = 10^3\,m$ to determine the number of kilometers in 1 mi.

Solution: You need to use two unity conversion factors to obtain the result; i.e.,

$$1\,mi = (5280\,\text{ft})\overbrace{\left(\frac{0.3048\,m}{1\,ft}\right)\left(\frac{1\,km}{10^3\,m}\right)}^{\text{unity conversion factors}} = \boxed{1.609\,344\,km}$$

note: You select unity conversion factors so that the units cancel, just as algebraic quantities do.

EXAMPLE 1-2: Convert the speed 60 mi/h into units of ft/s.

Solution: You know that 1 mi = 5280 ft, 1 h = 60 min, and 1 min = 60 s; thus you set up three unity conversion factors, or

$$60 \text{ mi/h} = \left(\frac{60 \text{ mi}}{1 \text{ h}}\right) \overbrace{\left(\frac{5280 \text{ ft}}{1 \text{ mi}}\right)\left(\frac{1 \text{ h}}{60 \text{ min}}\right)\left(\frac{1 \text{ min}}{60 \text{ s}}\right)}^{\text{unity conversion factors}} = \boxed{88 \text{ ft/s}}$$

1-2. Vector and Scalar Quantities

Some quantities, such as temperature and time, are completely specified by a number and units, so they have *magnitude only* and are called **scalar quantities**. Other quantities, such as displacement and velocity, involve more than magnitude and are called *vector quantities*:

• A **vector quantity** has both magnitude and direction.

For example, the velocity of a particle may be described by a vector **v**, which might be 2.5 m s^{-1} west or 6.7 m s^{-1} up, where the magnitudes of **v** are 2.5 m s^{-1} and 6.7 m s^{-1} and the directions are west and up, respectively.

Because it involves direction, a vector is graphed as a directed line segment, or arrow, using Cartesian coordinates. The arrow is drawn to scale, so its magnitude (length) is a scalar quantity. Thus, for example, the velocity of a particle, **v**, has a magnitude |**v**|, or v. The direction of the arrow is measured by the angle θ that the arrow makes with the chosen axis. Graphing any vector **V** this way gives us two ways to specify a vector: (1) We can give its magnitude |**V**| and its direction angle θ, or (2) we can give its components V_x and V_y (and V_z in a three-dimensional case) to specify a curve on a graph.

Naturally enough, we can't evaluate vectors by simple arithmetic addition or multiplication—we have to use vector methods.

1-3. Vector Addition

Vector addition can be done in two basic ways: (1) graphical methods and (2) component methods.

A. Graphical methods of vector addition

1. Head-to-tail method

We can add vectors by placing the tail of the added vector arrow to the head of the preceding one, keeping their directions constant. Then we find the *sum*, or **resultant vector**, by drawing another arrow from the tail of the first to the head of the last, as shown in Figure 1-1a and b. Vector addition is *commutative*, because the order in which the vectors are added doesn't matter. Thus, as we see from Figure 1-1,

$$\mathbf{A} + \mathbf{B} + \mathbf{C} = \mathbf{C} + \mathbf{B} + \mathbf{A} = \mathbf{R}$$

2. Parallelogram method

Any *two* vectors **A** and **B** can also be added by the parallelogram method, in which **A** and **B** are drawn from a common origin and used as adjacent sides in a parallelogram. The diagonal drawn from the common origin is the sum, or resultant **R** = **A** + **B**, as shown in Figure 1-1c.

3. Difference between two vectors

We find the *difference* between two vectors by adding the *negative* of the vector to be subtracted. The **negative of a vector** is obtained by interchanging the head and tail of the vector arrow, as shown for **B** in Figure 1-2.

note: Negative vectors reverse direction only—their magnitudes remain constant (and positive).

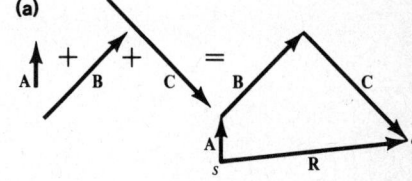
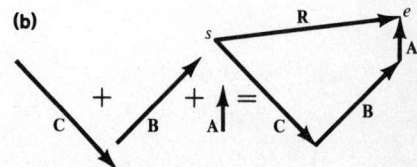
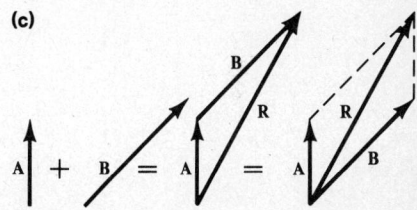

Figure 1-1. Vector sums: (a) **A** + **B** + **C** = **R**; (b) **C** + **B** + **A** = **R**; (c) **A** + **B** = **R**. (The points s and e indicate the start and end points.)

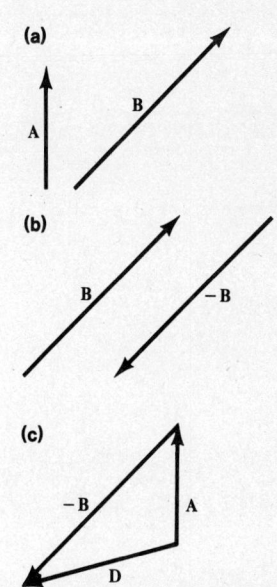

Figure 1-2. (a) The vectors **A** and **B**; (b) the vector **B** and its negative, **−B**; (c) the vector difference **D** = **A** − **B**.

EXAMPLE 1-3: Vector **A** has a magnitude of 1.5 units and direction $\theta = 140°$; vector **B** has a magnitude of 2.5 units and direction $\theta = 60°$. Determine the magnitudes (**a**) of their resultant **R** = **A** + **B** and (**b**) of their differences **D** = **A** − **B** graphically.

Solution: Plot **A** and **B** using Cartesian coordinates. Then measure the units on the graph with an appropriate scale, as in Figure 1-3a. (Note that if $\theta = 140°$, then $\phi = 40°$.)

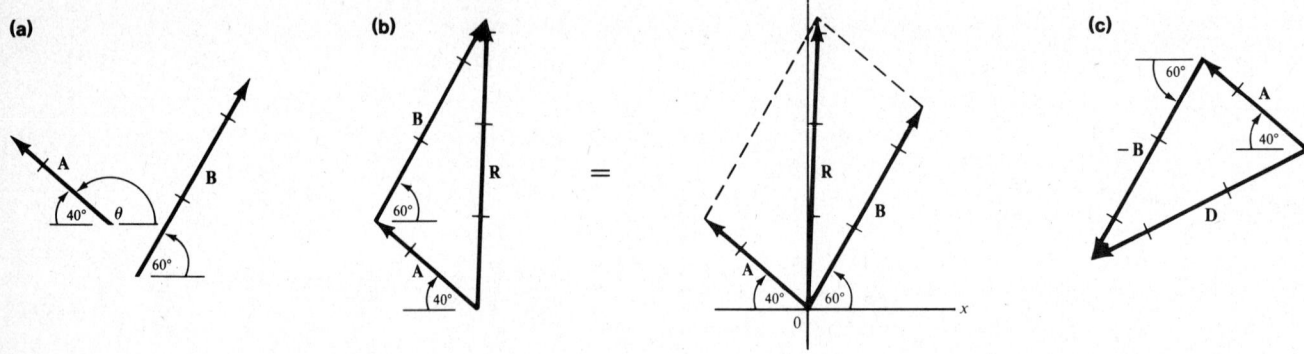

Figure 1-3

(**a**) To find the vector sum, place the tail of **B** to the head of **A** and draw the resultant **R** from the tail of **A** to the head of **B**. Or, draw **A** and **B** from a common origin, so that **A** and **B** form adjacent sides of a parallelogram, the length of whose diagonal **R** is R. To find the magnitude of **R** measure **R**: $R = |\mathbf{R}| = 3.1$ units. (See Figure 1-3b.)

(**b**) To find the vector difference, reverse the direction of **B** to get −**B**, place the tail of −**B** to the head of **A**, and draw the resultant **D** from the tail of **A** to the head of −**B**. Then measure **D**: $D = |\mathbf{D}| = 2.7$ units. (See Figure 1-3c.)

B. Component method of vector addition

The component method of vector addition is an analytic method that permits us to calculate (rather than measure) resultants in two or three dimensions. In order to use this method, we have to resolve vectors into their *scalar* and *vector components*.

The **scalar components** of a vector are obtained by constructing perpendiculars from the tip of the vector to the appropriate coordinate axes, as shown in Figure 1-4.

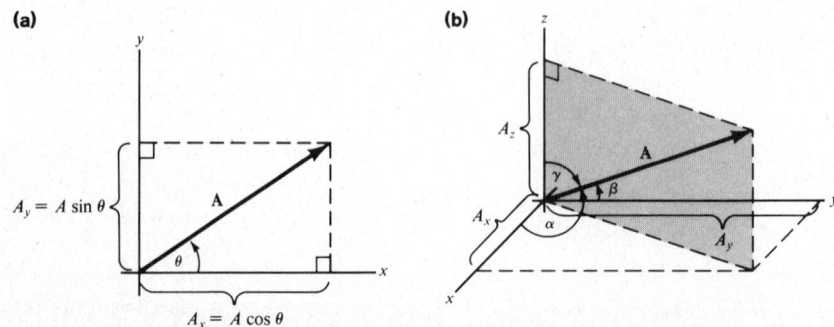

Figure 1-4. (a) The vector **A** that lies in the *x, y* plane has two scalar components: A_x and A_y. (b) The vector **R** that lies in three-dimensional space has three scalar components: R_x, R_y, and R_z.

In two dimensions there are two scalar components of any vector **A** (Figure 1-4a). These components are A_x and A_y:

$$A_x = A \cos\theta \quad \text{and} \quad A_y = A \sin\theta \qquad \textbf{(1-1)}$$

where θ is the direction angle of **A** (usually chosen to be the angle between **A** and the $+x$ axis). The magnitude A and the direction angle θ of **A** are

$$|\mathbf{A}| = A = \sqrt{A_x^2 + A_y^2} \qquad \textbf{(1-2a)}$$

$$\theta = \arctan\left(\frac{|A_y|}{|A_x|}\right) \quad \left[\text{or} \quad \theta = \tan^{-1}\left(\frac{|A_y|}{|A_x|}\right)\right] \qquad \textbf{(1-2b)}$$

In three dimensions there are three scalar components of any vector **A** (Figure 1-4b), which are obtained in terms of their direction cosines:

$$A_x = A \cos\alpha, \qquad A_y = A \cos\beta, \qquad A_z = A \cos\gamma$$

where α, β, and γ are the direction angles from the x, y, and z axes, respectively. Then the magnitude and direction angles of **A** are

$$A = \sqrt{A_x^2 + A_y^2 + A_z^2} \qquad \textbf{(1-3a)}$$

$$\alpha = \arccos\left(\frac{|A_x|}{A}\right) \quad \beta = \arccos\left(\frac{|A_y|}{A}\right) \quad \gamma = \arccos\left(\frac{|A_z|}{A}\right) \qquad \textbf{(1-3b)}$$

The **vector components** of a vector are the products of the scalar components and their respective *unit vectors*, $\hat{\mathbf{x}}$, $\hat{\mathbf{y}}$, and $\hat{\mathbf{z}}$ (also denoted by **i**, **j**, and **k**). A **unit vector** has a magnitude of unity and is directed along one of the coordinate axes, as shown in Figure 1-5. The vector components of **A** can be written as

$$\mathbf{A}_x = A_x\hat{\mathbf{x}} \quad [\text{or } A_x\mathbf{i}]$$

$$\mathbf{A}_y = A_y\hat{\mathbf{y}} \quad [\text{or } A_y\mathbf{j}]$$

$$\mathbf{A}_z = A_z\hat{\mathbf{z}} \quad [\text{or } A_z\mathbf{k}]$$

- Any vector can be specified by the sum of its vector components:

$$\mathbf{A} = \mathbf{A}_x + \mathbf{A}_y + \mathbf{A}_z$$
$$= A_x\hat{\mathbf{x}} + A_y\hat{\mathbf{y}} + A_z\hat{\mathbf{z}} \quad [\text{or } \mathbf{A} = A_x\mathbf{i} + A_y\mathbf{j} + A_z\mathbf{k}]$$

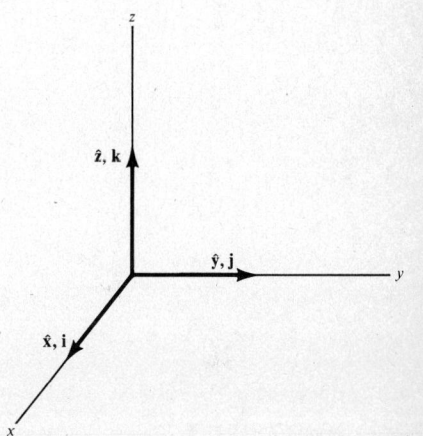

Figure 1-5. Unit vectors in three dimensions. Each unit vector has a magnitude of 1 unit.

Now we can use either scalar or vector components to add vectors.

$$\mathbf{R} = \mathbf{A} + \mathbf{B}$$

Using scalar components	Using vector components
(1) Resolve **A** and **B** into their scalar components:	(1) Express **A** and **B** as sums of their vector components:
$$A_x, A_y$$ $$B_x, B_y$$	$$\mathbf{A} = A_x\hat{\mathbf{x}} + A_y\hat{\mathbf{y}}$$ $$\mathbf{B} = B_x\hat{\mathbf{x}} + B_y\hat{\mathbf{y}}$$
(2) Add like components to get the scalar components of **R**:	(2) Add like components to get the vector components of **R**:
$$R_x = A_x + B_x$$ $$R_y = A_y + B_y$$	$$\mathbf{R} = (A_x\hat{\mathbf{x}} + B_x\hat{\mathbf{x}}) + (A_y\hat{\mathbf{y}} + B_y\hat{\mathbf{y}})$$ $$= (A_x + B_x)\hat{\mathbf{x}} + (A_y + B_y)\hat{\mathbf{y}}$$

Once the scalar components of the resultant vector **R** are found, its magnitude and direction angle(s) can be found from Eqs. (1-2)—or Eqs. (1-3) in the three-dimensional case.

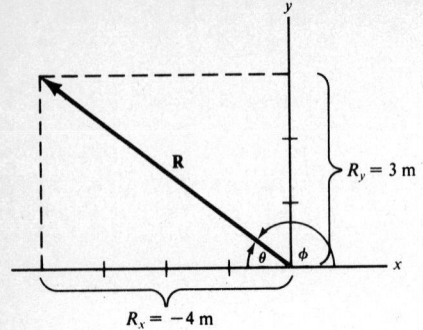

$R_x = -4$ m

Figure 1-6

EXAMPLE 1-4: Vectors **A**, **B**, and **C** are specified by their scalar components as $A_x = 3$ m, $A_y = 2$ m; $B_x = -5$ m, $B_y = -4$ m; and $C_x = -2$ m, $C_y = 5$ m. Determine **(a)** the scalar components of the resultant **R** and **(b)** its magnitude and direction.

Solution:

(a) Obtain the scalar components of the resultant by adding corresponding scalar components:

$$R_x = A_x + B_x + C_x$$
$$= 3 \text{ m} - 5 \text{ m} - 2 \text{ m} = \boxed{-4 \text{ m}}$$
$$R_y = A_y + B_y + C_y$$
$$= 2 \text{ m} - 4 \text{ m} + 5 \text{ m} = \boxed{3 \text{ m}}$$

The resultant is shown in Figure 1-6.

(b) Use Eqs. (1-2) to obtain the magnitude and direction:

$$R = \sqrt{R_x^2 + R_y^2}$$
$$= \sqrt{(-4 \text{ m})^2 + (3 \text{ m})^2} = \boxed{5 \text{ m}}$$

$$\theta = \arctan\left(\frac{|R_y|}{|R_x|}\right)$$
$$= \arctan\left(\frac{|3 \text{ m}|}{|-4 \text{ m}|}\right) = \boxed{36.9°}$$

Note that $\phi = 180° - \theta = 143.1°$, as shown in Figure 1-6.

EXAMPLE 1-5: Use the component method to add the vectors **A** and **B** shown in Figure 1-7, where $A = 3$ m, $\theta = 20°$; $B = 5$ m, $\phi = 150°$.

Solution: Put the tails of both vectors at the origin of a coordinate system and obtain their components (Figure 1-8a, b).

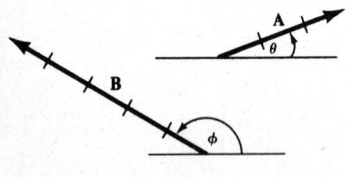

Figure 1-7

(a)

(b)

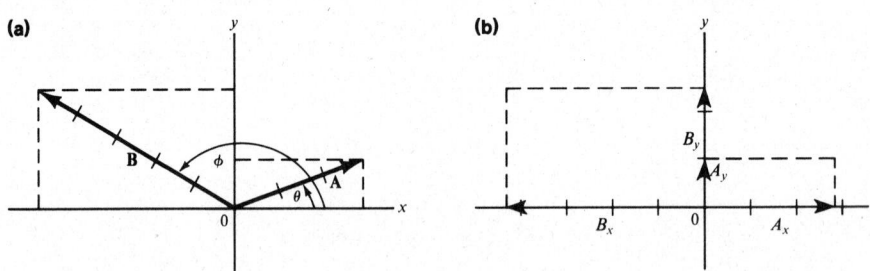

Figure 1-8. (a) and (b)

Add corresponding components, being careful in dealing with the negative components:

$$A_x = A\cos\theta = (3 \text{ m})\cos 20° = 2.819 \text{ m}$$
$$A_y = A\sin\theta = (3 \text{ m})\sin 20° = 1.026 \text{ m}$$
$$B_x = B\cos\phi = (5 \text{ m})\cos 150° = -4.330 \text{ m}$$
$$B_y = B\sin\phi = (5 \text{ m})\sin 150° = 2.500 \text{ m}$$

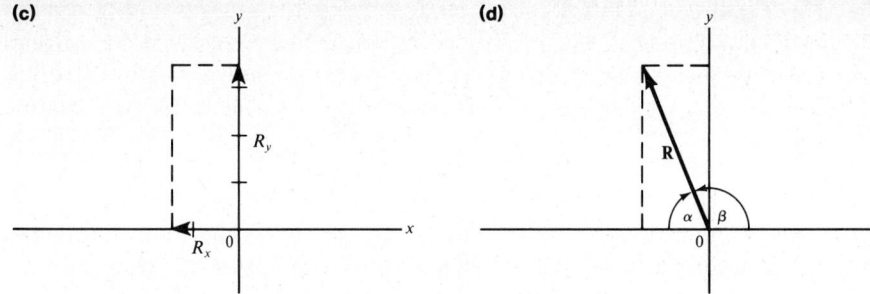

Figure 1-8. (c) and (d)

So the components of the resultant (Figure 1-8c) are

$$R_x = (2.819 - 4.330) \text{ m} = -1.511 \text{ m}$$
$$R_y = (1.026 + 2.500) \text{ m} = 3.526 \text{ m}$$

Finally, use Eqs. (1-2) to obtain the magnitude and direction of the resultant vector (Figure 1-8d):

$$R = \sqrt{R_x^2 + R_y^2} = \sqrt{(-1.511 \text{ m})^2 + (3.526 \text{ m})^2} = \boxed{3.836 \text{ m}}$$

$$\alpha = \arctan\left(\frac{|R_y|}{|R_x|}\right) = \arctan\left(\frac{|3.526 \text{ m}|}{|-1.511 \text{ m}|}\right) = \boxed{66.8°}$$

note: It's conventional to express the direction relative to the positive x axis, so

$$\beta = 180° - \alpha = 180° - 66.8° = \boxed{113.2°}$$

EXAMPLE 1-6: Add the vectors **A** and **B**, where $\mathbf{A} = (3\hat{x} + 5\hat{y} + 2\hat{z})$ m and $\mathbf{B} = (2\hat{x} - 2\hat{y} - 7\hat{z})$ m; i.e., find the resultant $\mathbf{R} = \mathbf{A} + \mathbf{B}$, determine R, and find its direction angles α, β, and γ.

Solution: Obtain the components of **R** by adding the corresponding components of **A** and **B**:

$$\mathbf{R} = \mathbf{A} + \mathbf{B} = [(3\hat{x} + 2\hat{x}) + (5\hat{y} - 2\hat{y}) + (2\hat{z} - 7\hat{z})] \text{ m} = (5\hat{x} + 3\hat{y} - 5\hat{z}) \text{ m}$$

Use Eq. (1-3a) to obtain the magnitude:

$$R = \sqrt{R_x^2 + R_y^2 + R_z^2} = \sqrt{5^2 + 3^2 + (-5)^2} \text{ m} = \boxed{7.68 \text{ m}}$$

Solve Eqs. (1-3b) to obtain the direction angles:

$$R_x = R\cos\alpha \qquad \alpha = \arccos\left(\frac{|R_x|}{R}\right) = \arccos\left(\frac{5 \text{ m}}{7.68 \text{ m}}\right) = \boxed{49.4°}$$

$$R_y = R\cos\beta \qquad \beta = \arccos\left(\frac{|R_y|}{R}\right) = \arccos\left(\frac{3 \text{ m}}{7.68 \text{ m}}\right) = \boxed{67.0°}$$

$$R_z = R\cos\gamma \qquad \gamma = \arccos\left(\frac{|R_z|}{R}\right) = \arccos\left(\frac{5 \text{ m}}{7.68 \text{ m}}\right) = \boxed{49.4°}$$

1-4. Vector Multiplication

There are three important ways to obtain the product of two quantities when one or both of the quantities is a vector.

A. Multiplication of a vector by a scalar

- The product of a scalar s and a vector **A** is a new vector:

$$\mathbf{B} = s\mathbf{A} \qquad\qquad \textbf{(1-4)}$$

The magnitude of **B** is $B = |s|A$; its direction is the same as that of **A** if s is positive and opposite that of **A** if s is negative. For example, if $\mathbf{B} = (-1)\mathbf{A} = -\mathbf{A}$, then $B = |-1|A = A$ and the direction of **B** is opposite that of **A**; but if $\mathbf{B} = (2)\mathbf{A} = 2\mathbf{A}$, then $B = |2|A = 2A$ and the directions of **A** and **B** are the same.

EXAMPLE 1-7: Momentum is an example of scalar multiplication. [A detailed treatment of momentum is given in Sections 8-1 and 8-2.] The momentum **p** of a particle of mass m moving with velocity **v** is $\mathbf{p} = m\mathbf{v}$. Because mass m, which is a scalar, can never be negative, **p** always has the same direction as **v**.

If a mass, $m = 1.7$ kg, is moving west at 3 m s^{-1}, what is the momentum of the mass?

Solution: From the momentum equation $\mathbf{p} = m\mathbf{v}$, you know that the magnitude of the momentum is

$$p = mv = (1.7 \text{ kg})(3 \text{ m s}^{-1}) = \boxed{5.1 \text{ kg m s}^{-1}}$$

and the direction of **p** is given as west.

B. Dot (or scalar) product of two vectors

- The **dot product** (also called the **scalar product**) of two vectors **A** and **B** is a scalar quantity:

$$s = \mathbf{A} \cdot \mathbf{B} \tag{1-5}$$

You can calculate s in two ways: (1) by using the magnitudes of the vectors and the angle θ between them (Figure 1-9); or (2) by using the scalar components of the vectors:

$$s = AB \cos \theta \tag{1-6a}$$

$$s = A_x B_x + A_y B_y + A_z B_z \tag{1-6b}$$

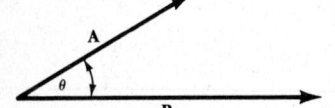

Figure 1-9. The angle between **A** and **B** is θ.

EXAMPLE 1-8: Work done by a force is an example of the dot product of two vectors. [A detailed treatment of work is given in Sections 5-1 and 5-2.] The work W done by the force **F** that drags a body a distance **D** is $W = \mathbf{F} \cdot \mathbf{D}$.

If $|\mathbf{F}| = 24$ lb, $|\mathbf{D}| = 3$ ft, and $\theta = 30°$, what is the amount of work done by the force dragging the block shown in Figure 1-10?

Solution: Use Eq. (1-6a):

$$W = FD \cos \theta = (24 \text{ lb})(3 \text{ ft})\cos 30° = \boxed{62.3 \text{ ft} \cdot \text{lb}}$$

Figure 1-10

C. Vector (or cross) product of two vectors

- The **vector product** (also called the **cross product**) of two vectors **A** and **B** is a vector:

$$\mathbf{V} = \mathbf{A} \times \mathbf{B} \tag{1-7}$$

We can calculate **V** in two ways:

(1) Use the magnitudes of the vectors and the angle between them:

$$\mathbf{V} = AB \sin \theta \tag{1-8a}$$

where the direction of **V** can be determined by using the *right-hand rule,* as shown in Figure 1-11. Place the fingers of your right hand in the direction of the first vector (**A** in this case) and close them toward your palm to align them with the second vector (**B** in this case). Then stick out your thumb: Your extended thumb points in the direction of **V**.

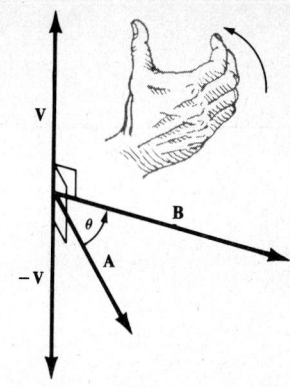

caution: It should be clear that $\mathbf{B} \times \mathbf{A} = -\mathbf{V}$; i.e., cross-multiplication of vectors is *not* commutative.

(2) Use the scalar components with the determinant:

$$\mathbf{V} = \mathbf{A} \times \mathbf{B} = \begin{vmatrix} \hat{\mathbf{x}} & \hat{\mathbf{y}} & \hat{\mathbf{z}} \\ A_x & A_y & A_z \\ B_x & B_y & B_z \end{vmatrix}$$

to obtain

$$\mathbf{V} = (A_y B_z - A_z B_y)\hat{\mathbf{x}} + (A_z B_x - A_x B_z)\hat{\mathbf{y}} + (A_x B_y - A_y B_x)\hat{\mathbf{z}} \quad \textbf{(1-8b)}$$

Figure 1-11. Using the right-hand rule: The vector **V** is perpendicular to both **A** and **B**.

EXAMPLE 1-9: Torque τ due to a force is an example of the cross product of two vectors. [A detailed treatment of torque is given in Section 9-2.] The torque due to the force **F** acting at point p a distance **r** from an axis is given by $\tau = \mathbf{r} \times \mathbf{F}$. In Figure 1-12, $F = 42$ N, $r = 0.8$ m, and $\theta = 50°$. (Assume that **r** and **F** are in the plane of the page.) Determine the torque due to the given force.

Solution: The magnitudes of the vectors and the angle between them are given, so we can use Eq. (1-8a):

$$\tau = rF\sin\theta$$
$$= (0.8 \text{ m})(42 \text{ N})\sin 50° = \boxed{25.7 \text{ N m}}$$

The direction of τ, obtained from the right-hand rule, is perpendicular to the surface of the page. In other words, the axis is perpendicular to the plane of the page, so the torque due to **F** is parallel to this axis and directed out of the page.

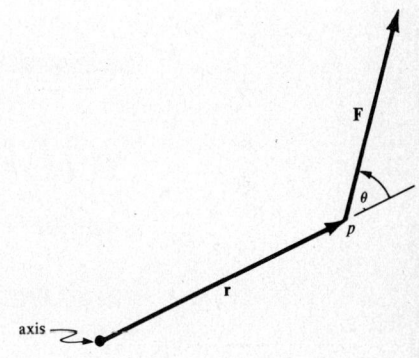

Figure 1-12

SUMMARY

1. The base metric units of the base quantities length, time, and mass are meter, second, kilogram in SI and centimeter, second, gram in cgs.
2. Conversion of a quantity from one system of units to another is accomplished by multiplying the quantity by a unity conversion factor obtained from an appropriate conversion relationship.
3. Vector quantities can be added in two basic ways: (1) graphical methods, and (2) the components method.
4. The three important ways of accomplishing vector multiplication are

(a) multiplication of a vector **A** by a scalar s: $\mathbf{B} = s\mathbf{A}$
(b) forming the dot scalar product: $s = AB\cos\theta = A_x B_x + A_y B_y + A_z B_z$
(c) forming the cross product or vector product:

$$\mathbf{V} = AB\sin\theta$$

$$\mathbf{V} = \mathbf{A} \times \mathbf{B} = \begin{vmatrix} \hat{\mathbf{x}} & \hat{\mathbf{y}} & \hat{\mathbf{z}} \\ A_x & A_y & A_z \\ B_x & B_y & B_z \end{vmatrix} = (A_y B_z - A_z B_y)\hat{\mathbf{x}} + (A_z B_x - A_x B_z)\hat{\mathbf{y}} + (A_x B_y - A_y B_x)\hat{\mathbf{z}}$$

with the direction of **V** given by the right-hand rule.

RAISE YOUR GRADES

Can you explain . . . ?

- ☑ the two main differences between the metric and British systems of units
- ☑ how to convert a quantity from one system of units to another
- ☑ how to add vectors by the graphical method
- ☑ how to add vectors by the component method
- ☑ how to multiply a scalar by a vector and the concept of a *negative vector*
- ☑ how to obtain the dot product of two vectors using either their magnitudes and included angle or their components
- ☑ how to obtain the cross product of two vectors using either their magnitudes and included angle or their components

SOLVED PROBLEMS

Quantities and Units

PROBLEM 1-1:   One acre has an area of 43 560 ft^2. Determine the area of an acre in square meters.

Solution: You can use the conversion relationship, 1 ft = 0.3048 m. Square both sides of this conversion relationship to obtain 1 ft^2 = 9.290 304 × 10^{-2} m^2, which you use as a unity conversion factor.

$$1 \text{ acre} = (43\,560 \text{ ft}^2)\underbrace{\left(\frac{9.290\,304 \times 10^{-2} \text{ m}^2}{1 \text{ ft}^2}\right)}_{\text{unity conversion factor}} = \boxed{4.047 \times 10^3 \text{ m}^2}$$

PROBLEM 1-2:   One hectare = 10^4 m^2. How many acres are there in one hectare?

Solution: You can start with the answer to Problem 1-1, or 1 acre = 4.047 × 10^3 m^2. Then

$$1 \text{ hectare} = (10^4 \text{ m}^2)\left(\frac{1 \text{ acre}}{4.047 \times 10^3 \text{ m}^2}\right) = \boxed{2.471 \text{ acres}}$$

PROBLEM 1-3:   One mile = 5280 ft. How many acres are there in a square mile?

Solution: Square the given conversion relationship and use a unity conversion factor to obtain

$$1 \text{ mi}^2 = (5280 \text{ ft})^2 = 2.787\,84 \times 10^7 \text{ ft}^2$$

$$= (2.787\,84 \times 10^7 \text{ ft}^2)\underbrace{\left(\frac{1 \text{ acre}}{43\,560 \text{ ft}^2}\right)}_{\text{unity conversion factor}} = \boxed{640 \text{ acres}}$$

PROBLEM 1-4:   The spark plug gap for an engine is 0.035 in. Determine the gap in millimeters.

Solution: Use 1 in. = 2.54 cm = 25.4 mm as a unity conversion factor:

$$0.035 \text{ in.} = (0.035 \text{ in.})\underbrace{\left(\frac{25.4 \text{ mm}}{1 \text{ in.}}\right)}_{\substack{\text{unity} \\ \text{conversion} \\ \text{factor}}} = \boxed{0.889 \text{ mm}}$$

PROBLEM 1-5: The engine in Problem 1-4 has a displacement of 252 in^3. Determine its displacement in cubic centimeters and liters (1 L = 10^3 cm^3).

Solution: Use 1 in. = 2.54 cm and cube both sides to obtain 1 in^3 = 16.39 cm^3. Then use this result as a unity conversion factor:

$$252 \text{ in}^3 = (252 \text{ in}^3)\underbrace{\left(\frac{16.39 \text{ cm}^3}{1 \text{ in}^3}\right)}_{\substack{\text{unity} \\ \text{conversion} \\ \text{factor}}} = \boxed{4130 \text{ cm}^3 = 4.13 \text{ L}}$$

Vector Addition

PROBLEM 1-6: Determine and graph the scalar components of the vector **D**, where D is 4.2 m and $\theta = 132°$.

Solution: Use Eqs. (1-1):

$D_x = (4.2 \text{ m})(\cos \phi) = (4.2 \text{ m})(\cos 132°)$

$\quad = -(4.2 \text{ m})(0.6691) = \boxed{-2.81 \text{ m}}$

$D_y = (4.2 \text{ m})(\sin \phi) = (4.2 \text{ m})(\sin 132°)$

$\quad = (4.2 \text{ m})(0.7431) = \boxed{3.12 \text{ m}}$

The scalar components are shown in Figure 1-13.

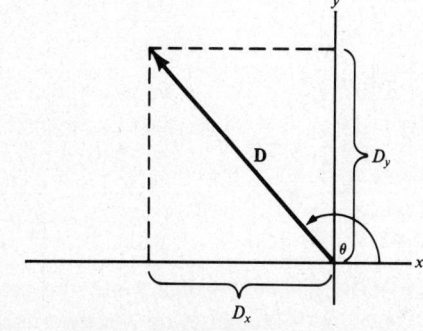

Figure 1-13

PROBLEM 1-7: Calculate (a) the sum **S** = **A** + **B** and (b) the difference **D** = **B** − **A**, where **A** = $(3\hat{\mathbf{x}} + 5\hat{\mathbf{y}})$ and **B** = $(-5\hat{\mathbf{x}} + 2\hat{\mathbf{y}})$. Express the resultants in terms of their magnitudes and their direction angles measured with respect to the $+x$ axis.

Solution:

(a) In terms of unit vectors

$$\mathbf{S} = \mathbf{A} + \mathbf{B} = [(3-5)\hat{\mathbf{x}} + (5+2)\hat{\mathbf{y}}] = \boxed{(-2\hat{\mathbf{x}} + 7\mathbf{y})}$$

So, by Eqs. (1-2a) and (1-2b),

$$\mathbf{S} = \sqrt{S_x^2 + S_y^2} = \sqrt{(-2)^2 + 7^2} \text{ m} = \boxed{7.28 \text{ m}} \quad \text{and} \quad \theta = \arctan\frac{|S_y|}{|S_x|} = \frac{|7|}{|-2|} = \boxed{106°}$$

(b)

$$\mathbf{D} = \mathbf{B} - \mathbf{A} = [(-5-3)\hat{\mathbf{x}} + (2-5)\hat{\mathbf{y}}] = \boxed{(-8\hat{\mathbf{x}} - 3\hat{\mathbf{y}})}$$

So $\quad \mathbf{D} = \sqrt{(-8)^2 + (-3)^2} \text{ m} = \boxed{8.54 \text{ m}} \quad \text{and} \quad \theta = \arctan\frac{|-3|}{|-8|} = \boxed{-20.6°}$

Vector Multiplication

PROBLEM 1-8: Determine the dot product s between the vectors **A** and **B**, where $A = 63$ m, $B = 4$ m, and the angle θ between them is 48°.

Solution: Use Eq. (1-6a):

$$s = \mathbf{A} \cdot \mathbf{B} = AB\cos\theta = (63 \text{ m})(4 \text{ m})\cos 48° = \boxed{168.6 \text{ m}^2}$$

PROBLEM 1-9: $\mathbf{E} = (3\hat{\mathbf{x}} + 2\hat{\mathbf{y}} + 4\hat{\mathbf{z}})$ m and $\mathbf{F} = (5\hat{\mathbf{x}} + \hat{\mathbf{y}} + 3\hat{\mathbf{z}})$ m. **(a)** Determine the dot product between vectors $\mathbf{E}$ and $\mathbf{F}$ and **(b)** obtain the angle θ between them.

Solution:

(a) Use Eq. (1-6b):

$$s = E_x F_x + E_y F_y + E_z F_z = [(3)(5) + (2)(1) + (4)(-3)] \text{ m}^2 = \boxed{5 \text{ m}^2}$$

(b) Solve Eq. (1-6a) for θ and obtain the magnitudes of $\mathbf{E}$ and $\mathbf{F}$ by using Eq. (1-2a):

$$s = EF\cos\theta$$

$$\theta = \arccos\left(\frac{s}{EF}\right)$$

where

$$E = \sqrt{3^2 + 2^2 + 4^2} \text{ m} = 5.385 \text{ m} \qquad \text{and} \qquad F = \sqrt{5^2 + 1^2 + (-3)^2} \text{ m} = 5.916 \text{ m}$$

So

$$\theta = \arccos\left[\frac{5 \text{ m}^2}{(5.385 \text{ m})(5.916 \text{ m})}\right] = \boxed{81.0°}$$

PROBLEM 1-10:   Calculate the magnitude V of the cross product of the vectors $\mathbf{A}$ and $\mathbf{B}$, where $A = 24$ m, $B = 48$ m, and the angle between them is $\theta = 37°$.

Solution: Use Eq. (1-8a):

$$V = AB\sin\theta = (24 \text{ m})(48 \text{ m})\sin 37° = \boxed{693 \text{ m}^2}$$

PROBLEM 1-11:   Calculate the cross product for vectors $\mathbf{G} = (3\hat{\mathbf{x}} + 5\hat{\mathbf{y}} + 2\hat{\mathbf{z}})$ m and $\mathbf{H} = (2\hat{\mathbf{x}} - 2\hat{\mathbf{y}} - 7\hat{\mathbf{z}})$ m and obtain the angle θ between them.

Solution: You can use Eq. (1-8b) to obtain the cross product:

$$
\begin{aligned}
\mathbf{G} \times \mathbf{H} &= (G_y H_z - G_z H_y)\hat{\mathbf{x}} + (G_z H_x - G_x H_z)\hat{\mathbf{y}} + (G_x H_y - G_y H_x)\hat{\mathbf{z}} \\
&= \{[(5)(-7) - (2)(-2)]\hat{\mathbf{x}} + [(2)(2) - (3)(-7)]\hat{\mathbf{y}} + [(3)(-2) - (5)(2)]\hat{\mathbf{z}}\} \text{ m}^2 \\
&= \boxed{(-31\hat{\mathbf{x}} + 25\hat{\mathbf{y}} - 16\hat{\mathbf{z}}) \text{ m}^2}
\end{aligned}
$$

To get θ, use Eq. (1-8a):

$$|\mathbf{G} \times \mathbf{H}| = GH\sin\theta$$

$$\theta = \arcsin\left(\frac{|\mathbf{G} \times \mathbf{H}|}{GH}\right)$$

$$= \arcsin\left[\frac{\sqrt{(-31)^2 + (25)^2 + (-16)^2} \text{ m}^2}{(\sqrt{3^2 + 5^2 + 2^2} \text{ m})(\sqrt{2^2 + (-2)^2 + (-7)^2} \text{ m})}\right] = \boxed{67.2°}$$

Supplementary Exercises

EXERCISE 1-1:   Assume that a and b are physical units. If the conversion relationship between them is $3a = 7b$, how many a are $42b$?

EXERCISE 1-2:   For the same a and b as in Exercise 1-1, determine the number of square b there are in $87a^2$.

EXERCISE 1-3:   Find the horsepower required to operate a 1300-watt toaster (1 hp = 746 W).

EXERCISE 1-4:   A human hair is 0.0035 in. in diameter. What is its diameter in meters?

EXERCISE 1-5:   Aunt Ester lived 96 years, 2 weeks, and 4 days. Determine her life span in seconds.

EXERCISE 1-6:   Determine and graph the vector components of the vector $\mathbf{F}$, where F is 63.2 N and $\theta = -36°$.

For Exercises 1-7 through 1-10, the vectors referred to are: $\mathbf{A} = (-3\hat{\mathbf{x}} + 4\hat{\mathbf{y}})$ and $\mathbf{B} = (-5\hat{\mathbf{x}} - 2\hat{\mathbf{y}})$.

EXERCISE 1-7:   Determine (**a**) the sum $\mathbf{S} = \mathbf{A} + \mathbf{B}$ and (**b**) the difference $\mathbf{D} = \mathbf{B} - \mathbf{A}$.

EXERCISE 1-8:   Obtain the dot product, $d = \mathbf{A} \cdot \mathbf{B}$.

EXERCISE 1-9:   Obtain the vector product, $\mathbf{V} = \mathbf{A} \times \mathbf{B}$.

EXERCISE 1-10:   What is the angle between $\mathbf{A}$ and $\mathbf{B}$?

Answers to Supplementary Exercises

1-1: $18a$

1-2: $473.7b^2$

1-3:   1.743 hp

1-4: 8.89×10^{-5} m

1-5: 3.03×10^9 s

1-6: $\mathbf{F}_x = (51.1 \text{ N})\hat{\mathbf{i}}$; $\mathbf{F}_y = (-37.1 \text{ N})\hat{\mathbf{j}}$
See Figure 1-14

1-7:   (**a**) $\mathbf{S} = (-8\hat{\mathbf{x}} + 2\hat{\mathbf{y}})$ m
$= 8.25$ m
$\theta_S = 166°$

(**b**) $\mathbf{D} = (-2\hat{\mathbf{x}} - 6\hat{\mathbf{y}})$ m
$= 6.32$ m
$\theta_D = 251.6°$
Both direction angles are relative to the positive x-axis.

1-8: 7 m^2

1-9: $+(26 \text{ m}^2)\hat{\mathbf{z}}$

1-10: $74.9°$

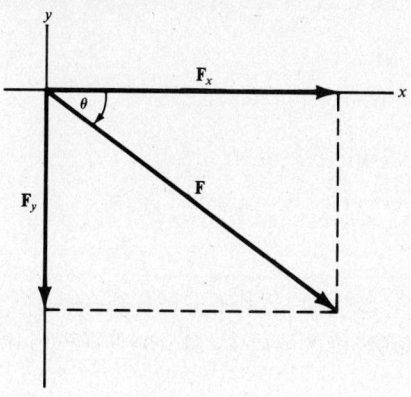

Figure 1-14

2 MOTION IN ONE DIMENSION

THIS CHAPTER IS ABOUT

☑ **Displacement, Speed, and Average Velocity**
☑ **Instantaneous Velocity**
☑ **Acceleration**
☑ **Constant Accelerated Motion: Kinematic Equations**
☑ **Bodies in Free Fall**
☑ **Applications of the Calculus for the Kinematic Equations**

agreement: Motion in one dimension is motion along a straight line. In this chapter we'll consider the motion of a particle along the x axis of the Cartesian coordinate system.

2-1. Displacement, Speed, and Average Velocity

A. Displacement

The change in position, i.e., the **displacement** Δx, of a particle along the x axis is obtained by subtracting the initial position x_1 from the final position x_2.

DISPLACEMENT $\qquad\qquad \Delta x = x_2 - x_1 \qquad\qquad$ **(2-1)**

A positive displacement indicates that the particle moves in the positive x direction, and a negative displacement indicates that the particle moves in the negative x direction. Displacement has the dimension of length.

EXAMPLE 2-1: A particle starts at point $x_1 = 3$ m and arrives at point $x_2 = -5$ m. Determine its displacement.

Solution: From Eq. (2-1) the displacement of the particle is

$$\Delta x = x_2 - x_1 = (-5 \text{ m}) - (3 \text{ m}) = \boxed{-8 \text{ m}}$$

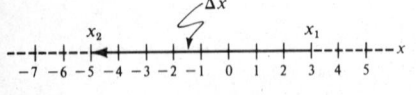

Figure 2-1

So the change in the particle's position is 8 m to the left of its starting point on the x axis, as shown in Figure 2-1.

B. Speed

Speed is the time it takes a moving object to travel a specific distance. Speed has the dimension of length per unit time, measured in meters per second, m s^{-1} (and km h^{-1}), in the metric system and feet per second, ft/s (and mi/h), in the British system. The average speed $\bar{v}_s$ of a moving body is determined by dividing the total path traversed, D, by the time interval, t, required to travel that path: $\bar{v}_s = D/t$.

C. Average velocity

The **average velocity** $\bar{v}$ of a particle moving in a straight line is its displacement divided by the time interval:

AVERAGE VELOCITY $$\bar{v} = \frac{x_2 - x_1}{t_2 - t_1} = \frac{\Delta x}{\Delta t} \qquad (2\text{-}2)$$

The quantities t_1 and t_2 are the instants when the particle is at points x_1 and x_2, respectively.

EXAMPLE 2-2: A particle traveling in a straight line along the x axis is at point $x_1 = 3$ m at instant $t_1 = 6$ s and arrives at $x_2 = -5$ m at $t_2 = 9$ s. Determine its average velocity and its average speed.

Solution: The time interval is $\Delta t = t_2 - t_1 = (9 - 6)$ s $= 3$ s. So, from Eq. (2-2), the average velocity is

$$\bar{v} = \frac{\Delta x}{\Delta t} = \frac{-8 \text{ m}}{3 \text{ s}} = \boxed{-2.67 \text{ m s}^{-1}}$$

As in Example 2-1, the negative sign indicates that the particle moved in the $-x$ direction. The particle traveled $|-8|$ m in 3 s, so its average speed is

$$\bar{v}_s = \frac{|-8| \text{ m}}{3 \text{ s}} = |-2.67| = 2.67 \text{ m s}^{-1}$$

---

note: Velocity always involves a direction and is defined in terms of *displacement*. Speed involves no direction and is the absolute value, or *magnitude*, of the velocity, i.e., $\bar{v}_s = |\bar{v}|$. [Velocity is, in fact, a vector quantity and can be written as **v**. But in one dimension there is no direction angle, so we'll use v to signify velocity.]

2-2. Instantaneous Velocity

The velocity of a particle may vary with time as shown, for example, in Figure 2-2. The **instantaneous velocity** v of a particle is its velocity at a given instant in time, which can be specified as a limit:

INSTANTANEOUS VELOCITY $$v = \lim_{\Delta t \to 0} \frac{\Delta x}{\Delta t} = \frac{dx}{dt} \qquad (2\text{-}3)$$

In other words, the instantaneous velocity is the limiting value of the average velocity as Δt approaches zero.

2-3. Acceleration

Acceleration is the time rate of change of the velocity of a body. Average acceleration is $\bar{a} = \Delta v/\Delta t$, so **instantaneous acceleration** is

INSTANTANEOUS ACCELERATION $$a = \lim_{\Delta t \to 0} \frac{\Delta v}{\Delta t} = \frac{dv}{dt} = \frac{d^2 x}{dt^2} \qquad (2\text{-}4)$$

The dimensions of acceleration are length per time squared, so its units may be meters per second squared (m s^{-2}) in the SI system and feet per second squared (ft s^{-2}) in the British system. A positive acceleration indicates that the body's velocity is increasing, whereas a negative acceleration (i.e., *deceleration*) indicates that the body is slowing down.

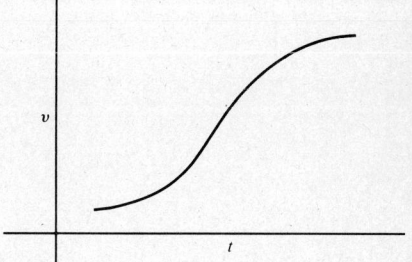

Figure 2-2. Velocity as a continuous function of time.

2-4. Constant Accelerated Motion: Kinematic Equations

When a body moves in a straight line on the x axis with constant acceleration a, its position x is

$$x = v_0 t + \tfrac{1}{2} a t^2 \qquad \text{(2-5)}$$

and its instantaneous velocity v is

$$v = v_0 + at \qquad \text{(2-6)}$$

where we assume that the body is at the origin at $t = 0$, so that $x_0 = 0$. Equations (2-5) and (2-6) are sufficient to describe the linear motion of a uniformly accelerating body. We can manipulate algebraically the two results to get two additional equations, which can be useful:

$$v^2 = v_0^2 + 2ax \qquad \text{(2-7)}$$

$$x = vt - \tfrac{1}{2} a t^2 \qquad \text{(2-8)}$$

Kinematic Equations for Straight-Line Motion

$$\left. \begin{array}{l} x = v_0 t + \tfrac{1}{2} a t^2 \\[4pt] v = v_0 + at \\[4pt] v^2 = v_0^2 + 2ax \\[4pt] x = vt - \tfrac{1}{2} a t^2 \end{array} \right\} \quad \begin{array}{l} a = \text{constant} \\[8pt] x_0 = 0 \end{array}$$

note: If the body is *not* at the origin at $t = 0$, Eq. (2-5) becomes

$$x = x_0 + v_0 t + \tfrac{1}{2} a t^2 \qquad \text{(2-5a)}$$

and Eq. (2-7) becomes

$$v^2 = v_0^2 + 2a(x - x_0) \qquad \text{(2-7a)}$$

Equations (2-5) through (2-8) are known as the **kinematic equations for straight-line motion with constant acceleration.**

EXAMPLE 2-3: The velocity of a toy train traveling on a straight track is graphed in Figure 2-3 for four intervals of time (I, II, III, and IV). **(a)** What is the acceleration of the train during interval I? **(b)** How far does it travel during interval I? **(c)** How far does it travel during interval II? **(d)** What is the acceleration of the train during interval III? **(e)** How far does it travel during interval III? **(f)** What is the acceleration of the train during interval IV? **(g)** How far does it travel during interval IV?

caution: You are asked to find *incremental*, not cumulative, values.

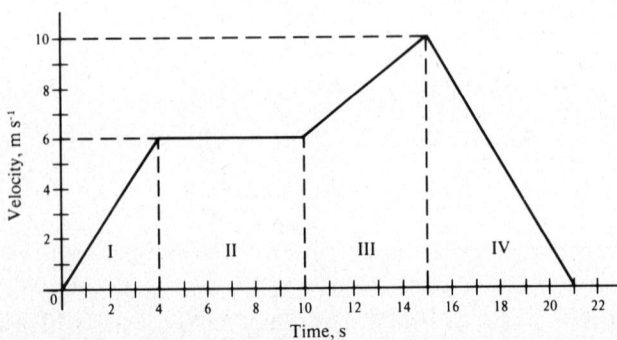

Figure 2-3

Solution:

(a) You can see from Figure 2-3 that $v_0 = 0$, $v = 6 \text{ m s}^{-1}$, and $t = (4 - 0) = 4$ s; so solve Eq. (2-6) for a to find the acceleration:

$$v = v_0 + at$$

$$a = \frac{v - v_0}{t} = \frac{6 \text{ m s}^{-1} - 0}{4 \text{ s}} = \boxed{1.5 \text{ m s}^{-2}}$$

(b) Substitute $a = 1.5 \text{ m s}^{-2}$ into Eq. (2-5) to find the distance traveled, x:

$$x = v_0 t + \tfrac{1}{2}at^2 = 0 + \tfrac{1}{2}(1.5 \text{ m s}^{-2})(4 \text{ s})^2 = \boxed{12 \text{ m}}$$

(c) In interval II, $v_0 = 6 \text{ m s}^{-1}$, which is the velocity at the beginning of the interval; $t = (10 - 4) = 6$ s, which is the duration of the interval; and $a = 0$. Use these values in Eq. (2-5) to find the distance traveled:

$$x = v_0 t + \tfrac{1}{2}at^2 = (6 \text{ m s}^{-1})(6 \text{ s}) + 0 = \boxed{36 \text{ m}}$$

(d) Solve Eq. (2-6) for a, with $t = (15 - 10) = 5$ s:

$$a = \frac{v - v_0}{t} = \frac{(10 - 6) \text{ m s}^{-1}}{5 \text{ s}} = \boxed{0.8 \text{ m s}^{-2}}$$

(e) Use Eq. (2-5):

$$x = v_0 t + \tfrac{1}{2}at^2 = (6 \text{ m s}^{-1})(5 \text{ s}) + \tfrac{1}{2}(0.8 \text{ m s}^{-2})(5 \text{ s})^2 = \boxed{40 \text{ m}}$$

(f) Solve Eq. (2-6) for a:

$$a = \frac{v - v_0}{t} = \frac{0 - 10 \text{ m s}^{-1}}{6 \text{ s}} = \boxed{-1.67 \text{ m s}^{-2}}$$

note: The negative sign indicates that the train is slowing down (decelerating).

(g) Use Eq. (2-5):

$$x = v_0 t + \tfrac{1}{2}at^2 = (10 \text{ m s}^{-1})(6 \text{ s}) + \tfrac{1}{2}(-1.67 \text{ m s}^{-2})(6 \text{ s})^2 = \boxed{30 \text{ m}}$$

EXAMPLE 2-4: A small car travels 80 m, during which its velocity uniformly increases from 20 m s^{-1} to 25 m s^{-1}. (a) Determine the acceleration of the car. (b) How long does it take to travel the 80 m?

Solution:

(a) Here, you don't know the time, so you solve Eq. (2-7) for a:

$$v^2 = v_0^2 + 2ax$$

$$a = \frac{v^2 - v_0^2}{2x} = \frac{(25 \text{ m s}^{-1})^2 - (20 \text{ m s}^{-1})^2}{2(80 \text{ m})} = \boxed{1.41 \text{ m s}^{-2}}$$

(b) Now you can solve Eq. (2-6) for t:

$$v = v_0 + at$$

$$t = \frac{v - v_0}{a} = \frac{(25 - 20) \text{ m s}^{-1}}{1.41 \text{ m s}^{-2}} = \boxed{3.55 \text{ s}}$$

EXAMPLE 2-5: A bicyclist increases her speed uniformly at the rate of 0.8 m s^{-2} for 10 s. At the end of this time interval her speed is 12 m s^{-1}. (a) How far does she travel in these 10 s? (b) What is her speed at the beginning of the 10-s interval?

Solution:

(a) Here, you don't know the initial speed (|velocity|), so you use Eq. (2-8):

$$x = vt - \tfrac{1}{2}at^2 = (12 \text{ m s}^{-1})(10 \text{ s}) - \tfrac{1}{2}(0.8 \text{ m s}^{-2})(10 \text{ s})^2 = \boxed{80 \text{ m}}$$

(b) Solve Eq. (2-7) for v_0:

$$v^2 = v_0^2 + 2ax$$

$$v_0 = \sqrt{v^2 - 2ax} = \sqrt{(12 \text{ m s}^{-1})^2 - 2(0.8 \text{ m s}^{-2})(80 \text{ m})} = \boxed{4 \text{ m s}^{-1}}$$

2-5. Bodies in Free Fall

Objects that do not experience much air resistance fall with the acceleration due to **gravity g**. Near the earth's surface this acceleration, $a = g$, is directed downward and has the following approximate values:

$$\text{SI system} \qquad g = 9.8 \text{ m s}^{-2}$$
$$\text{British system} \qquad g = 32 \text{ ft s}^2$$

So you can use the kinematic equations, Eqs. (2-5) through (2-8), for objects in free fall.

- $a = -g$ when the direction of x is specified as positive upward, $a = +g$ when x is specified as positive downward.

EXAMPLE 2-6: A ball is thrown vertically upward with a velocity of 8 m s^{-1}. **(a)** How high does it rise above the point of release? **(b)** How long does it take to reach this height? **(c)** How fast is it going when it is at the halfway point coming down? **(d)** How long does it take for the ball to complete the trip and return to the point of release?

Solution: Draw a sketch of the ball's path, letting the positive direction be upward, as in Figure 2-4.

(a) You don't know the time, so you solve Eq. (2-7) for $x = h$, with $v = 0$ at the point $x = h$ and $a = -g$:

$$v^2 = v_0^2 + 2ax$$

$$0 = v_0^2 + 2(-g)h$$

$$h = \frac{v_0^2}{2g} = \frac{(8 \text{ m s}^{-1})^2}{2(9.8 \text{ m s}^{-2})} = \boxed{3.26 \text{ m}}$$

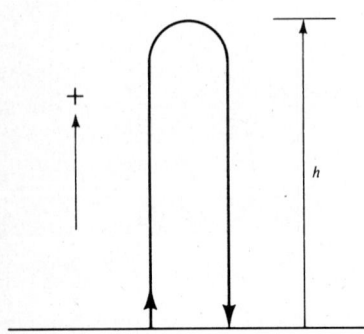

Figure 2-4

(b) Now you can find the time by solving Eq. (2-6) for t, with $v = 0$ and $a = -g$:

$$v = v_0 + at$$

$$0 = v_0 + (-g)t$$

$$t = \frac{v_0}{g} = \frac{8 \text{ m s}^{-1}}{9.8 \text{ m s}^{-2}} = \boxed{0.82 \text{ s}}$$

(c) Solve Eq. (2-7) for v, with $x = h/2$ and $a = -g$:

$$v^2 = v_0^2 + 2ax$$

$$v = \sqrt{v_0^2 + 2(-g)(h/2)} = \sqrt{(8 \text{ m s}^{-1})^2 - 2(9.8 \text{ m s}^{-2})((3.26/2) \text{ m})}$$

$$= \boxed{\pm 5.66 \text{ m s}^{-1}}$$

note: The *speed* of the ball is $|\pm 5.66| = 5.66 \text{ m s}^{-1}$ whether the ball is traveling up or down. The *velocity* is $\mathbf{v} = -5.66 \text{ m s}^{-1}$ coming down.

(d) Solve Eq. (2-5) for t, with $x = 0$ and $a = -g$:

$$x = v_0 t + \tfrac{1}{2}at^2 = (v_0 - \tfrac{1}{2}gt)t$$

This equation has two roots: One is $t = 0$, which is the trivial solution that indicates the ball was just released, and the other is

$$t = \frac{2v_0}{g} = \frac{2(8 \text{ m s}^{-1})}{9.8 \text{ m s}^{-2}} = \boxed{1.63 \text{ s}}$$

Note that this solution is just twice the time determined in part (b) for the ball to rise to $x = h$.

EXAMPLE 2-7: A small projectile is shot vertically upward from a point high above the ground; its initial velocity is 15 m s^{-1}. (a) What will the velocity of the projectile be when it is 20 m below the release point? (b) How long after release will it be 20 m below the release point?

Solution: Sketch the situation, as in Figure 2-5. Let the positive direction be upward.

(a) Solve Eq. (2-7) for v, with $x = -20$ m and $a = -g$:

$$v^2 = v_0^2 + 2ax$$

$$v = \sqrt{v_0^2 + 2(-g)x} = \sqrt{(15 \text{ m s}^{-1})^2 - 2(9.8 \text{ m s}^{-2})(-20 \text{ m})} = \boxed{\pm 24.8 \text{ m s}^{-1}}$$

note: Because the projectile is moving *downward* at this point, the velocity of the projectile should be the *negative* solution, or -24.8 m s^{-1}.

(b) Solve Eq. (2-6) for t, with $a = -g$:

$$v = v_0 + at$$

$$t = \frac{v - v_0}{-g} = \frac{-24.8 \text{ m s}^{-1} - 15 \text{ m s}^{-1}}{-9.8 \text{ m s}^{-2}} = \boxed{4.06 \text{ s}}$$

note: We don't switch from $-g$ to $+g$ just because the projectile is going down. We decided at the beginning that the positive direction was up—and *up* is up, regardless of the projectile's direction.

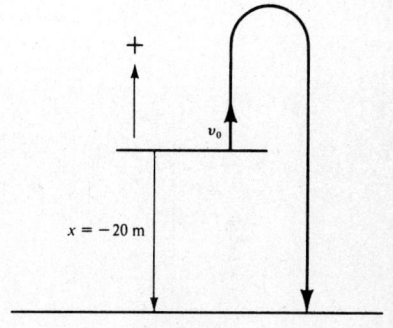

Figure 2-5

2-6. Applications of the Calculus for the Kinematic Equations

If the *position* of a body is known as a function of time, its velocity and acceleration can be obtained by *differentiation*:

$$x = f(t) \tag{2-9a}$$

$$v = \frac{dx}{dt} = \frac{df(t)}{dt} \tag{2-9b}$$

$$a = \frac{dv}{dt} = \frac{d^2x}{dt^2} = \frac{d^2f(t)}{dt^2} \tag{2-9c}$$

If the *acceleration* of a body is known as a function of time, the velocity and position of the body can be obtained by *integration*:

$$a = F(t) \tag{2-10a}$$

$$v = \int a \, dt + C_1 = \int F(t) \, dt + C_1 \tag{2-10b}$$

$$x = \int v \, dt + C_2 \tag{2-10c}$$

The constants C_1 and C_2 can be determined from the initial conditions of the problem.

EXAMPLE 2-8: The position of a body moving along the x axis can be described by the function $x = pt^3 - qt^2 + st$. Determine the velocity and acceleration of the object, assuming that p, q, and s are constants.

Solution: From Eq. (2-9b), the velocity is

$$v = \frac{dx}{dt} = \boxed{3pt^2 - 2qt + s}$$

and from Eq. (2-9c), the acceleration is

$$a = \frac{dv}{dt} = \boxed{6pt - 2q}$$

note: In the SI system, p has units of $m\,s^{-3}$, q has units of $m\,s^{-2}$, and s has units of $m\,s^{-1}$.

EXAMPLE 2-9: A particle is subject to a constant acceleration k. Determine its velocity and position by integration. Initially, at $t = 0$, $v = v_0$ and $x = x_0$.

Solution: Use Eqs. (2-10b, c). For velocity,

$$v = \int a\, dt + C_1 = \int k\, dt + C_1 = kt + C_1$$

At $t = 0$, $v = v_0 = 0 + C_1$. Thus $v = kt + v_0$, which can be written as

$$v = v_0 + kt = \boxed{v_0 + at}$$

For position,

$$x = \int v\, dt + C_2 = \int (v_0 + at)\, dt + C_2 = v_0 t + \tfrac{1}{2}at^2 + C_2$$

At $t = 0$, $x = x_0 = 0 + 0 + C_2$. Thus

$$x = \boxed{x_0 + v_0 t + \tfrac{1}{2}at^2}$$

SUMMARY

1. The displacement of a particle from position x_1 to position x_2 is

$$\Delta x = x_2 - x_1.$$

2. The average velocity of a particle is

$$\bar{v} = \Delta x / \Delta t$$

where $\Delta x = x_2 - x_1$ and $\Delta t = t_2 - t_1$.

3. The instantaneous velocity and acceleration of a particle are given by

$$v = dx/dt \quad \text{and} \quad a = dv/dt = d^2x/dt^2$$

4. The kinematic equations that describe the straight-line motion of a particle subject to a constant acceleration are

$$\left.\begin{aligned} x &= v_0 t + \tfrac{1}{2}at^2 \\ v &= v_0 + at \\ v^2 &= v_0^2 + 2ax \\ x &= vt - \tfrac{1}{2}at^2 \end{aligned}\right\} \quad \begin{aligned} a &= \text{constant} \\ x_0 &= 0 \end{aligned}$$

5. A particle in "free fall" is subject to the constant acceleration due to gravity, $g = 9.8\ m\,s^{-2}$ ($= 32\ ft\,s^{-2}$), directed downward.

RAISE YOUR GRADES

Can you explain . . . ?

- ☑ the subsequent motion of a particle that, at one instant, has zero velocity but constant acceleration
- ☑ the subsequent motion of a particle that has a negative acceleration
- ☑ the motion of a particle that has a negative velocity
- ☑ the subsequent motion of a particle that, at one instant, has a negative velocity but a positive acceleration
- ☑ the acceleration of a vertically thrown ball at the top of its path

SOLVED PROBLEMS

Constant Accelerated Motion

PROBLEM 2-1: A car accelerates at the constant rate of 1.8 m s^{-2} along a straight road. The velocity of the car is 12 m s^{-1} when it is 16 m past an intersection. Determine the velocity of the car as it passes the intersection.

Solution: You know that $a = 1.8$ m s^{-2} and that $v = 12$ m s^{-1} at $x = 16$ m, but you don't know the time. You want to determine v_0, so you solve Eq. (2-7) for v_0:

$$v^2 = v_0^2 + 2ax$$

$$v_0 = \sqrt{v^2 - 2ax} = \sqrt{(12 \text{ m s}^{-1})^2 - 2(1.8 \text{ m s}^{-2})(16 \text{ m})} = \boxed{\pm 9.3 \text{ m s}^{-1}}$$

The car is still *accelerating* when it passes the intersection, so its velocity is $v = 9.3$ m s^{-1}.

PROBLEM 2-2: A train moving along a straight track at 22 m s^{-1} accelerates uniformly to 28 m s^{-1}. The change in velocity takes 18 s. **(a)** Determine the acceleration for this time interval. **(b)** How far does the train travel during the period of acceleration?

Solution: You know that $v_0 = 22$ m s^{-1}, $v = 28$ m s^{-1}, and $t = 18$ s.

(a) To find the acceleration, solve Eq. (2-6) for a:

$$v = v_0 + at$$

$$a = \frac{v - v_0}{t} = \frac{28 \text{ m s}^{-1} - 22 \text{ m s}^{-1}}{18 \text{ s}} = \boxed{0.33 \text{ m s}^{-2}}$$

(b) To find the distance traveled, use Eq. (2-5):

$$x = v_0 t + \tfrac{1}{2}at^2 = (22 \text{ m s}^{-1})(18 \text{ s}) + \tfrac{1}{2}(0.33 \text{ m s}^{-2})(18 \text{ s})^2 = \boxed{450 \text{ m}}$$

PROBLEM 2-3: A car is decelerating at the constant rate of 3 m s^{-2}. At the instant the car passes a sign, its velocity is 18 m s^{-1}. **(a)** How far beyond the sign does the car stop? **(b)** How many seconds after the car passes the sign will the car come to a stop? **(c)** How far does the car travel in 4 s after it passes the sign?

Solution: You know that $a = -3 \text{ m s}^{-2}$, and you can set $v_0 = 18 \text{ m s}^{-1}$ at $t = 0$.

(a) Solve Eq. (2-7) for x:

$$v^2 = v_0^2 + 2ax$$

$$x = \frac{v^2 - v_0^2}{2a} = \frac{0 - (18 \text{ m s}^{-1})^2}{2(-3 \text{ m s}^{-2})} = \boxed{54 \text{ m}}$$

(b) Solve Eq. (2-6) for t:

$$v = v_0 + at$$

$$t = \frac{v - v_0}{a} = \frac{0 - 18 \text{ m s}^{-1}}{-3 \text{ m s}^{-2}} = \boxed{6 \text{ s}}$$

(c) Use Eq. (2-5):

$$x = v_0 t + \tfrac{1}{2}at^2 = (18 \text{ m s}^{-1})(4 \text{ s}) + \tfrac{1}{2}(-3 \text{ m s}^{-2})(4 \text{ s})^2 = \boxed{48 \text{ m}}$$

PROBLEM 2-4:   Starting from rest, a truck accelerates uniformly at the rate of 3 m s^{-2}. A car starts from rest from the same location 12 s later, but accelerates at the rate of 5 m s^{-2}. How far from the starting point does the car pass the truck?

Solution: You'll have to do a bit more algebra in this problem than in the preceding ones. Let a_t be the acceleration of the truck, t be the time the truck travels, a_c be the acceleration of the car, $t - T$ (where $T = 12$ s), be the time the car travels, and x be the distance traveled when the car passes the truck.

Use Eq. (2-5), $x = v_0 t + \tfrac{1}{2}at^2$, for both car and truck with $v_0 = 0$ for both. For the truck,

$$x = 0 + \tfrac{1}{2}a_t t^2$$

and for the car,

$$x = 0 + \tfrac{1}{2}a_c(t - T)^2$$

Because both car and truck travel the same distance, equate the two expressions and solve the resulting quadratic equation for t:

$$\tfrac{1}{2}a_t t^2 = \tfrac{1}{2}a_c(t - T)^2 = \tfrac{1}{2}a_c(t^2 - 2Tt + T^2)$$

$$(a_c - a_t)t^2 - 2a_c Tt + a_c T^2 = 0$$

You'll find it convenient to substitute the numerical values for a_c, a_t, and T at this point; so

$$(5 - 3)t^2 - (2)(5)(12)t + 5(12^2) = 0$$

$$2t^2 - 120t + 720 = 0$$

$$t^2 - 60t + 360 = 0$$

Complete the solution using the quadratic formula:

$$t = \frac{60 \pm \sqrt{3600 - 1440}}{2} = 6.762 \text{ s} \qquad \text{and} \qquad 53.24 \text{ s}$$

The root $t = 6.762$ s must be discarded because the car starts 12 s after the truck. You therefore substitute $t = 53.24$ s into Eq. (2-5) to find x:

$$x = 0 + \tfrac{1}{2}a_t t^2 = \tfrac{1}{2}(3 \text{ m s}^{-2})(53.24 \text{ s})^2 = \boxed{4.25 \times 10^3 \text{ m}}$$

PROBLEM 2-5:   The minimum stopping distance for a certain car moving at 90 km h^{-1} is 36 m. **(a)** Determine the acceleration of the car. **(b)** How long does it take to stop? Express your answers in SI base units (meters and seconds).

Solution: First, convert the given velocity into SI units, i.e.,

$$\left(\frac{90 \text{ km}}{1 \text{ h}}\right)\left(\frac{1 \times 10^3 \text{ m}}{1 \text{ km}}\right)\left(\frac{1 \text{ h}}{60 \text{ min}}\right)\left(\frac{1 \text{ min}}{60 \text{ s}}\right) = 25 \text{ m s}^{-1}$$

(a) Solve Eq. (2-7) for a:

$$v^2 = v_0^2 + 2ax$$

$$a = \frac{v^2 - v_0^2}{2x} = \frac{0 - (25 \text{ m s}^{-1})^2}{2(36 \text{ m})} = \boxed{-8.68 \text{ m s}^{-2}}$$

(b) Solve Eq. (2-6) for t and use the result from part (a):

$$v = v_0 + at$$

$$t = \frac{v - v_0}{a} = \frac{0 - 25 \text{ m s}^{-1}}{-8.68 \text{ m s}^{-2}} = \boxed{2.88 \text{ s}}$$

PROBLEM 2-6: A bicyclist accelerates at 1.5 m s^{-2} for 6 s. During this interval she travels 33 m. Determine her velocity at the end of this interval.

Solution: Note that you don't know the velocity at the *beginning* of the interval, so you solve Eq. (2-8) for v:

$$x = vt - \frac{1}{2}at^2$$

$$v = \frac{x}{t} + \frac{1}{2}at = \frac{33 \text{ m}}{6 \text{ s}} + \frac{1}{2}(1.5 \text{ m s}^{-2})(6 \text{ s}) = \boxed{10 \text{ m s}^{-1}}$$

Bodies in Free Fall

PROBLEM 2-7: A small stone is tossed upward with an initial velocity of 7 m s^{-1} and lands on a shelf 2 m above the point of release. **(a)** Determine the velocity of the stone as it strikes the shelf. **(b)** How long was it in flight?

Solution: Sketch the situation, as in Figure 2-6, letting the positive direction be upward. You know that $v_0 = +7 \text{ m s}^{-1}$, $a = -g = -9.8 \text{ m s}^{-2}$, and $x = +2 \text{ m}$. Now use the appropriate kinematic equation to get the answer for each part.

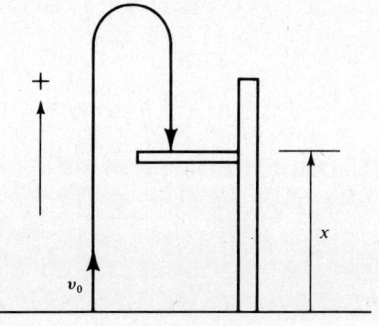

Figure 2-6

(a) Solve Eq. (2-7) for v:

$$v^2 = v_0^2 + 2ax$$

$$v = \pm\sqrt{(7 \text{ m s}^{-1})^2 + 2(-9.8 \text{ m s}^{-2})(2 \text{ m})}$$

$$= \pm 3.13 \text{ m s}^{-1}$$

You should select the negative solution because the stone is moving downward at the instant it touches. Thus

$$v = \boxed{-3.13 \text{ m s}^{-1}}$$

(b) Solve Eq. (2-6) for t:

$$v = v_0 + at$$

$$t = \frac{v - v_0}{a} = \frac{-3.13 \text{ m s}^{-1} - 7 \text{ m s}^{-1}}{-9.8 \text{ m s}^{-2}} = \boxed{1.03 \text{ s}}$$

PROBLEM 2-8: Clay pigeons can be thrown vertically with an initial velocity of 6 m s^{-1}. How much time must elapse between throws if a second pigeon is to collide with the first 1.2 m above the point of release?

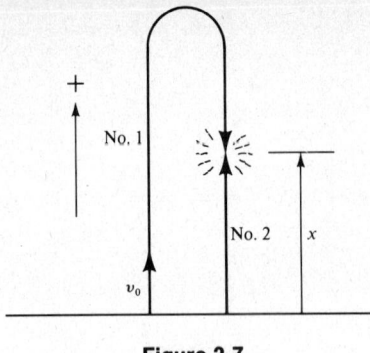

Figure 2-7

Solution: Sketch the situation, as in Figure 2-7, letting the positive direction be upward. You know that $v_0 == +6\ \mathrm{m\,s^{-1}}$, $a = -g = -9.8\ \mathrm{m\,s^{-2}}$, and $x = 1.2$ m. Rearrange Eq. (2-5) and solve for t by using the quadratic formula:

$$x = v_0 t + \tfrac{1}{2}at^2 = v_0 t - \tfrac{1}{2}gt^2$$

$$\frac{g}{2}t^2 - v_0 t + x = 0$$

$$4.9t^2 - 6t + 1.2 = 0$$

$$t = \frac{6 \pm \sqrt{36 - 23.52}}{9.8}$$

$$= 0.973\ \mathrm{s} \qquad \text{and} \qquad 0.252\ \mathrm{s}$$

The two roots correspond to the times of travel of the first and second clay pigeons, respectively. The difference between them is the time between the release of the two clay pigeons.

$$t = t_1 - t_2 = (0.973\ \mathrm{s}) - (0.252\ \mathrm{s}) = \boxed{0.721\ \mathrm{s}}$$

Applications of the Calculus for the Kinematic Equations

PROBLEM 2-9: The position of a jet-powered toy car on a straight track during its first 3 s of motion is $x = kt^3$. It subsequently slows down and comes to a stop. At the end of the 2nd second it has traveled 5.6 m. **(a)** Evaluate the constant k. **(b)** How fast is the car moving at the end of the 3rd second? **(c)** Determine the acceleration of the car at the end of the 2nd second.

Solution:

(a) Solve $x = kt^3$ for k and evaluate: $k = \dfrac{x}{t^3} = \dfrac{5.6\ \mathrm{m}}{(2\ \mathrm{s})^3} = \boxed{0.7\ \mathrm{m\,s^{-3}}}$

(b) Differentiate to obtain the velocity function and then evaluate:

$$v = \frac{dx}{dt} = 3kt^2 = 3(0.7\ \mathrm{m\,s^{-3}})(3\ \mathrm{s})^2 = \boxed{18.9\ \mathrm{m\,s^{-1}}}$$

(c) Differentiate the velocity function to obtain the acceleration function and then evaluate it:

$$a = \frac{dv}{dt} = 6\,kt = 6(0.7\ \mathrm{m\,s^{-3}})(2\ \mathrm{s}) = \boxed{8.4\ \mathrm{m\,s^{-2}}}$$

Note that the acceleration of the car is not constant during its first 3 s of motion.

PROBLEM 2-10: The acceleration of a small projectile down a long straight tube between the first and fifth seconds of its travel is $a = b/t^2$. At the end of the first second it has traveled 2 m, its velocity is $3\ \mathrm{m\,s^{-1}}$, and its acceleration is $4\ \mathrm{m\,s^{-2}}$. Determine the velocity and position of the projectile as functions of time. Evaluate all constants.

Solution: You can determine the constant b from the information given:

$$b = at^2 = (4\ \mathrm{m\,s^{-2}})(1\ \mathrm{s})^2 = \boxed{4\ \mathrm{m}}$$

Integrate the acceleration function to obtain the velocity:

$$v = \int a\,dt + C_1 = \int \frac{b}{t^2}\,dt + C_1 = -\frac{b}{t} + C_1$$

Evaluate C_1 at $t = 1$ s:

$$C_1 = v + \frac{b}{t} = 3\ \mathrm{m\,s^{-1}} + \frac{4\ \mathrm{m}}{1\ \mathrm{s}} = \boxed{7\ \mathrm{m\,s^{-1}}}$$

Now integrate the velocity function to obtain the position function:

$$x = \int v\, dt + C_2 = \int\left(-\frac{b}{t} + C_1\right) dt + C_2 = -b(\ln t) + C_1 t + C_2$$

Finally, evaluate C_2 at $t = 1$ s:

$$C_2 = x + b(\ln t) - C_1 t = 2\text{ m} + (4\text{ m})(\ln 1) - (7\text{ m s}^{-1})(1\text{ s}) = \boxed{-5\text{ m}}$$

Supplementary Exercises

EXERCISE 2-1: A car accelerates down a straight road at the rate of 1.8 m s^{-2}. It takes the car 14 s to travel the distance between two marker flags that are 300 m apart. How fast was the car moving when it passed the first flag?

EXERCISE 2-2: From a standing start, a skier accelerates down a long straight hill. She travels a distance of 120 m with a constant acceleration of 0.8 m s^{-2}. How long did it take her to cover that distance?

EXERCISE 2-3: The speed of a small racing airplane increases from 60 m s^{-1} to 75 m s^{-1} in 3.5 s. Determine the acceleration of the plane during this interval.

EXERCISE 2-4: The speed of a falling object at a given instant is 44 m s^{-1}. What was its speed 2 s earlier?

EXERCISE 2-5: A speeding motorist slows from 35 m s^{-1} to 20 m s^{-1} in a distance of 80 m after seeing a police car. What was the acceleration of his car?

EXERCISE 2-6: An object in vertical free fall passes the bottom of a 1.6-m high window outside a school building at 8 m s^{-1}. What was its speed when it passes the top of the window?

EXERCISE 2-7: Determine the constant acceleration of a sports car that takes 4 s to travel 84 m while accelerating. The car is moving at 25 m s^{-1} at the end of the 4-s interval.

EXERCISE 2-8: A small projectile is shot vertically upward. How much time elapses between the instant it is released until it is moving downward at 20 m s^{-1} at a point 1.6 m above the release point?

EXERCISE 2-9: The position of a toy car on a track during its first 10 s of motion is given by $x = Xe^{kt}$, where $X = 2$ m and $k = 0.3\text{ s}^{-1}$. How fast is the car moving after it has traveled for 4 s?

EXERCISE 2-10: During the first 8 s of its motion, the velocity of a small sphere projected down a long tube is given by $v = V^{kt}$, where $V = 1.2\text{ m s}^{-1}$ and $k = 0.4\text{ s}^{-1}$. At $t = 0$ the sphere is located at $x = 20$ m. Determine **(a)** the position of the sphere as a function of time during its first 8 s of motion and **(b)** its position at $t = 2$ s.

Answers to Supplementary Exercises

2-1: 8.83 m s^{-1}

2-2: 17.3 s

2-3: 4.29 m s^{-2}

2-4: 24.4 m s^{-1}

2-5: -5.16 m s^{-2}

2-6: 5.71 m s^{-1}

2-7: 2 m s^{-2}

2-8: 4.16 s

2-9: 1.99 m s^{-1}

2-10: **(a)** $x = \dfrac{X + V^{kt}}{k \ln V}$, where

$X = 6.29$ m

(b) $x = 22.16$ m

3 MOTION IN TWO DIMENSIONS

THIS CHAPTER IS ABOUT

- ☑ **Displacement, Velocity, and Acceleration**
- ☑ **Motion with Constant Acceleration: Kinematic Equations**
- ☑ **Projectile Motion**
- ☑ **Uniform Circular Motion**
- ☑ **Accelerated Circular Motion**
- ☑ **Relative Velocity**
- ☑ **Rotational Motion**

agreement: Motion in two dimensions is motion in a plane. In this chapter, we'll consider the motion of a particle in the xy plane.

3-1. Displacement, Velocity, and Acceleration

A. Displacement

Displacement ($\Delta \mathbf{R}$) in the xy plane is the vector difference between two position vectors $\mathbf{R}_1$ and $\mathbf{R}_2$ that locate the initial $P_1(x_1, y_1)$ and final $P_2(x_2, y_2)$ positions of a particle. The particle can move along any path from point $P_1(x_1, y_1)$ to point $P_2(x_2, y_2)$, but the magnitude of the displacement is always the straight-line distance between points P_1 and P_2. The displacement (Figure 3-1a) is written as

DISPLACEMENT (in two dimensions)
$$\Delta \mathbf{R} = \mathbf{R}_2 - \mathbf{R}_1 \tag{3-1a}$$

The scalar components of $\Delta \mathbf{R}$ (Figure 3-1b) are

$$\Delta R_x = x_2 - x_1 \quad \text{and} \quad \Delta R_y = y_2 - y_1 \tag{3-1b}$$

The magnitude of $\Delta \mathbf{R}$ is

$$\Delta R = \sqrt{(\Delta R_x)^2 + (\Delta R_y)^2} \tag{3-2}$$

The angle θ between $\Delta \mathbf{R}$ and the x axis can be obtained from

$$\theta = \arctan \frac{|\Delta R_y|}{|\Delta R_x|} \tag{3-3}$$

Expressed in terms of unit vectors, the displacement is

$$\Delta \mathbf{R} = R_x \hat{\mathbf{x}} + R_y \hat{\mathbf{y}} \tag{3-4}$$

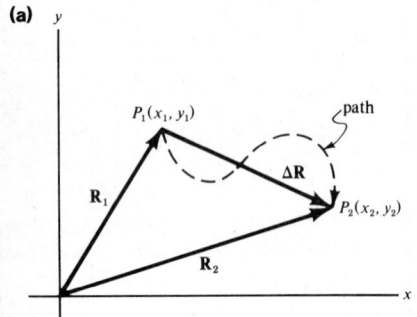
(a)

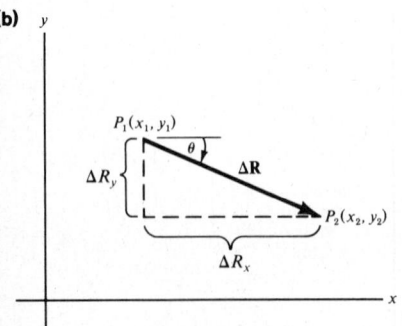
(b)

Figure 3-1. (a) Displacement is the vector difference between the final and initial position vectors. (b) ΔR_x and ΔR_y are the components of $\Delta \mathbf{R}$.

EXAMPLE 3-1: A somewhat inebriated person starts at point P_1 at $x_1 = 2$ m and $y_1 = 3$ m and wanders to point P_2 at $x_2 = 10$ m and $y_2 = 8$ m. Determine the magnitude and direction of his displacement.

Solution: Sketch the situation, as in Figure 3-2.

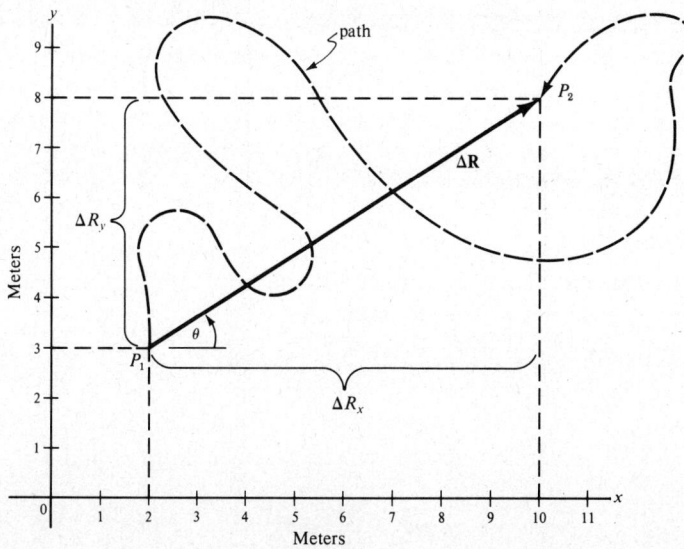

Figure 3-2

Use Eq. (3-1b) to get the components of $\Delta \mathbf{R}$:

$$\Delta R_x = x_2 - x_1 \qquad\qquad \Delta R_y = y_2 - y_1$$
$$= (10 - 2) \text{ m} = 8 \text{ m} \qquad = (8 - 3) \text{ m} = 5 \text{ m}$$

Use Eq. (3-2) to obtain the magnitude,

$$\Delta R = \sqrt{(\Delta R_x)^2 + (\Delta R_y)^2} = \sqrt{(8 \text{ m})^2 + (5 \text{ m})^2} = \boxed{9.43 \text{ m}}$$

and Eq. (3-3) to get the direction,

$$\theta = \arctan \frac{|\Delta R_y|}{|\Delta R_x|} = \arctan\left(\frac{5 \text{ m}}{8 \text{ m}}\right) = \boxed{32.0°}$$

B. Velocity

The **average velocity** for a particle moving in the xy plane is a vector quantity, which can be determined by dividing the displacement ($\Delta \mathbf{R}$) of the particle by the time ($\Delta t = t_2 - t_1$) required for it to go from P_1 to P_2:

$$\mathbf{v}_{av} = \frac{\Delta \mathbf{R}}{\Delta t} \qquad\qquad (3\text{-}5)$$

Note that the average velocity is *not* the same as the **average speed** v_{av}, which is a scalar quantity:

$$v_{av} = \frac{\text{Length of path between } P_1 \text{ and } P_2}{\Delta t}$$

The **instantaneous velocity** (usually referred to as simply *the velocity*) at a point along the path is given by the limit

**INSTANTANEOUS
VELOCITY
(in two dimensions)**
$$\mathbf{v} = \lim_{\Delta t \to 0} \frac{\Delta \mathbf{R}}{\Delta t} = \frac{d\mathbf{R}}{dt} \qquad\qquad (3\text{-}6)$$

where $\mathbf{R}$ is the position vector for the particle.

• The **speed** of the particle is the magnitude of its (instantaneous) velocity.

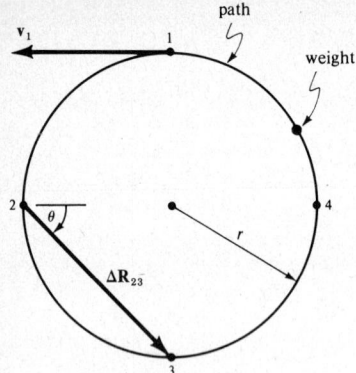

Figure 3-3

EXAMPLE 3-2: A child swings a weight on the end of a string around a horizontal circular path of radius 0.9 m. The weight moves at a constant speed, making 2 complete revolutions each second.

Determine (**a**) the average speed of the weight, (**b**) its average velocity between points 2 and 3, and (**c**) its instantaneous velocity at point 1, where the arc between two consecutive points is $\pi r/2$, as shown in Figure 3-3.

Solution:

(**a**) Because the weight moves at constant speed, its average speed is equal to its instantaneous speed at all points around the circular path. And since the weight completes 2 revolutions in 1 s, the time interval for one complete revolution is $\Delta t = 0.5$ s. So the speed is

$$v_{av} = \frac{\text{Circumference}}{\Delta t} = \frac{2\pi r}{\Delta t} = \frac{2\pi(0.9 \text{ m})}{0.5 \text{ s}} = \boxed{11.3 \text{ m s}^{-1}}$$

(**b**) Use Eq. (3-5) to get the average velocity between points 2 and 3:

$$\mathbf{v}_{av} = \frac{\Delta \mathbf{R}}{\Delta t} = \frac{\Delta \mathbf{R}_{23}}{\Delta t}$$

In this case, $\Delta R_{23} = \sqrt{2r^2} = \sqrt{2}r$ and $\Delta t = (\frac{1}{4})(0.5 \text{ s}) = 0.125$ s, so

$$|\mathbf{v}_{av}| = \frac{\Delta R_{23}}{\Delta t} = \frac{\sqrt{2}(0.9 \text{ m})}{0.125 \text{ s}} = \boxed{10.2 \text{ m s}^{-1}}$$

and the direction of $\mathbf{v}_{av}$ is the same as that of $\Delta \mathbf{R}_{23}$, which is, of course, $-45°$.

(**c**) The magnitude of $\mathbf{v}_1$ is the same as v_{av}, i.e.,

$$|\mathbf{v}_1| = v_{av} = \boxed{11.3 \text{ m s}^{-1}}$$

and its direction is shown in Figure 3-3.

C. Acceleration

The **average acceleration** of a particle between any two points 1 and 2 in the plane is the change in the velocity of the particle $\Delta \mathbf{v} = \mathbf{v}_2 - \mathbf{v}_1$, divided by the time interval $\Delta t = t_2 - t_1$:

AVERAGE ACCELERATION
$$\mathbf{a}_{av} = \frac{\mathbf{v}_2 - \mathbf{v}_1}{t_2 - t_1} = \frac{\Delta \mathbf{v}}{\Delta t} \tag{3-7}$$

The **instantaneous acceleration** at a point in the plane is given by

INSTANTANEOUS ACCELERATION
$$\mathbf{a} = \lim_{\Delta t \to 0} \frac{\Delta \mathbf{v}}{\Delta t} = \frac{d\mathbf{v}}{dt} \tag{3-8}$$

note: The velocity of a particle may change in magnitude only, in direction only, or in both magnitude and direction. If any one of these possibilities occurs, the particle experiences acceleration.

In terms of scalar components and unit vectors, the velocity and acceleration can be written as

$$\mathbf{v} = \frac{dR_x}{dt}\hat{\mathbf{x}} + \frac{dR_y}{dt}\hat{\mathbf{y}} \tag{3-9}$$

$$\mathbf{a} = \frac{dv_x}{dt}\hat{\mathbf{x}} + \frac{dv_y}{dt}\hat{\mathbf{y}} \tag{3-10}$$

EXAMPLE 3-3: **(a)** Can a particle have a velocity but not an acceleration? **(b)** Can a particle have an acceleration but not a velocity? **(c)** Can a particle move with constant speed and also be accelerated?

Solution: In a word, *yes*—to all three.

(a) If the velocity of the particle is constant, its acceleration is zero.
(b) If an object (particle) is released from rest at a height above the ground, its velocity is zero at the instant of release, but it has acceleration, which is downward and equal to the acceleration due to gravity.
(c) The velocity of a particle moving around a circle at constant speed is continuously changing because its direction is continuously changing—and so is undergoing continuous acceleration.

3-2. Motion with Constant Acceleration: Kinematic Equations

To obtain the velocity of a particle undergoing a constant acceleration, integrate Eq. (3-8):

$$\mathbf{v} = \mathbf{v}_0 + \mathbf{a}t \qquad (3\text{-}11)$$

where $\mathbf{v}_0$ is the velocity of the particle at $t = 0$. [Compare with Eq. (2-6), $v = v_0$ $v = v_0 + at$.] A second integration gives

$$\mathbf{R} = \mathbf{R}_0 + \mathbf{v}_0 t + \tfrac{1}{2}\mathbf{a}t^2 \qquad (3\text{-}12)$$

where $\mathbf{R}_0$ is the position vector of the particle at $t = 0$. [Compare with Eq. (2-5a), $x = x_0 + v_0 t + \tfrac{1}{2}at^2$.]

EXAMPLE 3-4: Prove that the motion of a particle subjected to a constant acceleration is confined to a plane.

Solution: Use Eq. (3-12) to obtain the displacement of the particle during the time interval t:

$$\Delta\mathbf{R} = \mathbf{R} - \mathbf{R}_0 = \mathbf{v}_0 t + \tfrac{1}{2}\mathbf{a}t^2$$

i.e., the vector $\Delta\mathbf{R}$ is the sum of two other vectors, $\mathbf{v}_0 t$ and $\tfrac{1}{2}\mathbf{a}t^2$, as shown in Figure 3-4. These three vectors determine the plane in which the particle's motion takes place. Common practice identifies this plane as the xy plane, with x as the horizontal axis and y as the vertical axis.

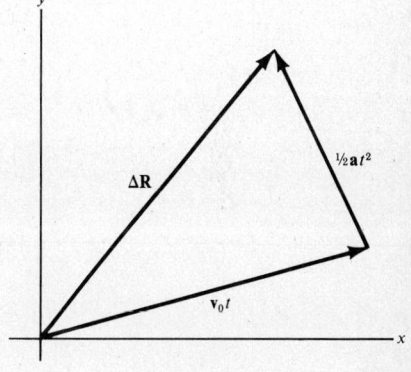

Figure 3-4

3-3. Projectile Motion

We'll neglect air friction in developing equations for the motion of a projectile that, we assume, takes place in the xy plane. The acceleration involved is that due to gravity, is constant, and acts vertically downward, i.e., $a_y = -g$. Consequently, $a_x = 0$, and the equations describing the motion of a projectile are given by

$$x = x_0 + v_{0x}t \qquad (3\text{-}13\text{a})$$

$$y = y_0 + v_{0y}t - \tfrac{1}{2}gt^2 \qquad (3\text{-}13\text{b})$$

$$v_y = v_{0y} - gt \qquad (3\text{-}13\text{c})$$

where v_{0x} is the x component of the velocity (v_{0x} is constant because $a_x = 0$), v_{0y} is the y velocity component at $t = 0$, and (x_0, y_0) is its initial position.

EXAMPLE 3-5: A golfer tees off from an elevated tee, $H = 3$ m (Figure 3-5a); the ball rises initially at an angle of elevation $\theta = 35°$ and speed $v_0 = 28$ m s^{-1}. (a) How far from the tee does the ball strike the ground? (b) What is the velocity of the ball (magnitude and direction) just before it lands?

(a)

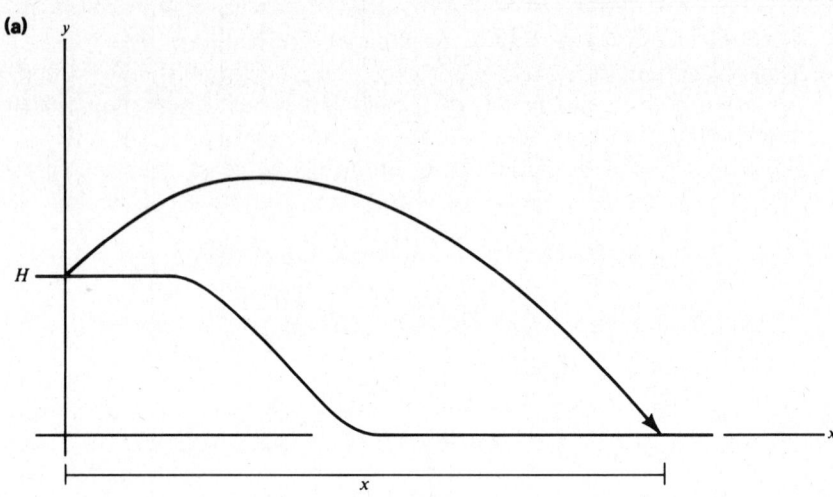

(b)   **(c)**

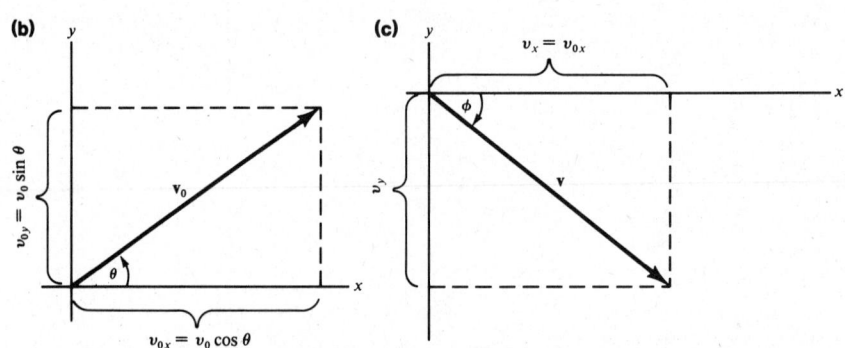

Figure 3-5

Solution:

(a) We use Eq. (3-13a) to obtain the horizontal distance traveled from the tee, with $x_0 = 0$, $v_{0x} = v_0 \cos \theta$ (Figure 3-5b), and $t =$ the time of flight:

$$x = x_0 + v_{0x}t = 0 + (v_0 \cos \theta)t$$

This equation has two unknowns, x and t. We can eliminate one unknown by solving Eq. (3-13b) for t, using the quadratic formula:

$$y = y_0 + v_{0y}t - \tfrac{1}{2}gt^2$$

$$0 = H + (v_0 \sin \theta)t - \tfrac{1}{2}gt^2$$

We can write this equation in standard form as

$$\tfrac{1}{2}gt^2 - (v_0 \sin \theta)t - H = 0$$

Substituting numbers, we get

$$4.9t^2 - 16.06t - 3 = 0$$

$$t = \frac{16.06 \pm \sqrt{258 + 58.8}}{9.8} = 3.455 \text{ s} \quad \text{and} \quad -0.177 \text{ s}$$

We disregard the negative root (because it has no physical significance) and evaluate x:

$$x = (v_0 \cos \theta)t = (28 \text{ m s}^{-1})(\cos 35°)(3.455 \text{ s}) = \boxed{79.24 \text{ m}}$$

(b) We determine **v** at the instant of impact by first calculating its scalar components (Figure 3-5b,c):

$$v_x = v_{0x} = v_0 \cos \theta$$
$$= (28 \text{ m s}^{-1}) \cos 35° = 22.94 \text{ m s}^{-1}$$

$$v_y = v_{0y} - gt = v_0 \sin \theta - gt \qquad [\text{Eq. (3-13c)}]$$
$$= (28 \text{ m s}^{-1}) \sin 35° - (9.8 \text{ m s}^{-2})(3.455 \text{ s})$$
$$= -17.80 \text{ m s}^{-1}$$

and then by evaluating v and ϕ, so that

$$v = \sqrt{v_x^2 + v_y^2}$$
$$= \sqrt{(22.94 \text{ m s}^{-1})^2 + (-17.80 \text{ m s}^{-1})^2} = \boxed{29.0 \text{ m s}^{-1}}$$

$$\phi = \arctan \frac{|v_y|}{|v_x|}$$

$$= \arctan \frac{|-17.80 \text{ m s}^{-1}|}{|22.94 \text{ m s}^{-1}|} = \boxed{37.8°}$$

3-4. Uniform Circular Motion

A particle moving around a circle (radius r) at a constant speed (v) is undergoing *uniform circular motion*. During this motion the particle experiences a constant (in magnitude) acceleration called the **centripetal acceleration** because its direction is toward the center of the circle.

CENTRIPETAL ACCELERATION $\qquad a_c = \dfrac{v^2}{r} \qquad\qquad$ (3-14)

EXAMPLE 3-6: Prove that the centripetal acceleration of a particle undergoing uniform circular motion is given by Eq. (3-14). Sketch the situation, as in Figure 3-6, exaggerating the separation between points 1 and 2 for clarity.

Solution: Initially, the particle is at point 1 and its position vector is $\mathbf{R}_0$. A short time later the particle is at point 2, its position vector is $\mathbf{R}$, where $|\mathbf{R}| = |\mathbf{R}_0| = R = r$ (the radius of the circle), and its displacement is therefore $\Delta\mathbf{R} = \mathbf{R} - \mathbf{R}_0$. Apply Eq. (3-12), as in Example 3-4. In this case $v_0 = v$, the constant speed of the particle. The time interval t involved is very small (the limit $t \to 0$ will ultimately be used), so **v** is perpendicular to $\mathbf{R}_0$ at point 1. In addition, the line of action of **a** must be parallel to **R** at all points because $v = |\mathbf{v}|$ is constant. Apply the Pythagorean theorem to the right triangle in Figure 3-6; i.e.,

$$(\tfrac{1}{2}at^2 + R)^2 = (vt)^2 + R_0^2$$
$$\tfrac{1}{4}a^2t^4 + at^2R + R^2 = v^2t^2 + R_0^2$$
$$(\tfrac{1}{4}a^2t^2)\cancel{t^2} = (v^2 - aR)\cancel{t^2} + R_0^2 - R^2$$
$$\tfrac{1}{4}a^2t^2 = (v^2 - aR) + 0$$

Take the limit as the time interval $t \to 0$ to obtain $0 = v^2 - aR$, and solve for a:

$$a = \frac{v^2}{R} = \frac{v^2}{r}$$

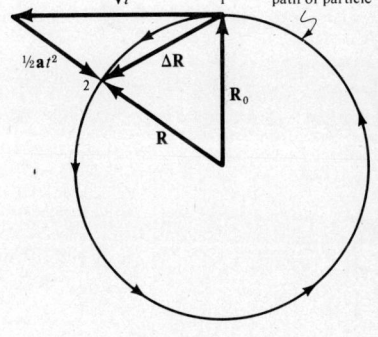

Figure 3-6.

3-5. Accelerated Circular Motion

A particle moving around a circle could change speed. If it does, there will be a corresponding acceleration in the direction of motion, which is always tangent to

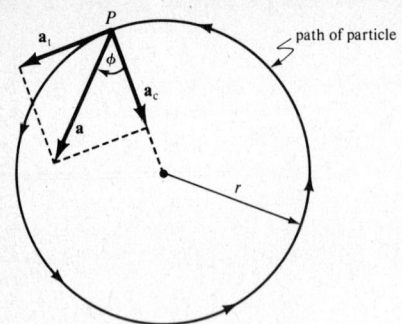

Figure 3-7. Path of a particle undergoing tangential acceleration as it moves in a circular path. The total acceleration is the vector sum of the tangential and centripetal accelerations.

the circle. This acceleration is called the **tangential acceleration** a_t and may be positive or negative depending on whether the speed of the particle is increasing or decreasing. Figure 3-7 shows a particle at point P, with an instantaneous speed v, a tangential acceleration a_t, and a centripetal acceleration $a_c = v^2/r$. The **total acceleration** of the particle is

TOTAL ACCELERATION $\qquad\qquad \mathbf{a} = \mathbf{a}_t + \mathbf{a}_c \qquad\qquad$ (3-15)

where

$$a = \sqrt{a_t^2 + a_c^2} \qquad\qquad (3\text{-}16)$$

and

$$\phi = \arctan\frac{|a_t|}{|a_c|} \qquad\qquad (3\text{-}17)$$

EXAMPLE 3-7: A small test rocket flies in a horizontal circle of radius 450 m. For a short time its speed is given by the formula $v = (4.2\ \mathrm{m\,s^{-2}})t + 6\ \mathrm{m\,s^{-1}}$. Determine at the instant $t = 12$ s the rocket's (**a**) tangential acceleration a_t, (**b**) centripetal acceleration a_c, and (**c**) total acceleration a.

Solution:

(**a**) Differentiate the expression for the speed of the rocket to obtain its tangential acceleration:

$$a_t = \frac{dv}{dt} = 4.2\ \mathrm{m\,s^{-2}}$$

Evidently, the tangential acceleration of the rocket is constant during the time the given expression is valid.

(**b**) Evaluate the speed of the rocket at $t = 12$ s and then calculate its centripetal acceleration using Eq. (3-14).

$$v = (4.2\ \mathrm{m\,s^{-2}})(12\ \mathrm{s}) + 6\ \mathrm{m\,s^{-1}} = 56.4\ \mathrm{m\,s^{-1}}$$

and

$$a_c = \frac{v^2}{r} = \frac{(56.4\ \mathrm{m\,s^{-1}})^2}{450\ \mathrm{m}} = \boxed{7.07\ \mathrm{m\,s^{-2}}}$$

(**c**) Use Eqs. (3-16) and (3-17) to calculate the magnitude and direction of the total acceleration. The magnitude is

$$a = \sqrt{a_t^2 + a_c^2}$$
$$= \sqrt{(4.2\ \mathrm{m\,s^{-2}})^2 + (7.07\ \mathrm{m\,s^{-2}})^2}$$
$$= \boxed{8.22\ \mathrm{m\,s^{-2}}}$$

And the direction is

$$\phi = \arctan\frac{|a_t|}{|a_c|}$$
$$= \arctan\left(\frac{4.2\ \mathrm{m\,s^{-2}}}{7.07\ \mathrm{m\,s^{-2}}}\right)$$
$$= \boxed{30.7^\circ}$$

3-6. Relative Velocity

We can determine the relative velocity of an object (particle) with respect to two different coordinate systems and of two objects with respect to each other. Let's begin by considering one particle and two coordinate systems. As shown in Figure 3-8a, $\mathbf{R}_{PA}$ is the position vector locating the particle P with respect to coordinate system A, $\mathbf{R}_{PB}$ is the position vector locating the particle P with respect to coordinate system B, and $\mathbf{R}_{BA}$ is the position vector locating coordinate system B with respect to A. From the figure it's obvious that

$$\mathbf{R}_{PA} = \mathbf{R}_{PB} + \mathbf{R}_{BA} \qquad (3\text{-}18)$$

We differentiate Eq. (3-18) with respect to time to obtain

$$\mathbf{V}_{PA} = \mathbf{V}_{PB} + \mathbf{V}_{BA} \qquad (3\text{-}19)$$

where

$$\mathbf{V}_{PA} = \frac{d\mathbf{R}_{PA}}{dt}, \qquad \mathbf{V}_{PB} = \frac{d\mathbf{R}_{PB}}{dt}, \qquad \text{and} \qquad \mathbf{V}_{BA} = \frac{d\mathbf{R}_{BA}}{dt}$$

Be sure to draw the velocity diagram correctly (Figure 3-8b). Draw velocity vectors from their reference points to the object, with the name (or position) of the object (P) at the tips of the arrows and the names of the references (A and B) at the tails. Then complete the diagram by drawing a vector that connects the reference names (A to B).

Finding the relative velocity of two moving objects is illustrated in Example 3-8. In this case the velocity vectors have a common reference point.

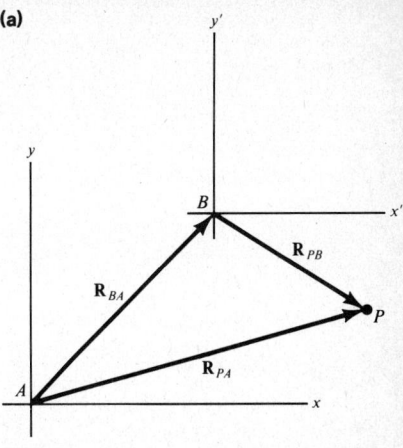

(a)

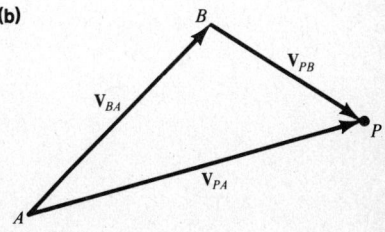

(b)

Figure 3-8. (a) Location of a particle in relation to two coordinate systems. (b) Velocity diagram for the particle with respect to the two coordinate systems.

EXAMPLE 3-8: An air traffic controller observes two planes on the radar screen. Plane K is moving at 240 km h^{-1}, 50° north of west, and plane J is moving at 180 km h^{-1} directly west (Figure 3-9a). Determine the velocity of plane J with respect to plane K.

Solution: Draw the relative velocity diagram (Figure 3-9b), connecting K and J with a *downward* vector, because we want $\mathbf{V}_{JK}$, the velocity of J (tip of vector) with respect to K (tail of vector). Solve for the magnitude of $\mathbf{V}_{JK}$ using the law of cosines:

$$V_{JK} = \sqrt{V_{KO}^2 + V_{JO}^2 - 2V_{KO}V_{JO}\cos\theta}$$
$$= \sqrt{(240)^2 + (180)^2 - 2(240)(180)\cos 50°}\ \text{km h}^{-1}$$
$$= \boxed{185.6\ \text{km h}^{-1}}$$

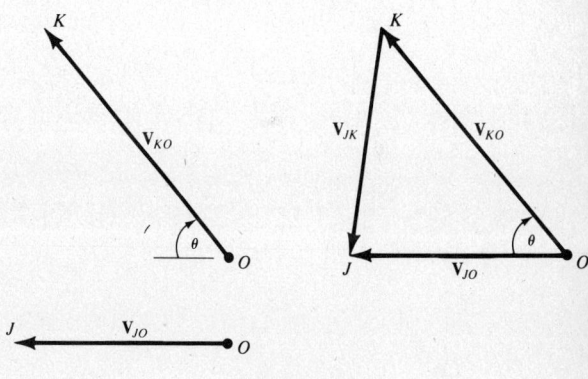

(a)

(b)

Figure 3-9

3-7. Rotational Motion

Bodies that rotate about an axis can be described in terms of the angle through which they rotate and their associated *angular velocity* and *angular acceleration*. **Angular displacement** θ (Figure 3-10) is measured in *radians* (rad) and is determined by the relationship

ANGULAR DISPLACEMENT $\qquad \theta = \dfrac{s}{r} \qquad (3\text{-}20)$

One radian is an angle of about 57.3°. We can obtain the conversion relationship between radians and degrees from Eq. (3-20). One complete rotation, 1 rev, corresponds to 360°, and $s = 2\pi r$, so

$$360° = \theta_{1\,\text{rev}} = \frac{2\pi r}{r} = 2\pi\ \text{rad}$$

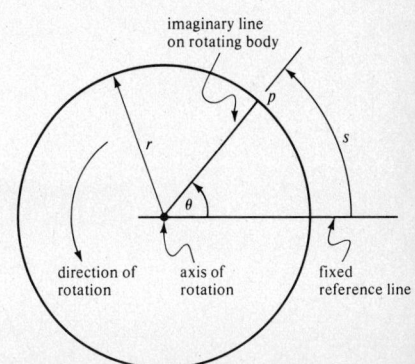

Figure 3-10. Rotation of a disk of radius r. Point p is on the edge of the disk.

or

$$\pi \, \text{rad} = 180° \quad \text{and} \quad 1 \, \text{rad} = \frac{180°}{\pi} \cong 57.3°$$

If the disk in Figure 3-10 is rotating, we can obtain the distance traversed by point p on its rim by using Eq. (2-5), i.e.,

$$s = v_0 t + \tfrac{1}{2} a_t t^2$$

Divide by r to obtain

$$\frac{s}{r} = \frac{v_0}{r} t + \frac{1}{2}\left(\frac{a_t}{r}\right)t^2$$

$$\theta = \omega_0 t + \tfrac{1}{2}\alpha t^2 \tag{3-21}$$

where $\omega_0 = v_0/r$, the *initial* angular velocity, is measured in radians per second (rad s^{-1}), and $\alpha = a_t/r$, the **angular acceleration**, is measured in rad s^{-2}. In general, the **angular velocity** is $\omega = v/r$, so we can show that

$$\omega = \omega_0 + \alpha t \tag{3-22a}$$

and

$$\omega^2 = \omega_0^2 + 2\alpha\theta \tag{3-22b}$$

by dividing $v = v_0 + at$ [Eq. (2-6)] by r and $v^2 = v_0^2 + 2ax$ [Eq. (2-7)] by r^2, respectively. The centripetal acceleration can also be expressed as $a_c = v^2/r = \omega^2 r$.

EXAMPLE 3-9: The disk shown in Figure 3-10 is rotating at $30 \, \text{rad s}^{-1}$ and then accelerates at the rate $\alpha = 2 \, \text{rad s}^{-2}$ for 12 s. Determine (a) the angular velocity of the disk at the end of the 12-s interval; (b) the angle through which the disk rotates during this interval; and (c) the speed of point p at the end of the interval. The disk has a radius of 16 cm.

Solution:

(a) Use Eq. (3-22a) to determine the angular velocity:

$$\omega = \omega_0 + \alpha t = (30 \, \text{rad s}^{-1}) + (2 \, \text{rad s}^{-2})(12 \, \text{s}) = \boxed{54 \, \text{rad s}^{-1}}$$

(b) Use Eq. (3-21) to obtain the angular displacement:

$$\theta = \omega_0 t + \tfrac{1}{2}\alpha t^2 = (30 \, \text{rad s}^{-1})(12 \, \text{s}) + \tfrac{1}{2}(2 \, \text{rad s}^{-2})(12 \, \text{s})^2 = \boxed{504 \, \text{rad}}$$

We could also express this result as 80.2 rev or 2.89×10^4 degrees.

(c) Use $\omega = v/r$ and solve for v:

$$v = \omega r = (54 \, \text{rad s}^{-1})(0.16 \, \text{m}) = \boxed{8.64 \, \text{m s}^{-1}}$$

SUMMARY

1. The displacement of a particle from position 1 to position 2, specified by position vectors $\mathbf{R}_1$ and $\mathbf{R}_2$, is $\Delta\mathbf{R} = \mathbf{R}_2 - \mathbf{R}_1$.
2. The velocity and acceleration for a particle are given by $\mathbf{v} = d\mathbf{R}/dt$ and $\mathbf{a} = d\mathbf{v}/dt$.
3. The motion of a particle subject to a constant acceleration takes place in a plane. The equations for the velocity and position of the particle as a function of time are

$$\mathbf{v} = \mathbf{v}_0 + \mathbf{a}t \quad \text{and} \quad \mathbf{R} = \mathbf{R}_0 + \mathbf{v}_0 t + \tfrac{1}{2}\mathbf{a}t^2$$

4. The equations that describe the motion of a projectile that moves in a vertical plane in which friction with the air can be neglected are

$$x = x_0 + v_{0x}t$$

$$y = y_0 + v_{0y}t - \tfrac{1}{2}gt^2$$

$$v_y = v_{0y} - gt$$

5. A particle that moves in a circle of radius r is subject to a centripetal acceleration

$$a_c = v^2/r = \omega^2 r$$

6. The relationship between the velocity of a particle relative to coordinate system A, the velocity of the particle relative to coordinate system B, and the velocity of system B relative to system A is

$$\mathbf{V}_{PA} = \mathbf{V}_{PB} + \mathbf{V}_{BA}$$

7. The equations describing rotational motion are

$$\theta = \omega_0 t + \tfrac{1}{2}\alpha t^2 \qquad \omega = \omega_0 + \alpha t \qquad \omega^2 = \omega_0^2 + 2\alpha\theta$$

8. The conversion relationships between angular and linear quantities are

$$\theta = \frac{s}{r} \qquad \omega = \frac{v}{r} \qquad \alpha = \frac{a_t}{r}$$

RAISE YOUR GRADES

Can you explain . . . ?

☑ the difference between speed and average speed
☑ the difference between speed and velocity
☑ the difference between velocity and acceleration
☑ the centripetal acceleration of a particle moving at constant speed around a circle
☑ the total acceleration of a particle that moves with a changing speed around a circle
☑ the motion of a projectile (neglect air friction)
☑ the effect of air friction on a projectile
☑ the velocities of a particle relative to two coordinate systems moving with respect to one another
☑ the relationship between angular and linear descriptions of rotational motion

SOLVED PROBLEMS

Displacement, Velocity, and Acceleration

PROBLEM 3-1: A ball bearing rolls around on a horizontal surface in what appears to be a random manner. At $t_1 = 2$ s, it is at the position $x_1 = 2$ m, $y_1 = 3$ m; at $t_2 = 5$ s, it is at the position $x_2 = -3$ m, $y_2 = 7$ m (Figure 3-11). Determine (**a**) the displacement of the bearing and (**b**) its average velocity during this interval.

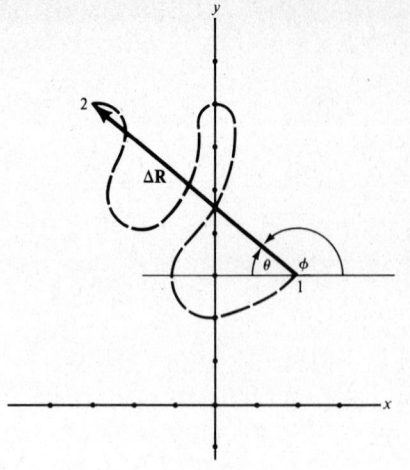

Figure 3-11

Solution:

(a) Use Eqs. (3-1b) to calculate the scalar components of the displacement:

$$\Delta R_x = x_2 - x_1$$
$$= -3 \text{ m} - 2 \text{ m} = -5 \text{ m}$$
$$\Delta R_y = y_2 - y_1$$
$$= 7 \text{ m} - 3 \text{ m} = 4 \text{ m}$$

The magnitude of $\Delta \mathbf{R}$ is

$$\Delta R = \sqrt{(\Delta R_x)^2 + (\Delta R_y)^2} \qquad \text{[Eq. (3-2)]}$$
$$= \sqrt{(-5 \text{ m})^2 + (4 \text{ m})^2} = \boxed{6.40 \text{ m}}$$

The direction of $\Delta \mathbf{R}$ is

$$\theta = \arctan \frac{|\Delta R_y|}{|\Delta R_x|} \qquad \text{[Eq. (3-3)]}$$

$$= \arctan \frac{|4 \text{ m}|}{|-5 \text{ m}|} = 38.7°$$

With respect to the positive x axis, the direction is

$$\phi = 180° - \theta = 180° - 38.7° = \boxed{141.3°}$$

(b) Equation (3-5) gives the average velocity. In this case

$$\Delta t = t_2 - t_1 = (5 - 2) \text{ s} = 3 \text{ s}$$

and

$$\mathbf{v}_{av} = \frac{\Delta \mathbf{R}}{\Delta t} = \frac{6.40 \text{ m}}{3 \text{ s}} = \boxed{2.13 \text{ m s}^{-1}}$$

The direction is the same as that of $\Delta \mathbf{R}$.

Motion with Constant Acceleration

PROBLEM 3-2: A small object moves in the $+x$ direction with a speed of 3 m s^{-1}. It is subjected to an acceleration of 1.6 m s^{-2} acting in the $+y$ direction for 2.5 s (Figure 3-12). Determine its velocity at the end of this time interval.

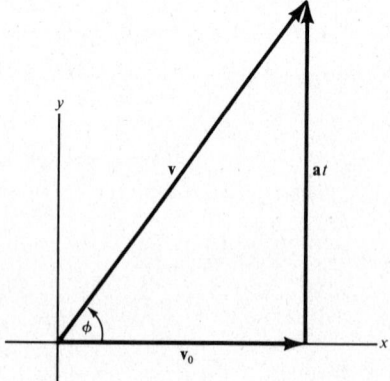

Figure 3-12

Solution: The velocity is the sum of two velocities: $\mathbf{v} = \mathbf{v}_0 + \mathbf{a}t$ [Eq. (3-11)]. The magnitudes are $v_0 = 3 \text{ m s}^{-1}$ and $at = (1.6 \text{ m s}^{-2})(2.5 \text{ s}) = 4 \text{ m s}^{-1}$.

$$v = \sqrt{(v_0)^2 + (at)^2} = \sqrt{(v_x)^2 + (v_y)^2}$$
$$= \sqrt{3^2 + 4^2} \text{ m s}^{-1} = \boxed{5 \text{ m s}^{-1}}$$

$$\phi = \arctan \frac{|v_y|}{|v_x|}$$

$$= \arctan \left(\frac{4 \text{ m s}^{-1}}{3 \text{ m s}^{-1}} \right) = \boxed{53.1°}$$

PROBLEM 3-3: The object in Problem 3-2 is at the origin at $t = 0$. Determine its displacement at $t = 2.5$ s. Determine the magnitude and direction of the displacement vector.

Solution: Use Eq. (3-12). Because $\mathbf{R}_0 = 0$, the displacement is the same as the position vector, so

$$\mathbf{R} = \mathbf{R}_0 + \mathbf{v}_0 t + \tfrac{1}{2}\mathbf{a}t^2$$

where

$$R_x = v_0 t = (3 \text{ m s}^{-1})(2.5 \text{ s}) = 7.5 \text{ m} \qquad \text{and} \qquad R_y = \tfrac{1}{2}at^2 = \tfrac{1}{2}(1.6 \text{ m s}^{-2})(2.5)^2 = 5 \text{ m}$$

Solving for R, you get

$$R = \sqrt{R_x^2 + R_y^2} = \sqrt{(7.5)^2 + (5)^2} \text{ m} = \boxed{9.01 \text{ m}}$$

The direction of $\mathbf{R}$ measured with respect to the $+x$ axis is θ, so

$$\theta = \arctan\frac{|R_y|}{|R_x|} = \arctan\left(\frac{5 \text{ m}}{7.5 \text{ m}}\right) = \boxed{33.7°}$$

Projectile Motion

PROBLEM 3-4: An airplane flying level at 252 km h^{-1} releases supplies from an altitude of 860 m. (You should be able to show that $252 \text{ km h}^{-1} = 70 \text{ m s}^{-1}$.) Determine the horizontal distance the supplies travel before striking the ground.

Solution: Solve Eq. (3-13b) for t, the time of flight. The quantities in the equation are $y = 0$, $y_0 = 860$ m, $v_{0y} = 0$, and $g = 9.8 \text{ m s}^{-2}$.

$$y = y_0 + v_{0y}t - \tfrac{1}{2}gt^2$$

$$t = \sqrt{\frac{2y_0}{g}} = \sqrt{\frac{2(860 \text{ m})}{9.8 \text{ m s}^{-2}}} = 13.25 \text{ s}$$

Use Eq. (3-13a) to determine x. In this case, $x_0 = 0$ and v_{0x} is the same as the horizontal speed of the airplane.

$$x = x_0 + v_{0x}t = 0 + (70 \text{ m s}^{-1})(13.25 \text{ s}) = \boxed{927.5 \text{ m}}$$

PROBLEM 3-5: A baseball is hit with an initial velocity of 22 m s^{-1} at an angle of $34°$ above the horizontal. What is the horizontal distance the ball travels if it is caught at the same height at which it was hit?

Solution: Solve Eq. (3-13b) for the time of flight. In this case, $y = y_0 = 0$ and $v_{0y} = v_0 \sin \theta$, where $v_0 = 22 \text{ m s}^{-1}$ and $\theta = 34°$.

$$y = y_0 + v_{0y}t - \tfrac{1}{2}gt^2$$
$$0 = 0 + (v_0 \sin \theta)t - \tfrac{1}{2}gt^2 = (v_0 \sin \theta - \tfrac{1}{2}gt)t$$

One root of this equation is $t = 0$, which corresponds to the instant the ball is hit. The other root is

$$t = \frac{2v_0 \sin \theta}{g} = \frac{2(22 \text{ m s}^{-1})\sin 34°}{9.8 \text{ m s}^{-2}} = 2.511 \text{ s}$$

Use this result in Eq. (3-13a) with $x_0 = 0$ and $v_{0x} = v_0 \cos \theta$.

$$x = x_0 + v_{0x}t = 0 + (v_0 \cos \theta)t = (22 \text{ m s}^{-1})(\cos 34°)(2.511 \text{ s}) = \boxed{45.8 \text{ m}}$$

PROBLEM 3-6: Determine the velocity of the baseball in Problem 3-5 0.8 s after it is hit.

Solution: The x component of the velocity of the ball is constant and equal to

$$v_x = v_0 \cos \theta = (22 \text{ m s}^{-1})\cos 34° = 18.24 \text{ m s}^{-1}$$

The vertical component of the velocity is obtained using Eq. (3-13c).

$$v_y = v_{0y} - gt = v_0 \sin \theta - gt = (22 \text{ m s}^{-1})\sin 34° - (9.8 \text{ m s}^{-2})(0.8 \text{ s}) = 4.46 \text{ m s}^{-1}$$

The magnitude of the velocity is

$$v = \sqrt{v_x^2 + v_y^2} = \sqrt{(18.24)^2 + (4.46)^2} \text{ m s}^{-1} = \boxed{18.8 \text{ m s}^{-1}}$$

The direction above the horizontal is

$$\phi = \arctan\frac{|v_y|}{|v_x|} = \arctan\left(\frac{4.46 \text{ m s}^{-1}}{18.24 \text{ m s}^{-1}}\right) = \boxed{13.7°}$$

Uniform Circular Motion

PROBLEM 3-7: Traveling at 57.6 km h^{-1}, an automobile rounds a circular curve that has a radius of 40 m. Determine the centripetal acceleration of the auto.

Solution: First convert 57.6 km h^{-1} to m s^{-1}.

$$v = \left(\frac{57.6 \times 10^3 \text{ m}}{\text{h}}\right)\left(\frac{1 \text{ h}}{3.6 \times 10^3 \text{ s}}\right) = 16 \text{ m s}^{-1}$$

Now use Eq. (3-14) to calculate the centripetal acceleration:

$$a_c = \frac{v^2}{r} = \frac{(16 \text{ m s}^{-1})^2}{40 \text{ m}} = \boxed{6.4 \text{ m s}^{-2}}$$

PROBLEM 3-8: At what speed would the car in Problem 3-7 have to travel for its centripetal acceleration to equal g?

Solution: Set $a_c = g$ in Eq. (3-14) and solve for v.

$$a_c = g = \frac{v^2}{r}$$

$$v = \sqrt{gr} = \sqrt{(9.8 \text{ m s}^{-2})(40 \text{ m})} = \boxed{19.8 \text{ m s}^{-1}}$$

Now convert back to obtain a speed of 71.3 km h^{-1}.

Accelerated Circular Motion

PROBLEM 3-9: A fly lands on a rotating disk 12 cm from the axis of rotation. The rotational speed of the disk is increasing at such a rate that the spot the fly is standing on has a tangential acceleration of 0.6 m s^{-2}. At the instant the speed of the disk is 24 rev min^{-1}, determine the total acceleration of the fly, assuming that it stays in place.

Solution: You should first of all determine the tangential speed of the fly at the instant the disk is at 24 rev min^{-1}.

$$v = \left(\frac{24 \text{ rev}}{\text{min}}\right)\left(\frac{2\pi r}{\text{rev}}\right)\left(\frac{1 \text{ min}}{60 \text{ s}}\right) = 0.302 \text{ m s}^{-1}$$

[The two unity conversion factors used are derived from 1 rev = circumference = $2\pi r$, and 1 min = 60 s.] Calculate the centripetal acceleration using Eq. (3-14).

$$a_c = \frac{v^2}{r} = \frac{(0.302 \text{ m s}^{-1})^2}{0.12 \text{ m}} = 0.76 \text{ m s}^{-2}$$

Now use Eqs. (3-16) and (3-17) to calculate the magnitude and direction of the total acceleration.

$$a = \sqrt{a_t^2 + a_c^2}$$

$$= \sqrt{(0.6)^2 + (0.76)^2} \text{ m s}^{-2} = \boxed{0.968 \text{ m s}^{-2}} \qquad \phi = \arctan\frac{|a_t|}{|a_c|}$$

$$= \arctan\left(\frac{0.6 \text{ m s}^{-2}}{0.76 \text{ m s}^{-2}}\right) = \boxed{38.3°}$$

Relative Velocity

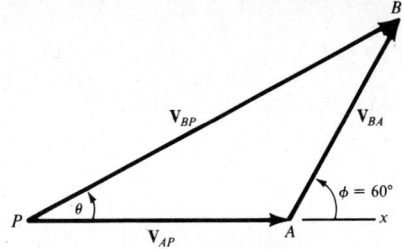

Figure 3-13.

PROBLEM 3-10: Two pucks are released simultaneously from the same point on a horizontal frictionless table. Puck A slides in the $+x$ direction at 3 m s^{-1}. Puck B slides at velocity $\mathbf{V}_{BP}$ ($\mathbf{V}_{BP}$ is the velocity of B relative to P, the point of release) at an angle θ with respect to the $+x$ axis. The velocity of puck B relative to A is 2.6 m s^{-1} and makes an angle of $60°$ with respect to the $+x$ axis. Draw to scale the velocity vector diagram.

Solution: Use a scale and protractor to draw the vector diagram. Remember to place the tip of the arrow toward the object and its tail toward the reference. First, draw $\mathbf{V}_{AP} = 3$ units along the $+x$ axis. Next, draw $\mathbf{V}_{BA} = 2.6$ units at an angle $60°$ above the $+x$ axis. Complete the diagram with $\mathbf{V}_{BP}$. The angle between $\mathbf{V}_{AP}$ and $\mathbf{V}_{BP}$ is θ. Your finished diagram should look like Figure 3-13.

PROBLEM 3-11: Use the law of cosines to determine the magnitude of $\mathbf{V}_{BP}$ in Problem 3-10.

Solution: The angle opposite $\mathbf{V}_{BP}$ is $\alpha = 180° - \phi = 120°$.

$$V_{BP}^2 = V_{AP}^2 + V_{BA}^2 - 2V_{AP}V_{BA}\cos\alpha = [(3)^2 + (2.6)^2 - 2(3)(2.6)\cos 120°]\ \text{m}^2\,\text{s}^{-2} = 23.56\ \text{m}^2\,\text{s}^{-2}$$

$$V_{BP} = \sqrt{23.56}\ \text{m s}^{-1} = \boxed{4.85\ \text{m s}^{-1}}$$

Rotational Motion

PROBLEM 3-12: The flywheel on the engine of a car has a diameter of 40 cm. The engine is revved up from 600 rev min^{-1} to 2400 rev min^{-1} uniformly in 1.6 s. Determine the angular acceleration of the wheel and the centripetal acceleration of a point on its rim at 2400 rev min^{-1}.

Solution: You should first convert the information given into a more useful form. The radius of the flywheel is $r = 40/2$ cm $= 20$ cm $= 0.2$ m. The initial and final angular velocities are

$$\omega_0 = \left(\frac{600\ \text{rev}}{\text{min}}\right)\left(\frac{2\pi\ \text{rad}}{\text{rev}}\right)\left(\frac{1\ \text{min}}{60\ \text{s}}\right) = 62.83\ \text{rad s}^{-1}$$

$$\omega = \left(\frac{2400\ \text{rev}}{\text{min}}\right)\left(\frac{2\pi\ \text{rad}}{\text{rev}}\right)\left(\frac{1\ \text{min}}{60\ \text{s}}\right) = 251.3\ \text{rad s}^{-1}$$

Use Eq. (3-22a) and solve for α.

$$\omega = \omega_0 + \alpha t$$

$$\alpha = \frac{\omega - \omega_0}{t} = \frac{(251.3 - 62.83)\ \text{rad s}^{-1}}{1.6\ \text{s}} = \boxed{118\ \text{rad s}^{-2}}$$

The centripetal acceleration is obtained from

$$a_c = \omega^2 r = (251.3\ \text{rad s}^{-1})^2(0.2\ \text{m}) = \boxed{1.26 \times 10^4\ \text{m s}^{-2}}$$

Note that, because *radians* are dimensionless and are not needed to describe the centripetal acceleration, they are dropped from this result.

Supplementary Exercises

EXERCISE 3-1: A teenager starts from home and walks 3 city blocks west, turns and walks 4 city blocks south, then turns and walks 2 blocks west, and finally walks 1 block north. (a) What is the displacement from home in city blocks? (b) Determine the angle of displacement with respect to west.

EXERCISE 3-2: A spacecraft travels at $150 \, \mathrm{m\,s^{-1}}$ due west. The commander fires thruster rockets that give the craft a constant acceleration of $5 \, \mathrm{m\,s^{-2}}$ due north. The boosters are fired for 15 s. Determine the velocity (magnitude and direction) of the spacecraft at the end of the burn.

EXERCISE 3-3: Determine the displacement of the spacecraft in Exercise 3-2 during the thruster burn.

EXERCISE 3-4: A military cannon fires a projectile with a velocity of $120 \, \mathrm{m\,s^{-1}}$ at an angle of 28° above the horizontal. (**a**) To what height does it rise? (**b**) How far from the cannon does it strike the ground, assuming that the terrain is level?

EXERCISE 3-5: The cannon in Exercise 3-4 shoots a projectile toward a cliff face 1000 m distant. How far above the base of the cliff, which is at the same level as the cannon, does the projectile strike?

EXERCISE 3-6: Determine the components of the velocity with which the projectile in Exercise 3-5 strikes the cliff face.

EXERCISE 3-7: Determine the centripetal acceleration of a point on the tip of a fan blade 15 cm from the axis of rotation when the fan rotates at $1700 \, \mathrm{rev\,min^{-1}}$.

EXERCISE 3-8: Determine the tangential speed of a seat on a Ferris wheel that will cause a person riding in the seat to experience a centripetal acceleration of g. The distance from the seat to the axis of rotation is 8 m.

EXERCISE 3-9: The speed of a car moving around a circular track that has a radius of 140 m is given by the expression $v = (1.2 \, \mathrm{m\,s^{-2}})t + 4 \, \mathrm{m\,s^{-1}}$, for a certain time interval. Determine the components of the total acceleration of the car at the instant $t = 9$ s.

EXERCISE 3-10: An airplane flies with a heading due east at $180 \, \mathrm{km\,h^{-1}}$ relative to the air. The wind is blowing at a steady rate of $90 \, \mathrm{km\,h^{-1}}$ in a direction 60° north of west. Determine the velocity of the airplane with respect to the ground.

EXERCISE 3-11: The wheels on a bicycle are 66 cm in diameter. A bicyclist increases her speed from $2 \, \mathrm{m\,s^{-1}}$ to $4.5 \, \mathrm{m\,s^{-1}}$ in traveling 40 m. Determine the angular acceleration of her bicycle wheels during this interval. [*Hint:* The angular velocity of a rolling wheel of radius r is $\omega = v/r$, where v is the linear speed of the wheel.]

Answers to Supplementary Exercises

3-1: (**a**) 5.83 city blocks
(**b**) 31° south of west

3-2: $v = 168 \, \mathrm{m\,s^{-1}}, \theta = 26.6°$ north of west

3-3: $R = 2.32 \times 10^3$ m, $\theta = 14°$ north of west

3-4: (**a**) 162 m
(**b**) 1.22×10^3 m

3-5: 95.2 m

3-6: $v_x = 106 \, \mathrm{m\,s^{-1}}, v_y = -36.2 \, \mathrm{m\,s^{-1}}$

3-7: $4.75 \times 10^3 \, \mathrm{m\,s^{-2}}$

3-8: $8.85 \, \mathrm{m\,s^{-1}}$

3-9: $a_t = 1.2 \, \mathrm{m\,s^{-2}}, a_c = 1.57 \, \mathrm{m\,s^{-2}}$

3-10: $v = 156 \, \mathrm{km\,h^{-1}}, \theta = 30°$ north of east

3-11: $0.62 \, \mathrm{rad\,s^{-2}}$

NEWTON'S LAWS OF MOTION

THIS CHAPTER IS ABOUT

☑ **Newton's Three Laws of Motion**
☑ **Force, Mass, and Weight**
☑ **Frictional Forces**
☑ **Statics—The First Condition of Equilibrium**
☑ **Dynamics—Motion of a Particle**
☑ **Motion Along a Curve**
☑ **Velocity-Dependent Forces**
☑ **Accelerated Frames of Reference**

4-1. Newton's Three Laws of Motion

• **First law:** A body will remain in its state of rest or of uniform motion in a straight line unless acted upon by an external unbalanced force.
• **Second law:** The acceleration of a body is directly proportional to the resultant external force acting on it and inversely proportional to its mass, or

$$\mathbf{a} = \frac{\Sigma \mathbf{F}}{m} \qquad \text{(4-1a)}$$

The more familiar form of this is

$$\Sigma \mathbf{F} = m\mathbf{a} \qquad \text{(4-1b)}$$

The resultant, or net force, acting on a body is the vector sum of all external forces. Note that the first law is the special case for which the resultant force is zero; consequently, the acceleration will be zero.

• **Third law:** Forces always act in pairs. If body A exerts a force on body B, body B exerts an equal, but opposite, force on body A.

When applying Newton's laws in the solution of problems, you have to be able to identify all the forces that act *on* an object. Forces exerted *by* the object don't affect its acceleration.

4-2. Force, Mass, and Weight

Newton's second law, Eqs. (4-1a) and (4-1b), defines the relationship between force and mass. The **weight** of an object is the force of gravity acting on its mass. Near the surface of the earth the force of gravity produces an acceleration $\mathbf{g}$ on a body, so the weight of a body of mass m is

$$\mathbf{w} = m\mathbf{g} \qquad \text{or} \qquad w = mg \quad \text{[magnitude only]} \qquad \text{(4-2)}$$

In the SI system mass is measured in kilograms (kg) and force in newtons (N). The newton is a derived unit, i.e.,

$$1\,\text{N} = 1\,\text{kg}\,\text{m}\,\text{s}^{-2}$$

In the British system, mass is measured in slugs and force in pounds (lb). The slug is a derived unit. The second law implies that

$$1 \text{ lb} = 1 \text{ slug} \cdot \text{ft/s}^2 \qquad \text{or} \qquad 1 \text{ slug} = 1 \text{ lb} \cdot \text{s}^2/\text{ft}$$

The conversion relationship between the systems is 1 lb = 4.448 N.

EXAMPLE 4-1: **(a)** Determine the weight of an 80-kg person. **(b)** Determine the weight of an 80-kg person on the surface of the moon, where the acceleration due to gravity is 1.6 m s^{-2}.

Solution:

(a) You should assume that the person is on the surface of the earth because no other location is specified. Use Eq. (4-2).

$$w = mg = (80 \text{ kg})(9.8 \text{ m s}^{-2}) = \boxed{784 \text{ N}}$$

(b) Use Eq. (4-2), but in this case $g = 1.6$ m s^{-2}.

$$w = mg = (80 \text{ kg})(1.6 \text{ m s}^{-2}) = \boxed{128 \text{ N}}$$

note: The mass of a body is an intrinsic property, but its weight depends on where the body is located in space.

EXAMPLE 4-2: What is the mass (in slugs) of a 130-lb crate?

Solution: Solve Eq. 4-2 for m.

$$w = mg$$

$$m = \frac{w}{g} = \frac{130 \text{ lb}}{32 \text{ ft/s}^2} = \boxed{4.06 \text{ slugs}}$$

4-3. Frictional Forces

A. Kinetic friction

When a body slides on a dry surface, interaction between the body and the surface results in a force of friction that acts in a direction opposite to the direction of motion of the body. This is the **kinetic friction, f$_k$**, which is also called the sliding friction. The *magnitude* of kinetic friction is

KINETIC FRICTION $\qquad\qquad f_k = \mu_k N$ $\qquad\qquad$ (4-3)

where μ_k is a dimensionless quantity called the **coefficient of kinetic friction** and N is the magnitude of the **normal force**. The normal force N is the force, perpendicular to the surface, that the surface exerts on the body. The coefficient of kinetic friction is an experimentally determined quantity; its value depends on the nature of the surfaces in contact.

EXAMPLE 4-3: A 4-kg (mass) block slides on a horizontal surface. The coefficient of friction between the block and surface is 0.36. Determine the force of friction and the acceleration of the block.

Solution: Use Eq. (4-3) to calculate the force of friction. First, however, you will have to determine the normal force. To help you do this, make a sketch and a

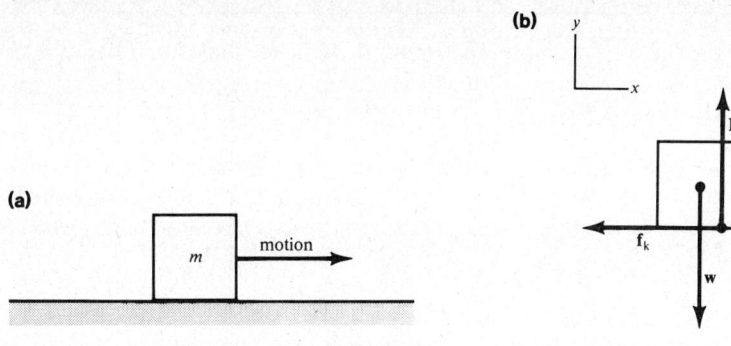

Figure 4-1

free-body diagram that shows *all* the forces acting on the block. Assume, as in Figure 4-1a, that the body is sliding to the right; $\mathbf{f_k}$ therefore is directed to the left.

The three forces that act on the block are shown on the free-body diagram, Figure 4-1b. Use Newton's second law, Eq. (4-1b),

$$\Sigma \mathbf{F} = m\mathbf{a}$$

in component form:

x component	*y* component
$\Sigma F_x = ma_x$	$\Sigma F_y = ma_y$
$-f_k = ma_x$	$N - w = ma_y = 0$

There is only one *x* force acting on the body, and it's directed along the negative *x* axis. There are two *y* forces acting on the body: *N* is positive, and $w = mg$ is negative. Their sum is zero because the block isn't accelerating in the *y* direction, or

$$N - w = 0 \quad \text{and} \quad N = w = mg$$

Use Eq. (4-3) to determine f_k, since *N* is now known.

$$f_k = \mu_k N = \mu_k mg = (0.36)(4 \text{ kg})(9.8 \text{ m s}^{-2}) = \boxed{14.1 \text{ N}}$$

Solve the *x* component equation for a_x.

$$a_x = -\frac{f_k}{m} = -\frac{14.1 \text{ N}}{4 \text{ kg}} = \boxed{-3.53 \text{ m s}^{-2}}$$

The negative sign indicates that the acceleration of the block is opposite its direction of motion and that the block is slowing down.

B. Static friction

The frictional force that acts on a body at rest is called **static friction** and may have a value that ranges from 0 to $\mu_s N$ when the body just begins to slide. Thus the force of static friction can be written as

STATIC FRICTION $\qquad\qquad f_s \leq \mu_s N \qquad\qquad\qquad$ (4-4)

where μ_s is called the **coefficient of static friction.** For any two surfaces, $\mu_s > \mu_k$. The direction of f_s is opposite the direction the body *tends* to move.

4-4. Statics—The First Condition of Equilibrium

Newton's first law describes a body that meets the first condition of equilibrium. This implies that the second law applied to that body gives $\Sigma \mathbf{F} = 0$, or in component form $\Sigma F_x = 0$, $\Sigma F_y = 0$, and $\Sigma F_z = 0$. In words, we say that the resultant, or net, force acting on the body is zero.

(a)

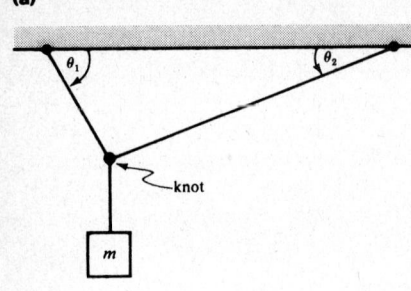

(b)

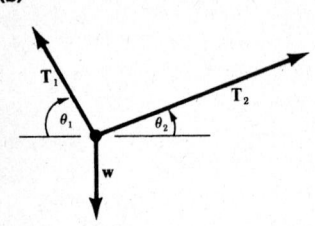

(c)

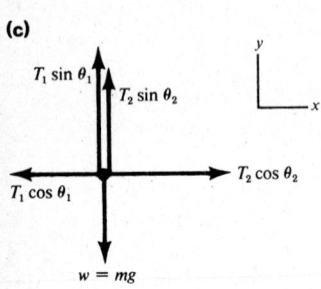

Figure 4-2

EXAMPLE 4-4: The block in Figure 4-2a has a mass of 6 kg. Calculate the tension in the supporting cords. Assume that the mass of the cords is negligible; $\theta_1 = 60°$ and $\theta_2 = 20°$.

Solution: Draw a free-body diagram for the knot (Figure 4-2b); T_1 and T_2 are the tensions in the cords. Resolve the forces into their x and y components (Figure 4-2c). Apply the first condition of equilibrium in component form.

x component	y component
$\Sigma F_x = 0$	$\Sigma F_y = 0$
$T_1 \sin \theta_1 + T_2 \sin \theta_2 - w = 0$	$T_2 \cos \theta_2 - T_1 \cos \theta_1 = 0$

Now solve the y component equation for T_1

$$T_1 = T_2 \left(\frac{\cos \theta_2}{\cos \theta_1} \right)$$

and substitute into the x component equation to solve for T_2:

$$T_2 \left(\frac{\cos \theta_2}{\cos \theta_1} \right) \sin \theta_1 + T_2 \sin \theta_2 = w = mg$$

Use the identity $\tan \theta = (\sin \theta)/(\cos \theta)$ to obtain

$$T_2 \cos \theta_2 (\tan \theta_1 + \tan \theta_2) = mg$$

Solve for T_2:

$$T_2 = \frac{mg}{\cos \theta_2 (\tan \theta_1 + \tan \theta_2)} = \frac{(6 \text{ kg})(9.8 \text{ m s}^{-2})}{\cos 20°(\tan 60° + \tan 20°)} = \boxed{29.8 \text{ N}}$$

Similarly,

$$T_1 = \frac{mg}{\cos \theta_1 (\tan \theta_1 + \tan \theta_2)} = \frac{(6 \text{ kg})(9.8 \text{ m s}^{-2})}{\cos 60°(\tan 60° + \tan 20°)} = \boxed{56.1 \text{ N}}$$

4-5. Dynamics—Motion of a Particle

A change in the state of motion of a body implies that the body accelerates. Newton's second law is used most frequently in component form to determine the acceleration that results from the net force acting on a body. A careful choice of a coordinate system often reduces the amount of math required.

EXAMPLE 4-5: A block of mass 3 kg is pulled along a flat surface with a cord, which makes an angle of $25°$ above the horizontal. The tension in the cord is 16 N, and the coefficient of kinetic friction between the block and the surface is 0.4. Determine the acceleration of the block.

Solution: Sketch the situation (Figure 4-3a) and draw a free-body diagram for the block, showing the four forces acting on it (Figure 4-3b). Next, define a coordinate system and resolve the forces into components (Figure 4-3c); then apply Newton's second law in component form.

(a)

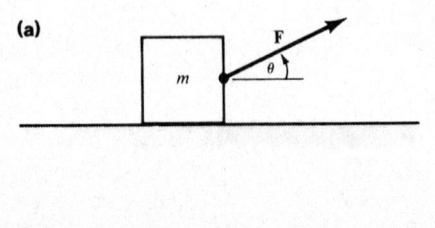

(b)

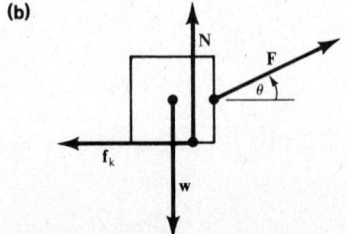

(c)

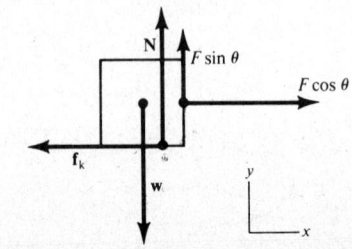

Figure 4-3

x component	y component
$\Sigma \mathbf{F}_x = ma_x$	$\Sigma \mathbf{F}_y = ma_y$
$F\cos\theta - f_k = ma_x$	$N + F\sin\theta - mg = ma_y = 0$

The y component equation is set equal to zero because $a_y = 0$; i.e., the block remains on the surface, and there is no acceleration in the y direction. Solve the y component equation for N and substitute in Eq. 4-3.

$$N = mg - F\sin\theta, \qquad f_k = \mu_k N, \qquad \text{and} \qquad f_k = \mu_k(mg - F\sin\theta)$$

Now substitute for f_k in the x-component equation,

$$F\cos\theta - \mu_k(mg - F\sin\theta) = ma_x$$

and solve for a_x.

$$a_x = \frac{F(\cos\theta + \mu_k\sin\theta)}{m} - \mu_k g$$

$$= \frac{(16\ \text{N})(\cos 25° + 0.4\sin 25°)}{3\ \text{kg}} - (0.4)(9.8\ \text{m s}^{-2}) = \boxed{1.82\ \text{m s}^{-2}}$$

4-6. Motion Along a Curve

When a particle moves along a curve, it is subject to a centripetal acceleration, even if it moves at a constant speed [Eq. (3-14)]. The force that produces this acceleration, called the **centripetal force**, is given by Newton's second law; i.e.,

CENTRIPETAL FORCE $$F_c = ma_c = m\frac{v^2}{r} \qquad (4\text{-}5)$$

and is directed toward the center of curvature. The **centrifugal force** is, by Newton's third law, the reaction to the centripetal force. When you solve problems involving centripetal force, an important early step is identification of the cause of the centripetal force.

EXAMPLE 4-6: A 2-kg mass swings in a horizontal circular orbit at the end of a cord of negligible mass that sweeps out an inverted cone of half angle $\alpha = 20°$. The cord has a length of 1.8 m, and the mass completes 42 rev min^{-1}. Determine **(a)** the cause of the centripetal force, **(b)** the magnitude of the centripetal force, and **(c)** the speed of the mass around the orbit.

Solution: Draw a sketch of the situation, a free-body diagram for the mass, showing the two forces acting on it, and, because the orbit is horizontal, the horizontal and vertical components of the tension in the cord (Figures 4-4a, b, and c, respectively).

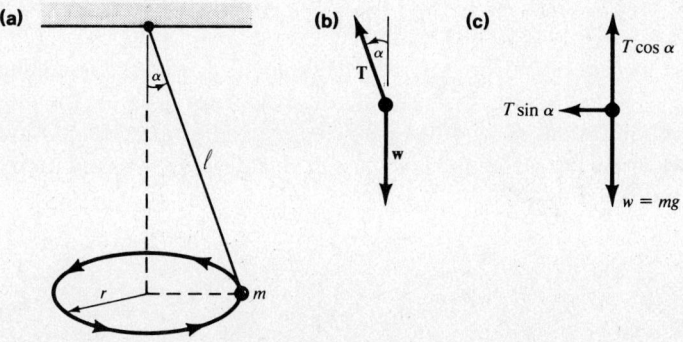

Figure 4-4

(a) The horizontal component of the tension in the cord is directed toward the center of the circular orbit and is therefore the centripetal force, so $F_c = T \sin \alpha$.

(b) Because the mass stays in the horizontal plane that contains the orbit, there is no net vertical acceleration; the net vertical force is zero, thus

$$T \cos \alpha - mg = 0$$

Solve for the tension in the cord, substitute into the result of (a), and evaluate.

$$T = \frac{mg}{\cos \alpha}$$

$$F_c = T \sin \alpha = \left(\frac{mg}{\cos \alpha}\right) \sin \alpha = mg \tan \alpha$$

$$= (2 \text{ kg})(9.8 \text{ m s}^{-2}) \tan 20° = \boxed{7.134 \text{ N}}$$

(c) Use Eq. (4-5) and solve for v. In this case $r = \ell \sin \alpha$:

$$F_c = \frac{v^2}{r} = \frac{v^2}{\ell \sin \alpha}$$

$$v = \sqrt{F_c \ell \sin \alpha}$$

$$= \sqrt{(7.134 \text{ N})(1.8 \text{ m}) \sin 20°} = \boxed{2.1 \text{ m s}^{-1}}$$

4-7. Velocity-Dependent Forces

Centripetal force is one example of a **velocity-dependent force**. Another is that of an object falling in a viscous medium. The frictional force exerted on such an object—a raindrop, for example—is proportional to the first power of its velocity, provided the velocity is not too great; that is,

$$f = bv \tag{4-6}$$

where b is the proportionality constant. This constant is a quantity that depends in general on the shape of the object and has units of kg s^{-1}. Its value must be determined experimentally. Because of the resistive force, a falling object will reach a constant **terminal velocity** if permitted to fall for a sufficient distance. The terminal velocity for raindrops ranges from about 4 to 8 m s^{-1}, depending on their size.

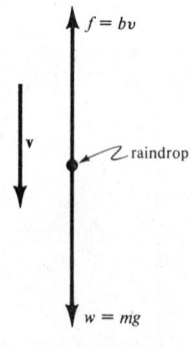

$f = bv$

v

raindrop

$w = mg$

Figure 4-5

EXAMPLE 4-7: (a) Determine the velocity as a function of time for a raindrop of mass $m = 4 \times 10^{-6}$ kg that reaches a terminal velocity of $v_t = 6$ m s^{-1}, for which the resistive force is given by Eq. (4-6). (b) Evaluate the constant b.

Solution:

(a) Two forces act on the raindrop: its weight, directed downward, and the resistive force, $f = bv$, acting upward. Because the raindrop is moving downward, it is convenient to let downward be the positive direction. Draw the free-body diagram for a raindrop of mass m, as in Figure 4-5, and apply Newton's second law, Eq. (4-1b).

$$\Sigma \mathbf{F} = m\mathbf{a}, \qquad w - f = ma, \qquad mg - bv = m\frac{dv}{dt}$$

where $a = \dfrac{dv}{dt}$ [Eq. (3-8)]. Separate the variables and integrate.

$$\int_0^v \frac{dv}{v - mg/b} = -\frac{b}{m}\int_0^t dt$$

$$\ln\left(\frac{v - mg/b}{-mg/b}\right) = -\frac{b}{m}t$$

Now solve for v.

$$v = (mg/b)(1 - e^{-bt/m}) = v_t(1 - e^{-bt/m})$$

(b) The velocity starts from zero and at sufficiently large t it reaches the constant terminal value mg/b. Solve for b and evaluate:

$$b = \frac{mg}{v_t}$$

$$= \frac{(4 \times 10^{-6}\ \text{kg})(9.8\ \text{m s}^{-2})}{6\ \text{m s}^{-1}} = \boxed{6.5 \times 10^{-6}\ \text{kg s}^{-1}}$$

4-8. Accelerated Frames of Reference

A coordinate system that is accelerated with respect to one that is not is said to be a **noninertial frame of reference**. The nonaccelerated coordinate system is said to be an **inertial frame of reference**. Newton's laws are defined for inertial frames. For noninertial frames new, **inertial forces** must be added to Newton's second law.

A. Linearly accelerated coordinate system

In Figure 4-6, A is an inertial coordinate system and B is a noninertial coordinate system accelerated at the rate a_{BA} with respect to A. A particle of mass m is located at point P. The relationship describing the position vectors is

$$\mathbf{R}_{PA} = \mathbf{R}_{BA} + \mathbf{R}_{PB}$$

We differentiate this equation twice with respect to time to obtain

$$\mathbf{a} = \mathbf{a}_{BA} + \mathbf{a}'$$

where $\mathbf{a}$ and $\mathbf{a}'$ are the accelerations of the mass at P measured with respect to coordinate systems A and B, respectively. Next, we multiply this equation through by m and subtract $m\mathbf{a}_{BA}$ from both sides:

$$m\mathbf{a} - m\mathbf{a}_{BA} = m\mathbf{a}' \qquad \textbf{(4-7)}$$

Equation (4-7) describes the motion of m relative to the linearly accelerated coordinate system. The quantity $m\mathbf{a}$ results from the proper application of Newton's second law to the inertial frame of reference; it is the vector sum of all the *real* forces; and $(-m\mathbf{a}_{BA})$ is the inertial force resulting from the acceleration of coordinate system B.

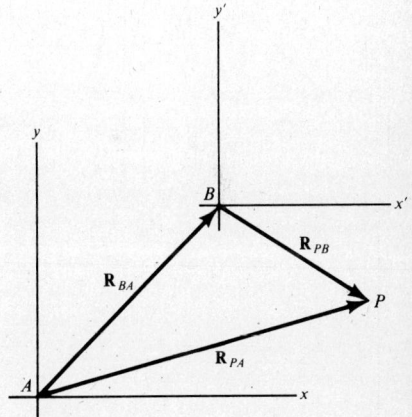

Figure 4-6. Coordinate system B is accelerated with respect to coordinate system A, an inertial frame of reference.

EXAMPLE 4-8: A 2-kg mass is connected by a cord to the ceiling of an elevator, which is accelerated upward at 3 m s^{-2}. **(a)** Determine the tension in the cord from the point of view of a stationary observer. **(b)** Determine the tension in the cord from the point of view of an observer riding in the elevator.

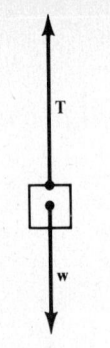

Figure 4-7

Solution:

(a) Draw a free body diagram for the mass, as in Figure 4-7, apply Newton's second law, Eq. (4-1b), and solve for T, the tension in the cord.

$$\Sigma \mathbf{F} = m\mathbf{a}$$

$$T - mg = ma$$

$$T = ma + mg = m(a + g) = (2 \text{ kg})(3 + 9.8) \text{ m s}^{-2} = \boxed{25.6 \text{ N}}$$

(b) The same free-body diagram applies. The sum of the *real* forces is $T - mg$, the inertial force caused by the acceleration of the elevator is ma_{elev}, and the acceleration in the accelerated system is $a' = 0$. Use Eq. (4-7) and solve for T.

$$(T - mg) - ma_{\text{elev}} = ma'$$

$$T - mg - ma_{\text{elev}} = 0$$

$$T = m(a_{\text{elev}} + g) = (2 \text{ kg})(3 + 9.8) \text{ m s}^{-2} = \boxed{25.6 \text{ N}}$$

B. Uniformly rotating coordinate system

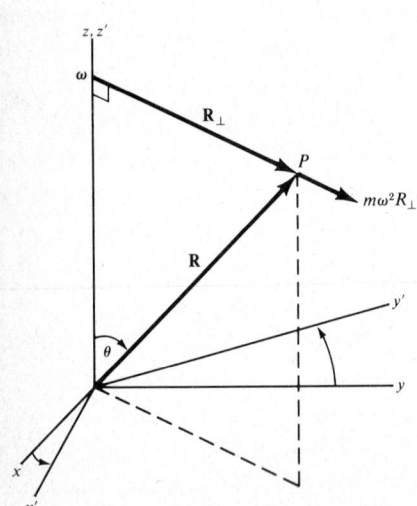

Figure 4-8. Coordinate system B is rotating about the common z,z' axis at a constant angular velocity ω with respect to the *fixed* inertial system A. The primed coordinates locate P in B; the nonprimed coordinates locate P in A.

In Figure 4-8, the z axes of the inertial and the rotating coordinate systems are coincident; the rotating system moves about this axis with uniform angular velocity ω. A particle of mass m is located at point P by the position vector $\mathbf{R}$, which is the same for both systems because their origins coincide and $\boldsymbol{\omega} = \omega \hat{\mathbf{z}}$.

There are two inertial forces, i.e.,

$$\mathbf{F}_{\text{Coriolis}} = -2m(\boldsymbol{\omega} \times \mathbf{v}') \qquad \text{(4-8a)}$$

$$\mathbf{F}_{\text{centrifugal}} = -m[\boldsymbol{\omega} \times (\boldsymbol{\omega} \times \mathbf{R})] \qquad \text{(4-8b)}$$

In magnitude, $F_{\text{centrifugal}} = m\omega^2 R_\perp$, and it is directed radially outward from the axis of rotation. $R_\perp = R \sin \theta$ and $\mathbf{v}'$ is the velocity of the particle measured relative to the rotating coordinate system. The relationship between $\mathbf{v}$ and $\mathbf{v}'$ is

$$\mathbf{v} = \mathbf{v}' + \boldsymbol{\omega} \times \mathbf{R} \qquad \text{(4-8c)}$$

Derivations of Eqs. (4-8a), (4-8b), and (4-8c) can be found in many physics textbooks.

Newton's second law modified for a rotating coordinate system is

$$\Sigma \mathbf{F}_{\text{real forces}} + \mathbf{F}_{\text{Coriolis}} + \mathbf{F}_{\text{centrifugal}} = m\mathbf{a}' \qquad \text{(4-9)}$$

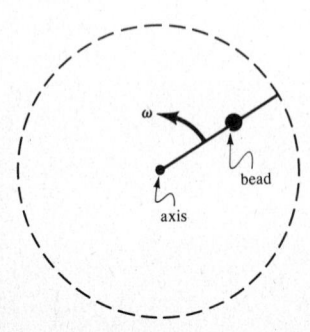

Figure 4-9

EXAMPLE 4-9: A bead of mass $m = 200$ g is free to slide on a straight, rigid wire that is rotating in a horizontal plane at constant angular velocity $\omega = 2.7 \text{ rad s}^{-1}$ (Figure 4-9). The length of the wire is 30 cm. At the instant the bead flies off the end of the wire its speed along the wire is 1.5 m s^{-1}. **(a)** Calculate the centrifugal force on the bead when it is 10 cm from the axis of rotation and just before it leaves the end of the wire. **(b)** Determine the Coriolis force on the bead just before it leaves the wire.

Solution:

(a) The magnitude of the centrifugal force is $F_{\text{centrifugal}} = m\omega^2 R_\perp$. At $R = R_\perp = 10 \text{ cm} = 0.1 \text{ m}$,

$$F_{\text{centrifugal}} = (0.2 \text{ kg})(2.7 \text{ rad s}^{-1})^2(0.1 \text{ m}) = \boxed{0.146 \text{ N}}$$

At $R = R_\perp = 30$ cm $= 0.3$ m,

$$F_{\text{centrifugal}} = (0.2 \text{ kg})(2.7 \text{ rad s}^{-1})^2(0.3 \text{ m}) = \boxed{0.437 \text{ N}}$$

In both cases, $\mathbf{F}_{\text{centrifugal}}$ is directed away from the axis of rotation. Note that the increase in acceleration of the bead is proportional to its distance from the axis of rotation of the wire.

(b) Use Eq. (4-8a). The z,z' axis of rotation is directed out of the page in Figure 4-9, so $\boldsymbol{\omega} = \omega\hat{\mathbf{z}}$ is also directed out of the page. The magnitude of F_{Coriolis} is $2m\omega v'$ because $\boldsymbol{\omega}$ and $\mathbf{v}'$ are perpendicular.

$$F_{\text{Coriolis}} = 2m\omega v' = 2(0.2 \text{ kg})(2.7 \text{ rad s}^{-1})(1.5 \text{ m s}^{-1}) = \boxed{1.62 \text{ N}}$$

The direction of $\mathbf{F}_{\text{Coriolis}}$ is clockwise, or opposite the direction of rotation. The force that the wire exerts on the bead is equal and opposite to $\mathbf{F}_{\text{Coriolis}}$.

SUMMARY

1. Newton's second law states that the net or resultant force on a particle, i.e., the vector sum of all forces acting on the particle, causes its acceleration:

$$\Sigma\mathbf{F} = m\mathbf{a}$$

where m is the mass of the particle.

2. Weight is the *force* of gravity acting on an object. It is related to the mass of the object by

$$w = mg$$

3. Kinetic (sliding) friction is proportional to the normal force acting on an object and is directed opposite to the motion of the object:

$$f_k = \mu_k N$$

4. Static friction applies to bodies at rest and can range from zero to a maximum value of $f_s = \mu_s N$, directed in opposition to the direction the body tends to move.

5. The first condition of equilibrium applies to bodies at rest or that move at constant velocity. Its mathematical form is that the net force acting on the body is zero:

$$\Sigma\mathbf{F} = 0$$

6. Objects falling *in air* do not continue to accelerate at the rate $g = 9.8$ m s^{-2}. They may, if given sufficient time, reach a constant (terminal) velocity.

7. In order to maintain the form of Newton's second law for an accelerated coordinate system, certain inertial forces must be added to account for the acceleration of the coordinate system. For the linearly accelerated coordinate system, the inertial force is $-m\mathbf{a}_{BA}$. For the coordinate system rotating at constant angular velocity, two inertial forces must be added: the *centrifugal force* and the *Coriolis force*.

RAISE YOUR GRADES
Can you explain . . . ?

☑ how a particle can move if there is no net force acting on it
☑ how an object can be weightless and yet have mass
☑ why objects weigh less on the moon than they do on Earth
☑ the reason for the direction of the force of friction
☑ how parachutes work
☑ why inertial forces are also called *fictional forces*

SOLVED PROBLEMS

Force, Mass, and Weight

PROBLEM 4-1:   **(a)** What is the acceleration due to gravity on the surface of planet X where a 230-kg space probe weighs 550 N? **(b)** What does it weigh on earth?

Solution:

(a) Use Eq. (4-2) and solve for g: $w = mg$

$$g = \frac{w}{m} = \frac{550 \text{ N}}{230 \text{ kg}} = \boxed{2.39 \text{ m s}^{-2}}$$

(b) On earth, $w = mg = (230 \text{ kg})(9.8 \text{ m s}^{-2}) = \boxed{2254 \text{ N}}$

Frictional Forces

PROBLEM 4-2:   A 3-kg block slides down a plane inclined at an angle of $36°$ above the horizontal. The coefficient of kinetic friction between block and plane is 0.26. Determine the acceleration of the block down the plane.

Solution: Make a sketch, draw a free-body diagram of the block, resolve the forces into components (where appropriate), and apply Newton's second law in component form (see Figure 4-10). The choice of coordinate system (Figure 4-10b), with the $+x$ axis directed down the plane and the $+y$ axis directed perpendicular to the surface of the plane, saves some math work. The weight of the block is the only force that has to be resolved into components (Figure 4-10c).

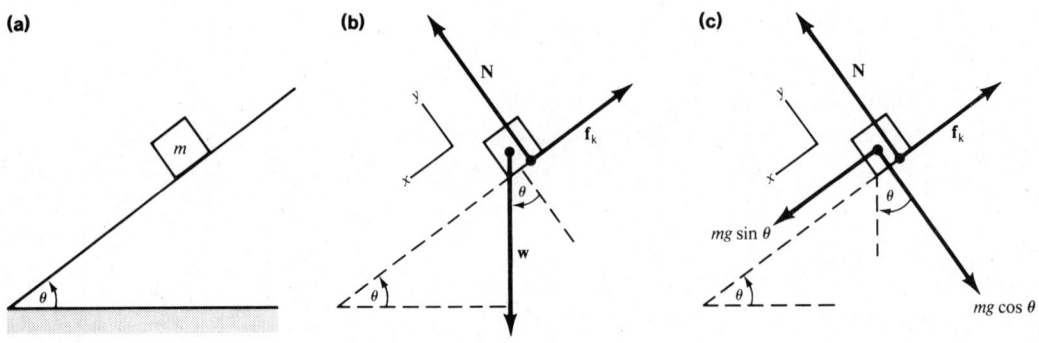

Figure 4-10

Use of Eq. (4-1b), Newton's second law, in component form yields

x component	y component
$\Sigma F_x = ma_x$	$\Sigma F_y = ma_y$
$mg \sin \theta - f_k = ma_x$	$N - mg \cos \theta = ma_y = 0$

Since there is no motion of the block in the y direction, $a_y = 0$. Solving the y component equation for N gives you $N = mg \cos \theta$. Substitute this expression for N in Eq. (4-3) and the resulting expression for f_k which becomes

$$mg \sin \theta - \mu_k mg \cos \theta = ma_x$$

Now solve for a_x, canceling m, to get

$$a_x = g(\sin \theta - \mu_k \cos \theta) = (9.8 \text{ m s}^{-2})(\sin 36° - 0.26 \cos 36°) = \boxed{3.7 \text{ m s}^{-2}}$$

Note that the acceleration of the block is independent of its mass; it depends only on the angle of inclination and the coefficient of kinetic friction.

Statics—The First Condition of Equilibrium

PROBLEM 4-3: A cord is attached to the block in Problem 4-2, passed over a frictionless pulley, and connected to a freely hanging weight. The coefficient of static friction between the block and plane is 0.32. Determine the maximum mass of the hanging weight required for both the block and the weight to remain at rest.

Solution: Make a sketch, draw a free-body diagram of the block, resolve the forces into components (where appropriate), and apply the first condition of equilibrium (see Figure 4-11).

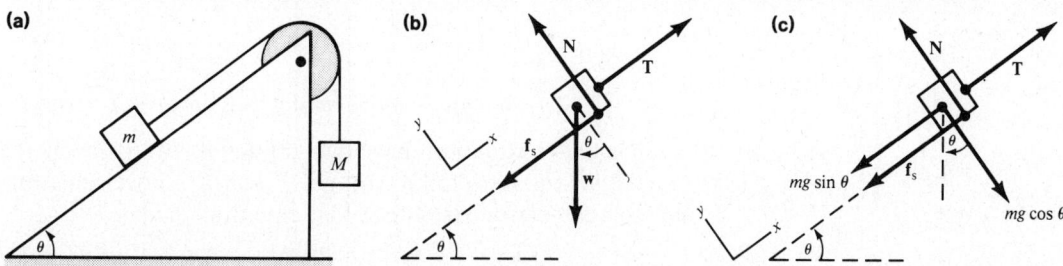

Figure 4-11

In this case the best choice of coordinate system is for the $+x$ axis to be directed up the plane. (You may want to convince yourself that you'll get the same result with the $+x$ axis directed down the plane. The correct result must be independent of the choice of a coordinate system.) Again, apply Newton's second law in component form.

x component	y component
ΣF_x	$\Sigma F_y = 0$
$T - mg \sin \theta - f_s = 0$	$N - mg \cos \theta = 0$

Because the system of weights is in equilibrium, the tension in the cord is equal to the weight of the hanging mass, so $T = Mg$. Because you want the maximum value for M, set $f_s = \mu_s N$, the maximum value of the force of static friction. Solve the y component equation for N and substitute these expressions for T, f_s, and N into the x component equation.

$$Mg - mg \sin \theta - \mu_s N = 0$$

$$Mg - mg \sin \theta - \mu_s mg \cos \theta = 0$$

Solve for M, canceling g.

$$M = m(\sin \theta - \mu_s \cos \theta) = (3 \text{ kg})(\sin 36° - 0.32 \cos 36°) = \boxed{0.987 \text{ kg}}$$

Note that M would be zero if $\sin \theta - \mu_s \cos \theta = 0$. If we solve this equation for μ_s,

$$\mu_s = \frac{\sin \theta}{\cos \theta} = \tan \theta \quad \text{or} \quad \theta = \arctan \mu_s = \arctan(0.32) = 17.7°$$

This means that, without the hanging weight and cord, the block wouldn't slide if the angle of inclination ranged from zero to 17.7°. If the angle exceeded 17.7°, the block would begin to slide down the plane.

Dynamics—Motion of a Particle

PROBLEM 4-4: In Figure 4-12, $m_1 = 3$ kg, $m_2 = 2$ kg, and $m_3 = 6$ kg; the pulley is frictionless, and the connecting cords can be considered weightless. **(a)** Calculate the acceleration of this system of weights when released from rest. **(b)** Determine the tension in the cord between m_1 and m_2 when the system is moving. (*Hint:* For part (a) consider m_1 and m_2 stuck together as a single mass, $m_1 + m_2$.)

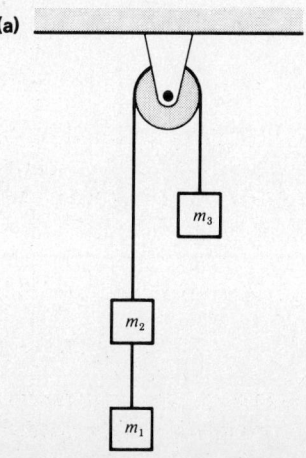

(a)

Figure 4-12a

(b)

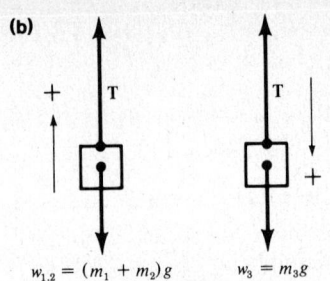

$$w_{1,2} = (m_1 + m_2)g \qquad w_3 = m_3g$$

(c)

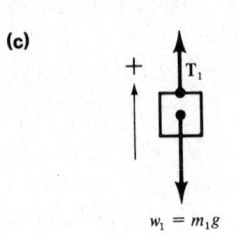

$$w_1 = m_1 g$$

Figure 4-12b and c

Solution:

(a) Because the pulley is frictionless, the tension T in the cord is the same on both sides of it. Assume that the cord doesn't stretch, so the *magnitude* of the acceleration a is the same for all three masses, but m_3 accelerates downward and m_1, m_2 accelerate upward. Apply Newton's second law to m_1, m_2 and to m_3. Let up be positive for m_1, m_2 and down be positive for m_3.

m_1, m_2	m_3
$\Sigma F = ma$	$\Sigma F = ma$
$T - (m_1 + m_2)g = (m_1 + m_2)a$	$m_3g - T = m_3a$

These two equations have two unknowns: T and a. Solving the m_3 equation for T; you get $T = m_3g - m_3a$; substitute this expression into the m_1, m_2 equation and solve for a.

$$m_3g - m_3a - (m_1 + m_2)g = (m_1 + m_2)a$$

$$a = g\left[\frac{m_3 - (m_1 + m_2)}{m_1 + m_2 + m_3}\right]$$

$$= (9.8 \text{ m s}^{-2})\left(\frac{[6 - (3 + 2)]\text{ kg}}{(3 + 2 + 6)\text{ kg}}\right)$$

$$= \boxed{0.89 \text{ m s}^{-2}}$$

(b) Apply Newton's second law to m_1 and solve for T_1, the tension in the cord between m_1 and m_2 (see Figure 4-12c).

$$\Sigma F = ma$$

$$T_1 - m_1g = m_1a$$

$$T_1 = m_1g + m_1a = m_1(g + a) = (3 \text{ kg})(9.8 + 0.89)\text{ m s}^{-2} = \boxed{32.1 \text{ N}}$$

The tension in the cord is greater than the weight of m_1, because m_1 is being *accelerated* upward.

PROBLEM 4-5: In Figure 4-13, $m_1 = 3$ kg, $m_2 = 2$ kg, $\theta = 65°$, and the coefficient of kinetic friction μ_k is 0.4. Assume that the pulleys are frictionless and that the mass of the connecting cord is negligible. Determine the acceleration of this system of masses.

(a)   **(b)**   **(c)**

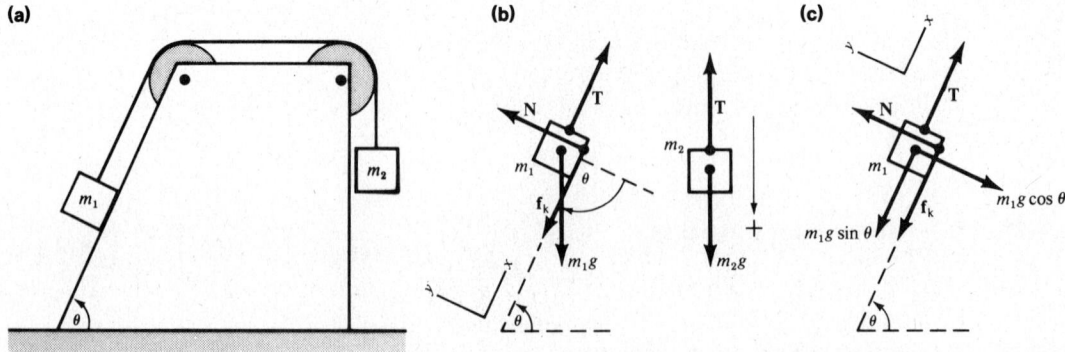

Figure 4-13

Solution: Because you don't know the direction this system will move by looking at it, *assume* a direction and proceed, e.g., that m_1 moves up the plane and, consequently, m_2 moves downward;

these become the positive directions. Draw free-body diagrams for m_1 and m_2, resolve forces into components (where appropriate), and apply Newton's second law.

m_1, x component	m_1, y component
$\Sigma F_x = ma$	$\Sigma F_y = 0$ (no motion in y direction)
$T - f_k - m_1 g \sin \theta = m_1 a$	$N - m_1 g \cos \theta = 0$

Solve the y component equation for N to get $N = m_1 g \cos \theta$ and then use this expression in the x component equation with $f_k = \mu_k N = \mu_k m_1 g \cos \theta$ to get

$$T - \mu_k m_1 g \cos \theta - m_1 g \sin \theta = m_1 a \qquad \text{(a)}$$

This equation has two unknowns: T and a. Apply the second law to m_2, keeping in mind that (because the pulleys are frictionless) the tension is the same on both sides of the pulleys, and the acceleration has the same magnitude for both m_1 and m_2.

$$\Sigma F = ma$$

$$m_2 g - T = m_2 a$$

Solve for T and substitute into Eq. (a).

$$T = m_2 g - m_2 a$$

$$m_2 g - m_2 a - \mu_k m_1 g \cos \theta - m_1 g \sin \theta = m_1 a$$

Now solve for a.

$$a = g \left[\frac{m_2 - m_1(\sin \theta + \mu_k \cos \theta)}{m_1 + m_2} \right]$$

$$= (9.8 \text{ m s}^{-2}) \left[\frac{2 - 3(\sin 65° + 0.4 \cos 65°) \text{ kg}}{(3 + 2) \text{ kg}} \right] = -2.403 \text{ m s}^{-2}$$

The negative sign indicates that the assumed direction of motion of the system of weights is *incorrect*. You now know that either the motion is in the *other direction* or that the masses *don't accelerate* when released. Let's proceed by assuming that the directions of motion of the weights are m_1 down the plane and m_2 upward.

Repeat the preceding analysis, which in this case gives

$$a = g \left[\frac{m_1(\sin \theta - \mu_k \cos \theta) - m_2}{m_1 + m_2} \right]$$

and which you should verify. Evaluating a, you obtain

$$a = (9.8 \text{ m s}^{-2}) \left[\frac{3(\sin 65° - 0.4 \cos 65°) - 2 \text{ kg}}{(3 + 2) \text{ kg}} \right] = \boxed{0.42 \text{ m s}^{-2}}$$

The acceleration is positive, so you should be confident that this is a correct result. An interesting further exercise is to show that for θ between 17° and 60° this system will remain at rest.

Motion Along a Curve

PROBLEM 4-6: A 1500-kg car rounds a level highway curve that has a radius of 40 m. Determine the force of static friction between the car wheels and the road surface when the speed of the car is 16 m s^{-1}.

Solution: The force of static friction provides the centripetal force to keep the car moving in its circular path. Use Eq. (4-5) and evaluate.

$$f = F_c = m \frac{v^2}{r} = (1500 \text{ kg}) \left[\frac{(16 \text{ m s}^{-1})^2}{40 \text{ m}} \right] = \boxed{9.6 \times 10^3 \text{ N}}$$

PROBLEM 4-7:   Determine the angle at which the road must be banked so that static friction isn't needed to hold the car in Problem 4-6 in its circular path.

Solution: Sketch the car on the banked road, draw a free-body diagram that shows the only two forces that act on the car, and resolve the normal force into horizontal (x) and vertical (y) components (see Figure 4-14).

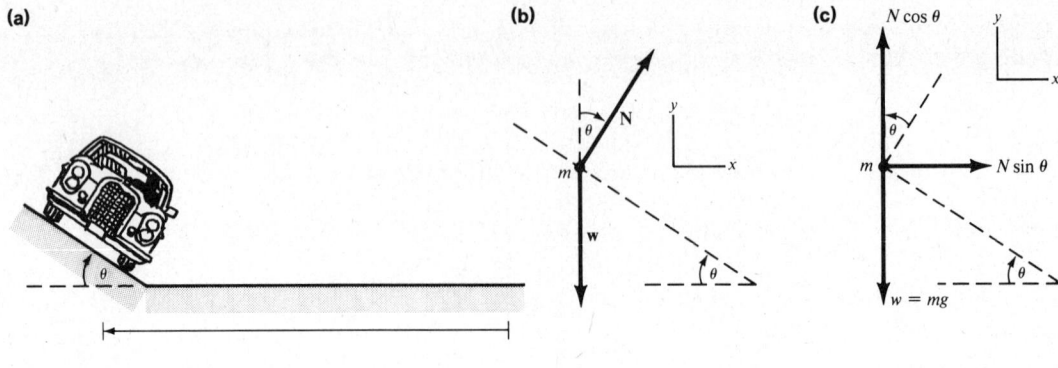

Figure 4-14

In this case the centripetal force is provided by the horizontal component of the normal force. Use Eq. (4-5),

$$N \sin \theta = F_c = m\frac{v^2}{r}$$

which has two unknowns: N and θ. First, apply Newton's second law in the vertical direction. The acceleration in the vertical direction is zero because there is no vertical motion of the car. Thus

$$\Sigma F_y = ma = 0$$
$$N \cos \theta - mg = 0$$

Then solve for N and substitute.

$$N = \frac{mg}{\cos \theta}$$

$$\left(\frac{mg}{\cos \theta}\right) \sin \theta = m\frac{v^2}{r}$$

The mass of the car cancels and $(\sin \theta)/(\cos \theta) = \tan \theta$, so $g \tan \theta = v^2/r$. Now solve for θ:

$$\theta = \arctan\left(\frac{v^2}{gr}\right) = \arctan\left[\frac{(16 \text{ m s}^{-1})^2}{(9.8 \text{ m s}^{-2})(40 \text{ m})}\right] = \boxed{33.2°}$$

Because the mass cancels, the result is the same for all vehicles, from large trucks to compacts, that travel this curve at 16 m s^{-1}.

Velocity-Dependent Forces

PROBLEM 4-8:   The friction force on an irregularly shaped object of mass 150 g is found to be proportional to its velocity. The proportionality constant is $b = 0.4$ kg s^{-1}. What is the terminal velocity for this object?

Solution: Solve the equation in Example 4-7b for v_t and evaluate.

$$v_t = \frac{mg}{b} = \frac{(0.15 \text{ kg})(9.8 \text{ m s}^{-2})}{0.4 \text{ kg s}^{-1}} = \boxed{3.68 \text{ m s}^{-1}}$$

Note that when an object is falling at its terminal velocity the frictional force on it equals its weight.

PROBLEM 4-9: For the object in Problem 4-8, determine how long it takes for the object to reach one-half its terminal velocity after being released from rest.

Solution: Use the equation derived in Example 4-7a, $v = v_t(1 - e^{-bt/m})$, set $v = \frac{1}{2}v_t$, and solve for t.

$$\tfrac{1}{2}v_t = v_t(1 - e^{-bt/m})$$

$$e^{bt/m} = 2$$

$$\frac{bt}{m} = \ln 2$$

$$t = \frac{m\ln 2}{b} = \frac{(0.15 \text{ kg})\ln 2}{0.4 \text{ kg s}^{-1}} = \boxed{0.26 \text{ s}}$$

Accelerated Frames of Reference

PROBLEM 4-10: A box rests on the flat bed of a truck that is moving at 80 km h^{-1}. The driver applies the brakes, the truck decelerates at the rate of 5 m s^{-2}, and the box begins to slide toward the cab with an acceleration of 2 m s^{-2}. Determine the coefficient of sliding friction between the box and truck bed.

Solution: Assume that the box has a mass m and that it and the truck are moving to the right. Draw a free body diagram for the box, as in Figure 4-15, and apply Newton's second law in component form.

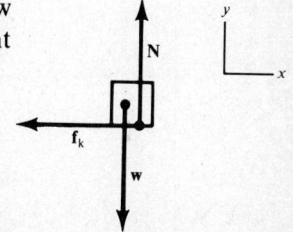

Figure 4-15

x component	y component	
$\Sigma F_x = ma$	$\Sigma F_y = 0$	(no motion in y direction)
$-f_k = ma$	$N - mg = 0$	

Solve the y component equation for $N = mg$ and substitute into the x component equation, where $f_k = \mu_k N$, to get $-\mu_k mg - ma$. Then use Eq. (4-7), $ma - ma_{BA} = ma'$, where $ma = -\mu_k mg$, a_{BA} is the acceleration of the truck with respect to the ground ($a_{BA} = -5$ m s^{-2}), and a' is the acceleration of the box with respect to the truck ($a' = 2$ m s^{-2}). Solve for μ_k and evaluate.

$$-\mu_k mg - ma_{BA} = ma'$$

$$\mu_k = -\frac{a' + a_{BA}}{g} = -\frac{(2 - 5) \text{ m s}^{-2}}{9.8 \text{ m s}^{-2}} = \boxed{0.31}$$

PROBLEM 4-11: A weight on the end of a cord swings at 30 re min^{-1} in a horizontal circle that has a radius of 40 cm. The cord sweeps out an inverted cone in space. Determine the angle between the cord and a vertical line.

Solution: Make a sketch of the cord and weight, draw a free-body diagram of the weight in a coordinate system that rotates with the weight, and resolve the forces into components (where appropriate), as in Figure 4-16.

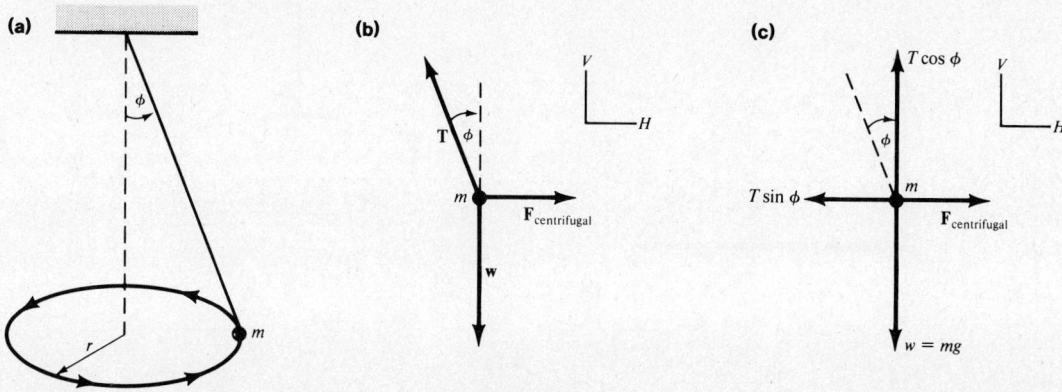

Figure 4-16

Let the mass of the weight be m and apply Eq. (4-9). In this case $F_{\text{Coriolis}} = 0$ and $a' = 0$ because the weight is at rest in the rotating coordinate system.

Horizontal component	Vertical component
$F_{\text{real}} + F_{\text{Coriolis}} + F_{\text{centrifugal}} = ma'$	$F_{\text{real}} + F_{\text{Coriolis}} + F_{\text{centrifugal}} = ma'$
$-T\sin\phi + 0 + m\omega^2 r = 0$	$(T\cos\phi - mg) + 0 + 0 = 0$

Solve the vertical-component equation for T, the tension in the cord, substitute that expression into the horizontal-component equation, and solve for ϕ.

$$-\left(\frac{mg}{\cos\phi}\right)\sin\phi + m\omega^2 r = 0 \quad \text{and} \quad \phi = \arctan\left(\frac{\omega^2 r}{g}\right)$$

where

$$\omega = \left(\frac{30 \text{ rev}}{\text{min}}\right)\left(\frac{2\pi \text{ rad}}{\text{rev}}\right)\left(\frac{1 \text{ min}}{60 \text{ s}}\right) = 3.14 \text{ rad s}^{-1}$$

So

$$\phi = \arctan\left[\frac{(3.14 \text{ rad s}^{-1})^2(0.4 \text{ m})}{9.8 \text{ m s}^{-2}}\right] = \boxed{21.9°}$$

Supplementary Exercises

EXERCISE 4-1:   Determine the tension in the slanted cord in Figure 4-17. Assume the rod is weightless, $m = 6$ kg, and $\phi = 36°$.

EXERCISE 4-2:   A 450-lb engine is lifted out of its mount in a car by means of an engine sling and cable hoist. The engine sling is inclined at $\theta = 10°$ relative to the horizontal, as shown in Figure 4-18. Determine the tension in the engine sling. [Interesting side exercise: Can you show that the engine weighs 2000 N and has a mass of 204 kg?]

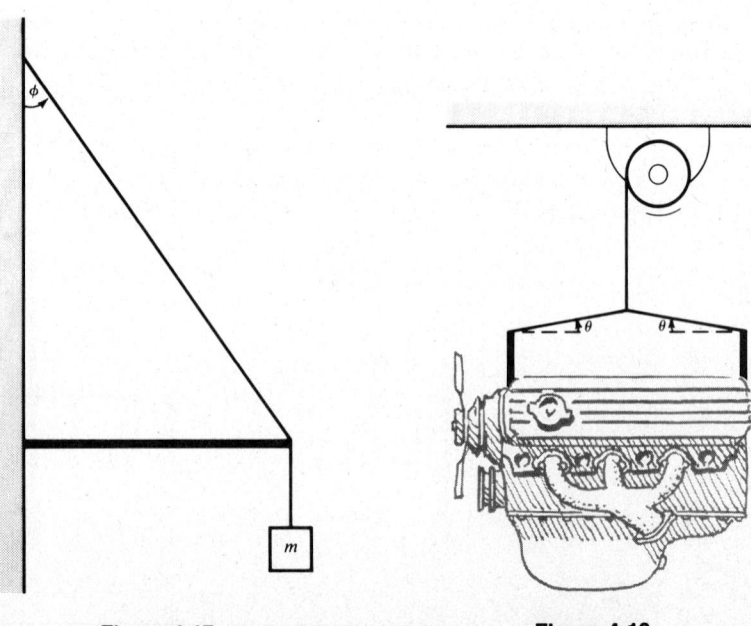

Figure 4-17   **Figure 4-18**

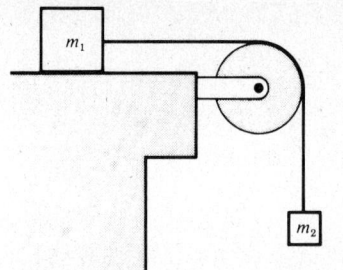

Figure 4-19

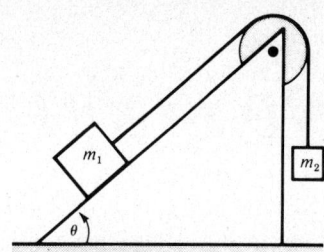

Figure 4-20

EXERCISE 4-3: The coefficient of kinetic friction between the 5-kg mass and the table surface (Figure 4-19) is $\mu_k = 0.24$. Determine the acceleration of the two masses when the hanging mass is 1.6 kg. Assume that the pulley is frictionless and that the cord has no mass.

EXERCISE 4-4: In Figure 4-20, $m_1 = 5$ kg, $m_2 = 4$ kg, $\theta = 40°$, and m_1 moves up the plane with an acceleration of $0.3 \, \text{m s}^{-2}$. Determine the coefficient of sliding friction between m_1 and the plane. Assume that the pulley is frictionless and that the cord has no mass.

EXERCISE 4-5: A small engine powers a model race car around a flat circular track. The car is attached to one end of a cord, which holds it in a circular orbit. The cord is 4 m long and at the other end is held in a frictionless pivot at the center of the circular track. The coefficient of static friction between the wheels of the car and the track is 0.8. Calculate the tension in the cord when the speed of the car is $7 \, \text{m s}^{-1}$.

EXERCISE 4-6: Mass m is whirled in a horizontal circle by means of a cord that passes through a frictionless pivot and is attached to a hanging mass M (see Figure 4-21). The distance from m to the pivot is ℓ, and this section of cord makes an angle θ with respect to the vertical. Determine the speed of m in its orbit in terms of $m, M, \ell, g,$ and θ.

EXERCISE 4-7: Derive an expression for the velocity as a function of time for a body of mass m that falls in a medium in which the frictional force is proportional to the square of the body's velocity, i.e., $f = kv^2$, where k is a constant that has units of kg m^{-1}. Assume that the initial condition is $v = 0$

at t = 0. $\left[\textit{Hint:} \quad \int \dfrac{dx}{p^2 - x^2} = \dfrac{1}{p} \tanh^{-1}\left(\dfrac{x}{p}\right). \right]$

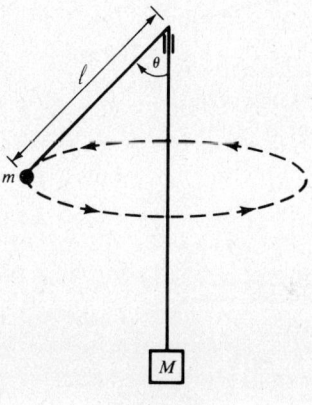

Figure 4-21

EXERCISE 4-8: Calculate the terminal velocity for a 4-kg body falling in a medium in which the frictional force is given by $f = kv^2$, where $k = 0.15 \, \text{kg m}^{-1}$.

EXERCISE 4-9: A small weight (mass m) hangs from a cord in a jetliner. Shortly after touchdown the plane decelerates and an observant passenger notices that during the deceleration the weight swings forward so that the cord makes an angle of 15° with respect to the vertical. Calculate the deceleration of the plane.

EXERCISE 4-10: Calculate the speed of the bead in Example 4-9 that is necessary for the Coriolis force on it to equal the bead's weight just before it leaves the end of the wire. As in Example 4-9, $\omega = 2.7 \, \text{rad s}^{-1}$.

Answers to Supplementary Exercises

4-1: 72.7 N   **4-2:** 1296 lb   **4-3:** $0.59 \, \text{m s}^{-2}$ **4-4:** $\mu_k = 0.133$ **4-5:** 6.6 N

4-6: $v = \sqrt{\dfrac{Mg\ell \sin^2 \theta}{m}}$ **4-7:** $v = \sqrt{\dfrac{mg}{k}} \tanh\left(\sqrt{\dfrac{kg}{m}}\, t\right)$ **4-8:** $v_1 = 16.2 \, \text{m s}^{-1}$

4-9: $2.6 \, \text{m s}^{-2}$ **4-10:** $v' = \dfrac{g}{2\omega} = 1.81 \, \text{m s}^{-1}$

5 WORK AND KINETIC ENERGY

THIS CHAPTER IS ABOUT
- ☑ **Work—The Dot Product**
- ☑ **Work Done by a Changing Force**
- ☑ **The Work–Energy Principle**
- ☑ **Power**

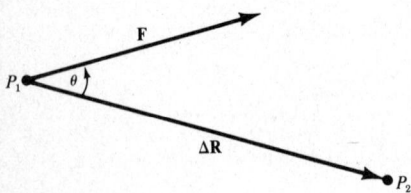

Figure 5-1. The force **F** moves a particle from P_1 to P_2. The component of **F** in the direction of the displacement is $F\cos\theta$.

(a)

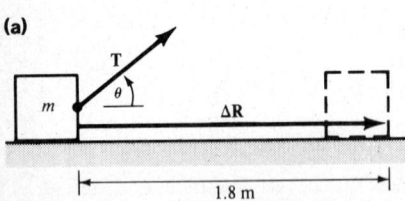

(b)

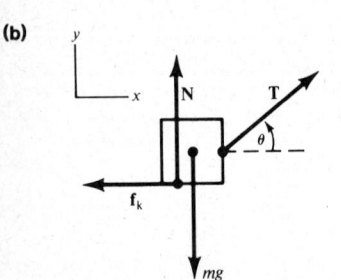

(c)

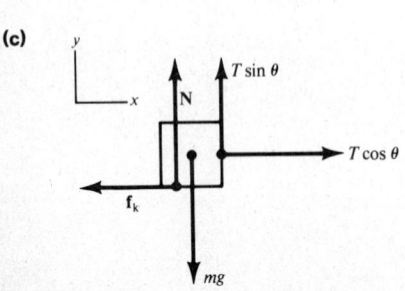

Figure 5-2

5-1. Work—The Dot Product

Work in physics involves both the force acting on a particle of mass m and its resulting displacement. For a constant force displacing a particle from point 1 to point 2 (Figure 5-1), the work done is

$$W = \mathbf{F} \cdot (\Delta\mathbf{R}) \tag{5-1a}$$

WORK
$$= F(\Delta R)\cos\theta \tag{5-1b}$$

$$= F_x(\Delta R_x) + F_y(\Delta R_y) + F_z(\Delta R_z) \tag{5-1c}$$

Work is a scalar quantity, and in the SI system of units it is measured in joules.

$$1 \text{ joule} = 1 \text{ newton-meter} = 1 \text{ (kilogram)} \cdot \text{(meter}^2/\text{second}^2)$$

$$1 \text{ J} = 1 \text{ N m} = 1 \text{ kg m}^2 \text{ s}^{-2}$$

In the British system, work is measured in foot-pounds, i.e., feet times pounds, or ft-lb.

$$1 \text{ ft-lb} = 1.356 \text{ J}$$

For work to be done by a force, it must be applied to the object during the displacement and the angle between the force and the displacement must be less than 90°.

EXAMPLE 5-1: A block of mass $m = 6$ kg is dragged across a horizontal surface a distance of 1.8 m by a cord that makes an angle of 38° above the horizontal, as shown in Figure 5-2. The tension in the cord is 80 N and the coefficient of kinetic friction between the block and surface is 0.4. Determine **(a)** which forces acting on the block do work, **(b)** the amount of work each of these forces does, and **(c)** the net work done on the block.

Solution: Make a sketch, draw a free body diagram to identify the forces acting on the block, resolve forces into components (where necessary), and evaluate the work done.

(a) Forces N and mg do no work because the angle between them and the displacement is 90°. Forces **T** and $\mathbf{f}_k$ do work.

(b) Use Eq. (5-1b) to calculate the work done by **T**.

$$W_T = T(\Delta R)\cos\theta = (80 \text{ N})(1.8 \text{ m})\cos 38° = \boxed{113.5 \text{ J}}$$

To calculate the work done against the force of friction you must first determine the force of friction. Use Eq. (4-3), $f_k = \mu_k N$, and the techniques presented in Section 4-3 to evaluate $N = mg - T\sin\theta$. Therefore $f_k = \mu_k(mg - T\sin\theta)$ and is directed opposite to $\Delta \mathbf{R}$ and the work done by f_k is

$$W_f = f_k(\Delta R)\cos 180°$$

$$= \mu_k(mg - T\sin\theta)(\Delta R)\cos 180°$$

$$= (0.4)[(6\text{ kg})(9.8\text{ m s}^{-2}) - (80\text{ N})\sin 38°](1.8\text{ m})(-1) = \boxed{-6.9\text{ J}}$$

Friction is a *dissipative force*. Work done against friction is converted into *thermal energy*, which explains the negative sign in this case.

(c) The net work done on the block is

$$W_{net} = W_T + W_f = (113.5 - 6.9)\text{ J} = \boxed{106.6\text{ J}}$$

5-2. Work Done by a Changing Force

When a force changes along a path or when the path is curved so that the angle between the force and path changes, the *line integral* in Eq. (5-2) must be evaluated to determine the work done by $\mathbf{F}$ in moving the particle from point 1 to point 2, i.e.,

$$W = \int_1^2 \mathbf{F} \cdot d\mathbf{R} \tag{5-2}$$

In component form this equation is

$$W = \int_1^2 F_x\,dx + \int_1^2 F_y\,dy + \int_1^2 F_z\,dz \tag{5-3}$$

where points 1 and 2 have coordinates (x_1, y_1, z_1) and (x_2, y_2, z_2), respectively, and it is presumed that the functional relationship between $\mathbf{F}$ and position is known.

EXAMPLE 5-2: A block is moved along the x axis from $x_1 = 1$ m to $x_2 = 3$ m by the force $\mathbf{F} = (ax^2 + b)\hat{\mathbf{x}}$, where $a = 2\text{ N m}^{-2}$ and $b = 1.8$ N. Determine the work done by this force.

Solution: Use Eq. (5-3). In this case $F_x = ax^2 + b$, $F_y = 0$, and $F_z = 0$, so

$$W = \int_1^2 F_x\,dx + 0 + 0 = \int_{x_1}^{x_2} (ax^2 + b)dx$$

$$= a\left(\frac{x^3}{3}\bigg|_{x_1}^{x_2}\right) + b\left(x\bigg|_{x_1}^{x_2}\right)$$

$$= \frac{a}{3}(x_2^3 - x_1^3) + b(x_2 - x_1)$$

$$= \left(\frac{2\text{ N m}^{-2}}{3}\right)[(3\text{ m})^3 - (1\text{ m})^3] + (1.8\text{ N})(3 - 1)\text{ m} = \boxed{20.9\text{ J}}$$

5-3. The Work–Energy Principle

Energy is that property, or capacity, of a system that enables it to do work. It is a scalar quantity and is measured in the same units as work. **Kinetic energy**, the energy of motion, of a particle of mass m moving with speed v is

KINETIC ENERGY $E_k = \tfrac{1}{2}mv^2$ (5-4)

EXAMPLE 5-3: Determine the kinetic energy of (a) a 500-g object moving at 3 m s^{-1} and (b) a 3000-kg automobile moving at 90 km h^{-1}.

Solution:

(a) Use Eq. (5-4).

$$E_k = \tfrac{1}{2}mv^2 = \tfrac{1}{2}(0.5 \text{ kg})(3 \text{ m s}^{-1})^2 = \boxed{2.25 \text{ J}}$$

(b) Use Eq. (5-4), with $v = 90 \text{ km h}^{-1} = 25 \text{ m s}^{-1}$.

$$E_k = \tfrac{1}{2}mv^2 = \tfrac{1}{2}(3 \times 10^3 \text{ kg})(25 \text{ m s}^{-1})^2 = \boxed{9.38 \times 10^5 \text{ J}}$$

The **work–energy principle** states that the net work done, i.e., the work done by the net or resultant force, is equal to the *change* in kinetic energy of the particle.

WORK–ENERGY PRINCIPLE

$$W_{net} = \Delta E_k \tag{5-5a}$$
$$= E_{k,\,final} - E_{k,\,initial} \tag{5-5b}$$
$$= \tfrac{1}{2}mv_2^2 - \tfrac{1}{2}mv_1^2 \tag{5-5c}$$

where v_1 is the speed of the particle before the work is done, and v_2 is its speed afterward.

EXAMPLE 5-4: If the block in Example 5-1 is moving at 0.8 m s^{-1} at one point, determine how fast it is moving after being dragged 1.8 m by the cord.

Solution: Use the work–energy principle, Eq. (5-5c), and solve for v_2. In this case $v_1 = 0.8 \text{ m s}^{-1}$ and from the result of Example 5-1, $W_{net} = 106.6$ J.

$$W_{net} = \tfrac{1}{2}mv_2^2 - \tfrac{1}{2}mv_1^2$$

$$v_2 = \sqrt{2W_{net}/m + v_1^2} = \sqrt{2(106.6 \text{ J})/(6 \text{ kg}) + (0.8 \text{ m s}^{-1})^2} = \boxed{6.01 \text{ m s}^{-1}}$$

5-4. Power

The time rate at which work is done on a system is called the **power** applied to that system:

POWER
$$P = \frac{dW}{dt} \tag{5-6}$$

Power is a scalar quantity and in the SI system of units is measured in watts.

$$1 \text{ watt} = 1 \text{ joule/second}$$
$$1 \text{ W} = 1 \text{ J s}^{-1}$$

In the British system, power is measured in foot-pounds/second, or ft-lb/s. The conversion relationship between the two systems most commonly used is the definition of **horsepower**.

$$1 \text{ horsepower} = 550 \text{ ft-lb/s} = 746 \text{ W} = 0.746 \text{ kW}$$

For approximate calculations you may wish to remember that one horsepower is approximately three-fourths of a kilowatt.

In Example 5-5, we will show that

$$P = \mathbf{F} \cdot \mathbf{v} \tag{5-7}$$

where **F** is the force applied to a system and, from Eq. (5-6),

$$W = \int P\,dt \qquad\qquad \textbf{(5-8)}$$

where it is assumed that P is a known function of time.

EXAMPLE 5-5: Starting with Eq. (5-6), $P = \dfrac{dW}{dt}$, prove that $P = \mathbf{F} \cdot \mathbf{v}$, Eq. (5-7).

Solution: From Eq. (5-1a), $W = F \cdot (\Delta \mathbf{R})$, you can conclude that $dW = \mathbf{F} \cdot d\mathbf{R}$, and from Eq. (3-6), $\mathbf{v} = \dfrac{d\mathbf{R}}{dt}$, you can also conclude that $d\mathbf{R} = \mathbf{v}\,dt$. Therefore $dW = \mathbf{F} \cdot \mathbf{v}\,dt$. Divide by dt to obtain the desired result, or $\dfrac{dW}{dt} = P = \mathbf{F} \cdot \mathbf{v}$.

EXAMPLE 5-6: A gallon of gasoline contains 1.3×10^8 J of chemical energy. At $90 \text{ km h}^{-1} = 25 \text{ m s}^{-1}$, a small car gets 48 km gal^{-1}. Determine the input power to the engine of the car. [*Hint:* Assume that the input power is constant during the consumption of one gallon of gasoline and that the speed of the car is also constant.]

Solution: Use Eq. (5-6) (for finite intervals $P = \Delta W/\Delta t$), in which $\Delta W = 1.3 \times 10^8$ J, the input energy from the gasoline, and Δt is the time interval during which the gallon of gasoline is consumed at 90 km h^{-1}. At constant speed, $x = v\,\Delta t$, so

$$\Delta t = \frac{x}{v} = \frac{48 \text{ km}}{90 \text{ km h}^{-1}} = \frac{48 \times 10^3 \text{ m}}{25 \text{ m s}^{-1}} = 1.92 \times 10^3 \text{ s}$$

and

$$P = \frac{\Delta W}{\Delta t} = \frac{1.3 \times 10^8 \text{ J}}{1.92 \times 10^3 \text{ s}} = \boxed{6.77 \times 10^4 \text{ W} = 67.7 \text{ kW}}$$

EXAMPLE 5-7: The car in Example 5-6 has an overall efficiency typical of small cars of 22% in converting the chemical energy in the gasoline into mechanical energy to maintain the speed of the car at 90 km h^{-1}. Determine the constant force that must be supplied to the car by its engine to maintain a speed of 90 km h^{-1}. This force is equal in magnitude to the effective force of moving friction acting on the car at this speed.

Solution: Efficiency, in this case, is the ratio of useful output power to input power.

$$\text{Efficiency} = \frac{\text{Output power}}{\text{Input power}} = \frac{P_{\text{out}}}{P_{\text{in}}} \quad \text{and} \quad P_{\text{out}} = (\text{eff})(P_{\text{in}})$$

Use Eq. (5-7), for which $\mathbf{F} \cdot \mathbf{v} = Fv$, because **F** and **v** are parallel in this case, and solve for F.

$$P = \mathbf{F} \cdot \mathbf{v} = Fv$$

$$F = \frac{P}{v} = \frac{P_{\text{out}}}{v} = \frac{(\text{eff})(P_{\text{in}})}{v} = \frac{(0.22)(6.77 \times 10^4 \text{ W})}{25 \text{ m s}^{-1}} = \boxed{596 \text{ N}}$$

EXAMPLE 5-8: The input power to a certain appliance is directly proportional to the time the appliance is in operation, or $P_{\text{in}} = Kt$, where the proportionality constant is $K = 24 \text{ W s}^{-1}$. Determine the input energy to this device when it is operated for one hour.

Solution: Use Eq. (5-8), with $T = 1$ h $= 3.6 \times 10^3$ s.

$$W = \int P \, dt = \int_0^T Kt \, dt = K \int_0^T t \, dt = \frac{1}{2} Kt^2 \Big|_0^T = \frac{1}{2} KT^2$$

$$= \frac{1}{2}(24 \text{ W s}^{-1})(3.6 \times 10^3 \text{ s})^2 = \boxed{1.56 \times 10^8 \text{ J}}$$

SUMMARY

1. The work done by a constant force displacing an object $\Delta \mathbf{R}$ is $W = \mathbf{F} \cdot (\Delta \mathbf{R})$.
2. When the force changes along the path, the work done is computed by the *line integral* $W = \int_1^2 \mathbf{F} \cdot d\mathbf{R}$, where 1 and 2 represent the initial and final positions of the object displaced.
3. The *kinetic energy* of an object of mass m is the work it is able to do by virtue of its motion and is given by $E_k = \frac{1}{2}mv^2$.
4. The *work–energy principle* states that the net work done on an object is equal to the change in its kinetic energy: $W_{net} = \Delta E_k$.
5. Power is the time rate at which work is done: $P = dW/dt$.

RAISE YOUR GRADES

Can you explain . . . ?

☑ how a force of 600 N acting on an object for 2 h could do no work
☑ why no work is done by the tension in the cord attached to a weight whirled in a horizontal circle at constant speed
☑ why the force of kinetic friction cannot increase the kinetic energy of a particle
☑ what will happen to the kinetic energy of a particle if the net work done on it is negative
☑ how the rate at which work done by a constant force can increase with time
☑ the conditions under which the power applied to a system could be negative
☑ why the kinetic energy of a particle can never be negative
☑ why the kinetic energy of a 210-g mass can never be 460 watts

SOLVED PROBLEMS

Work—The Dot Product

PROBLEM 5-1:   A 6-kg block is raised 2 m vertically at constant speed by a cord attached to it that passes over a frictionless pulley. How much work is done on the block by the tension in the cord?

Solution: Use Eq. (5-1b), in which F is the tension in the cord and is equal to the weight of the block because it is being lifted at a constant speed, ΔR is the displacement, and $\theta = 0°$.

$$W = F(\Delta R)\cos\theta = (mg)(\Delta R)\cos 0° = (6 \text{ kg})(9.8 \text{ m s}^{-2})(2 \text{ m})(1) = \boxed{118 \text{ J}}$$

PROBLEM 5-2: A 4-kg block is pulled up a frictionless plane inclined at an angle $\theta = 15°$ above the horizontal by a cord that makes an angle of $\phi = 20°$ above the surface of the plane. The tension in the cord is 20 N. Determine the net work done on the block to move it 2 m up the plane.

Solution: Make a sketch, draw a free-body diagram of the block, and resolve the forces acting on it into components parallel and perpendicular to the surface of the plane, as shown in Figure 5-3. Now use Eq. (5-1c), in which $\Delta R_y = 0$ and $\Delta R_z = 0$.

$$W = F_x(\Delta R_x) + F_y(\Delta R_y) + F_z(\Delta R_z) = (T\cos\phi - mg\sin\theta)(\Delta R) + 0 + 0$$

$$= [(20\text{ N})\cos 20° - (4\text{ kg})(9.8\text{ m s}^{-2})\sin 15°](2\text{ m}) = \boxed{17.3\text{ J}}$$

Note that forces N, $T\sin\phi$, and $mg\cos\theta$ do no work because they are perpendicular to the displacement.

(a)   **(b)**   **(c)**

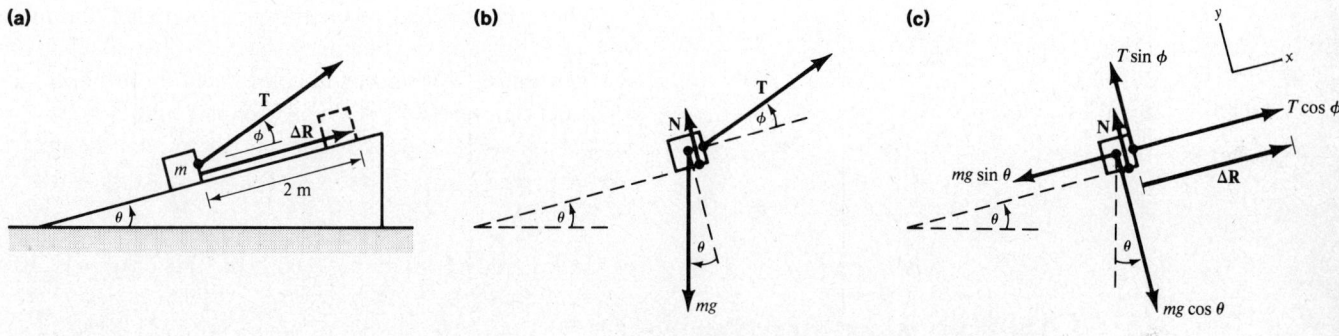

Figure 5-3

PROBLEM 5-3: A 3-kg block slides down a plane that is 2 m high (h) at one end and is inclined at an angle $\theta = 25°$. The coefficient of kinetic friction between the block and plane is 0.3. Determine the net work done on the block if it slides down the entire length of the plane.

Solution: Make a sketch, draw a free-body diagram of the block, and resolve the forces acting on it into components parallel and perpendicular to the plane (see Figure 5-4).

(a)   **(b)**   **(c)**

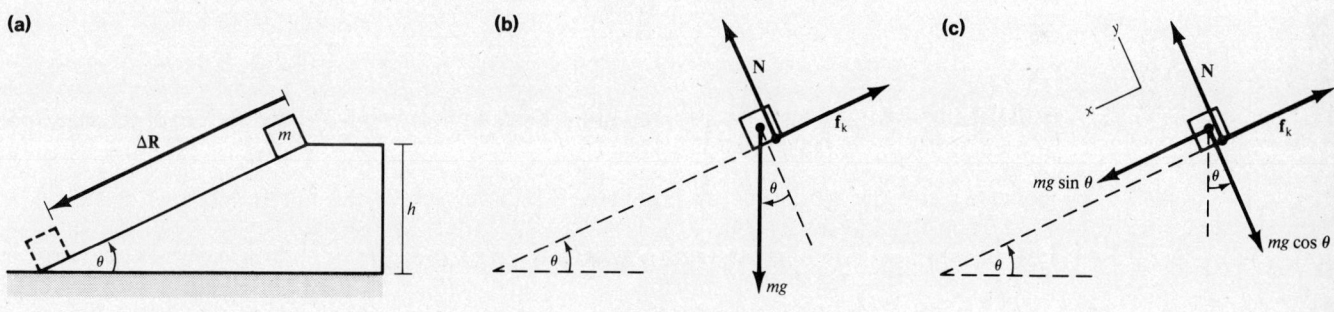

Figure 5-4

In this case you need to resolve only the weight. As in Problem 5-2, N and $mg\cos\theta$ do no work. Use Eq. (5-1c), in which $\Delta R_y = 0$ and $\Delta R_z = 0$.

$$W = F_x(\Delta R_x) + F_y(\Delta R_y) + F_z(\Delta R_z) = (mg\sin\theta - f_k)(\Delta R) + 0 + 0$$

$$= mg(\sin\theta - \mu_k\cos\theta)\left(\frac{h}{\sin\theta}\right) = mgh\left(1 - \frac{\mu_k}{\tan\theta}\right)$$

$$= (3\text{ kg})(9.8\text{ m s}^{-2})(2\text{ m})\left(\frac{1 - 0.3}{\tan 25°}\right) = \boxed{21\text{ J}}$$

Confirm the use of

$$f_k = \mu_k N = \mu_k mg\cos\theta, \qquad h = \Delta R\sin\theta, \qquad \text{and} \qquad \tan\theta = \frac{\sin\theta}{\cos\theta}$$

in the second and third steps of the solution.

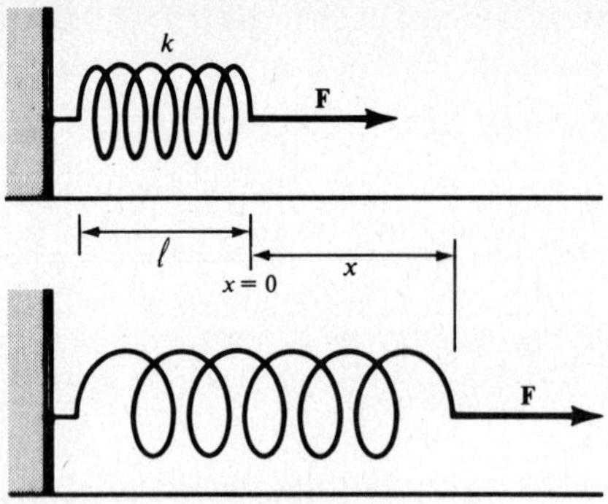

Figure 5-5

Work Done by a Changing Force

PROBLEM 5-4: The force required to stretch a spring a distance x is directly proportional to the distance stretched, or $F = kx$, where k is called the spring constant and is measured in $N\,m^{-1}$ in the SI system. The force that the spring exerts is, by Newton's third law, equal but opposite, or $F_s = -F = -kx$ (known as Hooke's law). How much work is required to stretch a spring, having a spring constant $k = 160\,N\,m^{-1}$, a distance of 20 cm?

Solution: Figure 5-5 shows the spring before and after it is stretched. Its unstretched length is ℓ. The force isn't constant, so you must use Eq. (5-2) to calculate the amount of work done; in component form it becomes Eq. (5-3), where $F_x = F = kx$, $F_y = 0$, and $F_z = 0$.

$$W = \int_1^2 F_x\,dx + \int_1^2 F_y\,dy + \int_1^2 F_z\,dz = \int_0^x kx\,dx + 0 + 0$$

$$= k\int_0^x x\,dx = \frac{1}{2}kx^2 = \frac{1}{2}(160\,N\,m^{-1})(0.2\,m)^2 = \boxed{3.2\,J}$$

The Work–Energy Principle

PROBLEM 5-5: Determine the speed of the block in Problem 5-2 at the end of the 2-m displacement up the plane if the block started from rest.

Solution: Use the work–energy principle, Eq. (5-5c), where the initial speed of the block is $v_1 = 0$, and solve for v_2.

$$W_{net} = \tfrac{1}{2}mv_2^2 - \tfrac{1}{2}mv_1^2$$

$$v_2 = \sqrt{2W_{net}/m + 0} = \sqrt{2(17.3\,J)/4\,kg} = \boxed{2.94\,m\,s^{-1}}$$

PROBLEM 5-6: Determine the speed of the block in Problem 5-3 at the bottom of the plane if it started at the top with a speed of $0.6\,m\,s^{-1}$.

Solution: Use the work–energy principle, Eq. (5-5c), where the initial speed of the block is $v_1 = 0.6\,m\,s^{-1}$, and solve for v_2.

$$W_{net} = \tfrac{1}{2}mv_2^2 - \tfrac{1}{2}mv_1^2$$

$$v_2 = \sqrt{2W_{net}/m + v_1^2} = \sqrt{2(21\,J)/3\,kg + (0.6\,m\,s^{-1})^2} = \boxed{3.79\,m\,s^{-1}}$$

PROBLEM 5-7: A particle of mass m moves in a straight line. Because of the net work (W_{net}) done on it, its speed increases from v_1 to v_2. Obtain an expression for the acceleration of the particle in terms of W_{net}, m, and x, the displacement of the particle during which the net work is done.

Solution: Use the work–energy principle, Eq. (5-5c), and Eq. (2-7), $v^2 = v_0^2 + 2ax$. In this case $v = v_2$ and $v_0 = v_1$, so Eq. (2-7) becomes $v_2^2 = v_1^2 + 2ax$, which can be written as $v_2^2 - v_1^2 = 2ax$. Now solve Eq. (5-5c) for $v_2^2 - v_1^2$.

$$W_{net} = \tfrac{1}{2}mv_2^2 - \tfrac{1}{2}mv_1^2$$

$$v_2^2 - v_1^2 = \frac{2W_{net}}{m}$$

Thus $2ax = 2W_{net}/m$ and

$$a = \boxed{W_{net}/mx}$$

Power

PROBLEM 5-8: Determine the average power needed to stretch the spring in Problem 5-4 if the stretching is to be done in 0.5 s.

Solution: Use Eq. (5-6), but in this case dW is the total work done, and dt is the total time required.

$$P = \frac{dW}{dt} = \frac{3.2\text{ J}}{0.5\text{ s}} = \boxed{6.4\text{ W}}$$

PROBLEM 5-9: A small engine can deliver 2 hp of power. How long will it take this engine to raise an 80-kg mass 70 m at constant speed? (See Figure 5-6.) The tension in the cord is equal to the weight of the mass because the mass is raised at a constant speed.

Solution: You may first want to convert the 2 hp into SI units. To do so, use the conversion relationship 1 hp = 746 W to obtain

$$P = (2\text{ hp})(746\text{ W/hp}) = 1.492\text{ kW}$$

Use Eq. (5-6), but in this case let $dW = W = Fh$, the work done by the engine in the time $dt = t$. Because the mass is raised at constant speed, F, the tension in the cord, is equal to the weight of the mass, $F = mg$. Solve for $dt = t$.

$$P = \frac{dW}{dt} = \frac{W}{t} = \frac{Fh}{t} = \frac{mgh}{t}$$

and

$$t = \frac{mgh}{P} = \frac{(80\text{ kg})(9.8\text{ m s}^{-2})(70\text{ m})}{1.492 \times 10^3\text{ W}} = \boxed{36.8\text{ s}}$$

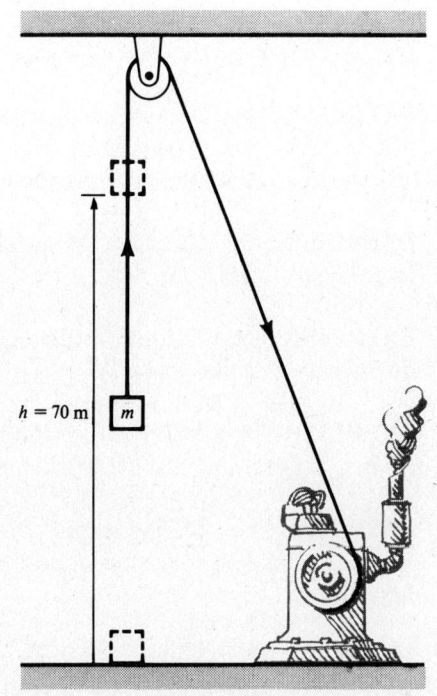

Figure 5-6

PROBLEM 5-10: At 55 mph, the combined force of air friction and rolling friction is 94 lb for a small car. Determine the horsepower its engine must deliver to maintain this speed.

Solution: First, convert the speed into units of ft/s.

$$v = (55\text{ mph})\left(\frac{88\text{ ft/s}}{60\text{ mph}}\right) = 80.67\text{ ft/s}$$

Now use Eq. 5-7.

$$P = \mathbf{F}\cdot\mathbf{v} = Fv\cos\theta$$

In this case $\theta = 0°$, so

$$P = (94\text{ lb})(80.67\text{ ft/s})\cos 0° = 7.583 \times 10^3\text{ ft-lb/s}$$

Convert this result to horsepower, using 1 hp = 550 ft-lb/s.

$$P = (7.583 \times 10^3\text{ ft-lb/s})\left(\frac{1\text{ hp}}{550\text{ ft-lb/s}}\right) = 13.8\text{ hp}$$

Supplementary Exercises

EXERCISE 5-1:   A horizontal force of 35 N drags a 6-kg mass a distance of 3 m across a horizontal surface. The coefficient of kinetic friction between the mass and the surface is 0.4. Calculate the net work done on the mass.

EXERCISE 5-2:   A force of 65 N parallel to the surface of a plane that is inclined at 32° above the horizontal drags a 6-kg mass a distance of 1.6 m up the plane. The coefficient of kinetic friction between the plane and the mass is 0.4. Calculate the net work done on the mass.

EXERCISE 5-3:   A bead, constrained to slide on a straight horizontal wire, is acted on by a force of 6 N that is directed 24° above the horizontal. The net work done on the bead in moving it 1.4 m is 4.2 J. Determine the force of friction acting on the bead.

EXERCISE 5-4:   Because of an attraction to the origin, a force given by $F = K/x^2$ is needed to move a particle of mass m from x_1 to x_2, where $x_2 > x_1$. How much work is done by this force? Assume that the motion is along the x axis.

EXERCISE 5-5:   Calculate the speed of the mass in Exercise 5-1 if it started from rest when the force begins to act on it.

EXERCISE 5-6:   The mass in Exercise 5-2 is moving at 0.5 m s^{-1} at the beginning of the 1.6-m displacement up the plane. What is its speed at the end of the plane?

EXERCISE 5-7:   A 3-kg mass is sliding on a surface at 1.5 m s^{-1}. A short time later it is moving at 2.4 m s^{-1}. Determine the net work done on the mass during this interval.

EXERCISE 5-8:   During its first 2 min of operation, a machine does work at the rate $P = Ct$, where the constant C has the value 20 W s^{-1}. Calculate the work done by this machine during its first 2 min.

EXERCISE 5-9:   To keep a bicycle moving at a constant speed against a net frictional force of 24 N, the cyclist must expend 60 W of power. Determine her speed on a level road.

EXERCISE 5-10:   How many gallons of water per minute can a $\frac{1}{2}$hp pump raise from a well 20 ft deep? One gallon of water weighs 8.3 lb.

Answers to Supplementary Exercises

5-1:   34.4 J

5-2:   22.2 J

5-3:   2.48 N

5-4: $W = K\left(\dfrac{1}{x_1} - \dfrac{1}{x_2}\right) = K\left(\dfrac{x_2 - x_1}{x_1 x_2}\right)$

5-5:   3.39 m s^{-1}

5-6:   2.77 m s^{-1}

5-7:   5.26 J

5-8: $W = \frac{1}{2}Ct^2 = 1.44 \times 10^5$ J

5-9:   2.5 m s^{-1}

5-10:   99.4 gal/min

6 CONSERVATIVE FORCES AND CONSERVATION OF ENERGY

THIS CHAPTER IS ABOUT

- ☑ **Conservative Forces**
- ☑ **Gravitational Potential Energy**
- ☑ **Elastic Potential Energy**
- ☑ **Conservation of Energy**

6-1. Conservative Forces

A **conservative force**, $\mathbf{F}_{con}$, has an associated *potential energy* function, E_p, a function of position; i.e.,

CONSERVATIVE FORCE
$$\mathbf{F}_{con} \cdot d\mathbf{R} = -dE_p \qquad (6\text{-}1)$$

The work done by the conservative force in proceeding from the initial position 1 to the final position 2 is

$$W_{con} = \int_1^2 \mathbf{F}_{con} \cdot d\mathbf{R} = -\int_1^2 dE_p = E_1 - E_2 = \Delta E_p \qquad (6\text{-}2)$$

Restating Eq. (6-2) in words, we say that the work done by a conservative force equals the *difference* in its associated potential energy function (initial minus final) and is independent of the path over which the integral is evaluated. **Potential energy** is the energy that a body has because of its position. Potential energy is measured in joules in the SI system and foot-pounds in the British system of units. A nonconservative force, on the other hand, has no associated potential energy function, and the work·done by such a force *does* depend on the path of integration. The force of kinetic friction is an example of a nonconservative force. In more advanced texts on mechanics, the authors prove that a force is conservative if its *curl* is zero. That is, the three following conditions must be met for a force to be conservative:

$$\frac{\partial F_z}{\partial y} - \frac{\partial F_y}{\partial z} = 0, \qquad \frac{\partial F_x}{\partial z} - \frac{\partial F_z}{\partial x} = 0, \quad \text{and} \quad \frac{\partial F_y}{\partial x} - \frac{\partial F_x}{\partial y} = 0 \qquad (6\text{-}3)$$

EXAMPLE 6-1: The potential energy function for a one-dimensional conservative force is $E_p = -Kx$, where K is a constant. Determine the force.

Solution: You should use Eq. (6-1),

$$\mathbf{F}_{con} \cdot d\mathbf{R} = -dE_p$$

In one dimension this reduces to $F_{x,con}dx = -dE_p$, or

$$F_{x,con} = -\frac{dE_p}{dx} = -(-K) = \boxed{K}$$

Thus the conservative force in this case is the constant force $F_{con} = K$, directed along the positive x axis.

EXAMPLE 6-2: Determine the potential energy function for the conservative one-dimensional force $F_{con} = C/x$, where C is a constant that has units of $N\,m^{-1}$.

Solution: You should use Eq. (6-2):

$$\int_1^2 \mathbf{F}_{con} \cdot d\mathbf{R} = -\int_1^2 dE_p$$

Limits are not specified in the problem, so you use the indefinite integral, or

$$\int \mathbf{F}_{con} \cdot d\mathbf{R} = -\int dE = -E$$

and

$$\int \frac{C}{x}\,dx = C\int \frac{dx}{x} = C \ln x = -E$$

Thus

$$E(x) = \boxed{-C \ln x}$$

Notice that this result specifies only that E is a function of x. An arbitrary constant term could be added on the right, but it would not change the functional dependence of E on x.

EXAMPLE 6-3: Determine the work done by the force in Example 6-2 in going from $x_1 = 2$ m to $x_2 = 5$ m for $C = 17$ J.

Solution: You should use Eq. (6-2).

$$W = \int_1^2 \mathbf{F} \cdot d\mathbf{R} = E_1 - E_2$$

$$= E(x_1) - E(x_2) = -C(\ln x_1 - \ln x_2) = C \ln\left(\frac{x_2}{x_1}\right)$$

$$= (17 \text{ J}) \ln\left(\frac{5 \text{ m}}{2 \text{ m}}\right) = \boxed{15.6 \text{ J}}$$

EXAMPLE 6-4: Determine whether the force $\mathbf{F} = axy\hat{\mathbf{x}} + bxy^2\hat{\mathbf{y}}$ is conservative, where a and b are constants.

Solution: You should use the test indicated by Eqs. (6-3).

$$F_x = axy \qquad\qquad \frac{\partial F_x}{\partial y} = ax$$

$$F_y = bxy^2 \qquad\qquad \frac{\partial F_y}{\partial x} = by^2$$

and

$$\frac{\partial F_x}{\partial y} \neq \frac{\partial F_y}{\partial x}$$

Because the necessary condition cannot be met, you can conclude that this force is not conservative.

6-2. Gravitational Potential Energy

The **gravitational potential energy** function of a body of mass m in a region where the acceleration due to gravity can be considered constant is

GRAVITATIONAL
   POTENTIAL $$E_{pg} = mgy \qquad \text{(6-4)}$$
   ENERGY

The y coordinate locates m relative to a fixed coordinate system in which the y axis is directed vertically upward. We can easily show that the conservative force associated with this gravitational potential energy function is the gravitational force $-mg$, using Eq. (6-1). We will consider the situation for which the acceleration due to gravity is a function of position in Chapter 11.

EXAMPLE 6-5: A 2-kg block is moved from P_1 at $x_1 = 2$ m, $y_1 = 1.5$ m to position P_2 at $x_2 = 3.5$ m, $y_2 = 5$ m. (The x axis is directed horizontally, and the y axis is directed vertically upward.) Determine (a) the potential energy of the block at these two positions; (b) the increase in the potential energy of the block; and (c) the work done by the force of gravity in moving the block.

Solution:

(a) You should use Eq. (6-4), $E_{pg} = mgy$. At P_1,

$$E_{pg1} = mgy_1 = (2 \text{ kg})(9.8 \text{ m s}^{-2})(1.5 \text{ m}) = \boxed{29.4 \text{ J}}$$

At P_2,

$$E_{pg2} = mgy_2 = (2 \text{ kg})(9.8 \text{ m s}^{-2})(5 \text{ m}) = \boxed{98.0 \text{ J}}$$

(b) The *increase* in E_{pg} is

$$\Delta E_{pg} = E_{pg2} - E_{pg1} = mgy_2 - mgy_1 = mg(y_2 - y_1)$$
$$= (2 \text{ kg})(9.8 \text{ m s}^{-2})(5 - 1.5)\text{ m} = \boxed{68.6 \text{ J}}$$

You can obtain the same result by subtraction:

$$\Delta E_{pg} = E_{pg2} - E_{pg1} = (98.0 - 29.4)\text{ J} = \boxed{68.6 \text{ J}}$$

(c) You should use Eq. (6-2).

$$W_{con} = \int_1^2 \mathbf{F}_{con} \cdot d\mathbf{R} = E_1 - E_2$$
$$= E_{pg1} - E_{pg2} = 29.4 \text{ J} - 98.0 \text{ J} = \boxed{-68.6 \text{ J}}$$

Note that the work done by the gravitational force is equal to the negative of the increase in the gravitational potential energy of the block.

6-3. Elastic Potential Energy

A Hooke's law type of force describes the **elastic potential energy** of a system: The force that restores the system to equilibrium is proportional to the negative of the displacement of the system from equilibrium. This type of force is conservative and therefore has an associated potential energy function. An elastic spring is an example of an object that behaves according to Hooke's law. (See Problem 5-4.) The force exerted by a spring stretched a distance x is

$$F_s = -kx \qquad \text{(6-5)}$$

where k is the spring constant measured in $N \text{ m}^{-1}$. The negative sign indicates that the direction of the force is opposite to that of the displacement.

The potential energy function for this force is

$$E_{pe} = \tfrac{1}{2}kx^2 \qquad \text{(6-6)}$$

We use the forms of Eqs. (6-5) and (6-6) to describe a number of elastic systems in physics.

EXAMPLE 6-6: An external force does 8 J of work to stretch a spring that has a spring constant $k = 2.4 \times 10^3 \, \text{N m}^{-1}$. **(a)** How far is the spring stretched? **(b)** Calculate the magnitude of the tension in the spring when it is stretched this distance.

Solution:

(a) The work done by the external force is equal to the elastic potential energy stored in the spring. You should use Eq. (6-6) and solve for x.

$$E_{\text{pe}} = \tfrac{1}{2}kx^2 \quad \text{and} \quad x = \sqrt{2E_{\text{pe}}/k}$$

$$x = \sqrt{2(8 \, \text{J})/(2.4 \times 10^3 \, \text{N m}^{-1})} = \boxed{8.16 \times 10^{-2} \, \text{m} = 8.16 \, \text{cm}}$$

(b) You should use Eq. (6-5) but ignore the negative sign because only the magnitude was requested.

$$F_s = -kx \quad \text{and} \quad |F_s| = kx$$

$$|F_s| = (2.4 \times 10^3 \, \text{N m}^{-1})(8.16 \times 10^{-2} \, \text{m}) = \boxed{196 \, \text{N}}$$

6-4. Conservation of Energy

A. Mechanical systems involving conservative forces only

One example of a *mechanical system* is a particle of mass m but this can be expanded to include many particles. The total **mechanical energy** of a system is the sum of the total potential energy plus the kinetic energy and is constant in time. Thus

MECHANICAL ENERGY $\qquad E = E_p + E_k \qquad$ (6-7)

The total potential energy of the system may be the sum of the elastic potential energy and the gravitational potential energy. We can also write Eq. (6-7) in terms of the initial and final configurations, 1 and 2, of the system, i.e.,

$$E_{p1} + E_{k1} = E_{p2} + E_{k2} \qquad (6\text{-}8)$$

B. Mechanical systems involving conservative and frictional forces

When a mechanical system is subject to both conservative and frictional forces,

$$E_{p2} + E_{k2} = E_{p1} + E_{k1} + W_f \qquad (6\text{-}9)$$

where W_f is the total work done against the force of friction and 1 and 2 represent the initial and final configurations of the system, respectively. You should keep in mind that, because the force of friction acts against the displacement, W_f in Eq. (6-9) is a negative quantity. Thus W_f is the quantity of mechanical energy transformed into thermal energy when the system goes from its initial to final state.

EXAMPLE 6-7: A small mass is released from a point 12 m above the ground. Use the conservation of energy principle to calculate its speed when it is 2 m above the ground. Neglect air friction.

Solution: It should be clear to you that the potential energy in this problem is gravitational. Let m represent the mass of the object and the coordinate system have its origin at the surface of the ground, i.e., $y = 0$ at the ground. You should use Eq. (6-8) with $v_1 = 0$ and solve for v_2.

$$E_{p1} + E_{k1} = E_{p2} + E_{k2} \quad \text{and} \quad mgy_1 + \tfrac{1}{2}mv_1^2 = mgy_2 + \tfrac{1}{2}mv_2^2$$

$$v_2 = \sqrt{2g(y_1 - y_2)} = \sqrt{2(9.8 \, \text{m s}^{-2})(12 - 2) \, \text{m}} = \boxed{14 \, \text{m s}^{-1}}$$

EXAMPLE 6-8: A 300-g block rests on a horizontal frictionless surface. The block is attached to one end of a spring that has a spring constant $k = 6\ \mathrm{N\,m}^{-1}$; the other end of the spring is fixed. The block is pulled a distance of 14 cm from its equilibrium position and then released from rest. (a) Plot a graph of the elastic potential energy of the block as a function of its position for $-14\ \mathrm{cm} \le x \le 14\ \mathrm{cm}$, where $x = 0$ is the equilibrium position of the block. Indicate on the graph the total energy of the system and, for a typical value of x, the kinetic and potential energies. (b) Determine the speed of the block when it is 3 cm from equilibrium.

Solution: (a) A sketch helps, so draw it first and then plot the graph. (See Figures 6-1a and b.) Note that the potential energy is positive for both positive and negative values of x. Note also that for any value of x between -14 cm and $+14$ cm the sum of the potential and kinetic energies always equals the total energy, $E = 5.88 \times 10^{-2}\mathrm{J}$. (b) Use Eq. (6-8) and solve for v_2; because the block is released from rest, $v_1 = 0$.

$$E_{p1} + E_{k1} = E_{p2} + E_{k2}$$

$$\tfrac{1}{2}kx_1^2 + \tfrac{1}{2}mv_1^2 = \tfrac{1}{2}kx_2^2 + \tfrac{1}{2}mv_2^2$$

$$v_2 = \pm\sqrt{\frac{k}{m}(x_1^2 - x_2^2)}$$

$$= \pm\sqrt{\left(\frac{6\ \mathrm{N\,m}^{-1}}{0.3\ \mathrm{kg}}\right)[(0.14\ \mathrm{m})^2 - (0.03\ \mathrm{m})^2]} = \boxed{\pm 0.612\ \mathrm{m\,s}^{-1}}$$

You should keep in mind that this result is valid for $x_2 = \pm 3$ cm and that the $\pm$ sign associated with v_2 implies that the block could be moving at $0.612\ \mathrm{m\,s}^{-1}$ in either the positive or negative x direction. Neither of these two conditions was specified in the problem statement.

EXAMPLE 6-9: In Figure 6-2, $y_1 = 5$ m, $y_2 = 1.5$ m, $\theta = 26°$, and the block (of mass m) is moving down the plane at $5\ \mathrm{m\,s}^{-1}$ when it is at y_1. The coefficient of kinetic friction between the block and plane is 0.6. Determine the speed of the block when it is at y_2.

Solution: You should use Eq. (6-9) and solve for v_2. The work done against friction is $W_f = -\mathbf{f} \cdot (\Delta\mathbf{R})$, where

$$f = \mu_k N = \mu_k mg\cos\theta \qquad \text{and} \qquad \Delta R = (y_1 - y_2)/\sin\theta$$

Thus

$$W_f = -\frac{\mu_k mg(y_1 - y_2)}{\tan\theta} \qquad \text{and} \qquad E_{p2} + E_{k2} = E_{p1} + E_{k1} + W_f$$

$$mgy_2 + \frac{1}{2}mv_2^2 = mgy_1 + \frac{1}{2}mv_1^2 - \frac{\mu_k mg(y_1 - y_2)}{\tan\theta}$$

$$v_2 = \sqrt{v_1^2 + 2g(y_1 - y_2)\left(1 - \frac{\mu_k}{\tan\theta}\right)}$$

$$= \sqrt{(5\ \mathrm{m\,s}^{-1})^2 + 2(9.8\ \mathrm{m\,s}^{-2})(5\ \mathrm{m} - 1.5\ \mathrm{m})\left(1 - \frac{0.6}{\tan 26°}\right)}$$

$$= \boxed{3.04\ \mathrm{m\,s}^{-1}}$$

Even though the block is sliding down the plane, the friction is sufficient to decrease its speed.

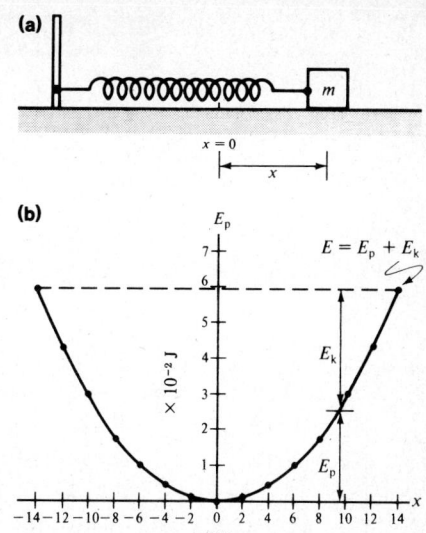

Figure 6-1. (a) Mass and spring on a frictionless surface. (b) Graph of potential energy as a function of position of the block.

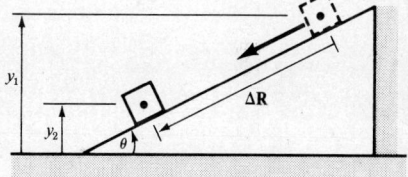

Figure 6-2

SUMMARY

1. The work done by a conservative force is equal to the change in its associated *potential energy* function, or $W_{con} = E_1 - E_2$, where E_1 and E_2 are the initial and final values of the potential energy, respectively.
2. The conservative force can be derived from its potential energy function by differentiation. In one dimension, $F_x = -\dfrac{dE_p}{dx}$. In three dimensions the components are

$$F_x = -\frac{\partial E_p}{\partial x}, \qquad F_y = -\frac{\partial E_p}{\partial y}, \qquad \text{and} \qquad F_z = -\frac{\partial E_p}{\partial z}$$

3. The *curl* of a conservative force is zero.
4. In a region where the acceleration due to gravity is constant the *gravitational potential energy function* for a mass m is $E_{pg} = mgy$, where y is the vertical position coordinate for the mass. The absolute value of the potential energy depends on the choice of coordinate system; however, the difference in the initial and final potential energies of a mass is the only relevant physical quantity.
5. An elastic system is described by a Hooke's law type of force, i.e., a force that is proportional to the negative of the displacement from equilibrium. For an elastic spring, Hooke's law is $F_s = -kx$, where k is the *spring constant*.
6. The elastic *potential energy function* for a spring is $F_{pe} = \frac{1}{2}kx^2$.
7. The total mechanical energy of a system involving only conservative forces is constant in time; i.e., $E = E_p + E_k$.
8. The work done against the force of friction by a nonconservative force is the quantity of mechanical energy transformed into thermal energy as a system goes from its initial to its final state. It is a negative quantity and appears in the conservation of energy equation as $E_{p2} + E_{k2} = E_{p1} + E_{k1} + W_f$.

RAISE YOUR GRADES

Can you explain . . . ?

☑ the difference between a conservative and nonconservative force
☑ why the force of friction is called a *dissipative* force
☑ why a one-dimensional force that depends only on the x coordinate is always conservative
☑ why the work done by a conservative force in going from an initial position to an intermediate position and then back to the original position over any path is always zero
☑ how it is possible for a ball to bounce to a height greater than its release point on its first bounce
☑ why a ball can never bounce to a height greater than its release point if it is released from rest
☑ why a ball bounces to decreasing heights
☑ why a force that depends only on time is not conservative

SOLVED PROBLEMS

Conservative Forces

PROBLEM 6-1: The potential energy function in a region near the origin is given by $E_p = Ke^{ax}$, where K and a are constants that have units of J and m^{-2}, respectively. Derive the conservative force associated with this potential energy function.

Solution: Use the one-dimensional case of Eq. (6-1).

$$F_{x,\,con} = -\frac{dE_p}{dx} = \boxed{-2axKe^{ax}}$$

You should verify that this result has units of force.

PROBLEM 6-2: Identify whether each force in (a)–(c) is conservative or nonconservative. The constant A has units of $N\,m^{-3}$.

(a) $\mathbf{F}_{x,\,y} = A(x^2y\hat{\mathbf{x}} + xy^2\hat{\mathbf{y}})$

(b) $\mathbf{F}_{x,\,y} = A(xy^2\hat{\mathbf{x}} + x^2y\hat{\mathbf{y}})$

(c) $\mathbf{F}_{x,\,y} = A(xy^2\hat{\mathbf{x}} - x^2y\hat{\mathbf{y}})$

Solution: Recall that a force is conservative if its *curl* is zero and use Eqs. (6-3). Because the given forces depend only on x and y, the third of Eqs. (6-3) is the only condition that must be met.

(a) $\dfrac{\partial F_y}{\partial x} = Ay^2$, $\qquad \dfrac{\partial F_x}{\partial y} = Ax^2$, $\quad$ and $\quad A(y^2 - x^2) \neq 0 \qquad \boxed{\text{Nonconservative}}$

(b) $\dfrac{\partial F_y}{\partial x} = 2Axy$, $\qquad \dfrac{\partial F_x}{\partial y} = 2Axy$, $\quad$ and $\quad 2Axy - 2Axy = 0 \qquad \boxed{\text{Conservative}}$

(c) $\dfrac{\partial F_y}{\partial x} = -2Axy$, $\qquad \dfrac{\partial F_x}{\partial y} = 2Axy$, $\quad$ and $\quad -2Axy - (2Axy) = -4Axy \neq 0$

$\boxed{\text{Nonconservative}}$

Gravitational Potential Energy

PROBLEM 6-3: A 2-kg mass is located 10 m above the ground. Determine the potential energy of this mass when the origin of the coordinate system is (a) 3 m above the ground and (b) 2 m below the ground.

Solution: In both cases use Eq. (6-4).

(a) In this case, $y = (10 - 3)$ m $= 7$ m:

$$E_{pg} = mgy = (2 \text{ kg})(9.8 \text{ m s}^{-2})(7 \text{ m}) = \boxed{137 \text{ J}}$$

(b) In this case, $y = [10 - (-2)]$ m $= 12$ m:

$$E_{pg} = mgy = (2 \text{ kg})(9.8 \text{ m s}^{-2})(12 \text{ m}) = \boxed{235 \text{ J}}$$

PROBLEM 6-4: Determine the change in the gravitational potential energy of the mass in Problem 6-3 when it is moved from its position 10 m above the ground to a position 6 m above the ground for both coordinate systems (a) and (b) in Problem 6-3.

Solution: You are asked to calculate $\Delta E_p = E_{p1} - E_{p2}$, where 1 and 2 represent the initial and final positions, respectively.

(a) $\Delta E_p = mgy_1 - mgy_2 = mg(y_1 - y_2) = (2 \text{ kg})(9.8 \text{ m s}^{-2})[7 - (6 - 3)] \text{ m} = \boxed{78.4 \text{ J}}$

(b) $\Delta E_p = mg(y_1 - y_2) = (2 \text{ kg})(9.8 \text{ m s}^{-2})\{12 - [6 - (-2)]\} \text{ m} = \boxed{78.4 \text{ J}}$

note: This problem illustrates two important points that you should keep in mind:

(1) *Changes* in the potential energy of a system are physically important. The value of the potential energy at a point depends on the location of the origin of the coordinate system.

(2) The results obtained in (a) and (b) represent a *decrease* in the potential energy. If the mass had been raised 4 m, its potential energy would have increased 78.4 J.

Elastic Potential Energy

PROBLEM 6-5: A force does 0.36 J of work in compressing a spring 12 cm. Determine the spring constant.

Solution: Use Eq. (6-6) and solve for k.

$$E_{pe} = \frac{1}{2}kx^2 \quad \text{and} \quad k = \frac{2E_{pe}}{x^2}$$

$$k = \frac{2(0.36 \text{ J})}{(0.12 \text{ m})^2} = \boxed{50 \text{ N m}^{-1}}$$

PROBLEM 6-6: One end of a spring is attached to the ceiling. A mass m is attached to its lower end (configuration 1) and permitted to descend slowly so that it comes to rest some distance below its initial position (configuration 2). Determine the change in potential energy for this system (if there is a change) in terms of m, g, and the spring constant k.

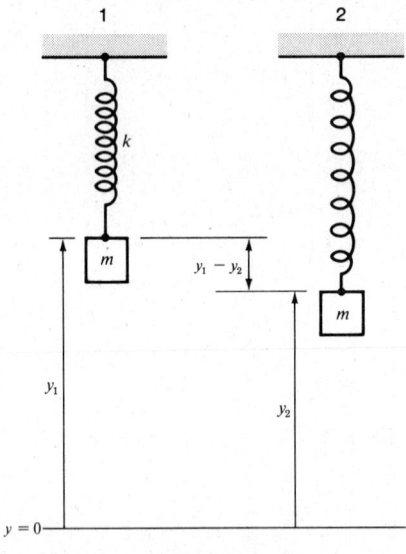

Figure 6-3

Solution: Draw a sketch (as in Figure 6-3), introduce a coordinate system, and evaluate the potential energy of this system in its initial and final configurations. Let the change in potential energy (if there is one) be $\Delta E_p = E_{p1} - E_{p2}$, where 1 and 2 represent the initial and final configurations, respectively. The potential energy of this system is the sum of the gravitational and elastic potential energies. Use Eqs. (6-4) and (6-6).

$$E_{p1} = E_{pg1} + E_{pe1} = mgy_1 + 0$$

and

$$E_{p2} = E_{pg2} + E_{pe2} = mgy_2 + \tfrac{1}{2}k(y_1 - y_2)^2$$
$$\Delta E_p = E_{p1} - E_{p2}$$

Thus

$$\Delta E_p = mg(y_1 - y_2) - \tfrac{1}{2}k(y_1 - y_2)^2$$

You can calculate the quantity $y_1 - y_2$ by using Hooke's law, Eq. (6-5). In configuration 2, the force that the spring exerts is equal and opposite in direction to the weight of m. The displacement is $x = (y_1 - y_2)$.

$$F_s = -kx, \qquad -mg = -k(y_1 - y_2),$$

and

$$y_1 - y_2 = \frac{mg}{k}$$

Thus

$$\Delta E_p = mg\left(\frac{mg}{k}\right) - \frac{1}{2}k\left(\frac{mg}{k}\right)^2 = \boxed{\frac{(mg)^2}{2k}}$$

You should again note that the decrease in potential energy is independent of the coordinate system.

Conservation of Energy

PROBLEM 6-7: In Figure 6-4, $m = 2.5$ kg, $M = 3.2$ kg, the pulley is frictionless, and the mass of the connecting cord can be neglected. What is the velocity of the masses after they have moved 4 m, if they were released from rest?

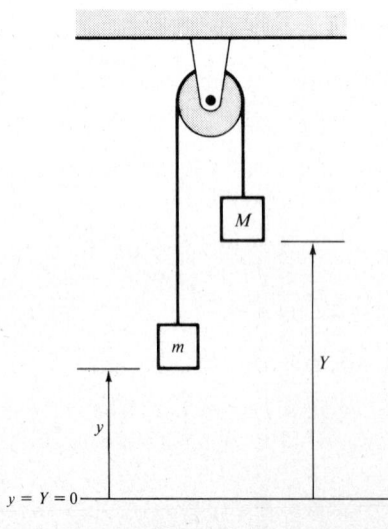

Figure 6-4

Solution: Because of the cord, both will have the same speed with m moving up and M moving down. Let 1 and 2 represent the initial and final configurations, respectively. You should use Eq. (6-8) because only conservative forces are involved.

$$E_{p1} + E_{k1} = E_{p2} + E_{k2}$$

and

$$mgy_1 + MgY_1 + \tfrac{1}{2}mv_1^2 + \tfrac{1}{2}MV_1^2 = mgy_2 + MgY_2 + \tfrac{1}{2}mv_2^2 + \tfrac{1}{2}MV_2^2$$

You can simplify this equation greatly: $v_1 = V_1 = 0$, and in magnitude $v_2 = V_2$. Therefore you can let $v_2 = V_2 = v$, or the quantity you are to calculate. Thus

$$mgy_1 + MgY_1 = mgy_2 + MgY_2 + \tfrac{1}{2}(m + M)v^2$$

and

$$mg(y_1 - y_2) + Mg(Y_1 - Y_2) = \tfrac{1}{2}(m + M)v^2$$

Then let $Y_1 - Y_2 = Y = 4$ m, and note that $y_1 - y_2 = -\Delta Y = -4$ m. Therefore

$$(M - m)g\,\Delta Y = \frac{1}{2}(m + M)v^2 \qquad \text{and} \qquad v = \sqrt{\frac{2(M - m)g\,\Delta Y}{m + M}} \qquad \text{(Independent of the coordinate system)}$$

$$v = \sqrt{\frac{2(3.2 - 2.5)\ \text{kg}\ (9.8\ \text{m s}^{-2})(4\ \text{m})}{(2.5 + 3.2)\ \text{kg}}} = \boxed{3.10\ \text{m s}^{-1}}$$

PROBLEM 6-8: In Figure 6-5 the spring constant is $k = 14\ \text{N m}^{-1}$, $m = 0.5$ kg, $y_1 = 0.8$ m, $y_2 = 0.1$ m, and $y_3 = 0.4$ m. In configuration 1, m rests on the floor and the spring is unstretched. In 2, the spring is stretched and attached to m and released. In 3, m is moving upward at speed v. Determine v.

Solution: Because only conservative forces are involved you can use Eq. (6-8). In this case the initial state is 2, the *final* state is 3, and $v_2 = 0$. Let $v_3 = v$, or the quantity to be determined. The initial and final y coordinates for m are y_2 and y_3, respectively.

$$E_{p2} + E_{k2} = E_{p3} + E_{k3}$$

and

$$mgy_2 + \tfrac{1}{2}k(y_1 - y_2)^2 + \tfrac{1}{2}mv_2^2$$
$$= mgy_3 + \tfrac{1}{2}k(y_1 - y_3)^2 + \tfrac{1}{2}mv_3^2$$

Thus

$$\tfrac{1}{2}k[(y_1 - y_2)^2 - (y_1 - y_3)^2] + mg(y_2 - y_3) = \tfrac{1}{2}mv^2$$

and

$$v = \sqrt{\left(\frac{k}{m}\right)[(y_1 - y_2)^2 - (y_1 - y_3)^2] + 2g(y_2 - y_3)}$$

$$= \sqrt{\left(\frac{14\ \text{N m}^{-1}}{0.5\ \text{kg}}\right)[(0.8 - 0.1)^2 - (0.8 - 0.4)^2]\ \text{m}^2 + 2(9.8\ \text{m s}^{-2})(0.1 - 0.4)} = \boxed{1.83\ \text{m s}^{-1}}$$

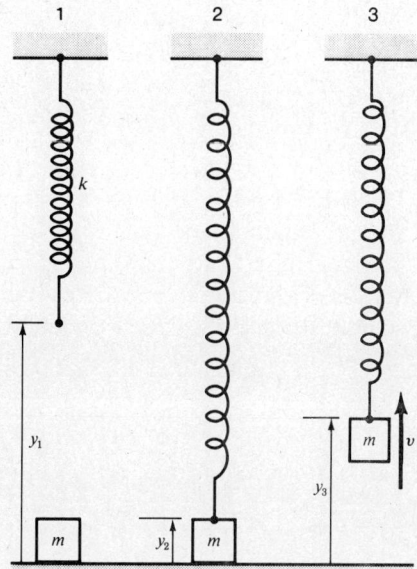

Figure 6-5

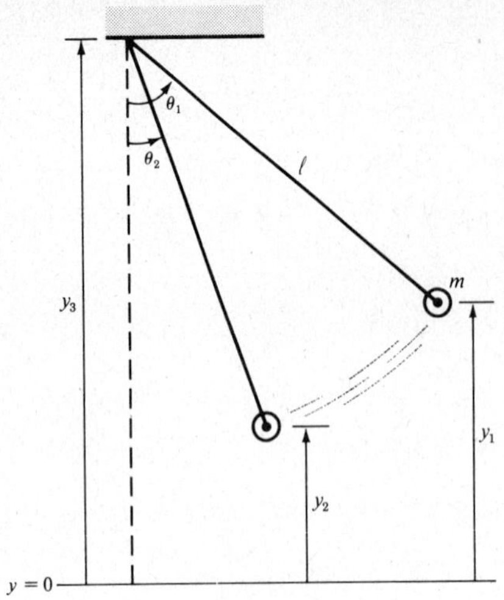

Figure 6-6

PROBLEM 6-9: One end of a cord of length $\ell = 1.2$ m is attached to a small mass, and the other end of the cord is attached to the ceiling. The mass is raised so that the cord makes an angle $\theta_1 = 50°$ relative to the vertical and then is released from rest. (See Figure 6-6.) How fast is the mass moving when the angle made by the cord is $\theta_2 = 20°$?

Solution: Because only conservative forces are involved you can use Eq. (6-8). Let 1 and 2 represent the initial and final configurations, respectively. Because the mass is released from rest, $v_1 = 0$. Solve for v_2.

$$E_{p1} + E_{k1} = E_{p2} + E_{k2}$$
$$mgy_1 + \tfrac{1}{2}mv_1^2 = mgy_2 + \tfrac{1}{2}mv_2^2$$
$$v_2 = \sqrt{2g(y_1 - y_2)}$$

From Figure 6-6 you should conclude that

$$y_1 = y_3 - \ell \cos \theta_1 \quad \text{and} \quad y_2 = y_3 - \ell \cos \theta_2$$

Therefore

$$y_1 - y_2 = (y_3 - \ell \cos \theta_1) - (y_3 - \ell \cos \theta_2) = \ell(\cos \theta_2 - \cos \theta_1)$$

Thus

$$v_2 = \sqrt{2g\ell(\cos \theta_2 - \cos \theta_1)} = \sqrt{2(9.8 \text{ m s}^{-2})(1.2 \text{ m})(\cos 20° - \cos 50°)} = \boxed{2.64 \text{ m s}^{-1}}$$

PROBLEM 6-10: A bead of mass m is released from rest and slides down a straight section of wire of length d, and then onto a curved section that has a radius of curvature r. A constant frictional force $\mathbf{f}$ acts on the bead as it moves down the straight section of wire, whereas the curved section of wire is frictionless. How far above the top of the curved section must the release point be so that the centripetal force on the bead is equal to its weight when it passes the top of the curved section; i.e., determine $y_1 - y_2$ in Figure 6-7.

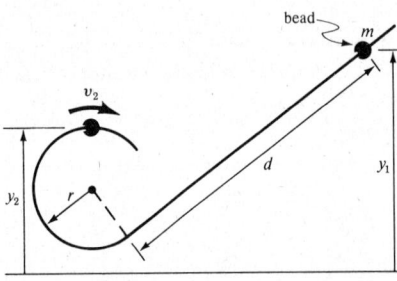

Figure 6-7

Solution: Because the nonconservative force of friction is involved in this problem you must use Eq. (6-9). Keep in mind that W_f is a negative quantity and that $v_1 = 0$ because the bead is released from rest.

$$E_{p2} + E_{k2} = E_{p1} + E_{k1} + W_f$$
$$mgy_2 + \tfrac{1}{2}mv_2^2 = mgy_1 + \tfrac{1}{2}mv_1^2 - \mathbf{f} \cdot \mathbf{d}$$
$$y_1 - y_2 = \frac{v_2^2}{2g} + \frac{\mathbf{f} \cdot \mathbf{d}}{mg}$$

The requirement that the centripetal force on the bead equal its weight at the top of the curved section of wire is $m(v_2^2/r) = mg$, which implies that $v_2^2/g = r$. Thus

$$y_1 - y_2 = \boxed{\frac{r}{2} + \frac{\mathbf{f} \cdot \mathbf{d}}{mg}}$$

Supplementary Exercises

EXERCISE 6-1: The one-dimensional form of the famous Yukawa potential is $E_p = Ke^{-ax}/x$, where K and a are constants that have units of J m and m^{-1}, respectively. Determine the conservative force associated with this potential.

EXERCISE 6-2: Determine the numerical values of r, s, and t required for the force

$$\mathbf{F}_{x,y} = K(x^2 y^r \hat{\mathbf{x}} + x^s y^t \hat{\mathbf{y}})$$

to be conservative. What are the units of K?

EXERCISE 6-3: In Figure 6-8, the small and large pulleys are attached to each other and pivot on a frictionless bearing; $m_1 = 2$ kg, $m_2 = 1.5$ kg, $r_1 = 8$ cm, $r_2 = 14$ cm, and $d = 80$ cm. When released from rest in this initial configuration, m_2 descends. Determine the change in potential energy of this system from its initial configuration to the state in which m_2 rests on the floor.

EXERCISE 6-4: In Figure 6-9, the pivot and pulley are frictionless, the rod and cords have negligible weight, each mass is $m = 2$ kg, and the length of the rod is $4\ell = 48$ cm. The initial and final configurations are $\theta_1 = 40°$ and $\theta_2 = 70°$, respectively. How much work is done by $\mathbf{F}$ in producing this change?

EXERCISE 6-5: Two springs with identical unstretched length but different spring constants, k_1 and k_2, are connected (**a**) in parallel; and (**b**) in series, as shown in Figure 6-10. Calculate the elastic potential energy stored when each system is stretched a distance x.

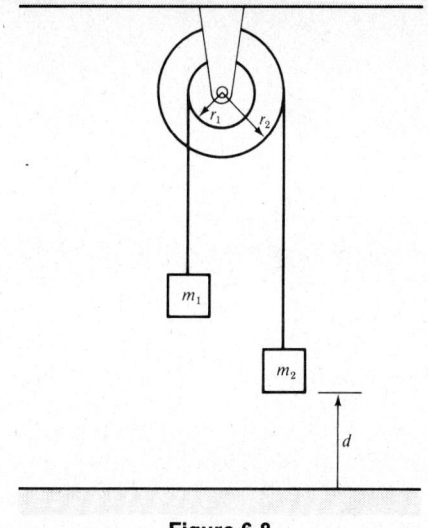

Figure 6-8

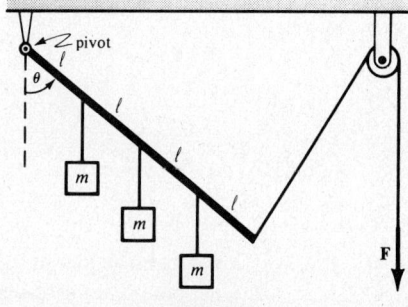

Figure 6-9

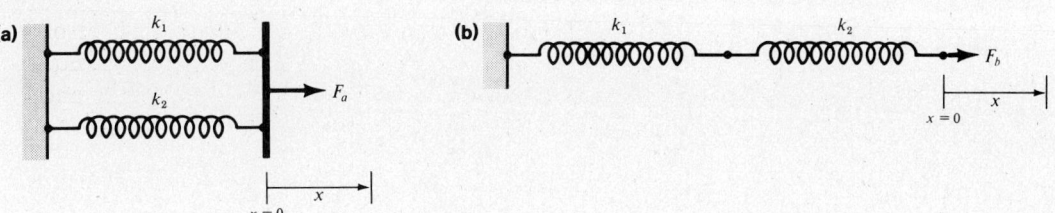

Figure 6-10

EXERCISE 6-6: One end of a coil spring ($k = 120$ N m^{-1}) that has an uncompressed length of 12 cm is placed on the floor. The spring is compressed 8 cm, and a small ball that has a mass of 200 g is placed on the spring and released. To what height above the floor does the ball rise?

EXERCISE 6-7: A small mass is swung in a circular orbit that has a radius of 0.8 m; the plane of the orbit is vertical. The speed of the mass at the top of the orbit is 4 m s^{-1}. Determine its speed at the bottom of the orbit.

EXERCISE 6-8: One end of a coil spring is attached to the ceiling. A 0.8-kg mass is attached to its lower end and let down slowly until it comes to rest 0.16 m below the point at which it was attached. How much below this point must the mass be pulled so that after it is released from rest it rises to the height at which it was attached?

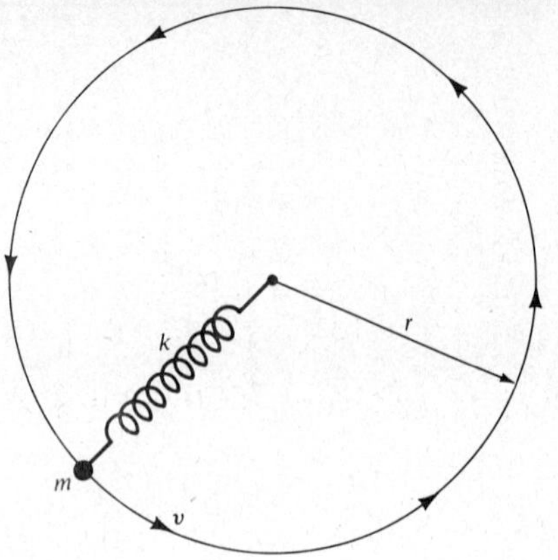

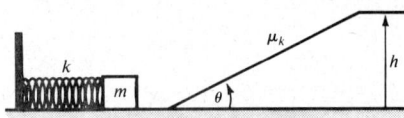

Figure 6-11

EXERCISE 6-9: A small mass m is attached to one end of a coil spring of unstretched length $\ell = 0.3$ m that has a spring constant $k = 50$ N m^{-1}. The mass is whirled in a horizontal circle at a speed that stretches the spring 0.1 m, as shown in Figure 6-11. The radius of the orbit is $r = \ell + \delta$, where δ is the elongation of the spring. Calculate the energy of this system, neglecting gravitational potential energy.

EXERCISE 6-10: A strong spring ($k = 1.5 \times 10^3$ N m^{-1}) is compressed a distance of 0.2 m. A block of mass $m = 1.2$ kg is placed against the spring and rests on a horizontal frictionless surface. When released the spring sends the block up a plane inclined at 26°. The coefficient of kinetic friction between block and plane is 0.3, and the horizontal surface at the top of the plane, 1.4 m above the lower surface, is frictionless. See Figure 6-12. Determine the speed of the block when it reaches the upper surface.

Figure 6-12

Answers to Supplementary Exercises

6-1: $Ke^{-ax}(1 + ax)/x^2$

6-2: $r = 3, s = 3$, and $t = 2$; K has units of N m^5.

6-3: $\Delta E_p = -2.8$ J; the negative sign indicates a decrease in the system's gravitational potential energy.

6-4: The work done by **F** is equal to the increase in gravitational potential for this system.

$$W = \Delta E_{pg} = 6\, mg\ell(\cos\theta_1 - \cos\theta_2) = 2.44 \text{ J}$$

6-5: (a) $E_{pe} = \frac{1}{2}(k_1 + k_2)x^2$;

(b) $E_{pe} = \frac{1}{2}\left(\dfrac{k_1 k_2}{k_1 + k_2}\right)(x^2)$

6-6: 0.236 m = 23.6 cm

6-7: 6.88 m s^{-1}

6-8: 0.16 m

6-9: The centripetal force is provided by the spring.

$$\frac{mv^2}{r} = k\delta \qquad \text{and} \qquad r = \ell + \delta$$

Thus

$$E_k = \frac{k\delta(\ell + \delta)}{2}, \qquad E_{pe} = \frac{k\delta^2}{2},$$

and

$$E = \frac{k\delta(\ell + 2\delta)}{2} = 1.5 \text{ J}$$

6-10: 2.38 m s^{-1}

7 *HARMONIC MOTION*

THIS CHAPTER IS ABOUT

☑ **Simple Harmonic Motion (SHM) and Hooke's Law**
☑ **Equations of SHM—The Mass–Spring Oscillator**
☑ **The Simple Pendulum**
☑ **Damped Harmonic Motion**
☑ **Forced Harmonic Motion**

7-1. Simple Harmonic Motion (SHM) and Hooke's Law

Systems that are governed by a Hooke's law type of force exhibit simple harmonic motion, SHM, about their equilibrium positions when displaced and released. Such a vibrating system is called a **simple harmonic oscillator**. A Hooke's law type of force occurs when a force that is directly proportional to the displacement from equilibrium acts to restore the system to equilibrium, or

$$F_r \propto - \text{Displacement from equilibrium} \qquad (7\text{-}1)$$

The negative sign indicates the *restoring* property of the force produced. A spring is an example of a device that obeys Hooke's Law, and the force produced by a stretched spring, as presented in Chapters 5 and 6, is expressed by Eq. (6-5), $F_s = -kx$.

7-2. Equations of SHM—The Mass–Spring Oscillator

The mass–spring oscillator system is almost always used as the basis for deriving the equations of SHM. If you thoroughly understand how this system works, you can transfer the equations derived to any other system governed by a Hooke's law type of force.

We begin by defining the system as a massless spring of spring constant k, with one end fixed, and a mass m, which is free to slide on a frictionless horizontal surface, attached to the other end. The location of m (position coordinate) is designated by x, and the equilibrium position of m is at $x = 0$, as shown in Figure 7-1. If m moves to positive x, the spring is stretched; if it moves to negative x, the spring is compressed. In either case the spring acts to restore m to its equilibrium position.

We apply Newton's second law to m in the form $F = ma = m\left(\dfrac{d^2x}{dt^2}\right)$, where F is the restoring force exerted by the spring, i.e., $F = F_s = -kx$, and solve for $\dfrac{d^2x}{dt^2}$.

$$m\frac{d^2x}{dt^2} = -kx \qquad (7\text{-}2a)$$

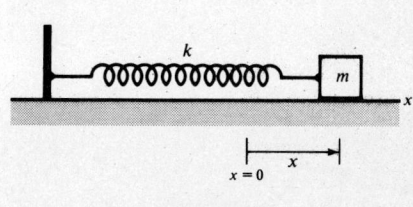

Figure 7-1

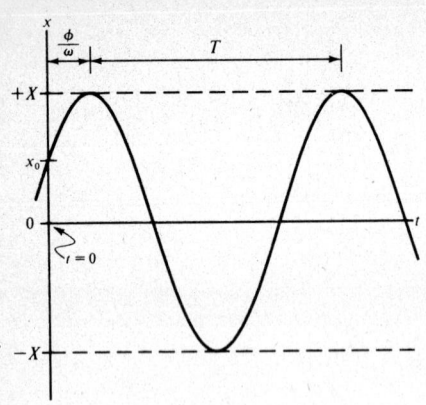

Figure 7-2

and

$$\frac{d^2x}{dt^2} = -\frac{k}{m}x \tag{7-2b}$$

Let $\omega^2 = k/m$, so that

$$\omega = \sqrt{\frac{k}{m}} \tag{7-3}$$

The term ω represents the *angular frequency* (not to be confused with angular velocity as defined in Chapter 3). You should verify that ω has units of s^{-1}. We can now write Eq. (7-2b) as

$$\frac{d^2x}{dt^2} = -\omega^2 x \tag{7-4}$$

You should also verify that the solution of Eq. (7-4) for x as a function of time is

$$x = X\cos(\omega t + \phi) \tag{7-5}$$

where X represents the **amplitude** of the motion and ϕ represents the **phase constant**, sometimes called the **phase angle**. These two constants depend on how the motion of the oscillator is initiated. A graph of Eq. (7-5) is shown in Figure 7-2; x_0 is the initial value of x, and T is the period of the motion.

The **period** of the motion, or the amount of time required for the motion to complete one cycle, is obtained from $\omega T = 2\pi$ rad.

PERIOD
$$T = \frac{2\pi}{\omega} \tag{7-6}$$

The **frequency** of the motion is

FREQUENCY
$$v = \frac{1}{T} = \frac{\omega}{2\pi} \tag{7-7}$$

Frequency is measured in **hertz** (Hz); 1 Hz is one cycle per second and has units of s^{-1}. The **angular frequency** is

ANGULAR FREQUENCY
$$\omega = 2\pi v \tag{7-8}$$

Using Eq. (7-3), we find that the period and frequency of the mass–spring oscillator are, respectively,

$$T = 2\pi\sqrt{\frac{m}{k}} \tag{7-9a}$$

$$v = \frac{1}{2\pi}\sqrt{\frac{k}{m}} \tag{7-9b}$$

We obtain the velocity of m from Eq. (7-5) by differentiating, i.e.,

$$v = \frac{dx}{dt} = -\omega X\sin(\omega t + \phi) \tag{7-10}$$

Its maximum absolute value, the *velocity amplitude*, is $v_{max} = \omega X$, and the acceleration, from Eq. (7-4), is

$$a = -\omega^2 x = -\omega^2 X\cos(\omega t + \phi)$$

The total energy of the oscillator is

$$E = E_k + E_{pe} = \tfrac{1}{2}mv^2 + \tfrac{1}{2}kx^2 = \tfrac{1}{2}kX^2 \tag{7-11}$$

note: Equations (7-2a)–(7-11) apply equally well to a mass hanging from a spring that oscillates up and down. The x coordinate locates the distance that m is from its equilibrium position, which is determined by hanging m on the end of the spring and lowering it slowly until it comes to rest.

The mass of a spring can be included by replacing m with the effective mass, m_{eff}, in Eqs. (7-2a), (7-2b), (7-3), (7-9a), and (7-9b). The effective mass is

$$m_{eff} = m + \tfrac{1}{3}m_s \qquad (7\text{-}12)$$

where m is the mass attached to the spring and m_s is the mass of the spring.

EXAMPLE 7-1: A 50-g mass is attached to one end of a spring ($k = 20 \text{ N m}^{-1}$), and the other end of the spring is attached to the ceiling. The mass is then pulled 8 cm below its equilibrium position and released from rest. Assume that the spring has negligible mass. Determine (**a**) the angular frequency of the oscillator; (**b**) the frequency; (**c**) the period of the motion; (**d**) the amplitude of the motion; (**e**) the phase constant of the motion; (**f**) the position of the mass 0.2 s after it is released; (**g**) the velocity (speed and direction) 0.2 s after it is released; (**h**) the speed when the mass is 6 cm above equilibrium and moving upward; and (**i**) the acceleration of the mass 0.2 s after release.

Solution:

(**a**) Use Eq. (7-3) to calculate the angular frequency.

$$\omega = \sqrt{\frac{k}{m}} = \sqrt{\frac{20 \text{ N m}^{-1}}{0.05 \text{ kg}}} = \boxed{20 \text{ Hz}}$$

(**b**) Use either Eq. (7-9b) or Eq. (7-7) to obtain the frequency.

$$v = \frac{1}{2\pi}\sqrt{\frac{k}{m}} = \frac{\omega}{2\pi} = \frac{20 \text{ Hz}}{2\pi} = \boxed{3.18 \text{ Hz}}$$

(**c**) Use Eq. (7-6) or Eq. (7-7) to obtain the period.

$$T = \frac{2\pi}{\omega} = v^{-1}$$

$$= (3.18 \text{ Hz})^{-1} = \boxed{0.314 \text{ s}}$$

(**d**) Use Eq. (7-11). At the instant m is released, $v = 0$. Solve for X.

$$\tfrac{1}{2}kX^2 = \tfrac{1}{2}mv^2 + \tfrac{1}{2}kx^2 = 0 + \tfrac{1}{2}kx^2$$

$$X^2 = x^2 = (-8 \text{ cm})^2 = (8 \times 10^{-2} \text{ m})^2 = 64 \times 10^{-4} \text{ m}^2$$

$$X = \sqrt{64 \times 10^{-4} \text{ m}^2} = \boxed{8 \times 10^{-2} \text{ m}}$$

The amplitude is a positive definite quantity: the magnitude of the maximum displacement of m. Note that $+x$ implies that m is above its equilibrium position and $-x$ implies that m is below it.

(**e**) Use Eq. (7-5), evaluated at $t = 0$, and solve for ϕ.

$$x = X\cos(\omega t + \phi)$$

$$-8 \times 10^{-2} \text{ m} = (8 \times 10^{-2} \text{ m})\cos(0 + \phi)$$

$$-1 = \cos\phi$$

$$\phi = \arccos(-1) = \pm\pi \text{ rad}$$

In this case it doesn't matter whether ϕ is $+\pi$ rad or $-\pi$ rad; the same correct

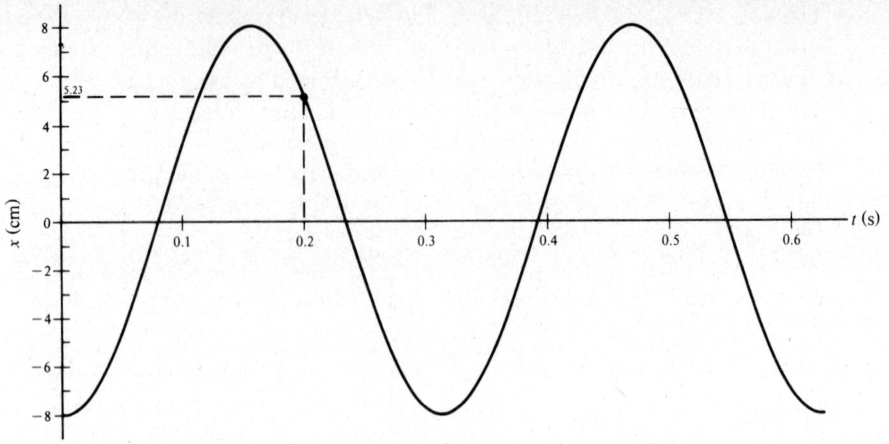

Figure 7-3

mathematical description results from either choice. Figure 7-3 shows the graph of the position of the mass as a function of time; $x = X \cos(\omega t + \phi)$, where $X = 8$ cm, $\omega = 20$ Hz, and $\phi = +\pi$ rad.

(f) Use Eq. (7-5), with $X = 8$ cm, $\omega = 20$ Hz, and $t = 0.2$ s; let $\phi = +\pi$ rad. The phase, $\omega t + \phi$, is in radian measure.

$$x = X \cos(\omega t \pm \phi) = (8 \text{ cm})\cos[(20 \text{ Hz})(0.2 \text{ s}) + \pi] = \boxed{+5.23 \text{ cm}}$$

(g) Use Eq. (7-10), with $\omega = 20$ Hz, $X = 8$ cm $= 8 \times 10^{-2}$ m, and $t = 0.2$ s; again, let $\phi = +\pi$ rad. As before, you get the same correct mathematical description of the velocity if you use $\phi = -\pi$ rad.

$$v = -\omega X \sin(\omega t + \phi)$$

$$= -(20 \text{ Hz})(8 \times 10^{-2} \text{ m})\sin[(20 \text{ Hz})(0.2 \text{ s}) + \pi] = \boxed{-1.21 \text{ m s}^{-1}}$$

The speed of the mass is 1.21 m s^{-1}, and the negative sign indicates that the mass is moving downward at this instant. Figure 7-4 shows the graph of the velocity of the mass as a function of time; $v = -\omega X \sin(\omega t + \phi)$, where $\omega = 20$ Hz, $X = 8$ cm, and $\phi = +\pi$ rad.

(h) Use Eq. (7-11) and solve for v.

$$E = \tfrac{1}{2}kX^2 = \tfrac{1}{2}mv^2 + \tfrac{1}{2}kx^2$$

$$v^2 = \frac{k}{m}(X^2 - x^2) = \omega^2(X^2 - x^2)$$

$$v = \pm\omega\sqrt{X^2 - x^2}$$

$$= \pm(20 \text{ Hz})\sqrt{(8 \times 10^{-2} \text{ m})^2 - (6 \times 10^{-2} \text{ m})^2} = \boxed{\pm 1.06 \text{ m s}^{-1}}$$

You should select the positive sign because you know that the mass is moving upward.

Figure 7-4

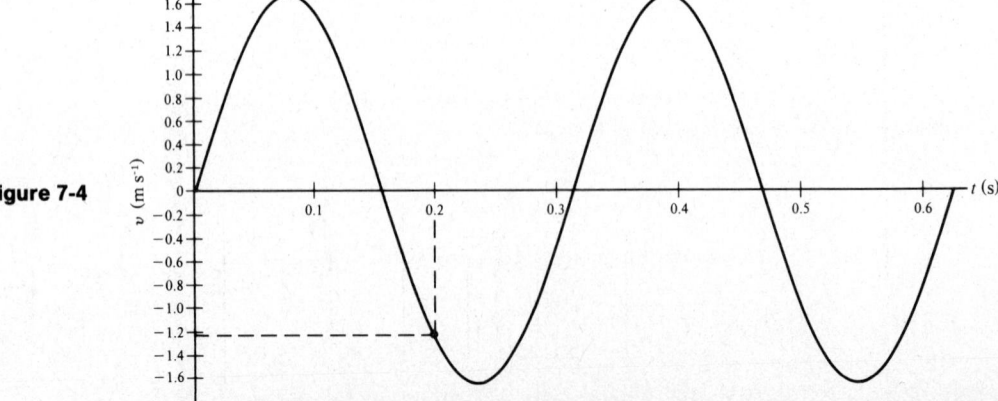

(i) Use Eqs. (7-4) and (7-5) to obtain the acceleration, with $X = 8 \times 10^{-2}$ m, $\omega = 20$ Hz, $t = 0.2$ s, and $\phi = +\pi$ rad.

$$a = -\omega^2 x = -\omega^2 X \cos(\omega t + \phi)$$

$$= -(20 \text{ Hz})^2 (8 \times 10^{-2} \text{ m})\cos[(20 \text{ Hz})(0.2 \text{ s}) + \pi] = \boxed{-20.9 \text{ m s}^{-1}}$$

The negative sign indicates that the acceleration is directed downward at this instant. As always in SHM, the acceleration of the mass is in the direction opposite that of its displacement. See Figures 7-3 and 7-5. Figure 7-5 shows the graph of the acceleration of the mass as a function of time; $a = -\omega^2 X \cos(\omega t + \phi)$, where $\omega = 20$ Hz, $X = 8$ cm, and $\phi = +\pi$ rad.

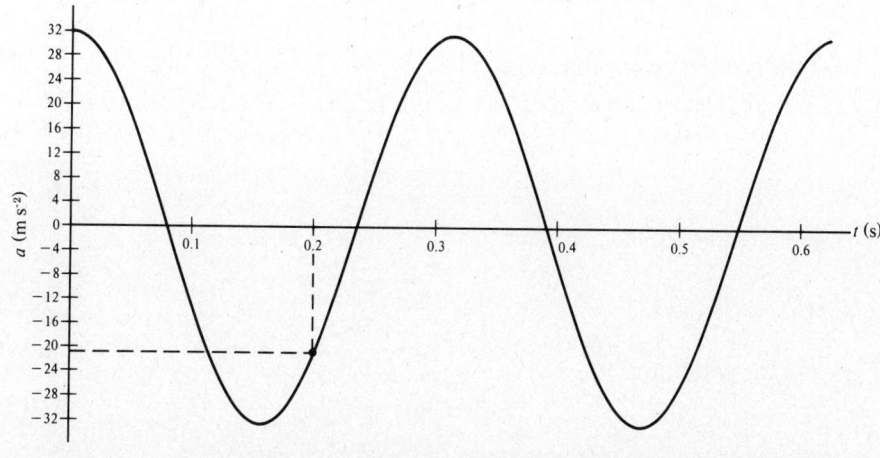

Figure 7-5

7-3. The Simple Pendulum

A simple pendulum consists of a mass m that is free to swing at the end of a cord of negligible mass. The length of a simple pendulum is measured from the pivot to the center of the mass and is designated ℓ. The coordinate that specifies the position of m is s (measured along the path), but a more useful coordinate is the angle θ, in radians, where

$$\theta = \frac{s}{\ell} \qquad (7\text{-}13)$$

See Figure 7-6. When we apply Newton's second law to m, we get

$$\Sigma F = ma$$

$$-mg \sin \theta = ma = m\frac{d^2 s}{dt^2} \qquad (7\text{-}14)$$

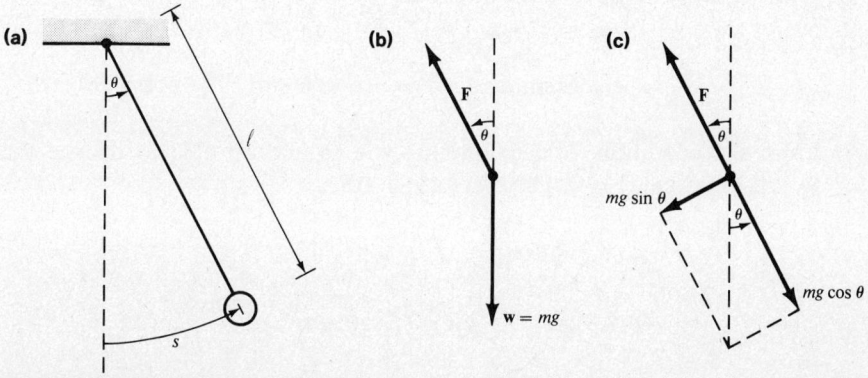

Figure 7-6. (a) A simple pendulum of length ℓ. (b) Free-body diagram of the forces acting on mass m; $\mathbf{w} = mg$, its weight, and $\mathbf{F}$, the tension in the cord. (c) Free-body diagram showing the weight resolved into components in the direction of motion and perpendicular to the direction of motion.

From Eq. (7-13), $s = \ell\theta$ and therefore $\dfrac{d^2s}{dt^2} = \ell\left(\dfrac{d^2\theta}{dt^2}\right)$. We can now write Eq. (7-14) as

$$\frac{d^2s}{dt^2} = \ell\left(\frac{d^2\theta}{dt^2}\right) = -g\sin\theta$$

or

$$\frac{d^2\theta}{dt^2} = -\frac{g}{\ell}\sin\theta \qquad (7\text{-}15)$$

You can find the exact solution of Eq. (7-15) in more advanced texts. We now turn to an approximate solution, which is valid when θ is small (up to about 0.2 rad, or about 10°).

A. Small-angle approximation

For small θ (measured in radians), $\sin\theta \cong \theta$ and Eq. (7-15) becomes

$$\frac{d^2\theta}{dt^2} = -\frac{g}{\ell}\theta \qquad (7\text{-}16)$$

Equation (7-16) has the same form as Eq. (7-2). In this case

$$\omega = \sqrt{\frac{g}{\ell}} \qquad (7\text{-}17)$$

and the solution is

$$\theta = \Theta\cos(\omega t + \phi) \qquad (7\text{-}18)$$

where Θ is the **angular amplitude**, the maximum value of θ.

EXAMPLE 7-2: A simple pendulum of length $\ell = 1.2$ m is released from rest from the position $\theta = 0.2$ rad. Determine (a) the period of its motion and (b) the speed of the mass when $t = 0.5$ s.

Solution:

(a) From Eq. (7-6), $T = 2\pi/\omega$, and from Eq. (7-17), $\omega = \sqrt{g/\ell}$. Combining these expressions, you get

$$T = 2\pi\sqrt{\frac{\ell}{g}} = 2\pi\sqrt{\frac{1.2\text{ m}}{9.8\text{ m s}^{-2}}} = \boxed{2.20\text{ s}}$$

(b) From Eq. (7-10), you can infer that the angular velocity of the mass is

$$\frac{d\theta}{dt} = -\omega\Theta\sin(\omega t + \phi)$$

so the velocity along its path would be $\dfrac{ds}{dt} = \ell\left(\dfrac{d\theta}{dt}\right)$, or

$$\frac{ds}{dt} = -\omega\ell\Theta\sin(\omega t + \phi) = -\sqrt{\frac{g}{\ell}}\,\ell\Theta\sin\left(\sqrt{\frac{g}{\ell}}\,t + \phi\right)$$

From the conditions stated initially, you should be able to deduce that $\Theta = 0.2$ rad and $\phi = 0$. Therefore at $t = 0.5$ s,

$$\frac{ds}{dt} = -\sqrt{g\ell}\,\Theta\sin\omega t$$

$$= -\sqrt{(9.8\text{ m s}^{-2})(1.2\text{ m})}\,(0.2\text{ rad})\sin\left(\sqrt{\frac{9.8\text{ m s}^{-2}}{1.2\text{ m}}}(0.5\text{ s})\right)$$

$$= -0.679\text{ m s}^{-1}$$

The negative sign should remind you that the motion of the mass is in the direction opposite that of its displacement, as is the case for all SHM.

B. Exact-solution result

One result of an exact solution of Eq. (7-15) is a series that describes how the period of the pendulum depends on its angular amplitude:

$$T(\Theta) = T\left\{1 + \left(\frac{1}{2}\right)^2 \sin^2\left(\frac{\Theta}{2}\right) + \left(\frac{1}{2}\right)^2\left(\frac{3}{4}\right)^2 \sin^4\left(\frac{\Theta}{2}\right) + \cdots\right\} \qquad \textbf{(7-19)}$$

where $T = 2\pi\sqrt{\ell/g}$ is the period obtained from combining Eqs. (7-6) and (7-17).

EXAMPLE 7-3: Determine the ratio $T(\Theta)/T$ for a pendulum that has an angular amplitude of $\pi/4$ rad (45°), using the first three terms of Eq. (7-19).

Solution: Use Eq. (7-19) with $\Theta = \pi/4$ rad.

$$\frac{T(\Theta)}{T} = 1 + \left(\frac{1}{2}\right)^2 \sin^2\left(\frac{\pi/4}{2}\right) + \left(\frac{1}{2}\right)^2\left(\frac{3}{4}\right)^2 \sin^4\left(\frac{\pi/4}{2}\right) + \cdots$$

$$= 1 + 3.661\,165 \times 10^{-2} + 3.344\,275 \times 10^{-3} + \cdots$$

$$= \boxed{1.039\,955}$$

The period is almost 4 percent longer when its angular amplitude is $\pi/4$ rad than when the amplitude is small.

7-4. Damped Harmonic Motion

When the mass of a mass–spring system moves in a viscous medium and the frictional force acting on it is proportional to the first power of its velocity, we have to modify Eq. (7-2) to include this force. As in Section 4-7, the frictional force, Eq. (4-6), is $f = bv = b\left(\dfrac{dx}{dt}\right)$. We include this force in Eq. (7-2a) and obtain

$$m\frac{d^2x}{dt^2} = -kx - b\left(\frac{dx}{dt}\right)$$

which we can rewrite to put it in a standard form as

$$\frac{d^2x}{dt^2} + \frac{b}{m}\frac{dx}{dt} + \frac{k}{m}x = 0 \qquad \textbf{(7-20)}$$

We can easily solve this second-order differential equation with constant coefficients by standard techniques.

$$x = Xe^{-\lambda t}\cos(\omega_b t + \phi) \qquad \textbf{(7-21)}$$

where

$$\lambda = \frac{b}{2m} \qquad \textbf{(7-22)}$$

and

$$\omega_b = \omega\sqrt{1 - \left(\frac{\lambda}{\omega}\right)^2} \qquad \textbf{(7-23)}$$

As before, $\omega = \sqrt{k/m}$.

EXAMPLE 7-4: The 50-g mass attached to the spring in Example 7-1 is immersed in a viscous fluid. The frictional force acting on the mass is $f = bv$, Eq. (4-6), with $b = 0.172 \text{ kg s}^{-1}$. The mass is pulled to a position 8 cm below its equilibrium position and released. Determine the displacement of the mass at the times $t = T, 2T, 3T, 4T,$ and $5T$, where T is the period of its motion.

Solution: You first have to determine the angular frequency ω_b, using Eq. (7-23), with

$$\lambda = \frac{b}{2m} = \frac{0.172 \text{ kg s}^{-1}}{2(0.05 \text{ kg})} = 1.72 \text{ s}^{-1}$$

and

$$\omega_b = \omega \sqrt{1 - \left(\frac{\lambda}{\omega}\right)^2} = (20 \text{ s}^{-1}) \sqrt{1 - \left(\frac{1.72 \text{ s}^{-1}}{20 \text{ s}^{-1}}\right)^2} = 19.93 \text{ s}^{-1}$$

Thus

$$T = \frac{2\pi}{\omega_b} = 0.3153 \text{ s}$$

To obtain the displacements, use Eq. (7-21), with $X = 8$ cm and $\phi = +\pi$ rad, evaluated at $t = T, 2T$, etc.

$$x = X^{-\lambda t} \cos(\omega_b t + \phi)$$
$$= -(8 \text{ cm})e^{-[(1.72 \text{ s}^{-1})(0.3153 \text{ s})]} \cos[(19.93 \text{ s}^{-1})(0.3153 \text{ s}) - \pi] = \boxed{-4.65 \text{ cm}}$$

The following tabulated results were obtained from calculations for six values of t. See Figure 7-7, also, which shows the graph of the position of the mass as a function of time; $x = Xe^{-\lambda t} \cos(\omega_b t + \phi)$, where $X = 8$ cm, $\omega_b = 1.72 \text{ s}^{-1}$, $\omega_b = 19.93$ Hz, and $\phi = +\pi$ rad.

x (in cm)	t	x (in cm)	t
−8.00	0	−1.58	$3T$
−4.65	T	−0.92	$4T$
−2.71	$2T$	−0.54	$5T$

Note that $\omega_b \cong \omega$, which is most often the case. Only when the frictional force is very large (a large value of b) does ω_b differ significantly from ω. Then the mass would undergo less than a full cycle of motion if displaced and released.

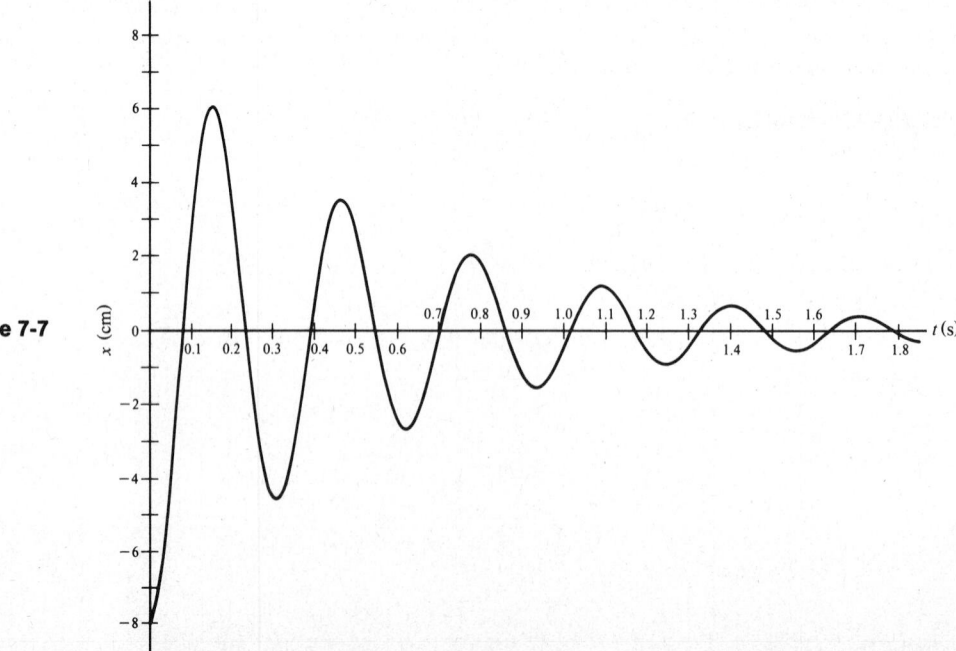

Figure 7-7

7-5. Forced Harmonic Motion

We now consider the damped mass–spring oscillator of Section 7-4, subject to a harmonic driving force, $F_t = F \cos \omega_e t$, where F is the amplitude of driving force and ω_e is its (external) angular frequency. In Figure 7-8, the driving force is provided by a "machine," and the damping force results from the motion of m in a viscous fluid such as heavy oil. When we include the driving force in Eq. (7-20),

$$\frac{d^2x}{dt^2} + \frac{b}{m}\frac{dx}{dt} + \frac{k}{m}x = \frac{F}{m}\cos\omega_e t \qquad (7\text{-}24)$$

The complete solution to Eq. (7-24) is presented in a more advanced course in mechanics and a first course in differential equations. Here we will deal only with the *steady state* solution, which is valid when $t > 10m/b$. In the steady state, we can still describe the position of m by an expression very much like Eq. (7-5), or

$$x = X_e \cos(\omega_e t - \phi_e) \qquad (7\text{-}25)$$

Note, however, that the angular frequency is that of the driving force. The phase constant and amplitude of the motion are, respectively,

$$\tan\phi_e = \frac{b\omega_e}{m(\omega^2 - \omega_e^2)} \qquad (7\text{-}26)$$

and

$$X_e = \frac{F}{\sqrt{m^2(\omega^2 - \omega_e^2)^2 + b^2\omega_e^2}} \qquad (7\text{-}27)$$

where, as before, $\omega = \sqrt{k/m}$, the *natural* angular frequency of the undamped oscillator. In terms of the *Q factor* (quality factor), $Q = \omega m/b$, and we can write the amplitude of the motion as

$$X_e = \frac{F/m}{\sqrt{(\omega^2 - \omega_e^2) + \omega^2\omega_e^2/Q^2}} \qquad (7\text{-}28)$$

Figure 7-9 is a graph of the amplitude of the motion as a function of the external driving angular frequency for two values of Q, 5.8 and 2.3. Quantities used to plot these graphs are $m = 50$ g, $k = 20$ N m^{-1}, $F = 0.344$ N, and $b = 0.172$ kg s^{-1} for $Q = 5.8$. The value of b for $Q = 2.3$ is determined in Example 7-6. Note that the largest value of X_e occurs at an angular frequency slightly less than the natural angular frequency. The angular frequency for which the amplitude is greatest is called the **resonance angular frequency**, given by

RESONANCE ANGULAR FREQUENCY
$$\omega_r = \sqrt{\omega^2 - \frac{b^2}{2m^2}} \qquad (7\text{-}29)$$

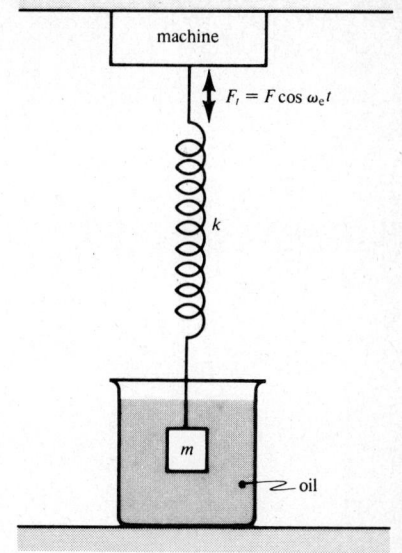

Figure 7-8

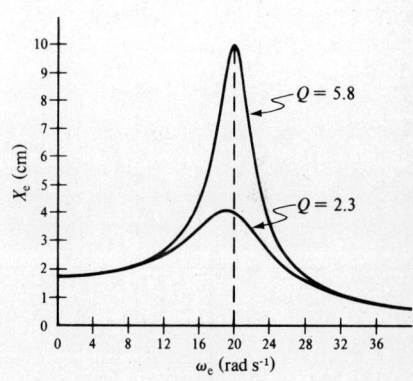

Figure 7-9

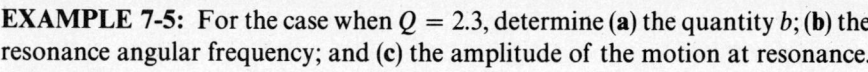

EXAMPLE 7-5: For the case when $Q = 2.3$, determine (a) the quantity b; (b) the resonance angular frequency; and (c) the amplitude of the motion at resonance.

Solution:

(a) Use the definition $Q = \omega m/b$ and solve for b.

$$b = \frac{\omega m}{Q} = \frac{(20 \text{ rad s}^{-1})(0.05 \text{ kg})}{2.3} = \boxed{0.435 \text{ kg s}^{-1}}$$

(b) Use Eq. (7-29) to obtain ω_r.

$$\omega_r = \sqrt{\omega^2 - \frac{b^2}{2m^2}} = \sqrt{(20 \text{ rad s}^{-1})^2 - \frac{(0.435 \text{ kg s}^{-1})^2}{2(0.05 \text{ kg})^2}} = \boxed{19.03 \text{ rad s}^{-1}}$$

(c) Use Eq. (7-28), with $\omega_e = \omega_r = 19.03$ rad s^{-1}.

$$X_e = \frac{F/m}{\sqrt{(\omega^2 - \omega_r^2)^2 + \omega^2\omega_r^2/Q^2}}$$

$$= \frac{(0.344 \text{ N})/(0.05 \text{ kg})}{\{[(20 \text{ rad s}^{-1})^2 - (19.03 \text{ rad s}^{-1})^2]^2 + (20 \text{ rad s}^{-1})^2(19.03 \text{ rad s}^{-1})^2/(2.3)^2\}^{1/2}}$$

$$= \boxed{4.05 \times 10^{-2} \text{ m} = 4.05 \text{ cm}}$$

EXAMPLE 7-6: (a) Obtain an expression for the resonance angular frequency in terms of ω and Q. (b) For what value of Q will $\omega_r = 0$? (c) For what value of b does $\omega_r = 0$?

Solution:

(a) Use the definition $Q = \omega m/b$, solve for b, and substitute the result into Eq. (7-29).

$$b = \frac{\omega m}{Q}$$

and

$$\omega_r = \sqrt{\omega^2 - \frac{b^2}{2m^2}} = \sqrt{\omega^2 - \frac{\omega^2}{2Q^2}} = \omega\sqrt{1 - \frac{1}{2Q^2}}$$

(b) Use the result from (a), let $\omega_r = 0$, and solve for Q.

$$\omega_r = 0 = \omega\sqrt{1 - \frac{1}{2Q^2}}$$

$$0 = 1 - \frac{1}{2Q^2}$$

$$Q = \boxed{\frac{1}{\sqrt{2}} = 0.707}$$

(c) Use the definition $Q = \omega m/b$ and solve for b.

$$b = \frac{\omega m}{Q} = (20 \text{ rad s}^{-1})(0.05 \text{ kg})(\sqrt{2}) = \boxed{\sqrt{2} \text{ kg s}^{-1} = 1.414 \text{ kg s}^{-1}}$$

SUMMARY

1. Hooke's law: The restoring force is proportional to the displacement from equilibrium; its direction is opposite that of the displacement: $F = -kx$.
2. The angular frequency of oscillation (the *natural angular frequency*) for an undamped mass–spring oscillator is $\omega = \sqrt{k/m}$.
3. The relationship between angular frequency and "ordinary" frequency is $\omega = 2\pi\nu$.
4. The position of the mass in a mass–spring oscillator is $x = X\cos(\omega t + \phi)$, where X is the amplitude of the motion and ϕ is the phase constant.
5. To account for the mass of the spring, m_s, use the effective mass, $m_{\text{eff}} = m + \frac{1}{3}m_s$, in mass–spring oscillator formulas.
6. The angular frequency of oscillation for a simple pendulum is $\omega = \sqrt{g/\ell}$, where ℓ is the length of the pendulum, and g is the acceleration due to gravity.

7. If a frictional force acts on the oscillator and is proportional to its velocity, $f = bv$, the position of the mass is described by

$$x = Xe^{-\lambda t}\cos(\omega_b t + \phi)$$

where $\lambda = b/2m$, and

$$\omega_b = \omega\sqrt{1 - (\lambda/\omega)^2}$$

8. If a damped oscillator is driven by a harmonic force at an angular frequency of ω_e, $F_t = F\cos\omega_e t$. The position of the mass and the amplitude of the motion in the steady state are, respectively,

$$x = X_e\cos(\omega_e t - \phi_e) \quad \text{and} \quad X_e = \frac{F/m}{\sqrt{(\omega^2 - \omega_e^2) + \omega^2\omega_e^2/Q^2}}$$

where $Q = \omega m/b$.

9. For a harmonically driven oscillator, the maximum amplitude of the driving angular frequency is equal to the resonance angular frequency ω_r, where

$$\omega_r = \sqrt{\omega^2 - \frac{b^2}{2m^2}}$$

RAISE YOUR GRADES

Can you explain . . . ?

☑ how to evaluate the phase constant
☑ the relationship between the angular frequency, the ordinary frequency, and the period
☑ the relationship between the kinetic, potential, and total energies for a harmonic oscillator
☑ the source of the restoring force for the simple pendulum
☑ the relationship between the frictional force coefficient b and the quality factor
☑ how to obtain the formula for the resonance angular frequency

SOLVED PROBLEMS

Simple Harmonic Motion and Hooke's Law

PROBLEM 7-1: An 80-g mass hangs on the lower end of a spring and undergoes SHM with a period of 0.3 s. The mass of the spring is 120 g. What is the value of the spring constant?

Solution: You should first find m_{eff} by using Eq. (7-12) and then solve Eq. (7-9a) for k.

$$m_{eff} = m + \tfrac{1}{3}m_s = 80\text{ g} + \tfrac{1}{3}(120\text{ g}) = 120\text{ g} = 0.12\text{ kg}$$

and

$$T = 2\pi\sqrt{\frac{m}{k}}$$

Thus

$$k = \frac{4\pi^2 m}{T^2} = \frac{4\pi^2(0.12\text{ kg})}{(0.3\text{ s})^2} = \boxed{52.6\text{ kg s}^{-2} = 52.6\text{ N m}^{-1}}$$

Equations of SHM—the Mass–Spring Oscillator

PROBLEM 7-2:   A 40-g mass is attached to the lower end of a spring of negligible mass and oscillates at a frequency of 2 Hz, with an amplitude of 8 cm. **(a)** Determine the spring constant. **(b)** How fast is the system moving when it is 2 cm from equilibrium? **(c)** How fast is it moving when it passes equilibrium?

Solution:

(a) Use Eq. (7-9b) and solve for k.

$$v = \frac{1}{2\pi}\sqrt{\frac{k}{m}}$$

$$k = (2\pi v)^2 m = [(2\pi)(2\text{ Hz})]^2(0.04\text{ kg}) = \boxed{6.32\text{ kg s}^{-1} = 6.32\text{ N m}^{-1}}$$

(b) Solve Eq. (7-11) for v, making use of Eq. (7-9b), i.e., $2\pi v = \sqrt{k/m}$. Thus

$$\tfrac{1}{2}kX^2 = \tfrac{1}{2}mv^2 + \tfrac{1}{2}kx^2$$

$$v^2 = \frac{k}{m}(X^2 - x^2)$$

$$v = \pm(2\pi v)\sqrt{X^2 - x^2} = \pm[(2\pi)(2\text{ Hz})]\sqrt{(0.08\text{ m})^2 - (0.02\text{ m})^2} = \boxed{\pm 0.973\text{ m s}^{-1}}$$

(c) Solve the same way as **(b)**, except that $x = 0$.

$$v = \pm(2\pi v)\sqrt{X^2} = \pm 2\pi v X = \pm(2\pi)(2\text{ Hz})(0.08\text{ m}) = \boxed{1.005\text{ m s}^{-1} = 100.5\text{ cm s}^{-1}}$$

PROBLEM 7-3:   The position of the mass in Problem 7-2 can be correctly described by Eq. (7-5). At $t = 0$, the mass is located at $x = 7.4$ cm **(a)** Determine ϕ. **(b)** Where is m when $t = 0.12$ s?

Solution:

(a) Solve Eq. (7-5) for ϕ when $t = 0$.

$$x = X\cos(\omega t + \phi) = X\cos\phi \qquad (\text{at } t = 0)$$

$$\phi = \arccos\left(\frac{x}{X}\right) = \arccos\left(\frac{7.4\text{ cm}}{8\text{ cm}}\right) = \boxed{0.39\text{ rad}}$$

(b) Use Eq. (7-5) with $\omega = 2\pi v$, or

$$x = X\cos(\omega t + \phi) = X\cos(2\pi v t + \phi)$$

$$= (8\text{ cm})\cos[(2\pi)(2\text{ Hz})(0.12\text{ s}) + 0.39\text{ rad}] = \boxed{-2.57\text{ cm}}$$

The negative sign tells you that the mass is below the equilibrium position at this instant.

PROBLEM 7-4:   **(a)** At what instant will the mass in Problem 7-2 first reach the point $x = -8$ cm? **(b)** At what instant will it again be at the point $x = -8$ cm?

Solution:

(a) Use the results from Problem 7-3 and solve Eq. (7-5) for t.

$$x = X\cos(\omega t + \phi) \qquad \text{and} \qquad \omega t + \phi = \arccos\left(\frac{x}{X}\right)$$

$$t = \frac{\arccos(x/X) - \phi}{\omega} = \frac{\arccos(x/X) - \phi}{2\pi v} = \frac{\arccos\left(\dfrac{-8\text{ cm}}{8\text{ cm}}\right) - 0.39\text{ rad}}{2\pi(2\text{ Hz})} = \boxed{0.219\text{ s}}$$

(b) You can determine the period of the motion from Eq. (7-7), i.e.,

$$T = v^{-1} = (2\text{ Hz})^{-1} = 0.5\text{ s}$$

Therefore the mass will again be at the position $x = -8$ cm, 0.5 s later, i.e., at the instant that

$$t = 0.219 \text{ s} + T = (0.219 + 0.5) \text{ s} = \boxed{0.719 \text{ s}}$$

The Simple Pendulum

PROBLEM 7-5: Determine the length of a simple pendulum that has a period of 1 s on the moon, where the surface acceleration due to gravity is 1.62 m s^{-2}.

Solution: Use Eq. (7-17) with $\omega = 2\pi v = 2\pi/T$, and solve for ℓ.

$$\omega = \frac{2\pi}{T} = \sqrt{\frac{g}{\ell}}$$

$$\ell = g\left(\frac{T}{2\pi}\right)^2 = (1.62 \text{ m s}^{-2})\left(\frac{1.0 \text{ s}}{2\pi}\right)^2 = \boxed{0.041 \text{ m} = 4.1 \text{ cm}}$$

PROBLEM 7-6: Use the first two terms in the exact solution to the simple pendulum equation [Eq. (7-19)] to determine the angular amplitude at which the period is 1 percent longer than that predicted by the small amplitude theory.

Solution: Solve Eq. (7-19) for Θ, with $T(\Theta)/T = 1.01$, but use only the first two terms.

$$T(\Theta) = T\left\{1 + \left(\frac{1}{2}\right)^2 \sin^2\left(\frac{\Theta}{2}\right) + \cdots\right\}$$

$$\frac{T(\Theta)}{T} - 1 = \frac{1}{4}\sin^2\left(\frac{\Theta}{2}\right)$$

$$\sin^2\left(\frac{\Theta}{2}\right) = 4(1.01 - 1) = 0.04$$

$$\sin\left(\frac{\Theta}{2}\right) = 0.2$$

$$\Theta = 2[\arcsin(0.2)] = \boxed{0.40 \text{ rad} = 23°}$$

Damped Harmonic Motion

PROBLEM 7-7: A damped mass–spring oscillator can be described by Eq. (7-21), with $\phi = 0$. During the first complete cycle the amplitude of the motion decreases by 5 percent. The spring constant and mass are, respectively, $k = 20 \text{ N m}^{-1}$ and $m = 60$ g. Assume that $\omega_b = \omega$ and determine the frictional force coefficient b.

Solution: This problem involves solving Eq. (7-21) for λ, then solving Eq. (7-22) for b. But first you need to eliminate some variables by using the information on the amplitude decrease. At $t = 0$, $\cos(\omega t + \phi) = 1$ (recall that $\phi = 0$), so, by Eq. (7-5), $x = X\cos(\omega t + \phi) = X$. So you know the oscillator is at its maximum amplitude X at $t = 0$, and that after the first complete cycle, that is, when $t = T = 2\pi/\omega$, $x = 0.95X$. Therefore we can write Eq. (7-21) for this problem as

$$0.95X = Xe^{-\lambda T}\left[\cos\left(\omega \frac{2\pi}{\omega}\right) + \phi\right]$$

The amplitude drops out and the cosine factor equals 1, so

$$e^{-\lambda T} = 0.95 \qquad \lambda = \frac{-\ln 0.95}{T}$$

We can find T by applying Eqs. (7-6) and (7-3).

$$T = \frac{2\pi}{\omega} = 2\pi\sqrt{\frac{m}{k}} = 2\pi\sqrt{\frac{0.06 \text{ kg}}{20 \text{ N m}^{-1}}} = 0.34414 \text{ s}$$

$$\lambda = \frac{-\ln 0.95}{0.34414 \text{ s}} = 0.14904 \text{ s}^{-1}$$

Finally, use Eq. (7-22)

$$\lambda = \frac{b}{2m} \qquad b = 2m\lambda = 2(0.06 \text{ kg})(0.14904 \text{ s}^{-1}) = \boxed{1.789 \times 10^{-2} \text{ kg/s}^{-1}}$$

A good exercise for you would be to verify the assumption $\omega_b \cong \omega$, using Eq. (7-23).

PROBLEM 7-8:   For the oscillator in Problem 7-7, how long will it take for the amplitude to decrease to 1 percent of its initial value? *Hint:* Consider only the exponential factor in Eq. (7-21).

Solution: Use only the exponential factor that reduces the amplitude as a function of time. Solve for t when $e^{-\lambda t} = 0.01$:

$$e^{\lambda t} = \frac{X}{x} = \frac{X}{0.01X} = 10^2$$

$$\lambda t = \ln(10^2)$$

$$t = \frac{\ln(10^2)}{\lambda} = \frac{\ln(10^2)}{0.149 \text{ s}^{-1}} = \boxed{30.9 \text{ s}}$$

Forced Harmonic Motion

PROBLEM 7-9:   The mass–spring oscillator in Problem 7-7 is driven harmonically at an angular frequency of 15 rad s^{-1}. Determine the phase constant between the driving force and the motion of the mass in the steady state.

Solution: Use Eq. (7-26) and solve for ϕ_e.

$$\tan \phi_e = \frac{b\omega_e}{m(\omega^2 - \omega_e^2)}$$

$$\phi_e = \arctan\left(\frac{b\omega_e}{m(\omega^2 - \omega_e^2)}\right)$$

$$= \arctan\left(\frac{(1.789 \times 10^{-2} \text{ kg s}^{-1})(15 \text{ rad s}^{-1})}{(0.06 \text{ kg})[(18.26 \text{ rad s}^{-1})^2 - (15 \text{ rad s}^{-1})^2]}\right) = \boxed{4.12 \times 10^{-2} \text{ rad} = 2.36°}$$

PROBLEM 7-10:   (a) Determine the quality factor for the oscillator in Problem 7-7. (b) Assume that the resonance angular frequency is ω and determine the force amplitude that produces an amplitude of the motion of 10 cm in the steady state at the driving angular frequency of ω_e; i.e., the oscillator is being driven at its resonant frequency.

Solution:

(a) Use the definition of the quality factor to obtain Q.

$$Q = \frac{\omega m}{b} = \frac{(18.26 \text{ rad s}^{-1})(0.06 \text{ kg})}{1.789 \times 10^{-2} \text{ kg s}^{-1}} = \boxed{61.2}$$

(b) Use Eq. (7-28), with $\omega_e = \omega_r \cong \omega$, and solve for F.

$$X_e = \frac{F/m}{\sqrt{(\omega^2 - \omega_e^2)^2 + \omega^2 \omega_e^2/Q^2}} = \frac{FQ}{m\omega^2}$$

$$F = \frac{X_e m \omega^2}{Q} = \frac{X_e m(k/m)}{Q} = \frac{X_e k}{Q} = \frac{(0.1 \text{ m})(20 \text{ N m}^{-1})}{61.2} = \boxed{3.27 \times 10^{-2} \text{ N}}$$

Another good exercise for you would be to verify the assumption that $\omega_r \cong \omega$, using Eq. (7-29).

Supplementary Exercises

EXERCISE 7-1: A mass of 80 g is attached to the lower end of a vertical spring that has a mass of 90 g and oscillates at a frequency of 3 Hz. At what frequency would it oscillate if the mass of the spring were negligible?

EXERCISE 7-2: A spring of negligible mass has a spring constant $k = 50 \text{ N m}^{-1}$. A mass of 60 g is attached to its lower end and set to oscillating with an amplitude of 8 cm. What is the shortest time for the mass to go from the 8-cm position to the -6-cm position?

EXERCISE 7-3: How fast is the mass in Exercise 7-2 moving when it gets to the -6-cm position?

EXERCISE 7-4: For the mass–spring oscillator in Exercise 7-2, the mass is located at $x = +4$ cm, and its velocity is 60 cm s^{-1}, moving downward ($v = -60$ cm s^{-1}), at $t = 0$. Determine the phase constant.

EXERCISE 7-5: Determine the length of a simple pendulum that has a period equal to that of the mass–spring oscillator in Exercise 7-2.

EXERCISE 7-6: A simple pendulum has a length of 80 cm. What is the speed of the pendulum bob as it passes the equilibrium position if its angular amplitude is 30°?

EXERCISE 7-7: The frictional force coefficient for a damped mass–spring oscillator is $b = 0.3 \text{ kg s}^{-1}$. The mass and spring constant are $m = 60$ g and $k = 50 \text{ N m}^{-1}$, respectively, and the mass is located at the 10-cm position above equilibrium at $t = 0$. Determine the position of m after one cycle.

EXERCISE 7-8: Determine the value of the frictional force coefficient for the mass–spring oscillator in Exercise 7-7 if the angular frequency of oscillation is 1 percent lower than ω.

EXERCISE 7-9: Determine the phase constant in the steady state between the driving angular frequency and the motion of the mass for the mass–spring oscillator in Exercise 7-7 when the driving angular frequency equals the resonance angular frequency, i.e., when $\omega_e = \omega_r$.

EXERCISE 7-10: Verify that, at the resonance angular frequency, the amplitude of the forced harmonic oscillator is

$$X_e = \frac{2QF}{b\omega\sqrt{(2Q)^2 - 1}}$$

where $Q = \omega m/b$. Note that for large values of Q, $X_e \cong F/b\omega$.

Answers to Supplemental Exercises

7-1: 3.52 Hz

7-2: 8.38×10^{-2} s

7-3: 1.53 m s^{-1}

7-4: $\tan(\omega t + \phi) = -v/\omega x$, $\quad \phi = 0.48$ rad $= 27.5°$

7-5: 1.18×10^{-2} m $= 1.18$ cm

7-6: 1.47 m s^{-1}

7-7: $+6.09$ cm

7-8: 0.489 kg s^{-1}

7-9: $\phi_e = 1.48$ rad $= 85°$

8 IMPULSE AND CONSERVATION OF MOMENTUM

THIS CHAPTER IS ABOUT

- ☑ **Impulsive Forces and Momentum**
- ☑ **Conservation of Linear Momentum**
- ☑ **Collisions: Elastic and Inelastic**
- ☑ **Center of Mass (CM)**
- ☑ **Collision of Two Bodies in CM Frame of Reference**
- ☑ **Rocket Propulsion**

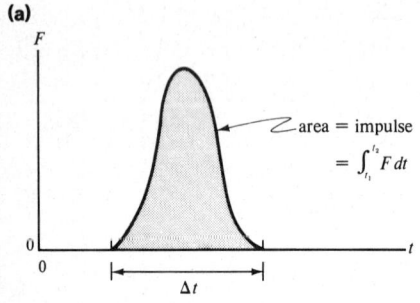

(a)

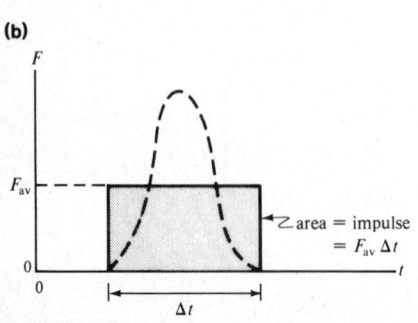

(b)

Figure 8-1. (a) Graph of a typical impulsive force. (b) Graph of the average impulsive force.

8-1. Impulsive Forces and Momentum

Suppose that a force acts for a short time on an object as, for example, when a tennis racket strikes a tennis ball. Newton's second law for this interaction is

$$\mathbf{F} = m\mathbf{a} = m\frac{d\mathbf{v}}{dt} = \frac{d}{dt}(m\mathbf{v}) \qquad \text{[Eq. (4-1b)]}$$

where $\mathbf{F}$ is the resultant force of the tennis racket acting on the tennis ball of mass m. Because the mass is constant, *in this case* it can be incorporated with the velocity. Equation (4-1b) is the form in which Newton originally formulated his second law. The quantity in parenthesis on the right-hand side is called the **linear momentum** or, more commonly, just the **momentum** of m. (We'll deal with angular momentum in Chapter 10.)

MOMENTUM $\qquad\qquad \mathbf{p} = m\mathbf{v}$ $\qquad\qquad$ **(8-1)**

No derived unit of momentum has yet been defined in the SI system. We therefore use $\mathrm{kg\,m\,s^{-1}}$ as the units for momentum. When we substitute $\mathbf{p}$ for $m\mathbf{v}$ in Eq. (4-1b), Newton's second law, as applied in this case, becomes

$$\mathbf{F} = \frac{d\mathbf{p}}{dt} \qquad\qquad \textbf{(8-2)}$$

We multiply by dt and integrate from t_1, the time at which $\mathbf{F}$ initially interacts with m, until t_2, when the interaction stops, to obtain

$$\int_{t_1}^{t_2} \mathbf{F}\,dt = \int_1^2 d\mathbf{p} = \mathbf{p}_2 - \mathbf{p}_1 = m\mathbf{v}_2 - m\mathbf{v}_1 \qquad \textbf{(8-3)}$$

The integral on the far left-hand side is called the **impulse**, and we can now reinterpret Newton's second law: *The impulse acting on an object produces a change in its momentum.* Figure 8-1a shows a graph of a typical impulsive force. The area under the curve equals the impulse. Note that impulse has units of $\mathrm{N\,s}$ in the SI system. The **average impulsive force**, F_{av}, is defined so that the rectangular area in Figure 8-1b is also equal to the impulse, or

AVERAGE IMPULSIVE
FORCE

$$F_{av}\,\Delta t = \int_{t_1}^{t_2} F\,dt \qquad\qquad \textbf{(8-4)}$$

where $\Delta t = t_2 - t_1$, i.e., the time interval.

EXAMPLE 8-1: A golfer hits a 50-g golf ball off the tee, and, at the end of the impulse, which lasts for 0.02 s, the ball is moving at 32 m s^{-1}. **(a)** What is the momentum of the ball at the end of the impulse? **(b)** Calculate the magnitude of the impulse. **(c)** Determine the average force the golf club exerted on the ball during the interaction.

Solution:

(a) Use the definition, Eq. (8-1), to determine the magnitude of **p**.

$$p = mv = (0.05\text{ kg})(32\text{ m s}^{-1}) = \boxed{1.6\text{ kg m s}^{-1}}$$

(b) Use Eq. (8-3), with zero initial momentum in this case.

$$\int_{t_1}^{t_2} F\,dt = mv_2 - mv_1 = (0.05\text{ kg})(32\text{ m s}^{-1}) - 0 = \boxed{1.6\text{ N s}}$$

(c) Use Eq. (8-4) and solve for F_{av}.

$$F_{av}\,\Delta t = \int_{t_1}^{t_2} F\,dt$$

$$F_{av} = \frac{\int_{t_1}^{t_2} F\,dt}{\Delta t} = \frac{1.6\text{ N s}}{0.02\text{ s}} = \boxed{80\text{ N}}$$

8-2. Conservation of Linear Momentum

Application of Newton's third law to a pair of interacting objects reveals that the impulse given to object A by B is equal and opposite to the impulse given to B by A. Thus the net impulse given for the pair is zero, and, consequently, the net change in momentum for the pair is zero; i.e.,

$$\mathbf{p}_{A1} + \mathbf{p}_{B1} = \mathbf{p}_{A2} + \mathbf{p}_{B2} \qquad\qquad \textbf{(8-5a)}$$

The momentum of the *system* before the interaction (1) is equal to the momentum of the *system* after the interaction (2). When we extend this statement to a system of many particles, upon which no *external* forces are acting, it is called the **principle of the conservation of linear momentum**. Its mathematical form is

PRINCIPLE OF THE
CONSERVATION OF
LINEAR MOMENTUM

$$\underset{\substack{i \\ \text{initial state}}}{\sum \mathbf{p}_i} = \underset{\substack{i \\ \text{final state}}}{\sum \mathbf{p}_i} \qquad\qquad \textbf{(8-5b)}$$

The sum is taken over all the particles in the system.

EXAMPLE 8-2: Two masses A and B, with $m_A = 2$ kg and $m_B = 3$ kg, slide on a straight frictionless track, with $v_A = 7$ m s^{-1} and $v_B = 4$ m s^{-1} before they make contact. Afterward, B is moving at $+6.4$ m s^{-1}. How fast and in what direction is A moving?

Solution: Use the conservation of linear momentum principle for two particles, Eq. (8-5a). Let 1 and 2 represent the velocities before and after the interaction, respectively. Consider the initial direction of A and B to be positive, i.e., one-dimensional motion along the x axis.

$$\mathbf{p}_{A1} + \mathbf{p}_{B1} = \mathbf{p}_{A2} + \mathbf{p}_{B2}$$

$$m_A v_{A1} + m_B v_{B1} = m_A v_{A2} + m_B v_{B2}$$

Solving for v_{A2}, you get

$$v_{A2} = v_{A1} + \frac{m_B}{m_A}(v_{B1} - v_{B2}) = 7 \text{ m s}^{-1} + \frac{3 \text{ kg}}{2 \text{ kg}}(4 - 6.4) \text{ m s}^{-1} = \boxed{3.4 \text{ m s}^{-1}}$$

EXAMPLE 8-3: As in Example 8-2, $m_A = 2$ kg, $m_B = 3$ kg, and $v_A = 7$ m s^{-1}, but in this case $v_B = 0.5$ m s^{-1}, and after the interaction B is moving at 5.7 m s^{-1} in the original direction. How fast and in what direction is A moving?

Solution: The solution is the same as that for Example 8-2.

$$v_{A2} = v_{A1} + \frac{m_B}{m_A}(v_{B1} - v_{B2}) = 7 \text{ m s}^{-1} + \frac{3 \text{ kg}}{2 \text{ kg}}(0.5 - 5.7) \text{ m s}^{-1} = \boxed{-0.8 \text{ m s}^{-1}}$$

Note that the negative sign tells you that A is moving in the opposite direction after the interaction.

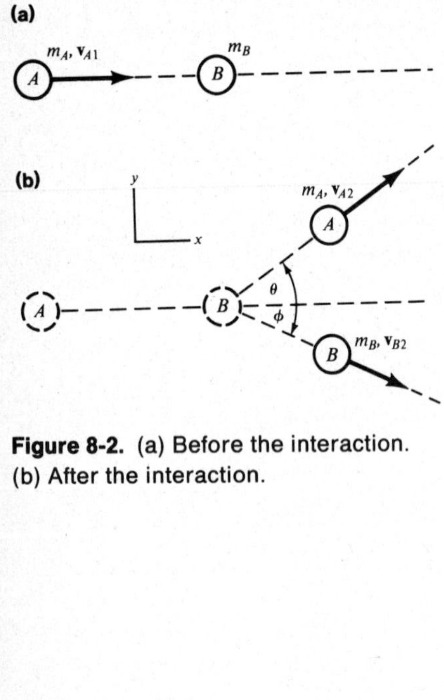

(a)

(b)

Figure 8-2. (a) Before the interaction. (b) After the interaction.

Now consider an interaction in two dimensions. Before the interaction m_A, with an initial velocity of v_{A1}, is approaching m_B, which is stationary, as shown in Figure 8-2a. Application of the principle of the conservation of linear momentum results in two equations: one for the x direction and one for the y direction, as shown in Figure 8-2b. For the x direction,

$$p_{A1_x} + p_{B1_x} = p_{A2_x} + p_{B2_x}$$

$$m_A v_{A1} + 0 = m_A v_{A2} \cos\theta + m_B v_{B2} \cos\phi \qquad \textbf{(8-6a)}$$

and for the y direction,

$$p_{A1_y} + p_{B1_y} = p_{A2_y} + p_{B2_y}$$

$$0 + 0 = m_A v_{A2} \sin\theta - m_B v_{B2} \sin\phi \qquad \textbf{(8-6b)}$$

EXAMPLE 8-4: Puck A ($m_A = 2$ kg) slides on a frictionless horizontal surface at 5 m s^{-1} and strikes puck B ($m_B = 3$ kg), which is stationary. After the impact, A is moving at 2.4 m s^{-1} in the direction $\theta = 35°$. (Refer back to Figure 8-2.) What is the velocity (magnitude and direction) of B?

Solution: You will find it somewhat easier to determine ϕ and then v_{B2}. Solve Eq. (8-6b) for $m_B v_{B2} \sin\phi$ and Eq. (8-6a) for $m_B v_{B2} \cos\phi$. Then divide one by the other to get $\tan\phi$.

$$\frac{m_B v_{B2} \sin\phi}{m_B v_{B2} \cos\phi} = \frac{m_A v_{A2} \sin\theta}{m_A v_{A1} - m_A v_{A2} \cos\theta}$$

$$\tan\phi = \frac{m_A v_{A2} \sin\theta}{m_A v_{A1} - m_A v_{A2} \cos\theta}$$

Thus

$$\phi = \arctan\left(\frac{v_{A2} \sin\theta}{v_{A1} - v_{A2} \cos\theta}\right)$$

$$= \arctan\left(\frac{(2.4 \text{ m s}^{-1})\sin 35°}{5 \text{ m s}^{-1} - (2.4 \text{ m s}^{-1})\cos 35°}\right) = \boxed{24.4°}$$

Now solve Eq. (8-6b) for v_{B2}.

$$v_{B2} = \frac{m_A v_{A2} \sin \theta}{m_B \sin \phi} = \frac{(2 \text{ kg})(2.4 \text{ m s}^{-1})\sin 35°}{(3 \text{ kg})\sin 24.4°} = \boxed{2.22 \text{ m s}^{-1}}$$

8-3. Collisions: Elastic and Inelastic

When the kinetic energy of a system before a collision is equal to the kinetic energy of the system afterward, we call the collision *completely elastic* or usually just *elastic*. When the kinetic energy of a system after a collision is less than before, we call the collision *inelastic*. In either case, keep in mind that linear momentum is always conserved during any collision. Newton discovered that the relative velocities of two objects after a collision are proportional to their relative velocities before the collision. This relationship is known as **Newton's rule**.

NEWTON'S RULE $\qquad v_{B2} - v_{A2} = e(v_{A1} - v_{B1})$ $\qquad\qquad$ **(8-7)**

where e is called the **coefficient of restitution**. The range of e is from 0 to 1, and if

$e = 1$, the collision is *completely elastic*;

$0 < e < 1$, the collision is *inelastic*; and

$e = 0$, the collision is *completely inelastic*.

In a completely inelastic collision the two bodies become attached in some way, so that they have the same velocity after the collision, i.e., $v_{B2} = v_{A2}$.

EXAMPLE 8-5: For a one-dimensional two-body collision, use the principle of the conservation of linear momentum and Eq. (8-7) to obtain expressions for v_{A2} and v_{B2}.

Solution: Solve Eq. (8-7) for v_{B2}.

$$v_{B2} = v_{A2} + e(v_{A1} - v_{B1})$$

Substitute this result into the conservation of momentum equation and solve for v_{A2}.

$$m_A v_{A1} + m_B v_{B1} = m_{A2} v_{A2} + m_B v_{B2} = m_{A2} v_{A2} + m_B[v_{A2} + e(v_{A1} - v_{B1})]$$
$$= (m_A + m_B)v_{A2} + em_B(v_{A1} - v_{B1})$$

$$v_{A2} = \frac{m_A v_{A1} + m_B v_{B1} - em_B(v_{A1} - v_{B1})}{m_A + m_B}$$

Similarly,

$$v_{B2} = \frac{m_A v_{A1} + m_B v_{B1} + em_A(v_{A1} - v_{B1})}{m_A + m_B}$$

EXAMPLE 8-6: Two masses ($m_A = 2$ kg and $m_B = 3$ kg) slide along a straight frictionless track with velocities $v_{A1} = 5 \text{ m s}^{-1}$ and $v_{B1} = 4 \text{ m s}^{-1}$. The coefficient of restitution e for the collision is 0.6. Determine the velocities of the two masses after the collision.

Solution: Use the results obtained in Example 8-5 and substitute to get

$$v_{A2} = \frac{(2 \text{ kg})(5 \text{ m s}^{-1}) + (3 \text{ kg})(4 \text{ m s}^{-1}) - (0.6)(3 \text{ kg})(5 - 4) \text{ m s}^{-1}}{(2 + 3) \text{ kg}}$$

$$= \boxed{4.04 \text{ m s}^{-1}}$$

$$v_{B2} = \frac{(2 \text{ kg})(5 \text{ m s}^{-1}) + (3 \text{ kg})(4 \text{ m s}^{-1}) + (0.6)(2 \text{ kg})(5 - 4) \text{ m s}^{-1}}{(2 + 3) \text{ kg}}$$

$$= \boxed{4.64 \text{ m s}^{-1}}$$

EXAMPLE 8-7: What are the velocities of the masses in Example 8-6 if the collision is completely elastic, i.e., if $e = 1$?

Solution:

$$v_{A2} = \frac{(2 \text{ kg})(5 \text{ m s}^{-1}) + (3 \text{ kg})(4 \text{ m s}^{-1}) - (1.0)(3 \text{ kg})(5 - 4) \text{ m s}^{-1}}{(2 + 3) \text{ kg}}$$

$$= \boxed{3.8 \text{ m s}^{-1}}$$

$$v_{B2} = \frac{(2 \text{ kg})(5 \text{ m s}^{-1}) + (3 \text{ kg})(4 \text{ m s}^{-1}) + (1.0)(2 \text{ kg})(5 - 4) \text{ m s}^{-1}}{(2 + 3) \text{ kg}}$$

$$= \boxed{4.8 \text{ m s}^{-1}}$$

EXAMPLE 8-8: If the collision of the masses in Example 8-6 is completely inelastic, i.e., if $e = 0$, **(a)** determine the common velocity of the masses after the collision. **(b)** How much kinetic energy (mechanical energy) is lost (i.e., converted into thermal energy)?

Solution:

(a) Use the results obtained in Example 8-5 to show that the common velocity $v = v_{A2} = v_{B2}$ is

$$v = \frac{m_A v_{A1} + m_B v_{B1}}{m_A + m_B} = \frac{(2 \text{ kg})(5 \text{ m s}^{-1}) + (3 \text{ kg})(4 \text{ m s}^{-1})}{(2 + 3) \text{ kg}} = \boxed{4.4 \text{ m s}^{-1}}$$

(b) The kinetic energy lost is $\Delta E_k = E_{k1} - E_{k2}$, or

$$\Delta E_k = \tfrac{1}{2} m_A v_{A1}^2 + \tfrac{1}{2} m_B v_{B1}^2 - \tfrac{1}{2}(m_A + m_B)v^2$$

$$= \tfrac{1}{2}[(2)(5)^2 + (3)(4)^2 - (2 + 3)(4.4)^2] \text{ J} = \boxed{0.6 \text{ J}}$$

8-4. Center of Mass (CM)

The **center of mass** (CM) has two important properties: (1) The total mass of the system, $M = m_A + m_B + m_C + \cdots$, is considered to be located at the point CM, which has coordinates (x_{CM}, y_{CM}, z_{CM}); and (2) the CM is considered to have the momentum of the system, or

$$\mathbf{p}_{CM} = \sum_i \mathbf{p}_i \qquad \text{and} \qquad M\mathbf{v}_{CM} = m_A \mathbf{v}_A + m_B \mathbf{v}_B + m_C \mathbf{v}_C + \cdots$$

where $\mathbf{v}_{CM}$ is the velocity of the CM and

$$\mathbf{v}_{CM} = \frac{m_A \mathbf{v}_A + m_B \mathbf{v}_B + m_C \mathbf{v}_C + \cdots}{m_A + m_B + m_C + \cdots} \qquad (8\text{-}8)$$

The coordinates for the CM are

$$x_{CM} = \frac{m_A x_A + m_B x_B + m_C x_C + \cdots}{m_A + m_B + m_C + \cdots} \qquad (8\text{-}9)$$

where x_A is the x coordinate for m_A, x_B is the x coordinate for m_B, etc. The y and z coordinates are obtained from similar expressions. The position vector for the CM is

$$\mathbf{R}_{CM} = x_{CM}\hat{\mathbf{x}} + y_{CM}\hat{\mathbf{y}} + z_{CM}\hat{\mathbf{z}}$$

EXAMPLE 8-9: Two pucks ($m_A = 2$ kg and $m_B = 3$ kg) slide on a frictionless horizontal surface. At a specific instant their positions and velocities are

$$\mathbf{R}_A = (3\hat{\mathbf{x}} + 4\hat{\mathbf{y}}) \text{ m} \qquad \text{and} \qquad \mathbf{v}_A = (2\hat{\mathbf{x}} - 5\hat{\mathbf{y}}) \text{ m s}^{-1}$$

$$\mathbf{R}_B = (2\hat{\mathbf{x}} - 3\hat{\mathbf{y}}) \text{ m} \qquad \text{and} \qquad \mathbf{v}_B = (-4\hat{\mathbf{x}} + 5\hat{\mathbf{y}}) \text{ m s}^{-1}$$

(a) Calculate the x and y coordinates of the CM. (b) Calculate the x and y components of $\mathbf{v}_{\text{CM}}$.

Solution:

(a) Use Eq. (8-9).

$$x_{\text{CM}} = \frac{m_A x_A + m_B x_B}{m_A + m_B} = \frac{(2 \text{ kg})(3 \text{ m}) + (3 \text{ kg})(2 \text{ m})}{(2 + 3) \text{ kg}} = \boxed{2.4 \text{ m}}$$

$$y_{\text{CM}} = \frac{m_A y_A + m_B y_B}{m_A + m_B} = \frac{(2 \text{ kg})(4 \text{ m}) + (3 \text{ kg})(-3 \text{ m})}{(2 + 3) \text{ kg}} = \boxed{-0.2 \text{ m}}$$

(b) Use Eq. (8-8) separately for each component of $\mathbf{v}_{\text{CM}}$.

$$v_{\text{CM}_x} = \frac{m_A v_{A_x} + m_B v_{B_x}}{m_A + m_B} = \frac{(2 \text{ kg})(2 \text{ m s}^{-1}) + (3 \text{ kg})(-4 \text{ m s}^{-1})}{(2 + 3) \text{ kg}} = \boxed{-1.6 \text{ m s}^{-1}}$$

$$v_{\text{CM}_y} = \frac{m_A v_{A_y} + m_B v_{B_y}}{m_A + m_B} = \frac{(2 \text{ kg})(-5 \text{ m s}^{-1}) + (3 \text{ kg})(5 \text{ m s}^{-1})}{(2 + 3) \text{ kg}} = \boxed{1.0 \text{ m s}^{-1}}$$

8-5. Collision of Two Bodies in CM Frame of Reference

The origin of the *CM coordinate system* is located at $\mathbf{R}_{\text{CM}}$ and has a velocity v_{CM} in the laboratory (LAB) frame of reference. In the CM frame of reference the momentum of the system is zero, and, as a consequence, the analysis of the collision of two bodies is simpler. We will use primed quantities to refer to CM coordinates and unprimed quantities to refer to LAB coordinates. Also, we continue to use 1 and 2 to refer to before and after a collision, respectively.

Recall from Section 3-6 the concept of velocity transformations from one coordinate system to another (Eq. 3-19). We can extend this concept to momentum and kinetic energy. From Eq. (3-19), for any body its velocity $v = v' + v_{\text{CM}}$; likewise, $E_k = E'_k + E_{k,\text{CM}}$, so we can express the kinetic energy of a body in the CM frame of reference before a collision as

$$E'_{k1} = E_{k1} - E_{k,\text{CM}}$$

$$\tfrac{1}{2}m_A(v'_{A1})^2 + \tfrac{1}{2}m_B(v'_{B1})^2 = \tfrac{1}{2}m_A v^2_{A1} + \tfrac{1}{2}m_B v^2_{B1} - \tfrac{1}{2}(m_A + m_B)v^2_{\text{CM}}$$

This means that the energy associated with the motion of the CM cannot contribute to a change in the velocity of two colliding bodies; it is unavailable during an interaction.

EXAMPLE 8-10: Two pucks are confined to a straight frictionless track; $m_A = 2$ kg and $m_B = 3$ kg; m_A is moving at 4 m s^{-1} and strikes m_B, which is stationary. After the collision, m_A is moving at 0.32 m s^{-1} in a direction opposite to its initial direction. Calculate the following: (a) v_{CM}; (b) the velocities of the pucks in the CM frame of reference before the collision; (c) the velocities of the pucks in the CM frame after the collision; and (d) the kinetic energy available for interaction of the pucks before the collision.

Solution:

(a) Use Eq. (8-8).

$$v_{\text{CM}} = \frac{m_A v_{A1} + m_B v_{B1}}{m_A + m_B} = \frac{(2 \text{ kg})(4 \text{ m s}^{-1}) + 0}{(2 + 3) \text{ kg}} = \boxed{1.6 \text{ m s}^{-1}}$$

(b) Use a velocity transformation.

$$v'_{A1} = v_{A1} - v_{CM} = (4 - 1.6)\ \mathrm{m\,s}^{-1} = \boxed{2.4\ \mathrm{m\,s}^{-1}}$$

$$v'_{B1} = v_{B1} - v_{CM} = 0 - 1.6\ \mathrm{m\,s}^{-1} = \boxed{-1.6\ \mathrm{m\,s}^{-1}}$$

You should verify that the momentum of the system is zero in these coordinates.

(c) Use the principle of conservation of linear momentum in CM coordinates and solve for v'_{B2}. You obtain v'_{A2} from a velocity transformation.

$$v'_{A2} = v_{A2} - v_{CM} = (0.32 - 1.6)\ \mathrm{m\,s}^{-1} = -1.28\ \mathrm{m\,s}^{-1}$$

$$m_A v'_{A1} + m_B v'_{B1} = m_A v'_{A2} + m_B v'_{B2} = 0$$

$$v'_{B2} = -\frac{m_A v'_{A2}}{m_B} = -\frac{(2\ \mathrm{kg})(-1.28\ \mathrm{m\,s}^{-1})}{3\ \mathrm{kg}}$$

$$= \boxed{+0.853\ \mathrm{m\,s}^{-1}}$$

(d)
$$E'_{k1} = E_{k1} - E_{k,CM} = \tfrac{1}{2}[m_A v^2_{A1} + m_B v^2_{B1} - (m_A + m_B)v^2_{CM}]$$
$$= \tfrac{1}{2}[(2)(4)^2 + 0 - (2+3)(1.6)^2]\ \mathrm{J} = \boxed{9.6\ \mathrm{J}}$$

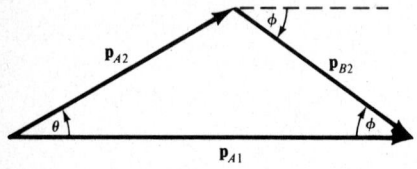

Figure 8-3

We now reconsider the collision of two bodies illustrated in Figure 8-2, which shows the relationships of their masses and velocities before and after in the LAB coordinate system. We begin by using the principle of the conservation of linear momentum, Eq. (8-5b).

$$\sum_i \mathbf{p}_i = \sum_i \mathbf{p}_i \qquad \mathbf{p}_{A1} = \mathbf{p}_{A2} + \mathbf{p}_{B2} \qquad \text{(LAB)}$$
initial state   final state

This vector triangle for the conservation of momentum is shown in Figure 8-3. We use the law of cosines to obtain the magnitude of $\mathbf{p}_{B2}$

$$(p_{B2})^2 = (p_{A1})^2 + (p_{A2})^2 - 2(p_{A1})(p_{A2})\cos\theta$$
$$(m_B v_{B2})^2 = (m_A v_{A1})^2 + (m_A v_{A2})^2 - 2(m_A v_{A1})(m_A v_{A2})\cos\theta$$

and solve for v_{B2}:

$$v_{B2} = \left(\frac{m_A}{m_B}\right)\sqrt{(v_{A1})^2 + (v_{A2})^2 - 2(v_{A1})(v_{A2})\cos\theta} \qquad \text{(8-10)}$$

To obtain ϕ, we use the law of sines.

$$\frac{\sin\phi}{p_{A2}} = \frac{\sin\theta}{p_{B2}}$$

$$\phi = \arcsin\left(\frac{p_{A2}}{p_{B2}}\sin\theta\right) = \arcsin\left(\frac{m_A v_{A2}}{m_B v_{B2}}\sin\theta\right) \qquad \text{(8-11)}$$

Next, we transform the velocity to the CM coordinate system. (See Section 3-6.) Figure 8-4 shows the velocity vector triangle for the transformation from the LAB to the CM coordinate system.

We calculate v_{CM}, using Eq. (8-8),

$$v_{CM} = \frac{m_A v_{A1} + 0}{m_A + m_B}$$

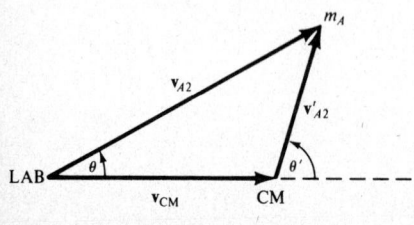

Figure 8-4

and v'_{A2}, using the law of cosines.

$$v'_{A2} = \sqrt{(v_{A2})^2 + (v_{CM})^2 - 2(v_{A2})(v_{CM})\cos\theta} \qquad \text{(8-12)}$$

We obtain θ' by applying the law of sines.

$$\frac{\sin(180° - \theta')}{v_{A2}} = \frac{\sin \theta'}{v_{A2}} = \frac{\sin \theta}{v'_{A2}}$$

$$\theta' = \arcsin\left(\frac{v_{A2}}{v'_{A2}} \sin \theta\right) \qquad \textbf{(8-13)}$$

These last two results give us the information necessary to construct the momentum vector diagram in the CM coordinate system for the collision of two bodies, as shown in Figure 8-5. Recall that in the CM coordinate system, $p'_{A1} = p'_{B1}$ and $p'_{A2} = p'_{B2}$; i.e., the magnitudes are equal. Therefore

$$v'_{B2} = \left(\frac{m_A}{m_B}\right)v'_{A2} \qquad \textbf{(8-14)}$$

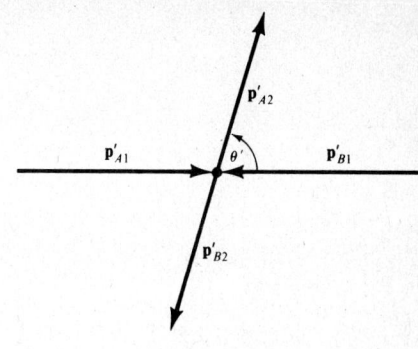

Figure 8-5

EXAMPLE 8-11: Use the data in Example 8-4, but in this case: (a) use Eqs. (8-10) and (8-11) to verify that you get the same results as in Example 8-4; (b) use Eqs. (8-12) and (8-13) to obtain the velocity of m_A in the CM frame after the collision; and (c) determine the velocity of m_B in the CM frame after the collision.

Solution:

(a) Use Eq. (8-10) to calculate v_{B2}.

$$v_{B2} = \left(\frac{m_A}{m_B}\right)\sqrt{(v_{A1})^2 + (v_{A2})^2 - 2(v_{A1})(v_{A2})\cos \theta}$$

$$= \left(\frac{2 \text{ kg}}{3 \text{ kg}}\right)\sqrt{(5)^2 + (2.4)^2 - 2(5)(2.4)\cos 35°} \text{ m s}^{-1}$$

$$= \boxed{2.22 \text{ m s}^{-1}}$$

Use Eq. (8-11) to find ϕ.

$$\phi = \arcsin\left(\frac{m_A v_{A2}}{m_B v_{B2}} \sin \theta\right)$$

$$= \arcsin\left(\frac{(2 \text{ kg})(2.4 \text{ m s}^{-1})}{(3 \text{ kg})(2.22 \text{ m s}^{-1})} \sin 35°\right) = \boxed{24.4°}$$

(b) You must first determine the velocity of the CM, for which you use Eq. (8-8).

$$v_{CM} = \frac{m_A v_{A1} + 0}{m_A + m_B} = \frac{(2 \text{ kg})(5 \text{ m s}^{-1})}{(2 + 3) \text{ kg}} = 2 \text{ m s}^{-1}$$

Then use Eq. (8-12) to calculate v'_{A2},

$$v'_{A2} = \sqrt{(v_{A2})^2 + (v_{CM})^2 - 2(v_{A2})(v_{CM})\cos \theta}$$

$$= \sqrt{(2.4)^2 + (2)^2 - 2(2.4)(2)\cos 35°} \text{ m s}^{-1} = \boxed{1.38 \text{ m s}^{-1}}$$

and Eq. (8-13) to obtain θ':

$$\theta' = \arcsin\left(\frac{v_{A2}}{v'_{A2}} \sin \theta\right) = \arcsin\left(\frac{2.4 \text{ m s}^{-1}}{1.38 \text{ m s}^{-1}} \sin 35°\right) = \boxed{88.6°}$$

(c) Eq. (8-14) gives you v'_{B2}.

$$v'_{B2} = \left(\frac{m_A}{m_B}\right)v'_{A2} = \frac{2 \text{ kg}}{3 \text{ kg}}(1.38 \text{ m s}^{-1}) = \boxed{0.92 \text{ m s}^{-1}}$$

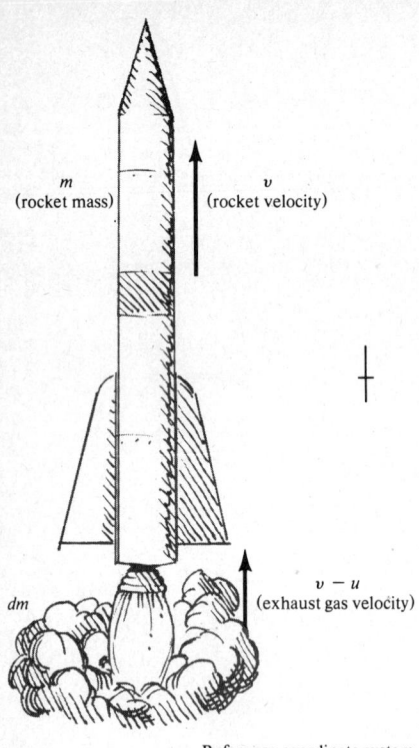

m
(rocket mass)

v
(rocket velocity)

$v - u$
(exhaust gas velocity)

dm

Reference coordinate system

Figure 8-6

8-6. Rocket Propulsion

A rocket of mass m moving with velocity v ejects a quantity of exhaust gas dm in a time interval dt at velocity u relative to the rocket. See Figure 8-6. The **rocket equation** is derived from the impulse delivered to the rocket by the exhaust gases.

ROCKET EQUATION
$$m\frac{dv}{dt} = -u\frac{dm}{dt} + F \qquad (8\text{-}15)$$

where F represents all *external* forces acting on the rocket. Near the earth's surface, $F = -mg$, the weight of the rocket when air friction is neglected. The quantity $-u\dfrac{dm}{dt}$ is called the **thrust**. Because the mass of the rocket is decreasing during a burn, $\dfrac{dm}{dt}$ is a negative quantity. If the initial mass of the rocket plus fuel is M and it burns and ejects mass at a constant rate r, its mass at time t later is $m = M - rt$. In free space, where F can be considered zero, integrating Eq. (8-15) results in

$$v_2 - v_1 = u\ln\left(\frac{m_1}{m_2}\right) \qquad (8\text{-}16)$$

which yields the increase in the velocity of the rocket after an amount of fuel $\Delta m = m_1 - m_2$ is burned and exhausted.

EXAMPLE 8-12: A rocket moving in free space at $1000\ \mathrm{m\,s^{-1}}$ has a mass, rocket plus fuel, of 1.2×10^4 kg. The engine is started and fuel is burned and exhausted at a rate of $50\ \mathrm{kg\ s^{-1}}$ for 40 s at a velocity of $1600\ \mathrm{m\,s^{-1}}$. **(a)** What is the velocity of the rocket after the burn? **(b)** Calculate the thrust.

Solution:

(a) Solve Eq. (8-16) for v_2, with

$$m_2 = m_1 - \Delta m = m_1 - rt$$
$$= 1.2 \times 10^4\ \mathrm{kg} - (50\ \mathrm{kg\,s^{-1}})(40\ \mathrm{s}) = 10^4\ \mathrm{kg}$$

Thus

$$v_2 = v_1 + u\ln\left(\frac{m_1}{m_2}\right)$$

$$= 10^3\ \mathrm{m\,s^{-1}} + (1600\ \mathrm{m\,s^{-1}})\ln\left(\frac{1.2 \times 10^4\ \mathrm{kg}}{1.0 \times 10^4\ \mathrm{kg}}\right) = \boxed{1.29 \times 10^3\ \mathrm{m\,s^{-1}}}$$

(b) The thrust is

$$F_{\mathrm{th}} = -u\frac{dm}{dt} = -(1600\ \mathrm{m\,s^{-1}})(-50\ \mathrm{kg\,s^{-1}}) = \boxed{8 \times 10^4\ \mathrm{N}}$$

SUMMARY

1. *Linear momentum* is the product of the mass of an object times its velocity: $\mathbf{p} = m\mathbf{v}$

2. The *impulse* of a force produces a change in the momentum of a body, as expressed by

$$\int_{t_1}^{t_2} \mathbf{F}\,dt = \mathbf{p}_2 - \mathbf{p}_1$$

3. The *average impulsive force* acting on a body is

$$F_{av} = \frac{\int_{t_1}^{t_2} F \, dt}{\Delta t}$$

where Δt is the duration of the impulse.

4. The *principle of the conservation of linear momentum* states that when no EXTERNAL forces act on a system of particles, the momentum of the system doesn't change.

5. The coefficient of restitution e, with a range of $0 \leq e \leq 1$, relates the relative velocities of two bodies before and after a collision. This relationship is *Newton's rule*, or

$$v_{B2} - v_{A2} = e(v_{A1} - v_{B1})$$

6. For a completely *elastic* collision, $e = 1$, and for a completely *inelastic* collision, $e = 0$.

7. The center of mass, CM, of a system of particles is a point at which the mass of the entire system appears to be located.

8. The origin of the CM coordinate system is located at the CM. The momentum of the system of particles is zero in this frame of reference.

9. The kinetic energy associated with the motion of the CM, $\frac{1}{2}Mv^2$, is not available for the interactions among the particles of a system.

10. The best way to solve a difficult collision problem is to transform the problem to CM coordinates, solve the problem, and then transform the solution back to LAB coordinates.

11. The rocket equation can be used to calculate the change in the velocity of the rocket when the velocity of the exhaust gases and the quantity of fuel burned are known.

RAISE YOUR GRADES

Can you explain . . . ?

☑ how to determine the value of the average impulsive force if you know the change in the velocity of the object and the duration of the impulse

☑ the relationship between Newton's third law and the principle of conservation of linear momentum

☑ how a system of two particles can have a total momentum of zero when both particles are moving

☑ the condition regarding external forces that must be met before the principle of conservation of linear momentum can be applied to a system of particles

☑ the relationship between the coefficient of restitution and the relative velocities between two bodies before and after a collision

☑ the relationship between the coefficient of restitution and the total kinetic energy before and after a collision between two bodies

☑ how to locate the CM for a system of particles

☑ how to transform a two-body collision problem from the LAB to the CM coordinate system

☑ the relationship between the kinetic energy of a two-body system as determined by LAB coordinates before a collision, the kinetic energy available for their interaction, and the kinetic energy associated with the CM

☑ how a rocket can accelerate in free space when there is nothing available for it to "push against"

SOLVED PROBLEMS

Impulsive Forces and Momentum

PROBLEM 8-1:   A 200-g hockey puck moving at 3 m s^{-1} is given a smack, increasing its speed to 24 m s^{-1} in the same direction. The striking stick was in contact with the puck for 0.08 s. **(a)** Determine the impulse delivered to the puck. **(b)** What was the average force on the puck during contact?

Solution:

(a) Use Eq. (8-3) and solve for $\int_{t_1}^{t_2} F \, dt$.

$$\int_{t_1}^{t_2} F \, dt = mv_2 - mv_1 = m(v_2 - v_1) = (0.2 \text{ kg})(24 - 3) \text{ m s}^{-1} = \boxed{4.2 \text{ N s}}$$

(b) Solve Eq. (8-4) for F_{av}.

$$F_{\text{av}} \, \Delta t = \int_{t_1}^{t_2} F \, dt$$

$$F_{\text{av}} = \frac{\int_{t_1}^{t_2} F \, dt}{\Delta t} = \frac{4.2 \text{ N s}}{0.08 \text{ s}} = \boxed{52.5 \text{ N}}$$

Conservation of Linear Momentum

PROBLEM 8-2:   A 20-g bullet moving horizontally at 250 m s^{-1} strikes a 60-g tin can resting on a post and passes through it. After being hit, the can has a horizontal velocity of 2 m s^{-1}. What is the velocity of the bullet?

Solution: Use the principle of the conservation of linear momentum, Eq. (8-5a), with A and B representing the bullet and can, respectively. Solve for v_{A2}.

$$\mathbf{p}_{A1} + \mathbf{p}_{B1} = \mathbf{p}_{A2} + \mathbf{p}_{B2}$$

$$m_A v_{A1} + 0 = m_A v_{A2} + m_B v_{B2}$$

$$v_{A2} = v_{A1} - \left(\frac{m_B}{m_A}\right) v_{B2} = 250 \text{ m s}^{-1} - \left(\frac{0.06 \text{ kg}}{0.02 \text{ kg}}\right)(2 \text{ m s}^{-1}) = \boxed{244 \text{ m s}^{-1}}$$

Collisions: Elastic and Inelastic

The next three problems deal with two pucks confined to a straight frictionless track.

PROBLEM 8-3:   Puck A ($m_A = 350$ g and $v_{A1} = 2 \text{ m s}^{-1}$) strikes puck B ($m_B = 200$ g and $v_{B1} = 1.5 \text{ m s}^{-1}$). After the collision, puck A has a velocity $v_{A2} = 1.71 \text{ m s}^{-1}$. **(a)** What is the velocity of puck B? **(b)** Calculate the coefficient of restitution. **(c)** How much kinetic energy is "lost" in the collision?

Solution:

(a) Use the principle of the conservation of linear momentum, Eq. (8-5a), and solve for v_{B2}.

$$\mathbf{p}_{A1} + \mathbf{p}_{B1} = \mathbf{p}_{A2} + \mathbf{p}_{B2}$$

$$m_A v_{A1} + m_B v_{B1} = m_A v_{A2} + m_B v_{B2}$$

$$v_{B2} = \frac{m_A}{m_B}(v_{A1} - v_{A2}) + v_{B1} = \left(\frac{0.35 \text{ kg}}{0.2 \text{ kg}}\right)(2 - 1.71) \text{ m s}^{-1} + 1.5 \text{ m s}^{-1} = \boxed{2.01 \text{ m s}^{-1}}$$

(b) Use Eq. (8-7) and solve for e.

$$v_{B2} - v_{A2} = e(v_{A1} - v_{B1})$$

$$e = \frac{v_{B2} - v_{A2}}{v_{A1} - v_{B1}} = \frac{(2.01 - 1.71) \text{ m s}^{-1}}{(2 - 1.5) \text{ m s}^{-1}} = \boxed{0.62}$$

The collision is inelastic.

(c) The kinetic energy lost is

$$\Delta E_{k,\text{lost}} = E_{k1} - E_{k2} = (\tfrac{1}{2}m_A v_{A1}^2 + \tfrac{1}{2}m_B v_{B1}^2) - (\tfrac{1}{2}m_A v_{A2}^2 + \tfrac{1}{2}m_B v_{B2}^2)$$
$$= \tfrac{1}{2}[m_A(v_{A1}^2 - v_{A2}^2) + m_B(v_{B1}^2 - v_{B2}^2)]$$
$$= \tfrac{1}{2}\{(0.35)[(2)^2 - (1.71)^2] + (0.2)[(1.5)^2 - (2.01)^2]\} \text{ J} = \boxed{1.02 \times 10^{-2} \text{ J}}$$

PROBLEM 8-4: Puck A ($m_A = 200$ g and $v_{A1} = 3 \text{ m s}^{-1}$) strikes puck B ($m_B = 350$ g and $v_{B1} = 2 \text{ m s}^{-1}$) in a completely elastic collision. What are the velocities of the pucks after the collision?

Solution: Use the result in Example 8-5, with $e = 1.0$.

$$v_{A2} = \frac{m_A v_{A1} + m_B v_{B1} - e m_B(v_{A1} - v_{B1})}{m_A + m_B} = \frac{(m_A - m_B)v_{A1} + 2m_B v_{B1}}{m_A + m_B}$$

$$= \frac{(0.2 - 0.35) \text{ kg } (3 \text{ m s}^{-1}) + 2(0.35)(2 \text{ m s}^{-1})}{(0.2 + 0.35) \text{ kg}} = \boxed{1.73 \text{ m s}^{-1}}$$

$$v_{B2} = \frac{m_A v_{A1} + m_B v_{B1} + e m_A(v_{A1} - v_{B1})}{m_A + m_B} = \frac{2m_A v_{A1} + (m_B - m_A)v_{B1}}{m_A + m_B}$$

$$= \frac{2(0.2 \text{ kg})(3 \text{ m s}^{-1}) + (0.35 - 0.2) \text{ kg } (2 \text{ m s}^{-1})}{(0.2 + 0.35) \text{ kg}} = \boxed{2.73 \text{ m s}^{-1}}$$

PROBLEM 8-5: Puck A ($m_A = 200$ g) moves to the right at 3 m s^{-1}, and puck B ($m_B = 350$ g) moves to the left at 2 m s^{-1}. The collision leaves them stuck together. **(a)** What is the common velocity of the pucks? **(b)** How much kinetic energy is "lost" in the collision?

Solution: You should recognize this as a completely inelastic collision, $e = 0$.

(a) To calculate the velocity after the collision, use the result in Example 8-5, with $e = 0$, or Eq. (8-5a), where v_2 is the common velocity after the collision.

$$m_A v_{A1} = m_B v_{B1} = m_A v_{A2} + m_B v_{B2} = (m_A + m_B)v_2$$

$$v_2 = \frac{m_A v_{A1} + m_B v_{B1}}{m_A + m_B} = \frac{(0.2 \text{ kg})(3 \text{ m s}^{-1}) + (0.35 \text{ kg})(-2 \text{ m s}^{-1})}{(0.2 + 0.35) \text{ kg}}$$

$$= \boxed{-0.182 \text{ m s}^{-1}}$$

The negative sign indicates that the pair is moving to the left.

(b) The kinetic energy "lost" is

$$E_{k,\text{lost}} = E_{k1} - E_{k2} = \tfrac{1}{2}m_A v_{A1}^2 + \tfrac{1}{2}m_B v_{B1}^2 - \tfrac{1}{2}(m_A + m_B)v_2^2$$
$$= \tfrac{1}{2}[(0.2)(3)^2 + (0.35)(-2)^2 - (0.2 + 0.35)(-0.182)^2] \text{ J} = \boxed{1.59 \text{ J}}$$

PROBLEM 8-6: Puck A ($m_A = 0.2$ kg) slides on a frictionless horizontal surface at 3 m s^{-1} and strikes puck B ($m_B = 0.35$ kg), which is stationary. After the collision, puck A is moving at 2 m s^{-1} in a direction that forms an angle of $60°$ to the original direction of movement. **(a)** Determine the velocity (magnitude and direction) of puck B after the collision (see Figure 8-2). **(b)** Is kinetic energy conserved in this collision?

Solution:

(a) Use the procedure outlined in Example 8-4.

$$\phi = \arctan\left(\frac{v_{A2}\sin\theta}{v_{A1} - v_{A2}\cos\theta}\right) = \arctan\left(\frac{(2\text{ m s}^{-1})\sin 60°}{3\text{ m s}^{-1} - (2\text{ m s}^{-1})\cos 60°}\right) = \boxed{40.9°}$$

$$v_{B2} = \frac{m_A v_{A2}\sin\theta}{m_B\sin\phi} = \frac{(0.2\text{ kg})(2\text{ m s}^{-1})\sin 60°}{(0.35\text{ kg})\sin 40.9°} = \boxed{1.51\text{ m s}^{-1}}$$

(b) In order to determine whether kinetic energy is conserved, you must calculate $E_{k,\text{lost}} = E_{k1} - E_{k2}$. Use the result in Problem 8-3c, with $v_{B1} = 0$.

$$E_{k,\text{lost}} = \tfrac{1}{2}[m_A(v_{A1}^2 - v_{A2}^2) - m_B v_{B2}^2] = \tfrac{1}{2}[(0.2)(3^2 - 2^2) - (0.35)(1.51)^2]\text{ J} = \boxed{0.1\text{ J}}$$

Because some kinetic energy is "lost," the collision is inelastic, and kinetic energy is not conserved.

PROBLEM 8-7:   Car A ($m_A = 1200$ kg) enters an intersection moving north at 6 m s^{-1}. Car B ($m_B = 1500$ kg) enters the same intersection going east at 5 m s^{-1}. They collide, and the entangled wreckage W careens off as a unit, as shown in Figure 8-7a. (Fortunately, no one was seriously injured.) Determine the velocity (magnitude and direction) of the wreckage immediately after the collision.

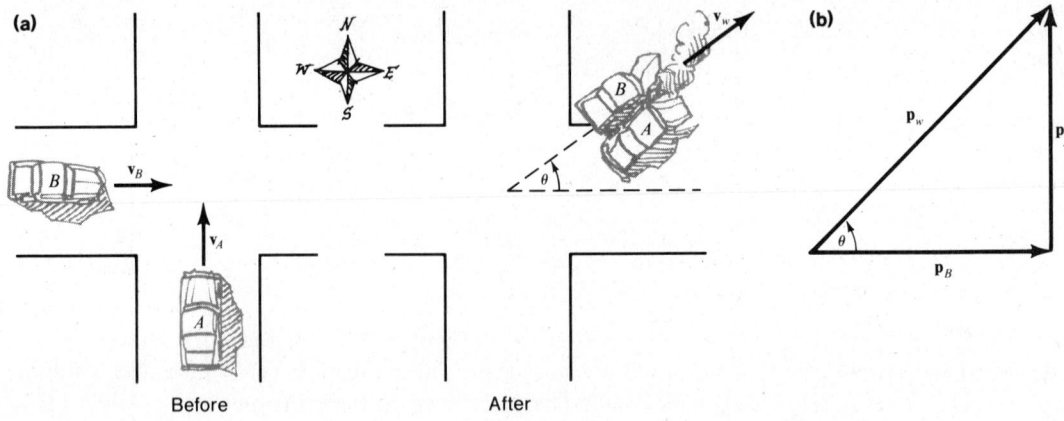

Figure 8-7

Solution: This two-dimensional collision is completely inelastic. Linear momentum is conserved, however, which implies that the momentum vector triangle is closed, as in Figure 8-7b.

$$\mathbf{p}_A + \mathbf{p}_B = \mathbf{p}_W \qquad \text{and} \qquad \tan\theta = \frac{p_A}{p_B}$$

First, solve for θ.

$$\theta = \arctan\left(\frac{p_A}{p_B}\right) = \arctan\left(\frac{m_A v_A}{m_B v_B}\right) = \arctan\left(\frac{(1200\text{ kg})(6\text{ m s}^{-1})}{(1500\text{ kg})(5\text{ m s}^{-1})}\right) = \boxed{43.8°}$$

Now calculate the magnitude of the momentum of the wreckage, using either the Pythagorean theorem or a trig function.

$$p_W = \frac{p_A}{\sin\theta}$$

$$(m_A + m_B)v_W = \frac{m_A v_A}{\sin\theta}$$

Finally, solve for v_W.

$$v_W = \frac{m_A v_A}{(m_A + m_B)\sin\theta} = \frac{(1200\text{ kg})(6\text{ m s}^{-1})}{(1200 + 1500)\text{ kg }(\sin 43.8°)} = \boxed{3.85\text{ m s}^{-1}}$$

Center of Mass: Collisions in CM

PROBLEM 8-8: Note that v_W is the velocity of the CM in Problem 8-7. Obtain the velocity before the collision of car A in the CM coordinate system. (See Section 3-6 and Example 8-10.)

Solution: The transformation of velocity to the CM system (primed quantity), as shown in Figure 8-8, is

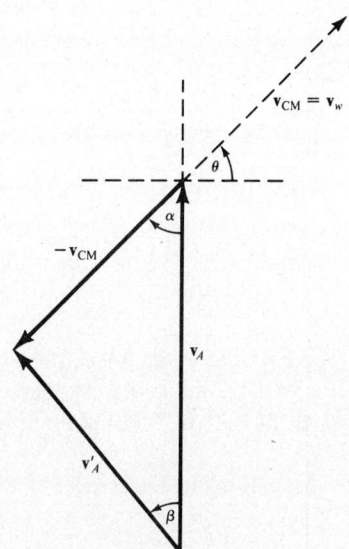

$$\mathbf{v}_A = \mathbf{v}'_A + \mathbf{v}_{CM}$$

$$\mathbf{v}'_A = \mathbf{v}_A - \mathbf{v}_{CM}$$

$$= \mathbf{v}_A - \mathbf{v}_W$$

Use the law of cosines to determine v'_A, where

$$\alpha = 90° - \theta.$$

$$v_A = \sqrt{v_A^2 + v_{CM}^2 - 2v_A v_{CM}\cos\alpha}$$

$$= \sqrt{6^2 + (3.85)^2 - 2(6)(3.85)\cos 46.2°} \text{ m s}^{-1}$$

$$= \boxed{4.34 \text{ m s}^{-1}}$$

You can calculate the direction of $\mathbf{v}'_A$ relative to the original direction of $\mathbf{v}_A$ by using the law of sines. Solve for β.

$$\frac{\sin\beta}{v_{CM}} = \frac{\sin\alpha}{v'_A}$$

$$\beta = \arcsin\left(\frac{v_{CM}}{v'_A}\sin\alpha\right)$$

$$= \arcsin\left(\frac{3.85 \text{ m s}^{-1}}{4.34 \text{ m s}^{-1}}\sin 46.2°\right) = \boxed{39.8°}$$

Figure 8-8

Rocket Propulsion

PROBLEM 8-9: Start with the rocket equation, Eq. (8-15), and obtain an expression for the velocity of the rocket after all its fuel is burned. The rocket's initial velocity is zero (at $t = 0$), and the burn lasts for a short time (t_b), so you can assume that g is constant. The only external force acting on the rocket is its weight, or $F = -mg$. The initial mass of the rocket plus fuel is M, the mass of the fuel is M_f, and the velocity of the exhaust gases relative to the rocket is u.

Solution:

$$m\frac{dv}{dt} = -u\frac{dm}{dt} + F = -u\frac{dm}{dt} - mg$$

Multiply by dt, divide by m, and integrate.

$$\int_0^{v_b} dv = u\int_M^{M-M_f} \frac{dm}{m} - g\int_0^{t_b} dt$$

$$\boxed{v_b = u\ln\left(\frac{M}{M - M_f}\right) - gt_b}$$

PROBLEM 8-10: Eighty percent of the total mass of a rocket resting on a launch pad is fuel. During launch all its fuel is burned in 25 s and is exhausted at 1500 m s^{-1}. Calculate the velocity attained by this rocket at burnout.

Solution: Use the result of Problem 8-9, with $M_f = 0.8M$. /

$$v_b = u \ln\left(\frac{M}{M - M_f}\right) - gt_b = u \ln\left(\frac{M}{M - 0.8M}\right) - gt_b$$

$$= (1500 \text{ m s}^{-1})\ln 5 - (9.8 \text{ m s}^{-2})(25 \text{ s}) = \boxed{2.17 \times 10^3 \text{ m s}^{-1}}$$

Supplementary Exercises

EXERCISE 8-1:   A 200-g baseball is pitched at 15 m s^{-1}. The average impulsive force delivered by the bat is 250 N, which gives the ball a velocity of 20 m s^{-1} in a direction opposite its initial direction. What is the duration of the impulse?

EXERCISE 8-2:   A 20-g bullet moving horizontally at 300 m s^{-1} strikes and becomes imbedded in a 1.4-kg block of wood at rest on a horizontal frictionless surface. (a) With what velocity does the pair move? (b) What percent of the initial kinetic energy is "lost" in the collision process?

The actions in Exercises 8-3, 8-4, 8-5, and 8-6 take place on a straight, horizontal, and frictionless track.

EXERCISE 8-3:   Puck A ($m_A = 200$ g) moves to the right at 3 m s^{-1}, and puck B ($m_B = 150$ g) moves to the left at 6 m s^{-1}. After the collision, puck A is moving to the left at 3.943 m s^{-1}. (a) What is the velocity and direction of puck B? (b) Find the value of the coefficient of restitution. (c) How much kinetic energy is "lost" in the collision?

EXERCISE 8-4:   Puck A ($m_A = 200$ g) moves to the right at 3 m s^{-1}, and puck B ($m_B = 150$ g) moves to the left at 6 m s^{-1}. The collision is completely elastic. Determine the velocities and directions of the pucks after the collision.

EXERCISE 8-5:   Use the data in Exercise 8-4, but this time assume the collision is completely inelastic. Determine (a) the common velocity and direction of the pucks, and (b) the kinetic energy "lost" in the collision.

EXERCISE 8-6:   A compressed spring of negligible mass separates pucks A and B. This combination is held by a short length of string attached between the pucks. The masses of the pucks are $m_A = 200$ g and $m_B = 150$ g. The combination is moving to the right at 3 m s^{-1} when the string breaks. Puck B is now moving to the right at 6 m s^{-1}. (a) What is the velocity and direction of puck A? (b) Calculate the kinetic energy "gained" in this interaction.

EXERCISE 8-7:   Puck A ($m_A = 50$ g), moving at 4 m s^{-1} on a horizontal frictionless surface, strikes stationary puck B ($m_B = 200$ g). After the collision, puck A is moving at 2 m s^{-1} in a direction 30° from its initial direction. Determine (a) the magnitude of the velocity of puck B; (b) its direction relative to the initial direction of puck A; and (c) the kinetic energy "lost" in the collision.

EXERCISE 8-8:   Car A ($m_A = 1600$ kg) is southbound and enters an intersection at 4 m s^{-1}. Car B ($m_B = 1400$ kg) is eastbound at 6 m s^{-1} and enters the same intersection. The cars collide, become attached, and the wreckage skids away as a unit. What is the velocity (magnitude and direction) of the wreckage as it begins to move?

EXERCISE 8-9:   For the conditions in Exercise 8-7, calculate (a) the velocity and direction of the CM; (b) the speed of puck A in the CM coordinate system after the collision; (c) the direction of the velocity of puck A in the CM system after the collision; and (d) the speed of puck B in the CM system. Assume that the velocity of puck A before the collision is along the $+x$ axis.

EXERCISE 8-10: It is necessary to increase the velocity of a rocket that is moving in a gravity-free region from 900 m s^{-1} to 1200 m s^{-1}. The burned rocket gases leave the engine at 1500 m s^{-1} relative to the rocket. The mass of the rocket plus fuel before the burn is 1.2×10^4 kg. Determine the mass of fuel that must be burned to get the increase in velocity.

Answers to Supplementary Exercises

8-1: 2.8×10^{-2} s

8-2: (a) 3.95 m s^{-1}; (b) 98.7%

8-3: (a) $v_{B2} = 3.257$ m s^{-1}, to the right; (b) 0.8; (c) 1.25 J

8-4: $v_{A2} = 4.71$ m s^{-1}, to the left; $v_{B2} = 4.29$ m s^{-1}, to the right

8-5: (a) 8.57 m s^{-1}, to the left; (b) 3.47 J

8-6: (a) 0.75 m s^{-1}, to the right; (b) 1.01 J

8-7 (a) 0.62 m s^{-1}; (b) 23.8°; (c) 0.26 J

8-8: 3.52 m s^{-1}, 37.3° south of east

8-9: (a) $v_{CM} = 0.8$ m s^{-1}, in the $+x$ direction; (b) $v'_{A2} = 3.33$ m s^{-1}; (c) $\theta' = 0.3°$ (relative, to the x axis); (d) $v'_{B2} = 0.83$ m s^{-1}, opposite the direction of v'_{A2}

8-10: 2.18×10^3 kg

EXAM 1 (Chapters 1 to 8)

1. Two forces, $\mathbf{F}_1 = (3\hat{\mathbf{x}} + 2\hat{\mathbf{y}})$ N and $\mathbf{F}_2 = (5\hat{\mathbf{x}} - 4\hat{\mathbf{y}})$ N, simultaneously act on a body.
   (a) Determine the resultant $\mathbf{R}$ of these forces. [Ch. 1]
   (b) What is the magnitude of $\mathbf{R}$? [Ch. 1]
   (c) Find the angle between $\mathbf{R}$ and the positive x axis. [Ch. 1]

2. (a) Evaluate $\mathbf{F}_1 \cdot \mathbf{F}_2$, where $\mathbf{F}_1$ and $\mathbf{F}_2$ are the forces in Question 1. [Ch. 1]
   (b) Find the angle ϕ between $\mathbf{F}_1$ and $\mathbf{F}_2$. [Ch. 1]

3. A toy car accelerates uniformly as it travels along a straight horizontal track. At t_1, its speed is 2.6 m s^{-1}; at t_2 its speed is 3.8 m s^{-1}. The time interval $t_2 - t_1$ is 3.4 s.
   (a) What is the acceleration of the car? [Ch. 2]
   (b) How far does the car travel in the interval $t_2 - t_1$? [Ch. 2]
   (c) If it started from rest, how long did it take to get to the speed 2.6 m s^{-1}? [Ch. 2]

4. A small stone is released from rest at the top of a building 10 m tall. Neglect air friction.
   (a) How far has it fallen when its speed is 8 m s^{-1}? [Ch. 2]
   (b) How fast is it moving just before it strikes the ground? [Ch. 2]
   (c) How much time does it take to travel the 10 m descent? [Ch. 2]

5. A car is accelerating uniformly at 2 m s^{-2}.
   (a) How long from the instant its speed is 4 m s^{-1} does the car take to travel 21 m? [Ch. 2]
   (b) How fast is the car moving after it travels the 21 m? [Ch. 2]

6. A baseball is hit so that its initial velocity is 20 m s^{-1} at 40° above the horizontal.
   (a) How far from the bat is the ball when it returns to the same height at which it was struck? [Ch. 3]
   (b) How long was the ball in flight? [Ch. 3]

7. (a) How fast is a toy car traveling around a circular track that has a radius of 4 m if the centripetal acceleration is 6 m s^{-2}? [Ch. 3]
   (b) What is the angular velocity of the line that joins the center of the circle and the car? [Ch. 3]

8. Blocks A and B have masses 2.4 kg and 4.7 kg, respectively, and are connected by a light cord. B, pulling A behind it, is dragged at constant speed on a horizontal surface for which the coefficient of friction is 0.6 for both blocks. What is the tension in the cord between the blocks? [Ch. 4]

9. A 4-kg block is pulled up a plane inclined at an angle 30° above the horizontal by a force of 50 N parallel to the plane. The coefficient of friction between block and plane is 0.6. Find the block's acceleration. [Ch. 4]

10. At what maximum speed can a 5-kg toy car travel around a flat circular track that has a diameter of 6 m if the coefficient of static friction between the car tires and the track is 0.8? [Ch. 4]

11. A constant force of 5 N is needed to push a 200-g plug straight up a tight-fitting tube at constant speed.
    (a) If the plug is pushed 40 cm, how much work is done by the force? [Ch. 5]
    (b) Determine the force of sliding friction. [Ch. 5]

12. (a) How much work must be done to increase the speed of a 3-kg object, sliding on a frictionless surface, from 6 m s^{-1} to 9 m s^{-1}? [Ch. 5]
    (b) If the speed increase takes 4 s, how much power is delivered? [Ch. 5]

13. A 0.4-kg mass is hung by a 80-cm length of light cord. The mass is pulled to the side so that the cord makes an angle of 40° with the vertical. What is the maximum speed of the mass after it is released from rest? [Ch. 6]

14. A 2-kg block slides at 3 m s^{-1} on a frictionless horizontal surface. It strikes, head on, the free end of a horizontally mounted coil spring of spring constant 450 N m^{-1}. By how much does the block compress the spring? **[Ch. 6]**

15. A 0.3-kg mass hangs from the bottom end of a spring, the top end of which is held fixed. After being disturbed, the mass vibrates up and down with a frequency of 4 Hz. The location of the mass, relative to its equilibrium position, is given by $x = (0.12 \text{ m}) \times \cos(8\pi t + 0.2\pi)$, where t is measured in seconds.

    (a) What is the amplitude of the motion? **[Ch. 7]**
    (b) What is the period of the motion? **[Ch. 7]**
    (c) What is the value of the spring constant? **[Ch. 7]**
    (d) What is the value of the phase constant? **[Ch. 7]**

16. At a location on the Earth's surface where the value of g is exactly 9.80 m s^{-2} a certain simple pendulum has a vibrational period of exactly 2.00 s. On top of a mountain the period of the pendulum is 2.01 s. Determine the value of g atop this mountain. **[Ch. 7]**

17. Two masses, 2 kg and 3 kg, are connected by a compressed spring and are moving along a straight, frictionless track at 1.2 m s^{-1}. The 3-kg mass is ahead of the 2-kg mass. After the spring is released the 3-kg mass is moving at 3 m s^{-1} in the same direction as before.

    (a) What is the velocity of the 2-kg mass? **[Ch. 8]**
    (b) How much work is done by the expanding spring? **[Ch. 6]**

18. A 300-g ball moving at 15 m s^{-1} is struck by a bat and now travels at 20 m s^{-1} in the opposite direction.

    (a) What is the impulse given to the ball by the bat? **[Ch. 8]**
    (b) The impulse lasts for 0.2 s. Determine the average force exerted on the ball by the bat. **[Ch. 8]**

Solutions to Exam 1

1. (a)
$$\mathbf{R} = \mathbf{F}_1 + \mathbf{F}_2 = (3\hat{\mathbf{x}} + 2\hat{\mathbf{y}}) \text{ N} + (5\hat{\mathbf{x}} - 4\hat{\mathbf{y}}) \text{ N} = \boxed{(8\hat{\mathbf{x}} - 2\hat{\mathbf{y}}) \text{ N}}$$

(b)
$$|\mathbf{R}| = \sqrt{R_x^2 + R_y^2} = \sqrt{8^2 + (-2)^2} \text{ N} = \boxed{8.246 \text{ N}}$$

(c) Let ϕ be the angle between $\mathbf{R}$ and the x axis.

$$\cos\phi = \frac{|R_x|}{|\mathbf{R}|} \qquad \phi = \arccos\left(\frac{|R_x|}{|\mathbf{R}|}\right) = \arccos\left(\frac{8 \text{ N}}{8.246 \text{ N}}\right) = \boxed{14.0° \text{ below the } x \text{ axis}}$$

2. (a)
$$\mathbf{F}_1 \cdot \mathbf{F}_2 = (F_{1x})(F_{2x}) + (F_{1y})(F_{2y}) = [(3)(5) + (2)(-4)]\text{N}^2 = \boxed{7 \text{ N}^2}$$

(b) Let θ be the angle between $\mathbf{F}_1$ and $\mathbf{F}_2$.

$$\mathbf{F}_1 \cdot \mathbf{F}_2 = (F_1)(F_2)\cos\theta$$

where

$$F_1 = \sqrt{F_{1x}^2 + F_{1y}^2} = \sqrt{3^2 + 2^2} \text{ N} = 3.61 \text{ N} \qquad \text{and} \qquad F_2 = \sqrt{F_{2x}^2 + F_{2y}^2} = \sqrt{5^2 + (-4)^2} = 6.40 \text{ N}$$

$$\theta = \arccos\left[\frac{\mathbf{F}_1 \cdot \mathbf{F}_2}{(F_1)(F_2)}\right] = \arccos\left[\frac{7 \text{ N}^2}{(3.61)(6.40) \text{ N}^2}\right] = \boxed{72.4°}$$

3. (a)
$$a = \frac{v_2 - v_1}{t_2 - t_1} = \frac{(3.8 - 2.6) \text{ m s}^{-1}}{3.4 \text{ s}} = 0.353 \text{ m s}^{-2}$$

(b)
$$x = v_0 t + \tfrac{1}{2}at^2 = (2.6 \text{ m s}^{-1})(3.4 \text{ s}) + \tfrac{1}{2}(0.353 \text{ m s}^{-2})(3.4 \text{ s})^2 = \boxed{10.9 \text{ m}}$$

(c) Use $v = v_0 + at$, where $v_0 = 0$, and solve for t.

$$t = \frac{v - v_0}{a} = \frac{(2.6 - 0)\ \text{m s}^{-1}}{0.353\ \text{m s}^{-2}} = \boxed{7.37\ \text{s}}$$

4. (a) Take the downward direction as positive, $a = 9.8\ \text{m s}^{-2}$, and $v_0 = 0$.

$$v^2 = v_0^2 + 2ax \qquad x = \frac{v^2 - v_0^2}{2a} = \frac{(8\ \text{m s}^{-1})^2 - 0}{2(9.8\ \text{m s}^{-2})} = \boxed{3.27\ \text{m}}$$

(b) $\qquad v^2 = v_0^2 + 2ax \qquad v = \sqrt{v_0^2 + 2ax} = \sqrt{0 + 2(9.8\ \text{m s}^{-2})(10\ \text{m})} = \boxed{14\ \text{m s}^{-1}}$

(c) $\qquad x = v_0t + \tfrac{1}{2}at^2 = \tfrac{1}{2}at^2 \qquad$ when $\qquad v_0 = 0$

$$t = \sqrt{\frac{2x}{a}} = \sqrt{\frac{2(10\ \text{m})}{9.8\ \text{m s}^{-2}}} = \boxed{1.43\ \text{s}}$$

5. (a) Use $x = v_0t + \tfrac{1}{2}at^2$ and solve for t with the quadratic formula.

$$t = \frac{-v_0 \pm \sqrt{v_0^2 + 4(a/2)x}}{2(a/2)} = \frac{-4\ \text{m s}^{-1} \pm \sqrt{16 + (4)(1)(21)}\ \text{m s}^{-1}}{2\ \text{m s}^{-2}}$$

$$t_+ = \boxed{3\ \text{s}} \qquad t_- = -7\ \text{s}$$

The negative solution is disregarded because a negative time interval doesn't apply to the physical context of this problem.

(b) $\qquad v = v_0 + at = 4\ \text{m s}^{-1} + (2\ \text{m s}^{-2})(3\ \text{s}) = \boxed{10\ \text{m s}^{-1}}$

6. Because the ball accelerates in the vertical direction only, the horizontal displacement of the ball is $x = v_{0x}t$ where $v_{0x} = v_0 \cos\theta$ and t is the time the ball is in flight. You must therefore determine the answer to part (b) of this problem first.

(b) Solve $y = y_0 + v_{0y}t - \tfrac{1}{2}gt^2$ for t where $v_{0y} = v_0 \sin\theta$. When the ball returns to the same height from which it was struck, $y = y_0$, so

$$0 = 0 + v_{0y}t - \tfrac{1}{2}gt^2 = v_0t \sin\theta - \tfrac{1}{2}gt^2$$

$$t = \frac{2v_0 \sin\theta}{g} = \frac{2(20\ \text{m s}^{-1})(\sin 40°)}{9.8\ \text{m s}^{-2}} = \boxed{2.62\ \text{s}}$$

(a) $\qquad x = v_{0x}t = (v_0 \cos\theta)t = (20\ \text{m s}^{-1})(\cos 40°)(2.62\ \text{s}) = \boxed{40.2\ \text{m}}$

7. (a) Use $a = v^2/r$ and solve for v.

$$v = \sqrt{ar} = \sqrt{(6\ \text{m s}^{-2})(4\ \text{m})} = \boxed{4.90\ \text{m s}^{-1}}$$

(b) $\qquad \omega = \dfrac{v}{r} = \dfrac{4.9\ \text{m s}^{-1}}{4\ \text{m}} = \boxed{1.22\ \text{rad s}^{-1}}$

8. Figure E-1 is a free-body diagram for block A. Because the block is moving at constant velocity it is in equilibrium and the net force on it is zero. So $N = w = mg$ and

$$T = f_k = \mu_k N = \mu_k mg = (0.6)(2.4\ \text{kg})(9.8\ \text{m s}^{-2}) = \boxed{14.1\ \text{N}}$$

9. Figure E-2 is a free-body diagram for the block. Define the x axis as parallel to the inclined plane, resolve the forces on the block into their x and y components, and use Newton's second law, $\Sigma F = ma$, on the forces in the x direction.

$$F - mg \sin\theta - f_k = ma$$

where

$$f_k = \mu_k N = \mu_k mg \cos\theta$$

$$a = \frac{F - mg \sin\theta - \mu_k mg \cos\theta}{m} = \frac{F}{m} - g(\sin\theta + \mu_k \cos\theta)$$

$$= \frac{50\ \text{N}}{4\ \text{kg}} - (9.8\ \text{m s}^{-2})[\sin 30° + (0.6)(\cos 30°)]$$

$$= \boxed{2.51\ \text{m s}^{-2}}$$

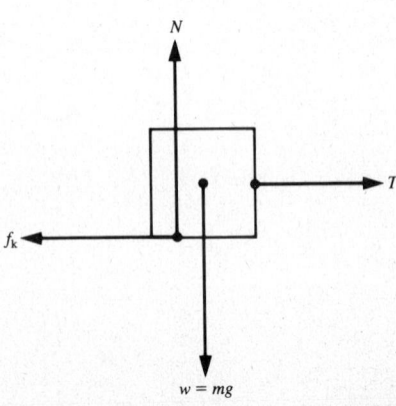

Figure E-1

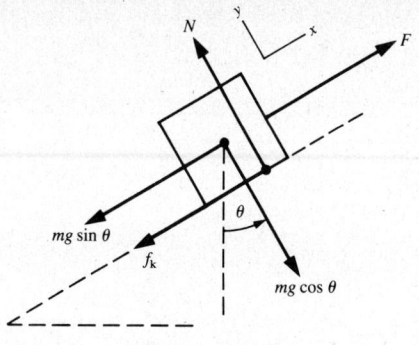

Figure E-2

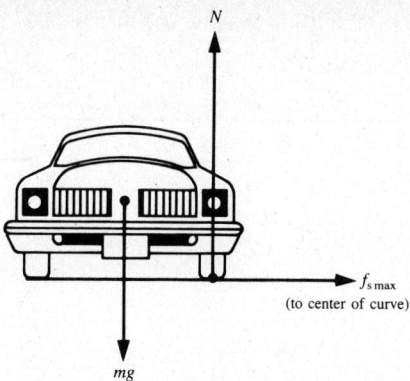

Figure E-3

10. Figure E-3 is a free-body diagram for the car. The maximum value of the force of static friction provides the centripetal force.

$$f_{s\,max} = \mu_s N = \mu_s mg = m\left(\frac{v^2}{r}\right)$$

Solve for v:

$$v = \sqrt{\mu_s g r} = \sqrt{(0.8)(9.8 \text{ m s}^{-2})(3 \text{ m})} = \boxed{4.85 \text{ m s}^{-1}}$$

11. **(a)**
$$W_{tot} = F\,\Delta R = (5 \text{ N})(0.4 \text{ m}) = \boxed{2 \text{ J}}$$

(b) The work done by the force is equal to the increase in the potential energy of the plug plus the work done against friction.

$$F\,\Delta R = \Delta E_p + W_{friction} = mg\,\Delta R + f_k\,\Delta R$$

$$f_k = F - mg = 5 \text{ N} - (0.2 \text{ kg})(9.8 \text{ m s}^{-2}) = \boxed{3.04 \text{ N}}$$

12. **(a)** The work done is equal to the object's increase in kinetic energy.

$$W_{net} = \Delta E_k = \tfrac{1}{2}mv_2^2 - \tfrac{1}{2}mv_1^2 = \tfrac{1}{2}m(v_2^2 - v_1^2)$$

$$= \tfrac{1}{2}(3 \text{ kg})[(9 \text{ m s}^{-1})^2 - (6 \text{ m s}^{-1})^2] = \boxed{67.5 \text{ J}}$$

(b)
$$P = \frac{dW}{dt} = \frac{67.5 \text{ J}}{4 \text{ s}} = \boxed{16.9 \text{ W}}$$

13. When the mass is pulled to the side, it acquires potential energy

$$E_p = mg(\ell - \ell\cos\phi) = mg\ell(1 - \cos\phi)$$

where ℓ is the length of the cord and ϕ is the angle the cord makes with the vertical when the mass is released. After its release, as the mass accelerates, its potential energy is converted to kinetic energy.

$$\Delta E_p = \Delta E_k \qquad mg\ell(1 - \cos\phi) = \tfrac{1}{2}mv^2 - 0$$

$$v = \sqrt{2g\ell(1 - \cos\phi)} = \sqrt{2(9.8 \text{ m s}^{-2})(0.8 \text{ m})(1 - \cos 40°)} = \boxed{1.92 \text{ m s}^{-1}}$$

14. Use the conservation of energy principle. Because friction is absent, all the kinetic energy of the block is converted into elastic potential energy in the spring: $\tfrac{1}{2}mv^2 = \tfrac{1}{2}kx^2$. Solve for x.

$$x = v\sqrt{\frac{m}{k}} = (3 \text{ m s}^{-1})\sqrt{\frac{2 \text{ kg}}{450 \text{ N m}^{-1}}} = \boxed{0.2 \text{ m}}$$

15. Use the equation for the position of a mass in simple harmonic motion, $x = X\cos(\omega t + \phi)$.

(a) The amplitude is $X = \boxed{0.12 \text{ m}}$

(b)
$$T = \frac{1}{v} = \frac{1}{4 \text{ Hz}} = \boxed{0.25 \text{ s}}$$

(c) The spring constant k and angular frequency ω are related by $\omega = \sqrt{k/m}$.

$$k = m\omega^2 = (0.3 \text{ kg})(8\pi \text{ s}^{-1})^2 = \boxed{189 \text{ N m}^{-1}}$$

(d) The phase constant is $\phi = \boxed{0.2\pi \text{ rad}}$

16. Use the expression for the period of a pendulum, $T = 2\pi\sqrt{\ell/g}$. The length ℓ is the same at both locations, so

$$\ell = \frac{g_1 T_1^2}{4\pi^2} = \frac{g_2 T_2^2}{4\pi^2}$$

$$g_2 = g_1\left(\frac{T_1^2}{T_2^2}\right) = (9.80 \text{ m s}^{-2})\left(\frac{2.00 \text{ s}}{2.01 \text{ s}}\right)^2 = \boxed{9.70 \text{ m s}^{-2}}$$

17. (a) Use the principle of conservation of linear momentum

$$m_A v_{A1} + m_B v_{B2} = m_A m_{A2} + m_B v_{B2}$$

with $v_{A1} = v_{B1} = v_1 = 1.2 \text{ m s}^{-1}$ and solve for v_{B2}.

$$v_{B2} = \frac{(m_A + m_B)v_1 - m_A v_{A2}}{m_B} = \frac{[(3+2)(1.2) - (3)(3)] \text{ kg m s}^{-1}}{2 \text{ kg}} = \boxed{-1.5 \text{ m s}^{-1}}$$

The negative sign indicates that this mass is moving opposite to its initial direction.

(b) The work done by the expanding spring is equal to the increase in kinetic energy of the system.

$$W_s = E_{k2} - E_{k1} = \tfrac{1}{2}(m_A v_{A2}^2 + m_B v_{B2}^2) - \tfrac{1}{2}(m_A + m_B)v_1^2$$
$$= \tfrac{1}{2}[(3 \text{ kg})(3 \text{ m s}^{-1})^2 + (2 \text{ kg})(-1.5 \text{ m s}^{-1})^2 - (5 \text{ kg})(1.2 \text{ m s}^{-1})^2] = \boxed{12.15 \text{ J}}$$

18 (a) The impulse given to the ball equals the ball's change in momentum.

$$F_{av}\Delta t = mv_2 - mv_1 = m(v_2 - v_1) = (0.3 \text{ kg})(-20 \text{ m s}^{-1} - 15 \text{ m s}^{-1}) = \boxed{-10.5 \text{ N s}}$$

The negative sign indicates that the impulse is directed opposite to the ball's initial velocity.

(b)
$$F_{av} = \frac{m(v_2 - v_1)}{\Delta t} = \frac{-10.5 \text{ N s}}{0.2} = \boxed{-52.5 \text{ N}}$$

9 RIGID BODIES I: STATICS

THIS CHAPTER IS ABOUT

☑ **Center of Mass**
☑ **Torque—The Vector Cross Product**
☑ **Equilibrium Conditions for Rigid Bodies**

9-1. Center of Mass

To locate the CM of a rigid continuous body, we apply the procedure of Section 8-4, except that we replace the sums with integrals that are to be evaluated over the entire body. The equivalent of Eq. (8-9) in this case is

$$x_{CM} = \frac{\int x \, dm}{M} \qquad (9\text{-}1)$$

where the mass of the body is

$$M = \int dm \qquad (9\text{-}2)$$

We obtain the coordinates y_{CM} and z_{CM} from expressions like Eq. (9-1), replacing x with y and z, respectively. For a *uniform, symmetrical* body, the CM is located at its geometric center, as for a square or rectangular sheet, a cricular hoop or disk, a solid or hollow sphere, etc. We use the symmetry of the object to calculate the coordinates whenever possible.

EXAMPLE 9-1: Locate the CM for the T-shaped sheet of uniform thickness shown in Figure 9-1a. The dimensions of the sheet are $a = 20$ cm, $b = 35$ cm, $c = 15$ cm, and $d = 40$ cm. The mass of the sheet is $M = 1.6$ kg.

Solution: First, you should derive a formula for the location of the CM, which by symmetry must be located on the center line L_c; then substitute values for the variables.

Because of the high degree of symmetry involved in this problem, you can solve it as though it were a two-body problem, similar to those problems in Chapter 8, and without having to use calculus. The CMs of the two segments, A and B, of the object are at their geometric centers, CM_A and CM_B, as shown in Figure 9-1b. The mass of each segment is proportional to its area. The mass per unit area for the object is M/S, where S is the total area of the sheet, or $S = ad + bc$. The masses of the two segments are $m_A = (M/S)S_A$ and $m_B = (M/S)S_B$. Applying Eq. (8-9), you get

$$x_{CM} = \frac{m_A x_A + m_B x_B}{m_A + m_B} = \frac{\left(\dfrac{M}{S}\right)\left[S_A\left(\dfrac{a}{2}\right) + S_B\left(a + \dfrac{b}{2}\right)\right]}{M} = \frac{(ad)\left(\dfrac{a}{2}\right) + (bc)\left(a + \dfrac{b}{2}\right)}{ad + bc}$$

$$= \frac{a^2 d + 2abc + b^2 c}{2(ad + bc)} = \frac{[(20)^2(40) + 2(20)(35)(15) + (35)^2(15)] \text{ cm}^3}{2[(20)(40) + (35)(15)] \text{ cm}^2} = \boxed{20.9 \text{ cm}}$$

Note that the location of the CM is independent of the mass of the sheet.

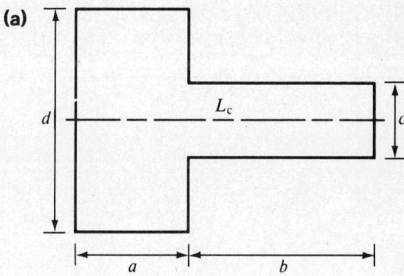

(a)

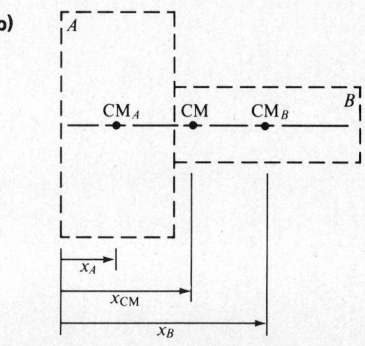

(b)

Figure 9-1. (a) T-shaped sheet. (b) The CM.

EXAMPLE 9-2: Locate the CM for the uniform sheet shown in Figure 9-2. The sheet is in the form of an isosceles triangle of base a, height h, and total mass M.

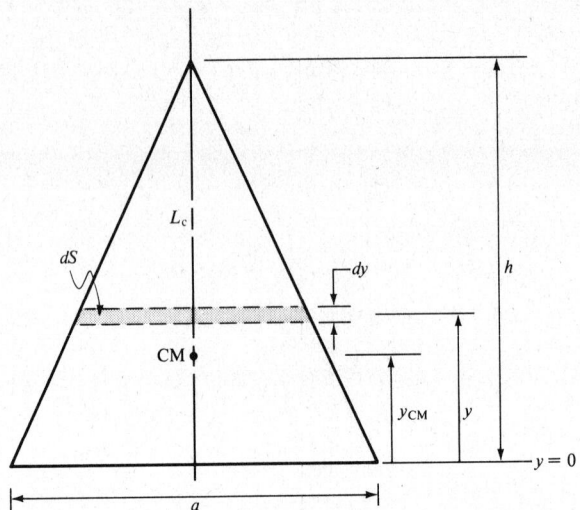

Figure 9-2

Solution: Because of symmetry, it should be clear that the CM is located on the center line, so we need to find only the y coordinate. In order to use Eq. (9-1), we need to find the mass, dm, of the shaded area, dS, shown in Figure 9-2.

$$dm = \sigma\, dS = \sigma\left[a\left(1 - \frac{y}{h}\right)\right] dy$$

where σ is the mass per unit area of the triangle, $\sigma = 2M/ah$, and $a(1 - y/h)$ is the length of the shaded area. From Eq. (9-1),

$$y_{CM} = \frac{\int y\, dm}{M} = \frac{\int_0^h y\sigma\left[a\left(1 - \frac{y}{h}\right)\right] dy}{M} = \frac{\sigma\int_0^h y\left[a\left(1 - \frac{y}{h}\right)\right] dy}{M}$$

$$= \frac{\frac{2M}{ah}\int_0^h\left(ay - \frac{ay^2}{h}\right) dy}{M} = \frac{2}{h}\left[\int_0^h y\, dy - \frac{1}{h}\int_0^h y^2\, dy\right]$$

$$= \left(\frac{2}{h}\right)\left[\frac{1}{2}y^2 - \left(\frac{1}{h}\right)\left(\frac{1}{3}y^3\right)\right]\Big|_0^h = \frac{2}{h}\left(\frac{h^2}{2} - \frac{h^3}{3h}\right) = \boxed{\frac{h}{3}}$$

Note that the location of the CM is independent of both the mass and length of the base.

9-2. Torque—The Vector Cross Product

Torque, which is also called the **moment of the force**, is responsible for rotation of a body about an axis, or for a body's tendency to rotate. In mechanical systems torque is caused by a force acting at a distance **r** from the axis and is designated by τ.

TORQUE $$\boldsymbol{\tau} = \mathbf{r} \times \mathbf{F} \tag{9-3a}$$

The magnitude of the torque is

$$\tau = rF \sin\theta \tag{9-3b}$$

which can also be written as $\tau = \ell F$, since $\ell = r\sin\theta$. The quantity ℓ is called the

moment arm or **lever arm** and is the perpendicular distance between the axis and the line of action of **F**. In Figure 9-3, when the force **F** is applied at point P, the wrench exerts a torque of magnitude $\tau = rF \sin \theta$ on the nut. The units of torque are N m in SI and lb ft in the British system. Although torque has the same basic units as energy, it is an entirely different concept. It would be *incorrect* to say that a torque of 4 N m is 4 J.

We don't use the vector property of torque in this chapter but we will in Chapter 10. In Figure 9-3, the direction of the torque produced by the wrench is out of the plane of the paper along the *axis*. In this chapter we will describe torque by stating the direction of the *rotation*, either clockwise (cw) or counterclockwise (ccw), associated with it. The angle θ is the acute angle between the vectors **r** and **F**.

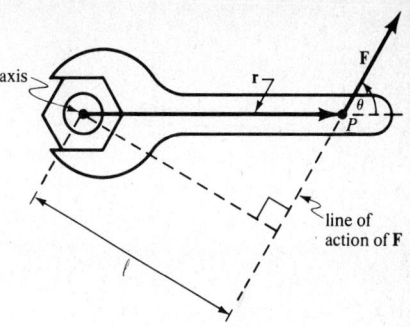

Figure 9-3

EXAMPLE 9-3: The force applied to the wrench in Figure 9-3 is 90 N. The distance from the axis to the point P is 24 cm and $\theta = 60°$. **(a)** What is the torque applied to the tight nut? **(b)** Determine the lever arm.

Solution:

(a) Use Eq. (9-3b).

$$\tau = rF \sin \theta = (0.24 \text{ m})(90 \text{ N})\sin 60° = \boxed{18.7 \text{ N m}}$$

The direction of rotation associated with this torque is counterclockwise.

(b)
$$\ell = r \sin \theta = (0.24 \text{ m})\sin 60° = \boxed{0.208 \text{ m}}$$

9-3. Equilibrium Conditions for Rigid Bodies

In Section 4-4, we stated the *condition for translational equilibrium*, or the *first* condition of equilibrium. That is, the net force acting on a body is zero, or $\Sigma \mathbf{F} = 0$. The **condition for rotational equilibrium**, or the *second* condition of equilibrium (important only for extended rigid bodies) is that the net external torque, evaluated about *any* axis, is zero.

CONDITION FOR ROTATIONAL EQUILIBRIUM $\Sigma \tau = 0$

In evaluating this sum we'll consider τ_{cw} to be negative and τ_{ccw} to be positive. The fact that we may select *any* axis in evaluating $\Sigma \tau = 0$ for a static rigid body means that it is *not* rotating. Problems involving rigid bodies in equilibrium in two dimensions can, in general, lead to three algebraic equations with three unknowns. Two of the equations represent $\Sigma \mathbf{F} = 0$, i.e., $\Sigma F_x = 0$ and $\Sigma F_y = 0$; the other is $\Sigma \tau = 0$.

EXAMPLE 9-4: A uniform rod 2 m in length that has a mass of 3 kg leans against a *smooth* wall at an angle $\theta = 60°$ with the ground. A mass of 2 kg hangs by a cord from a point $\frac{1}{4}\ell$ from the upper end of the rod. See Figure 9-4a. Determine the force (magnitude and direction) that the ground exerts on the rod.

Solution: Make a free-body diagram for the rod, as shown in Figure 9-4b. The components of the force exerted by the ground are R_x and R_y. The force exerted by the wall is **F**. Because the wall is smooth (frictionless), **F** has no component parallel to the surface of the wall; i.e., **F** must be perpendicular to the wall in this case. It should be clear from the information given that $\ell = 2$ m, $M = 2$ kg, $m = 3$ kg, $\theta = 60°$, and $\phi = 90° - \theta = 30°$. We apply the first condition of equilibrium, $\Sigma \mathbf{F} = 0$.

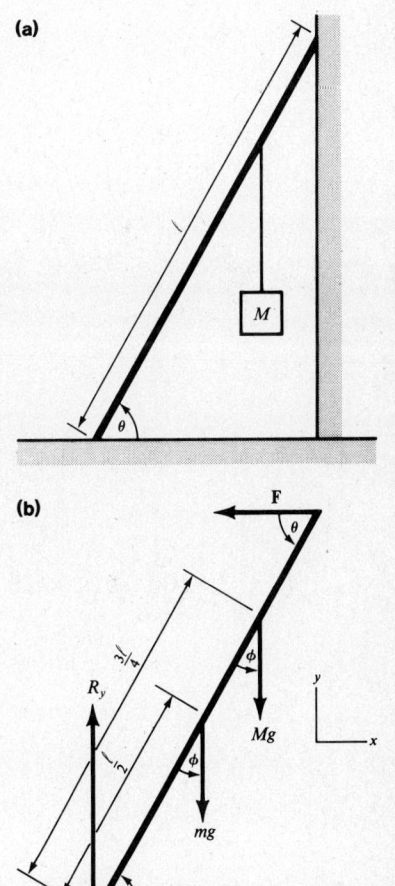

(a)

(b)

Figure 9-4. (a) Placement of the rod. (b) Free-body diagram for the rod.

$$\Sigma F_x = 0 \qquad\qquad \Sigma F_y = 0$$

$$R_x - F = 0 \qquad\qquad R_y - mg - Mg = 0$$

$$R_x = F \qquad\qquad R_y = (m + M)g = [(3 + 2)\,\text{kg}](9.8\,\text{m s}^{-2}) = 49\,\text{N}$$

The x component equation tells us the relationship between the two unknowns, R_x and F. We'll need to use the torque equation (second condition of equilibrium), $\Sigma\tau = 0$, to go on from here.

Since we can use any axis as a point about which to sum torques, the choice of a certain axis can make the mathematical part of the solution a bit easier. In this case let's select the bottom of the rod as the origin, with the ground as the x axis, because the two unknown forces R_x and R_y are applied at that point. Because the lever arms for both these forces are zero for this choice of axis, R_x and R_y don't exert any torque. Thus, about the lower end of the rod,

$$\Sigma\tau = 0 = -\frac{\ell}{2}mg\sin\phi - \frac{3\ell}{2}Mg\sin\phi + \ell F\sin\theta = 0$$

At this point, it is *important* for you to pause and make sure you understand the reason for each term in the preceding equation, including its sign (recall that τ_{cw} is negative and τ_{ccw} is positive). We now go on to solve this equation for the only unknown, F. Note that ℓ cancels.

$$F = \frac{(m + 3M)g\sin\phi}{2\sin\phi} = \frac{[3\,\text{kg} + 3(2\,\text{kg})](9.8\,\text{m s}^{-2})\sin 30°}{2\sin 60°} = 25.5\,\text{N}$$

From the x component equation,

$$R_x = F = 25.5\,\text{N}$$

The magnitude of **R** is

$$R = \sqrt{R_x^2 + R_y^2} = \sqrt{(25.5)^2 + (49)^2}\,\text{N} = \boxed{55.2\,\text{N}}$$

The angle between **R** and the x axis is

$$\alpha = \arctan\left(\frac{R_y}{R_x}\right) = \arctan\left(\frac{49\,\text{N}}{25.5\,\text{N}}\right) = \boxed{62.5°}$$

Note that this angle is *not* equal to θ. Thus the reaction force is not directed exactly along the rod.

EXAMPLE 9-5: In many situations R_x is provided by the force of static friction. Determine the minimum value of the coefficient of static friction necessary to prevent the bottom of the rod in Example 9-4 from slipping.

Solution: Use Eq. (4-4), or $f_s = \mu_s N$, where $f_s = R_x$ and $N = R_y$. Solve for μ_s.

$$R_x = \mu_s R_y$$

$$\mu_s = \frac{R_x}{R_y} = \frac{25.5\,\text{N}}{49\,\text{N}} = \boxed{0.52}$$

EXAMPLE 9-6: A cylinder that has a radius of 5 cm and a mass of 6 kg is held on a 26° inclined plane by a horizontal cord. **(a)** Determine the tension in the cord. **(b)** What is the minimum value of the coefficient of static friction needed to keep the cylinder from slipping?

Solution: Before proceeding with the solution to parts **(a)** and **(b)** of the problem, let's determine the length ℓ in Figure 9-5a. We will need this quantity, as well as ϕ, later in the solution. The length ℓ is the base of an isosceles triangle formed by the cord and the distance between the point P where the cylinder contacts the plane

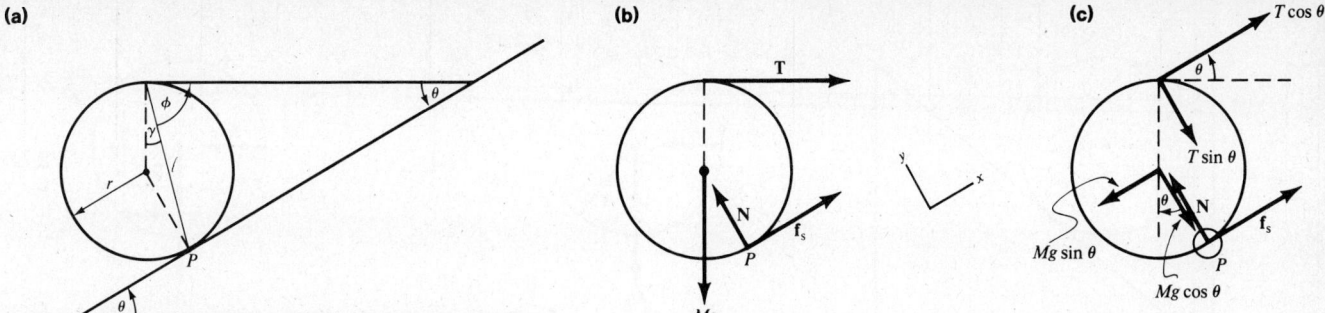

Figure 9-5. (a) Cylinder on an inclined plane. (b) Free-body diagram of cylinder and forces. (c) Force components.

and the point where the cord is attached to the plane. Therefore

$$\phi + \phi + \theta = 180°$$

$$\phi = \frac{180° - \theta}{2} = 90° - \frac{\theta}{2} = 90° - \frac{26°}{2} = 77°$$

$$\gamma = 90° - \phi = \frac{\theta}{2} = \frac{26°}{2} = 13°$$

$$\frac{\ell}{2} = r \cos \gamma$$

$$\ell = 2r \cos\left(\frac{\theta}{2}\right) = 2(5 \times 10^{-2} \text{ m})\cos 13° = 9.74 \times 10^{-2} \text{ m}$$

(a) Draw a free-body diagram of the cylinder, Figure 9-5b, and resolve the forces into x and y components, Figure 9-5c. Applying the first condition of equilibrium, $\Sigma\mathbf{F} = 0$, we get

$$\Sigma F_x = 0 \qquad\qquad \Sigma F_y = 0$$

$$f_s + T \cos \theta - Mg \sin \theta = 0 \qquad N - Mg \cos \theta - T \sin \theta = 0$$

These last two equations involve three unknowns: f_s, N, and T. We apply the second condition of equilibrium, using the point P as an axis,

$$\Sigma\tau = 0$$

$$rMg \sin \theta - \ell T \sin \phi = 0$$

and solve for T.

$$T = \frac{rMg \sin \theta}{\ell \sin \phi} = \frac{(5 \times 10^{-2} \text{ m})(6 \text{ kg})(9.8 \text{ m s}^{-2})\sin 26°}{(9.74 \times 10^{-2} \text{ m})\sin 77°} = \boxed{13.6 \text{ N}}$$

It's not evident from the conditions stated initially that the torque equation yields directly the tension in the cord.

(b) Use Eq. (4-4), $f_s = \mu_s N$, and solve for μ_s. (We obtain N and f_s from the x and y component equations.)

$$\mu_s = \frac{f_s}{N} = \frac{Mg \sin \theta - T \cos \theta}{Mg \cos \theta - T \sin \theta} = \frac{Mg \tan \theta - T}{Mg + T \tan \theta}$$

$$= \frac{(6 \text{ kg})(9.8 \text{ m s}^{-2})\tan 26° - 13.6 \text{ N}}{(6 \text{ kg})(9.8 \text{ m s}^{-2}) + (13.6 \text{ N})\tan 26°} = \boxed{0.23}$$

EXAMPLE 9-7: The 6-kg cylinder in Example 9-6 ($r = 5$ cm) rests on a 40° inclined plane and touches a wall, as shown in Figure 9-6a. How much force does the wall exert on the cylinder?

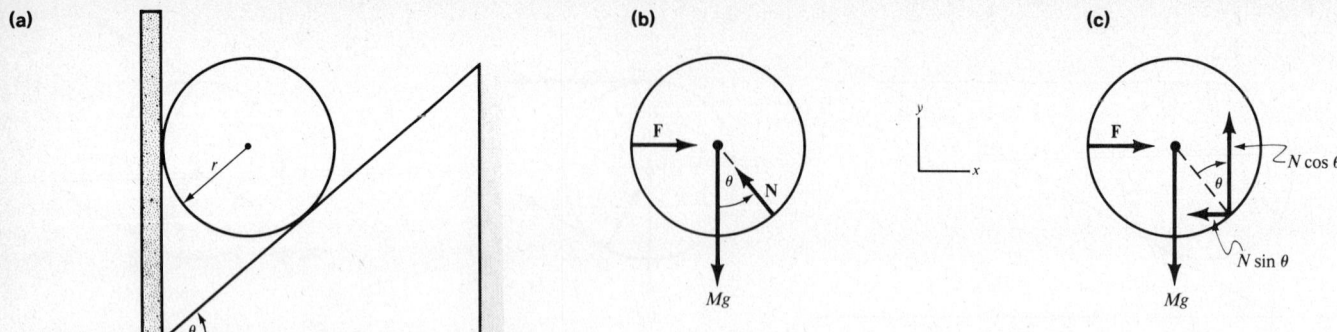

Figure 9-6. (a) Cylinder on an inclined plane. (b) Free-body diagram of cylinder and forces. (c) Normal force resolved into *x* and *y* components.

Solution: We draw a free-body diagram of the cylinder and resolve the normal force into *x* and *y* components, as shown in Figure 9-6b and c. The three forces acting on the cylinder are *concurrent*; i.e., their lines of action pass through a single point, and therefore no torques are acting on the cylinder. Thus the first condition of equilibrium is sufficient to determine *F*.

$$\Sigma F_x = 0 \qquad\qquad \Sigma F_y = 0$$

$$F - N\sin\theta = 0 \qquad\qquad N\cos\theta - Mg = 0$$

Solve the last two equations simultaneously for *F*.

$$F = Mg\tan\theta = (6\text{ kg})(9.8\text{ m s}^{-2})\tan 40° = \boxed{49.3\text{ N}}$$

SUMMARY

1. The CM of a continuous rigid body has coordinates

$$x_{CM} = \frac{\int x\,dm}{\int dm}, \qquad y_{CM} = \frac{\int y\,dm}{\int dm}, \qquad \text{and} \qquad z_{CM} = \frac{\int z\,dm}{\int dm}$$

where $M = \int dm$, the total mass of the body. All integrations are carried out over the entire body.

2. The *torque* exerted on a body by the force **F** acting at a point located **r** from an axis is given by the cross product $\boldsymbol{\tau} = \mathbf{r} \times \mathbf{F}$. Its magnitude is $rF\sin\theta$, where θ is the acute angle between **r** and **F**.

3. The *moment arm* or *lever arm* is $\ell = r\sin\theta$.

4. The *second condition of equilibrium* for a rigid body is that the net torque acting on the body be zero, i.e., $\Sigma\tau = 0$. The sum may be evaluated about *any* convenient axis.

RAISE YOUR GRADES

Can you explain . . . ?

☑ how to locate experimentally the CM of a flat object of irregular shape
☑ why no torque is exerted by a force that has a line of action that passes through the selected axis
☑ why a mechanic sometimes slips a length of pipe over the handle of a wrench to increase its length
☑ why it is incorrect to specify the torque rating of a bolt in kg m
☑ how to make a torque wrench from an ordinary wrench and a spring scale
☑ under what conditions the application of a force to the CM of a uniform rod results in a torque on the rod

SOLVED PROBLEMS

Center of Mass

PROBLEM 9-1: Locate the CM for the L-shaped piece of plywood of uniform thickness shown in Figure 9-7a.

(a)

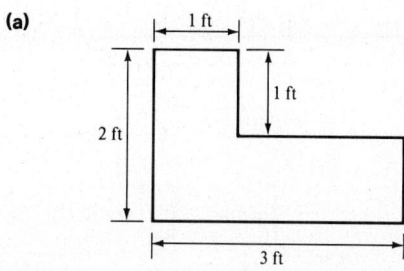

Solution: Divide the sheet as shown in Figure 9-7b. In this case, each division has one-half the mass M of the sheet, or $m_A = \frac{1}{2}M$ and $m_B = \frac{1}{2}M$. Use Eq. (8-9).

$$x_{CM} = \frac{m_A x_A + m_B x_B}{m_A + m_B}$$

$$= \frac{\frac{1}{2}M(0.5\text{ ft}) + \frac{1}{2}M(2\text{ ft})}{M}$$

$$= \frac{1}{2}(0.5 + 2)\text{ ft} = \boxed{1.25\text{ ft}}$$

$$y_{CM} = \frac{m_A y_A + m_B y_B}{m_A + m_B}$$

$$= \frac{\frac{1}{2}M(1\text{ ft}) + \frac{1}{2}M(0.5\text{ ft})}{M}$$

$$= \frac{1}{2}(1 + 0.5)\text{ ft} = \boxed{0.75\text{ ft}}$$

(b)

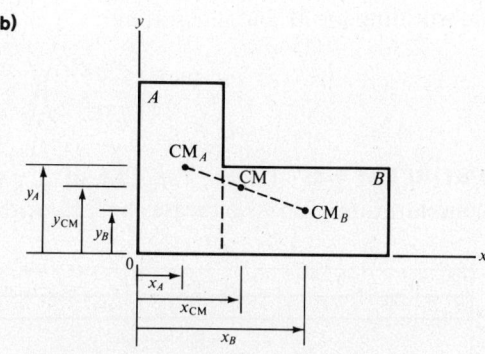

Figure 9-7. (a) L-shaped plywood sheet. (b) Coordinates for the parts of the sheet.

PROBLEM 9-2: Derive a formula for locating the CM of a generalized L-shaped uniform sheet, shown in Figure 9-8a.

Solution: Divide the sheet into two symmetrical parts, as shown in Figure 9-8b. The mass of each part is directly proportional to its area. The proportionality constant σ is the mass of the sheet divided by its total area, or $\sigma = M/[ad + b(c - d)]$. Use Eq. (8-9), with $m_A = \sigma b(c - d)$ and $m_B = \sigma ad$.

$$x_{CM} = \frac{m_A x_A + m_B x_B}{m_A + m_B}$$

$$= \frac{[\sigma b(c-d)](b/2) + (\sigma ad)(a/2)}{M}$$

$$= \boxed{\frac{b^2(c-d) + a^2 d}{2[ad + b(c-d)]}}$$

$$y_{CM} = \frac{m_A y_A + m_B y_B}{m_A + m_B}$$

$$= \frac{[\sigma b(c-d)][d + (c-d)/2] + (\sigma ad)(d/2)}{M}$$

$$= \boxed{\frac{d^2(a-b) + c^2 b}{2[ad + b(c-d)]}}$$

Note that the location of the CM is independent of the mass of this object.

(a)

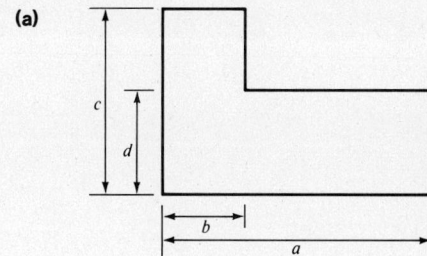

(b)

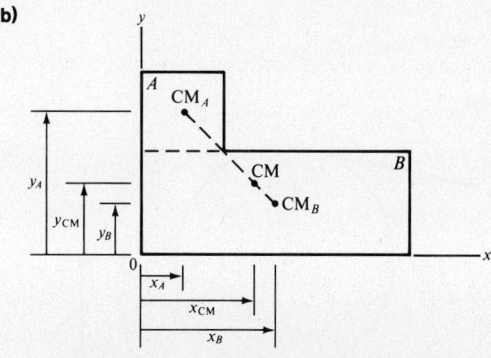

Figure 9-8. (a) L-shaped sheet. (b) Coordinates for the parts of the sheet.

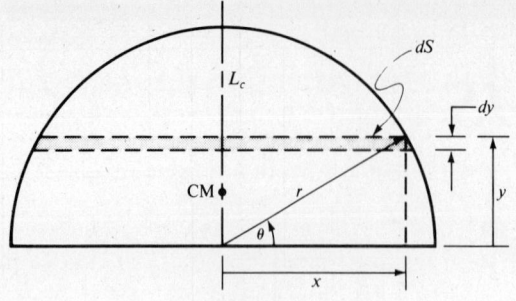

Figure 9-9

PROBLEM 9-3: Locate the CM of the uniform half disk of radius r shown in Figure 9-9.

Solution: Because of symmetry, you know that the CM is located on the center line L_c. Assume that the disk has a mass of M and, because it is uniform, has a mass per unit area of $\sigma = 2M/(\pi r^2)$. Use Eq. (9-1); $dm = \sigma\, dS = \sigma(2x)\, dy$, $x = r\cos\theta$, $y = r\sin\theta$, and $dy = r\cos\theta\, d\theta$. Thus

$$dm = 2\sigma r^2 \cos^2\theta\, d\theta$$

$$y_{CM} = \frac{\displaystyle\int y\, dm}{M} = \frac{\displaystyle\int_0^{\pi/2} (r\sin\theta)(2\sigma r^2 \cos^2\theta\, d\theta)}{M} = \frac{2\sigma r^3 \displaystyle\int_0^{\pi/2} \sin\theta\cos^2\theta\, d\theta}{M}$$

Substituting for M and integrating

$$\left(\frac{2\sigma r^3}{\sigma\pi r^2/2}\right)\left(-\frac{1}{3}\cos^3\theta\,\Big|_0^{\pi/2}\right) = \boxed{\frac{4r}{3\pi}}$$

PROBLEM 9-4: Locate the CM of a rod of length ℓ that has a nonuniform mass. The concentration of mass increases linearly with the distance from one end of the rod.

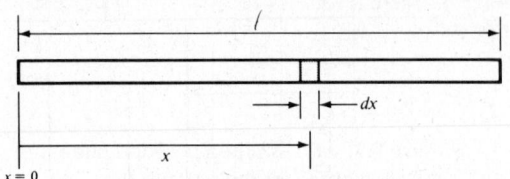

Figure 9-10

(a)

Solution: A sketch of the rod is shown in Figure 9-10. The mass of the element of length dx is $dm = Kx\, dx$, where K is a proportionality constant that has units of kg m^{-2}. Use Eq. (9-1).

$$x_{CM} = \frac{\displaystyle\int x\, dm}{\displaystyle\int dm} = \frac{K\displaystyle\int_0^\ell x^2\, dx}{K\displaystyle\int_0^\ell x\, dx} = \frac{\frac{1}{3}x^3\,\Big|_0^\ell}{\frac{1}{2}x^2\,\Big|_0^\ell} = \boxed{\frac{2\ell}{3}}$$

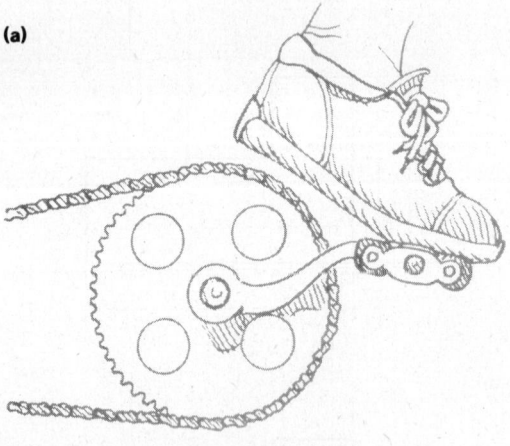

(b)

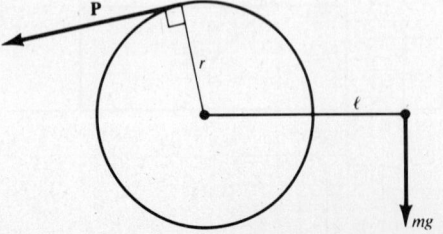

Figure 9-11

Torque

PROBLEM 9-5: The distance from the bicycle crank axis to the pedal pivot shown in Figure 9-11 is 26 cm. The diameter of the sprocket wheel is 22 cm. **(a)** What is the maximum torque that an 82-kg person can exert on the crank assembly? **(b)** What is the resulting tension in the chain?

Solution:

(a) The maximum force that the person can apply to the pedal is equal to his or her body weight, i.e., $w = mg$. Use Eq. (9-3b). The maximum torque results when $\theta = 90°$.

$$\tau = rF\sin\theta = \ell mg\sin 90°$$

$$= (0.26\text{ m})(82\text{ kg})(9.8\text{ m s}^{-2})\sin 90°$$

$$= \boxed{209\text{ N m}}$$

(b) This torque is transmitted to the chain as a force (the tension in the chain) and is represented by **P**. The radius of the sprocket wheel is 11 cm, represented by r. The angle between **P** and **r** is 90° in this case. Therefore the transmitted torque is $\tau = rP \sin 90°$. Solve for P.

$$P = \frac{\tau}{r} = \frac{209 \text{ N m}}{(0.11 \text{ m})} = \boxed{1.9 \times 10^3 \text{ N}}$$

Equilibrium Conditions for Rigid Bodies

PROBLEM 9-6: A nonuniform rod is suspended from two spring scales. The indications on the scales are $F_1 = 60$ N and $F_2 = 20$ N, and the length of the rod is 1.4 m, as shown in Figure 9-12a. Locate the CM of the rod from its heavier end.

Solution: Draw a free-body diagram of the rod, as in Figure 9-12b, and apply the conditions of equilibrium. You can consider the entire mass of the body, and consequently its weight, to be located at the CM. Only one component, y, appears in the equation for the first condition of equilibrium.

$$\Sigma F_y = 0$$

$$F_1 + F_2 - w = 0$$

Solving for w, you get

$$w = F_1 + F_2 = (60 + 20) \text{ N} = 80 \text{ N}$$

Now use the second condition of equilibrium and sum the torques about an axis at the left end of the rod.

$$\Sigma \tau = 0$$

$$-(x_{CM})(w)\sin 90° + (\ell)(F_2)\sin 90° = 0$$

and solve for x_{CM}.

$$x_{CM} = \left(\frac{F_2}{w}\right)\ell = \left(\frac{20 \text{ N}}{80 \text{ N}}\right)(1.4) \text{ m} = \boxed{0.35 \text{ m}}$$

PROBLEM 9-7: An "A" frame is constructed from two 5-kg, 2-m-long uniform rods, joined at the top by a frictionless pivot. A massless cord connects the rods 20 cm from their lower ends, the angle between the rods is 40°, and a 120-N weight is suspended from the pivot, as shown in Figure 9-13a. Find the tension in the cord when the frame rests on a frictionless surface.

Solution: Draw a free-body diagram for one of the rods showing the five forces that act on it (Fig. 9-13b). The reaction force **F** results

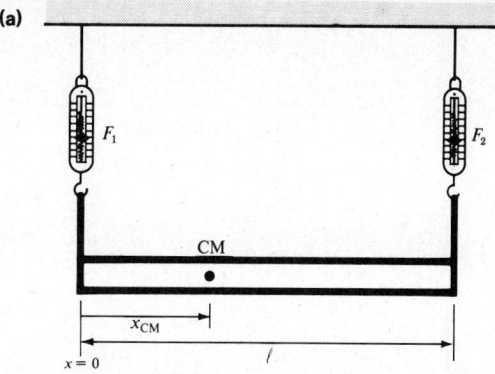
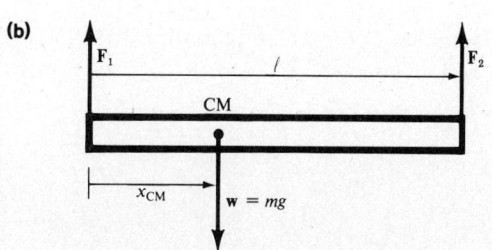

Figure 9-12. (a) Rod suspended from spring scales. (b) Free-body diagram for rod.

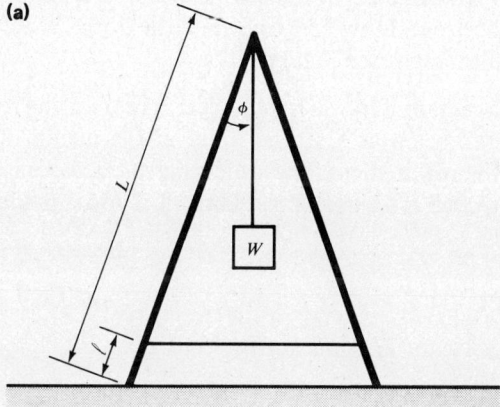
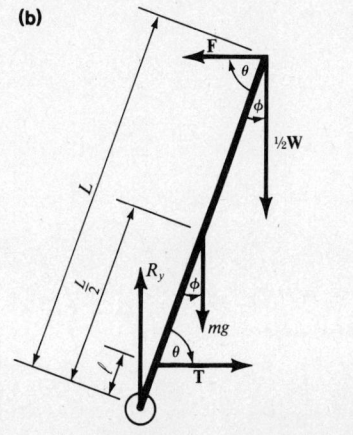

Figure 9-13. (a) "A" frame, weight, and cord. (b) Free-body diagram for one rod.

from the other rod, and, because of symmetry, $\frac{1}{2}W$ is the component of **W** supported by one rod; R_y is the reaction force of the horizontal surface on the rod. The known quantities are $L = 2$ m, $\ell = 0.2$ m, $m = 5$ kg, $W = 120$ N, $\phi = 20°$, and $\theta = 70°$. Apply the first condition of equilibrium to the rod.

$$\Sigma F_x = 0 \qquad\qquad \Sigma F_y = 0$$

$$F - T = 0 \qquad\qquad R_y - mg - W/2 = 0$$

Apply the second condition of equilibrium about an axis at the bottom of the rod.

$$\Sigma \tau = 0$$

$$-\ell T \sin \theta - \frac{L}{2} mg \sin \phi - L\left(\frac{W}{2}\right)\sin \phi + LF \sin \theta = 0$$

From the x component equation, you have $F = T$. Substitute in the torque equation and solve for T.

$$T(L - \ell)\sin \theta = \frac{L(mg + W)\sin \phi}{2}$$

$$T = \left[\frac{L(mg + W)}{2(L - \ell)}\right]\left(\frac{\sin \phi}{\sin \theta}\right)$$

Since $\sin \theta = \cos \phi$, because ϕ and θ are complementary,

$$T = \frac{L(mg + W)\tan \phi}{2(L - \ell)} = \frac{(2 \text{ m})[(5 \text{ kg})(9.8 \text{ m s}^{-2}) + 120 \text{ N}]\tan 20°}{2(2 - 0.2) \text{ m}} = \boxed{34.2 \text{ N}}$$

PROBLEM 9-8:   A lawn roller with a diameter of 50 cm and a mass of 180 kg is connected to a horizontal tow bar. Calculate the horizontal force necessary to pull the roller over an oak beam 8 cm high. See Figure 9-14a.

Solution: Draw a free-body diagram for the roller, as in Figure 9-14b. The reaction force **R** of the oak beam acts on the roller at the instant it breaks contact with the ground. At that instant the requirement for **F** will be greatest. Because you need only to determine F, you need only the second condition of equilibrium, if you sum the torques about an axis through the point of contact.

$$\Sigma \tau = 0$$

$$-rF \sin \theta + rMg \sin \phi = 0$$

Solve for F, using $\sin \theta = \cos \phi$.

$$F = Mg\frac{\sin \phi}{\cos \phi} = Mg \tan \phi$$

Figure 9-14. (a) Roller and wooden block. (b) Free-body diagram for roller. (c) Triangle for calculation of tan θ.

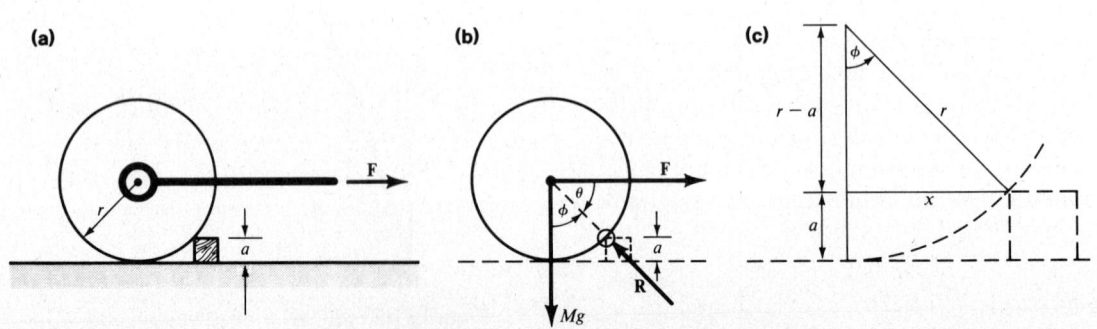

Figure 9-14c and the equation for a circle as applied to this problem, $r^2 = x^2 + (r - a)^2$, will help you calculate $\tan \phi$.

$$\tan \phi = \frac{x}{r - a} = \frac{\sqrt{r^2 - (r - a)^2}}{r - a} = \frac{\sqrt{a(2r - a)}}{r - a}$$

Thus

$$F = (Mg)\left(\frac{\sqrt{a(2r - a)}}{r - a}\right) = (180 \text{ kg})(9.8 \text{ m s}^{-2})\left\{\frac{\sqrt{(0.08)[(2)(0.5) - 0.08]} \text{ m}}{(0.5 - 0.08) \text{ m}}\right\}$$

$$= \boxed{1.14 \times 10^3 \text{ N}}$$

PROBLEM 9-9: Two rods of negligible weight having lengths of $a = 30$ cm and $b = 80$ cm are welded to form an L-shaped bracket. One end is attached to a frictionless pivot and the other end to a vertical cord. A weight of $W = 500$ N is attached to the mid-point of rod b. Find the tension in the cord when $\theta = 60°$. See Figure 9-15a.

Solution: Draw a free-body diagram for the welded rods, as in Figure 9-15b, where R is the reaction force at the pivot. You can greatly simplify this problem by recognizing the lever arms for the forces, i.e., x_2 for W and $(x_1 + x_2)$ for T. See Figure 9-15c. Using the geometry of the configuration, you can first obtain the value of x_2 and then that of x_1.

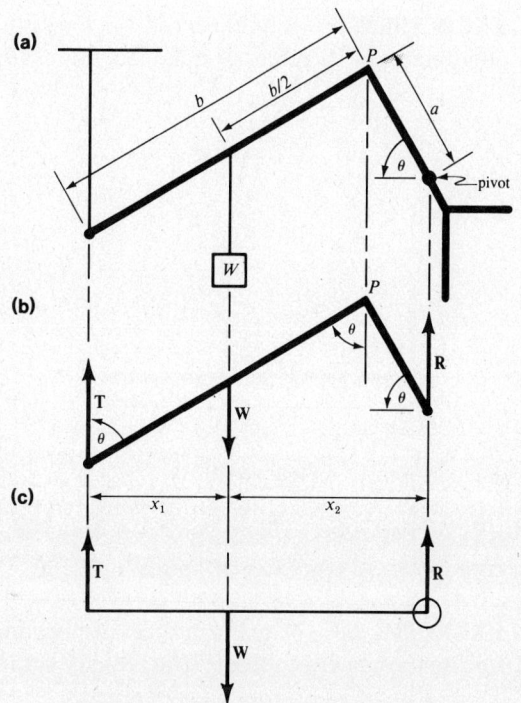

$$x_2 = a\cos\theta + \tfrac{1}{2}b\sin\theta$$
$$= (0.3 \text{ m})\cos 60° + \tfrac{1}{2}(0.8)\sin 60°$$
$$= 0.4964 \text{ m}$$

$$x_1 = \tfrac{1}{2}b\sin\theta = \tfrac{1}{2}(0.8 \text{ m})\sin 60° = 0.3464 \text{ m}$$

To find the tension in the cord, apply the second condition of equilibrium and sum torques about an axis through the pivot, or

$$\Sigma\tau = 0$$
$$x_2 W - (x_1 + x_2)T = 0$$

and solve for T.

$$T = W\frac{x_2}{x_1 + x_2}$$

$$= (500 \text{ N})\left[\frac{0.4964 \text{ m}}{(0.3464 + 0.4964) \text{ m}}\right]$$

$$= \boxed{294 \text{ N}}$$

Figure 9-15. (a) Bracket configuration. (b) Free-body diagram of welded rods. (c) Equivalent free-body diagram using lever arms.

PROBLEM 9-10: Determine the bending torque on the weld (point P in Figure 9-15) that tends to enlarge the right angle between the two rods in Problem 9-9. The *net* torque at P is zero because the rod is in equilibrium.

Solution: You should consider only the torques exerted by forces to the left of a vertical line drawn through P. (The torques due to forces from the right of P cancel those due to forces from the left.) Thus the bending torque at P is

$$\Sigma\tau_p = \tfrac{1}{2}bW\sin\theta - bT\sin\theta$$

$$= (\tfrac{1}{2}W - T)b\sin\theta$$

$$= [\tfrac{1}{2}(500\text{ N}) - 294\text{ N}](0.8\text{ m})\sin 60°$$

$$= \boxed{-30.5\text{ N m}}$$

Notice that the bending torque is clockwise, indicated by the negative sign.

Supplementary Exercises

EXERCISE 9-1:   A uniform rod 4 m long is bent 90° at a point 1 m from one end. Locate the x, y coordinates of its center of mass. See Figure 9-16.

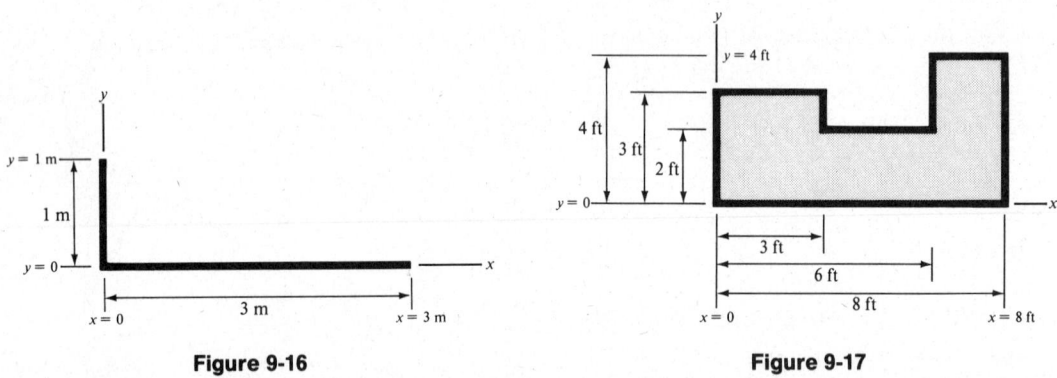

Figure 9-16   Figure 9-17

EXERCISE 9-2:   A uniform sheet of plywood is cut as shown in Figure 9-17. Locate the x, y coordinates of its center of mass.

EXERCISE 9-3:   A uniform sheet of plywood has a circular hole in it, as shown in Figure 9-18. Locate the center of mass. [*Hint:* It may help if you consider the hole as a "negative mass."]

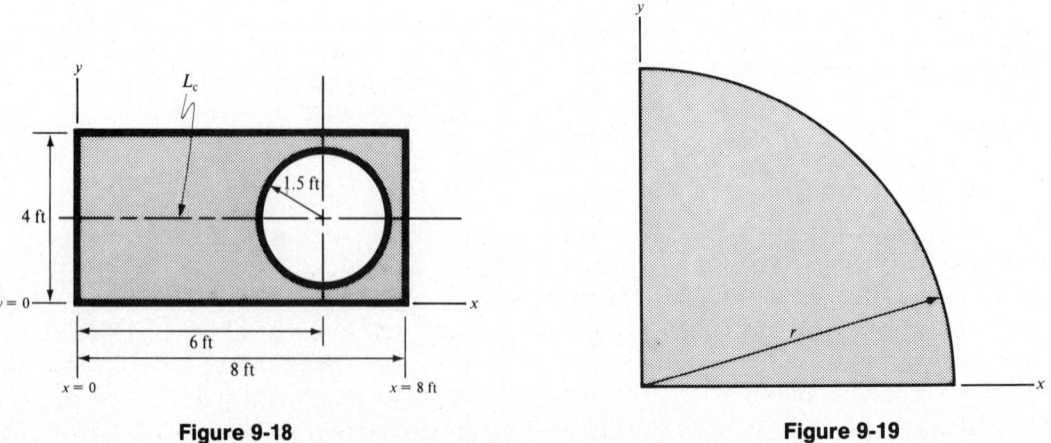

Figure 9-18   Figure 9-19

EXERCISE 9-4:   Locate the center of mass for a uniform sheet of aluminum cut as a quarter disk of radius r. See Figure 9-19.

EXERCISE 9-5: A uniform rod 60 cm long is pivoted at one end. The rod weighs 2 N and has a 3-N weight attached at its far end. A spring holds the rod, with the weight attached, at an angle of $\theta = 30°$ below the horizontal. How much torque is provided by the spring about the pivot? See Figure 9-20.

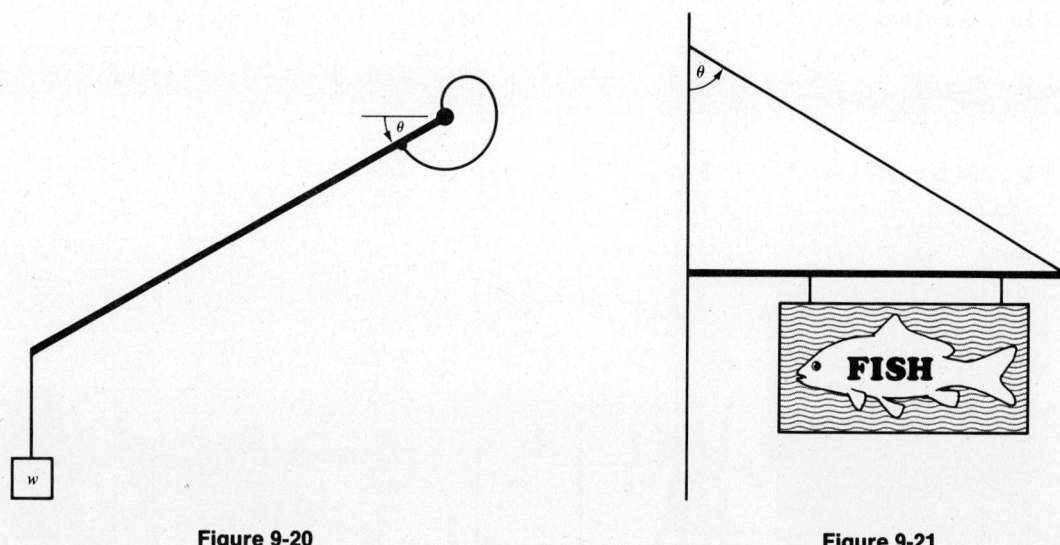

Figure 9-20　　　　　　**Figure 9-21**

EXERCISE 9-6: A sign made of a uniform sheet of wood has dimensions 1.4 m × 0.6 m and a mass of 20 kg. The edge of the sign is 0.5 m from the wall. The uniform rod that supports it has a mass of 12 kg and is 2 m long. Assume that the left end of the rod is pivoted at the wall. See Figure 9-21. Determine the tension in the supporting cable, which makes an angle of $\theta = 60°$ with the wall.

EXERCISE 9-7: A uniform rod 2 m long weighs 20 N, is pivoted at its lower end, and supports a 50-N weight by a cord at its upper end. The rod makes an angle $\theta = 50°$ with the horizontal, and the cable attached 1.5 m from the pivot makes an angle of $\phi = 65°$ with the rod. Find the tension in the cable. See Figure 9-22.

EXERCISE 9-8: Determine the x and y components of the reaction force provided by the pivot in Exercise 9-7.

EXERCISE 9-9: A uniform 30-ft ladder weighs 40 lb and leans against a smooth (frictionless) wall, making an angle of 30° with the wall. The coefficient of static friction between the bottom of the ladder and the ground is only 0.5. How far up the ladder can a 160-lb person climb before the ladder begins to slip?

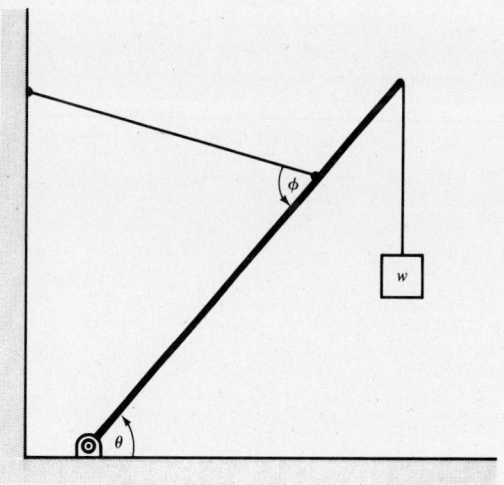

Figure 9-22

EXERCISE 9-10: A uniform ladder of length ℓ and weight w rests against a wall. The coefficient of static friction between the top of the ladder and the wall, and between the ladder and the ground upon which it rests, is μ_s. Calculate the tangent of the smallest angle the ladder can make with the ground before it will slip.

Answers for Supplementary Exercises

9-1: $x_{CM} = 1.125$ m, $y_{CM} = 0.125$ m

9-2: $x_{CM} = 4.196$ ft, $y_{CM} = 1.544$ ft

9-3: $x_{CM} = 3.433$ ft, $y_{CM} = 2$ ft

9-4: $x_{CM} = \dfrac{4r}{3\pi}$, $y_{CM} = \dfrac{4r}{3\pi}$

9-5: 2.08 N m, cw

9-6: 353 N

9-7: 56.7 N

9-8: $R_x = 54.8$ N, $R_y = 55.3$ N

9-9: 28.7 ft

9-10: $\dfrac{1 - \mu_s^2}{2\mu_s}$

10 RIGID BODIES II: DYNAMICS

THIS CHAPTER IS ABOUT

☑ **Moment of Inertia**
☑ **Kinetic Energy of Rotation**
☑ **Angular Momentum**
☑ **Torque and Angular Acceleration**
☑ **Conservation of Angular Momentum**

10-1. Moment of Inertia

The **moment of inertia** of a body is a measure of its resistance (inertia) to a change in its state of rotation. It is a quantity analogous to mass for linear motion. For a collection of particles having masses of $m_1, m_2, \ldots$, the moment of inertia, denoted by I, relative to an axis is

MOMENT OF INERTIA
$$I = m_1 r_1^2 + m_2 r_2^2 + m_3 r_3^2 + \cdots = \sum_i m_i r_i^2 \qquad \text{(10-1)}$$

where the sum includes all particles in the collection. (See Figure 10-1.) For a rigid body, the distances r to the mass particles from the axis, measured perpendicular to the axis, are fixed; therefore I is constant. It is measured in kg m^2 in SI units and slug ft^2 in British units.

When the mass of the body is distributed continuously throughout a volume, Eq. (10-1) becomes

$$I = \int r^2 \, dm \qquad \text{(10-2)}$$

where r is the perpendicular distance from the axis to the mass element dm, and the integration is carried out over the entire mass of the body. Occasionally, we find it desirable to express the moment of inertia of a body in terms of its entire mass M, or $M = \sum_i m_i$ for the collection of mass particles and $M = \int dm$ for the continuous body. In this case,

$$I = Mk^2 \qquad \text{(10-3)}$$

and k is called the **radius of gyration**, which is measured in meters or feet.

The **parallel axis theorem** allows us to calculate easily the moment of inertia of a body about an axis parallel to an axis that passes through the center of mass of the body. This theorem is stated as

PARALLEL AXIS THEOREM $\qquad I = I_c + Mb^2 \qquad \text{(10-4)}$

where I_c is the moment of inertia of the body relative to an axis passing through the center of mass, and b is the distance between the parallel axes.

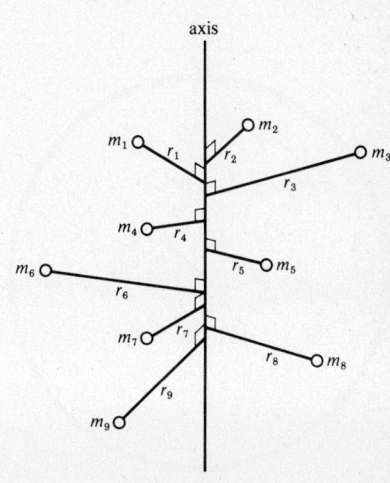

Figure 10-1

(a)

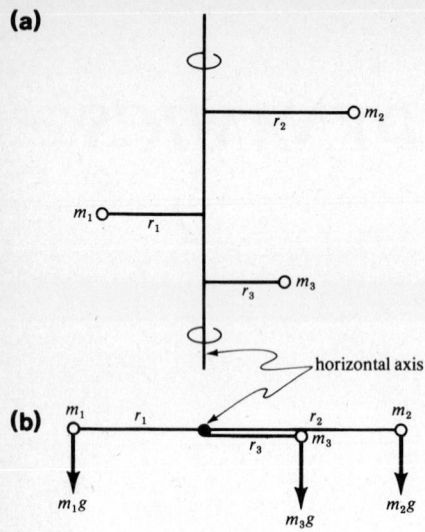

(b)

Figure 10-2

EXAMPLE 10-1: Figure 10-2a is a top view looking down on an object made of rigid wire of negligible mass and set in bearings to rotate about the horizontal axis. The mass particles are $m_1 = 400$ g, $m_2 = 100$ g, $m_3 = 300$ g, and the wire lengths are $r_2 = 0.20$ m and $r_3 = 0.10$ m. **(a)** Determine r_1 so that the net torque about the horizontal axis is zero. **(b)** Calculate the moment of inertia about the horizontal axis. **(c)** What is the radius of gyration for this object?

Solution:

(a) Figure 10-2b shows the object as viewed along its horizontal axis. The sum of the torques about the axis is to be zero, or

$$\Sigma\tau = 0$$
$$= m_1 g r_1 - m_2 g r_2 - m_3 g r_3$$

Solving for r_1, we get

$$r_1 = \frac{m_2 r_2 + m_3 r_3}{m_1} = \frac{(0.1 \text{ kg})(0.2 \text{ m}) + (0.3 \text{ kg})(0.1 \text{ m})}{0.4 \text{ kg}} = \boxed{0.125 \text{ m}}$$

(b) We use Eq. (10-1).

$$I = \sum_i m_i r_i^2 = m_1 r_1^2 + m_2 r_2^2 + m_3 r_3^2$$
$$= [(0.4)(0.125)^2 + (0.1)(0.2)^2 + (0.3)(0.1)^2] \text{ kg m}^2 = \boxed{1.325 \times 10^{-2} \text{ kg m}^2}$$

(c) We use Eq. (10-3) and solve for k.

$$I = Mk^2$$
$$k = \sqrt{\frac{I}{M}} = \sqrt{\frac{I}{m_1 + m_2 + m_3}} = \sqrt{\frac{1.325 \times 10^{-2} \text{ kg m}^2}{(0.4 + 0.1 + 0.3) \text{ kg}}} = \boxed{0.1287 \text{ m}}$$

Notice how we could interpret the radius of gyration: a single mass of 0.8 kg located 0.1287 m from the axis has the same moment of inertia as the three-mass object in this example.

EXAMPLE 10-2: **(a)** Calculate the moment of inertia for a uniform circular disk of mass M and radius R about an axis through its center of mass and perpendicular to the plane of the disk. **(b)** What is its radius of gyration?

Solution: A sketch of the uniform disk is shown in Figure 10-3.

(a) We must use Eq. (10-2) in this case. The mass per unit area is $\sigma = M/\pi R^2$, and, from the figure, the element of area is $dA = (r\,d\theta)(dr) = r\,dr\,d\theta$. The element of mass associated with dA is $dm = \sigma\,dA = \sigma r\,dr\,d\theta$.

$$I = \int r^2\,dm = \int r^2(\sigma r\,dr\,d\theta) = \sigma \int_0^{2\pi}\int_0^R r^3\,dr\,d\theta$$

The integration on θ is independent of r, so

$$I = \sigma \int_0^{2\pi} d\theta \int_0^R r^3\,dr = \sigma(2\pi)\left(\frac{R^4}{4}\right) = 2\pi\left(\frac{M}{\pi R^2}\right)\left(\frac{R^4}{4}\right) = \boxed{\frac{1}{2}MR^2}$$

(b) Solve Eq. (10-3) for k.

$$I = Mk^2$$
$$k = \sqrt{\frac{I}{M}} = \sqrt{\frac{\frac{1}{2}MR^2}{M}} = \frac{R}{\sqrt{2}} = \boxed{\frac{\sqrt{2}}{2}R}$$

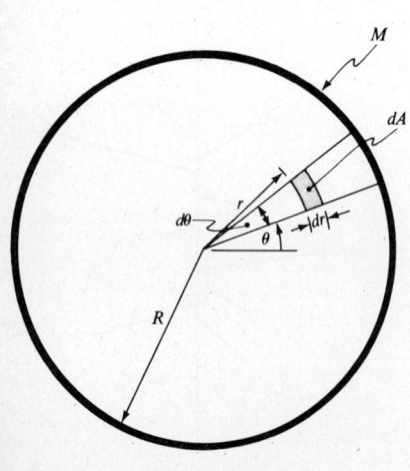

Figure 10-3

EXAMPLE 10-3: What is the moment of inertia of the disk in Example 10-2 when the axis is located on the edge of the disk, perpendicular to its surface?

Solution: We use the parallel axis theorem, Eq. (10-4), where $I_c = \frac{1}{2}MR^2$ and $b = R$.

$$I = I_c + Mb^2 = \frac{1}{2}MR^2 + MR^2 = \boxed{1.5MR^2}$$

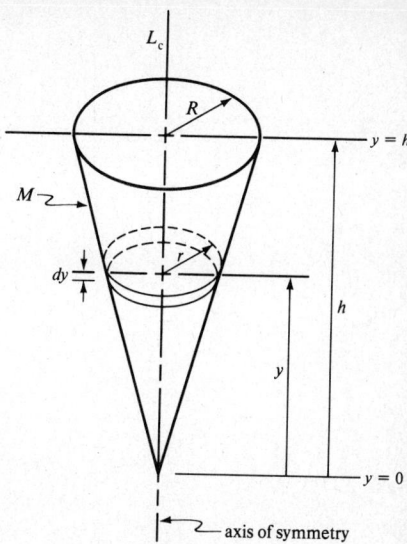

Figure 10-4

EXAMPLE 10-4: Use $I = \frac{1}{2}MR^2$ for a uniform disk about its center of mass to obtain the moment of inertia of a right circular cone of base radius R and height h about its axis of symmetry. Assume that the mass of the cone is uniformly distributed throughout its volume. The volume of a cone is $V = \frac{1}{3}$(area of base)(height). For the cone in this example, $V = \frac{1}{3}\pi R^2 h$. (See Figure 10-4).

Solution: In this case the disk (a section of the cone) at y has a moment of inertia $dI = \frac{1}{2}r^2\,dm$. Its mass is $dm = \rho\,dV = \rho(\pi r^2)\,dy$, where ρ is the mass per unit volume (the density) of the cone, or $\rho = M(\frac{1}{3}\pi R^2 h)^{-1}$, and dV is the

TABLE 10-1: Moment of Inertia of Selected Objects

Object	Diagram	Formula
Thin rod of length L, axis perpendicular to rod		$I = \frac{1}{12}ML^2$
Cylinder or disk, axis perpendicular to disk surface		$I = \frac{1}{2}MR^2$
Hoop (supported by spokes of negligible mass), axis perpendicular to plane of hoop		$I = MR^2$
Disk, axis along a diameter		$I = \frac{1}{4}MR^2$
Solid uniform sphere, axis along a diameter		$I = \frac{2}{5}MR^2$
Thin spherical shell, axis along a diameter		$I = \frac{2}{3}MR^2$
Rectangular sheet of dimensions $a \times b$, axis perpendicular to surface		$I = \frac{1}{12}M(a^2 + b)$

M is the total mass of the body and R is the radius of circular or spherical objects. The axis passes through the center of mass.

volume of the disk, or $dV = (\pi r^2)\,dy$. Thus

$$I = \int dI = \int \left(\frac{1}{2}r^2\right)\rho(\pi r^2)\,dy = \frac{1}{2}\rho\pi\int_0^h r^4\,dy$$

The relationship between r and y is proportional to that of R and h (similar triangles). Therefore $r/y = R/h$ and

$$I = \frac{1}{2}\rho\pi\int_0^h \left(\frac{R}{h}\right)^4 y^4\,dy = \frac{1}{2}\left(\frac{3M}{\pi R^2 h}\right)\pi\left(\frac{R}{h}\right)^4\left(\frac{1}{5}y^5\bigg|_0^h\right)$$

$$= \left(\frac{3MR^2}{2h^5}\right)\left(\frac{1}{5}h^5\right) = \boxed{\frac{3}{10}MR^2}$$

The moments of inertia for selected objects are shown in Table 10-1. You can extend the usefulness of this table by applying the parallel axis theorem.

10-2. Kinetic Energy of Rotation

If we let the system of mass particles shown in Figure 10-1 rotate about the axis with angular velocity ω, the kinetic energy associated with this rotation is the sum of kinetic energies of each of the particles; i.e.,

$$E_{k,\,rot} = \tfrac{1}{2}m_1 v_1^2 + \tfrac{1}{2}m_2 v_2^2 + \tfrac{1}{2}m_3 v_3^2 + \cdots = \tfrac{1}{2}\sum_i m_i v_i^2$$

From Section 3-7, for rotation about an axis, we know that $v = \omega r$, so $v_1 = \omega r_1$, $v_2 = \omega r_2, \ldots, v_i = \omega r_i$. Therefore

$$E_{k,\,rot} = \tfrac{1}{2}\Sigma m_i(\omega r_i)^2 = \tfrac{1}{2}(\Sigma m_i r_i^2)\omega^2$$

and Eq. (10-1) states that $I = \Sigma m_i r_i^2$, which is the moment of inertia about the axis of rotation. Thus

KINETIC ENERGY OF ROTATION
$$E_{k,\,rot} = \tfrac{1}{2}I\omega^2 \qquad\qquad\qquad (10\text{-}5a)$$

We use the parallel axis theorem, Eq. (10-4), in Eq. (10-5a) to obtain

$$E_{k,\,rot} = \tfrac{1}{2}(I_c + Mb^2)\omega^2 = \tfrac{1}{2}I_c\omega^2 + \tfrac{1}{2}Mb^2\omega^2$$

and

$$E_{k,\,rot} = \tfrac{1}{2}I_c\omega^2 + \tfrac{1}{2}Mv_{CM}^2 \qquad\qquad (10\text{-}5b)$$

We can express the kinetic energy associated with the rotation of a body about an axis at angular velocity ω in two ways: (1) with Eq. (10-5a), where I is the moment of inertia of the body about the axis of rotation; or (2) as a sum, as in Eq. (10-5b). The first term on the right-hand side of Eq. (10-5b) is the kinetic energy of rotation of the body about the center of mass. (I_c is the moment of inertia of the body about an axis that passes through the center of mass and is parallel to the axis of rotation.) The second term is the kinetic energy of translation of the center of mass of the body.

EXAMPLE 10-5: A small engine drives a flywheel up to 3000 rev min⁻¹ in the power unit of a lightweight experimental car ($m = 1200$ kg). The flywheel consists of a 10-kg thin disk with a radius of 0.4 m to which two 10-kg hoops, 0.4 m in radius, are welded. The flywheel spins about an axis through the center of mass, parallel to the plane of the disk, as shown in Figure 10-5. (a) Find the moment of inertia of the flywheel. (b) How much energy is stored in the spinning flywheel? (c) Assume that the car is moving on a level road at 10 m s⁻¹, with the flywheel spinning at 3000 rev min⁻¹. When the car reaches the base of a hill, the driver uses the clutch to disconnect the engine from the flywheel and the car

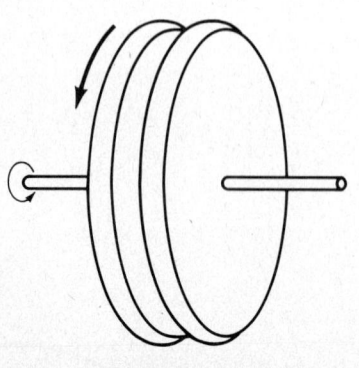

Figure 10-5

rolls up the hill. What elevation above the road does the car reach when it stops? Neglect friction.

Solution:

(a) Moment of inertia, like mass, is a scalar quantity, so the total moment of inertia is the sum of its components. Thus

$$I_{\text{flywheel}} = I_{\text{disk}} + I_{\text{hoop}} + I_{\text{hoop}}$$

In this case the mass of each hoop is equal to the mass of the disk ($M = 10$ kg), and all three have the same radius ($R = 0.4$ m), so

$$I = \tfrac{1}{2}MR^2 + MR^2 + MR^2 = 2.5MR^2 = (2.5)(10 \text{ kg})(0.4 \text{ m})^2 = \boxed{4.0 \text{ kg m}^2}$$

(b) We use Eq. (10-5a), with

$$\omega = (3000 \text{ rev min}^{-1})(2\pi \text{ rad rev}^{-1})\left(\frac{1}{60} \text{ min s}^{-1}\right) = 314 \text{ rad s}^{-1}$$

Thus

$$E_{\text{k, rot}} = \tfrac{1}{2}I\omega^2 = \tfrac{1}{2}(4 \text{ kg m}^2)(314 \text{ rad s}^{-1})^2 = \boxed{1.97 \times 10^5 \text{ J}}$$

(c) The total kinetic energy of the car at the instant it starts up the hill is the sum of the kinetic energy of translation and the kinetic rotational energy stored in the flywheel. This quantity of energy is transformed into gravitational potential energy as the car climbs the hill. We apply the work–energy principle, Eq. (5-5b), where $W_{\text{net}} = m(-g)h$, or the work done by the force of gravity,

$$W_{\text{net}} = E_{\text{k, final}} - E_{\text{k, init}} \qquad m(-g)h = 0 - (\tfrac{1}{2}mv^2 + E_{\text{k, rot}})$$

and solve for h to obtain

$$h = \frac{v^2}{2g} + \frac{E_{\text{k, rot}}}{mg} = \frac{(10 \text{ m s}^{-1})^2}{2(9.8 \text{ m s}^{-2})} + \frac{1.97 \times 10^5 \text{ J}}{(1500 \text{ kg})(9.8 \text{ m s}^{-2})} = \boxed{18.5 \text{ m}}$$

EXAMPLE 10-6: (a) Consider a pool ball of mass M and radius R. It starts from rest at height h and rolls down an incline without slipping. Derive an expression for its speed when it reaches the bottom of the incline. (b) Repeat (a) for a right circular cylinder of the same mass and radius as the pool ball. (c) In a race between the pool ball and cylinder, which one would win?

Solution:

(a) We use the work–energy principle, Eq. (5-5b), where $W_{\text{net}} = M(-g)(-h) = Mgh$ in this case and $v = \omega R$ (because there is no slipping). Solve for v.

$$W_{\text{net}} = E_{\text{k, final}} - E_{\text{k, initial}}$$

$$Mgh = \left(\frac{1}{2}Mv^2 + \frac{1}{2}I_b\omega^2\right) - 0$$

$$= \frac{1}{2}Mv^2 + \frac{1}{2}\left(\frac{2}{5}MR^2\right)\left(\frac{v^2}{R^2}\right) = \frac{1}{2}Mv^2\left(1 + \frac{2}{5}\right)$$

$$v = \sqrt{\frac{10}{7}gh}$$

Note that we used $I_b = \frac{2}{5}MR^2$ (solid sphere) from Table 10-1. Interestingly, the result is independent of the mass and radius of the ball. This result implies that all solid spheres would have identical speeds upon reaching the bottom of the incline.

(b) The analysis is the same as for (a), except that we use the moment of inertia of a cylinder, $I_c = \frac{1}{2}MR^2$ from Table 10-1. The result is $v = \sqrt{\frac{4}{3}gh}$.

(c) Because $\sqrt{10/7} > \sqrt{4/3}$, the pool ball would have the greater speed at the bottom of the incline and thus would win the race. In a race of this type, the object with the smallest moment of inertia would reach the bottom first. An object with a larger moment of inertia converts a larger proportion of its available potential energy into *translational* kinetic energy; an object with a smaller moment of inertia converts a larger proportion of its available potential energy into *rotational* kinetic energy.

10-3. Angular Momentum

The **angular momentum** of a particle of mass m at a distance $\mathbf{r}$ from the origin and moving with velocity $\mathbf{v}$ (and hence with linear momentum $\mathbf{p} = m\mathbf{v}$) is

ANGULAR MOMENTUM $$\mathbf{L} = \mathbf{r} \times \mathbf{p} \qquad (10\text{-}6)$$

which we can also write as $\mathbf{L} = m(\mathbf{r} \times \mathbf{v})$. The units of angular momentum are $\text{kg m}^2 \text{ s}^{-1}$, which we can also express in J s. Figure 10-6 shows a particle of mass m moving in the x, y plane, i.e., the plane of the page. The angular momentum for this particle is directed along the positive z axis, projecting upward from the page. Its magnitude at a specific instant is

$$L = rp \sin \phi = mvr \sin \phi.$$

The angular momentum of a system of particles is the vector sum of the momentum of all the particles, or

$$\mathbf{L}_{\text{system}} = \sum_i (\mathbf{r}_i \times \mathbf{p}_i)$$

A rigid body rotating about an *axis of symmetry that also passes through the center of mass* has an angular momentum of

$$\mathbf{L} = I\boldsymbol{\omega} \qquad (10\text{-}7)$$

where I is the moment of inertia of the body about that axis, and $\boldsymbol{\omega}$ is the **angular velocity vector**.

We can obtain the direction of $\boldsymbol{\omega}$ by using a right-hand rule: Grip the axis of rotation with the right hand, with your fingertips pointing in the direction of motion; the extended thumb indicates the direction of $\boldsymbol{\omega}$. Figure 10-7 shows an example of the right-hand rule for a disk of mass M and radius R, rotating about an axis of symmetry passing through its center of mass. In this case $\mathbf{L} = I\boldsymbol{\omega}$, where $I = \frac{1}{2}MR^2$. Later, we'll consider examples in which $\mathbf{L}$ and $\boldsymbol{\omega}$ aren't parallel.

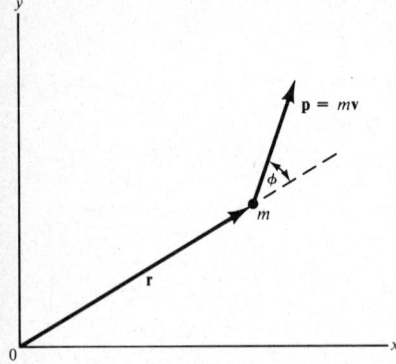

Figure 10-6

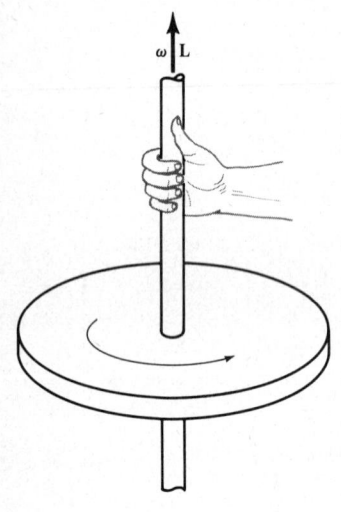

Figure 10-7

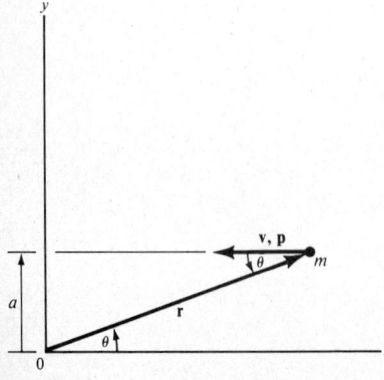

Figure 10-8

EXAMPLE 10-7: A particle of mass m moves at constant velocity $\mathbf{v}$ parallel to the x axis in the negative x direction at distance a above the x axis. **(a)** Prove that the angular momentum for this particle has the constant value $L = mva$. **(b)** Determine the direction of $\mathbf{L}$.

Solution:

(a) At a certain instant the particle is at the position shown in Figure 10-8. Use Eq. (10-6), with $p = mv$ and $\mathbf{L} = \mathbf{r} \times \mathbf{p}$. The magnitude of $\mathbf{L}$ is

$$L = rp \sin \theta = rmv \sin \theta$$

Note that $a = r \sin \theta$. Thus

$$L = \boxed{mva}$$

which is constant in time.

(b) We use the right-hand rule for the vector cross product to determine that $\mathbf{L}$ is directed in the positive z direction, projecting from the page in Figure 10-8.

EXAMPLE 10-8: A 4-kg disk, 30 cm in diameter, is spinning about an axis passing through its center and perpendicular to the plane of the disk. What angular momentum does the disk have when it spins at 600 rev min^{-1}?

Solution: Use Eq. (10-7), $\mathbf{L} = I\boldsymbol{\omega}$. Use the right-hand rule to find the direction of $\boldsymbol{\omega}$ from the direction of rotation. In this case, the direction of $\mathbf{L}$ is the same as that of $\boldsymbol{\omega}$, which is along the axis of rotation. To obtain the magnitude of $L = I\omega$, we use $I = \frac{1}{2}MR^2$ and

$$\omega = (600 \text{ rev min}^{-1})(2\pi \text{ rad rev}^{-1})(60 \text{ s min}^{-1})^{-1} = 20\pi \text{ rad s}^{-1}$$

$$L = I\omega = (\tfrac{1}{2}MR^2)\omega$$

$$= [\tfrac{1}{2}(4 \text{ kg})(0.15 \text{ m})^2](20\pi \text{ rad s}^{-1}) = \boxed{2.83 \text{ kg m}^2\text{ s}^{-1} = 2.83 \text{ J s}}$$

EXAMPLE 10-9: A dumbbell is made from a rod of negligible mass with a length of $2a$ to which a sphere of mass m is attached each end. The rod makes a fixed angle ϕ with respect to the z axis, about which it spins at its midpoint with angular velocity ω. (See Figure 10-9.) Derive an expression for the angular momentum of this system.

Solution: The angular momentum for this object is the angular momentum owing to the motion of both the upper and lower masses. Let's consider first the upper mass and use Eq. (10-6), $\mathbf{L} = \mathbf{r} \times \mathbf{p}$. If we use the midpoint of the rod as the origin of a coordinate system, $r = a$, $p = mv$, and $v = \omega a \sin \phi$. Note that r and p are perpendicular to each other, so that the magnitude of the angular momentum owing to the upper mass is

$$L_{\text{upper}} = (a)(m)(\omega a \sin \phi) = ma^2 \omega \sin \phi$$

When we use the right-hand rule, we see that the direction of L_{upper} is perpendicular to the rod and makes an angle θ with respect to the spin axis; $\theta + \phi = 90°$.

When we analyze the motion of the lower mass, we find that its contribution to the total angular momentum is equal in magnitude and has the same direction as the upper mass. Thus the total angular momentum is

$$L = L_{\text{upper}} + L_{\text{lower}} = \boxed{2ma^2 \omega \sin \phi}$$

Note that the magnitude of $\mathbf{L}$ is constant but that $\mathbf{L}$ itself is not constant in time because its direction changes continuously. We say that $\mathbf{L}$ *precesses* about the z axis with angular velocity ω. The z component of $\mathbf{L}$, however, is constant in time and has the value

$$L_z = L \cos \theta = L \sin \phi = 2ma^2 \omega \sin^2 \phi$$

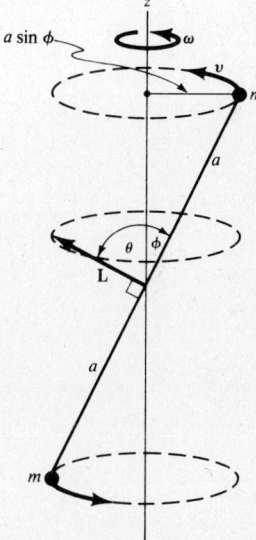

Figure 10-9

10-4. Torque and Angular Acceleration

A net external torque is required to change the angular momentum of a system. This change may be in magnitude, direction, or both, and is a statement of Newton's second law for rotational motion, i.e.,

$$\Sigma \boldsymbol{\tau} = \frac{d\mathbf{L}}{dt} \tag{10-8}$$

We'll consider two special cases, both involving rigid bodies rotating around an axis that passes through their center of mass: $\mathbf{L}$ changes in magnitude only, and $\mathbf{L}$ changes in direction only.

A. Change in magnitude

In this case **L** and **ω** are parallel, so we can use Eq. (10-7), $\mathbf{L} = I\boldsymbol{\omega}$, and substitute into Eq. (10-8) to obtain

$$\Sigma\boldsymbol{\tau} = \frac{d\mathbf{L}}{dt} = \frac{d(I\boldsymbol{\omega})}{dt} = I\frac{d\boldsymbol{\omega}}{dt} + \frac{dI}{dt}\boldsymbol{\omega}$$

For a rigid body $\dfrac{dI}{dt} = 0$, so that Eq. (10-8) becomes

$$\Sigma\boldsymbol{\tau} = I\frac{d\boldsymbol{\omega}}{dt} = I\boldsymbol{\alpha} \qquad \text{(10-9)}$$

where $\boldsymbol{\alpha}$ is the **angular acceleration vector**.

EXAMPLE 10-10: The engine in Example 10-5 can accelerate the flywheel from 0 to 3000 rev min^{-1} in 6 s. **(a)** How much torque does the engine exert on the flywheel? **(b)** How much power is delivered by the engine during this acceleration?

Solution: From Example 10-5, the moment of inertia of the flywheel is $I = 4.0 \text{ kg m}^2$, 3000 rev min^{-1} = 314 rad s^{-1}, and the kinetic energy of rotation stored in the flywheel spinning at 3000 rev min^{-1} is $E_{k,\text{rot}} = 1.97 \times 10^5$ J.

(a) Using Eq. (3-22a) to determine α,

$$\omega = \omega_0 + \alpha t$$

$$\alpha = \frac{\omega - \omega_0}{t} = \frac{314 \text{ rad s}^{-1} - 0}{6 \text{ s}} = 52.4 \text{ rad s}^{-2}$$

and Eq. (10-9) to determine the torque, we get

$$\Sigma\tau = I\alpha$$

$$\tau = (4.0 \text{ kg m}^2)(52.4 \text{ rad s}^{-2}) = \boxed{209 \text{ N m}}$$

(b) We obtain the power by using Eq. (5-6), so

$$P = \frac{dW}{dt} = \frac{1.97 \times 10^5 \text{ J}}{6 \text{ s}} = \boxed{3.28 \times 10^4 \text{ W}}$$

or in horsepower,

$$P = (3.28 \times 10^4 \text{ W})(746 \text{ W hp}^{-1})^{-1} = \boxed{44 \text{ hp}}$$

EXAMPLE 10-11: A disk of radius $R = 0.12$ m and mass $M = 3$ kg is supported by frictionless bearings. A lightweight cord passes over the rim of the disk and is attached to two blocks of $m_1 = 2.2$ kg and $m_2 = 0.8$ kg. (See Figure 10-10a). The friction between the cord and disk prevents the cord from slipping. What is the acceleration of this system if m_1 slides on a frictionless surface?

Figure 10-10

(a)

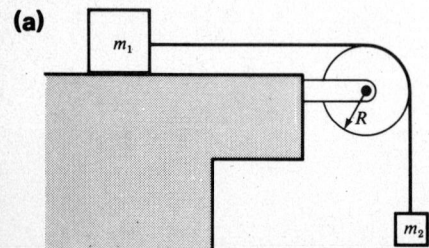

(b)

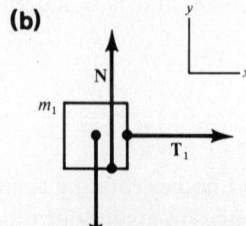

(c)

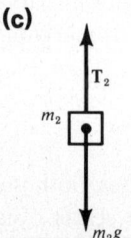

(d)

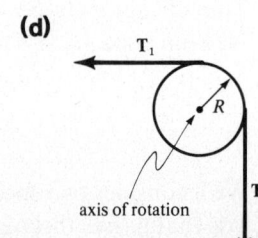

Solution: Draw free-body diagrams for m_1 (Figure 10-10b), m_2 (Figure 10-10c), and the disk (Figure 10-10d) and apply Newton's second law. Note that the acceleration of m_1 (in the positive x direction) is the same as the downward acceleration of m_2. This is the acceleration that we want to find; let's designate it a.

For m_1;

$$\Sigma F_x = m_1 a \qquad\qquad \Sigma F_y = m_1 a_y \qquad\qquad (a_y = 0)$$

$$T_1 = m_1 a \qquad\qquad N - m_1 g = 0$$

$$N = m_1 g \qquad\qquad \text{(This result isn't needed in this case.)}$$

Let's assign downward to be the positive direction, so for m_2,

$$\Sigma F = m_2 a$$

$$m_2 g - T_2 = m_2 a$$

Solve for T_2.

$$T_2 = m_2 g - m_2 a$$

For the disk, let's assign clockwise rotation to be positive, and $I = \frac{1}{2} M R^2$. We use Eq. (10-9) to calculate the magnitude.

$$\Sigma \tau = I \alpha \qquad\qquad T_2 R - T_1 R = I \left(\frac{a}{R}\right)$$

We substitute for T_1 and T_2 and solve for a.

$$(m_2 g - m_2 a)R - (m_1 a)R = I \left(\frac{a}{R}\right)$$

$$a = \frac{m_2 g}{\dfrac{I}{R^2} + m_1 + m_2} = \frac{m_2 g}{\dfrac{1}{2} M + m_1 + m_2} = \frac{(0.8\text{ kg})(9.8\text{ m s}^{-2})}{(1.5 + 2.2 + 0.8)\text{ kg}} = \boxed{1.74\text{ m s}^{-2}}$$

The direction of the torque, as given by the cross product $\boldsymbol{\tau} = \mathbf{r} \times \mathbf{F} = \mathbf{R} \times \Sigma m\mathbf{a}$, is along the axis of rotation perpendicular to the page in Figure 10-10d.

B. Change in direction

Example 10-9 is a case in which the magnitude of $\mathbf{L}$ is constant, but its direction changes.

EXAMPLE 10-12: Determine the torque exerted on the dumbbell in Example 10-9.

Solution: Figure 10-11 shows how the angular momentum vector precesses about the axis of rotation. In the time increment dt the component of $\mathbf{L}$ perpendicular to the axis of rotation, $L \sin \theta$, moves through the angle $\omega\, dt$. Thus the magnitude of $d\mathbf{L}$ is $|d\mathbf{L}| = (L \sin \theta)(\omega\, dt)$, and the magnitude of the torque, from Eq. (10-8), is

$$\tau = \frac{|d\mathbf{L}|}{dt} = \omega L \sin \theta$$

From Example 10-9, $L = 2ma^2 \omega \sin \phi$ and $\sin \theta = \cos \phi$. Therefore

$$\boxed{\tau = 2ma^2 \omega^2 \sin \phi \cos \phi}$$

The direction of τ is the same as $d\mathbf{L}$, or always perpendicular to both $\mathbf{L}$ and the line that joins the two masses. This torque is supplied by bearings that constrain

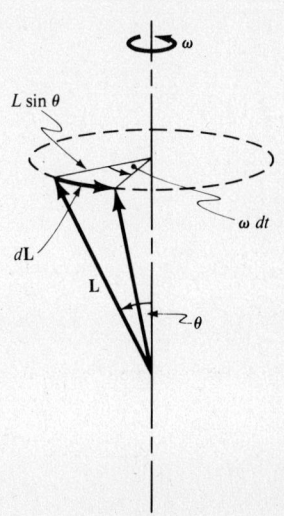

Figure 10-11

the dumbbell's rotation about ω. Note that when $\phi = 90°$ the line joining the masses is perpendicular to the axis of rotation, and the bearings no longer supply a torque. We then say that the object possesses *dynamic balance*.

10-5. Conservation of Angular Momentum

From Newton's second law of rotational motion, Eq. (10-8), we can deduce that when the net external torque is zero, $\frac{d\mathbf{L}}{dt} = 0$. This condition implies that the angular momentum must be constant, which leads to the statement of the **law of conservation of angular momentum**:

- If the net external torque acting on a system is zero, the system's angular momentum is constant.

(a)

(b)

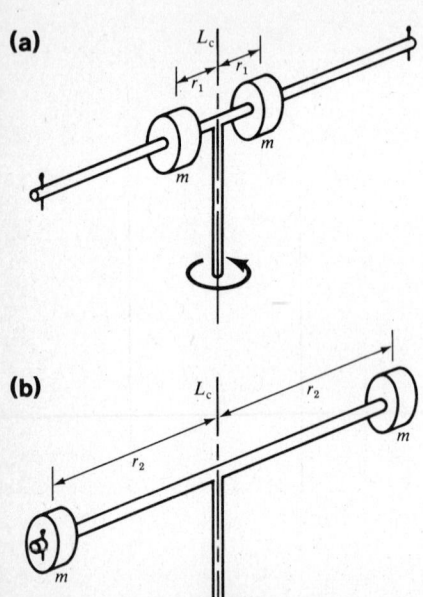

Figure 10-12

EXAMPLE 10-13: A rod 80 cm long is pivoted at its center and permitted to rotate freely about an axis perpendicular to its length. The moment of inertia of the rod about this axis is 5×10^{-2} kg m². Two weights, each of mass 0.5 kg, are placed on the rod; they can slide along the rod but not without friction. This system is set spinning at 12 rad s^{-1}, with the masses located initially at $r_1 = 10$ cm, as shown in Figure 10-12a. They slide along the rod until stopped by pins at the ends of the rod, which are located at $r_2 = 37$ cm, as shown in Figure 10-12b. (a) What is the angular velocity of this device when the weights are at r_2? (b) How much kinetic energy of rotation is lost during the motion of the masses?

Solution:

(a) Because no external torque acts on this device we can use the conservation of angular momentum principle: The angular momentum in configuration 1 is equal to the angular momentum in configuration 2. Because the device is symmetrical and rotating around an axis that passes through its center of mass, we can use Eq. (10-7), $\mathbf{L} = I\boldsymbol{\omega}$. So $I_1\omega_1 = I_2\omega_2$, and solve for ω_2.

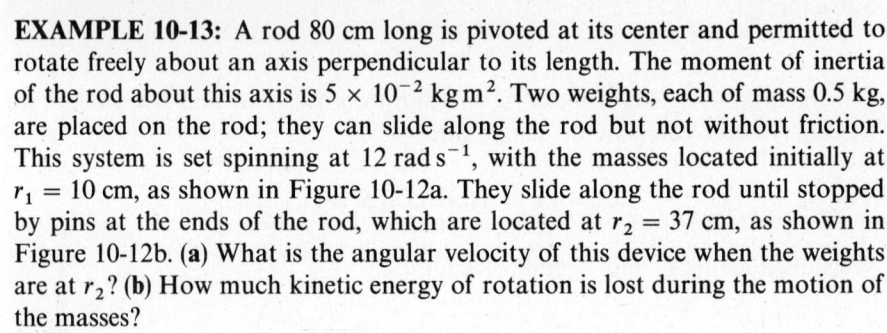

$$\omega_2 = \omega_1 \frac{I_1}{I_2} = \omega_1 \left(\frac{I_{\text{rod}} + 2mr_1^2}{I_{\text{rod}} + 2mr_2^2} \right)$$

$$= (12 \text{ rad s}^{-1}) \left(\frac{5 \times 10^{-2} \text{ kg m}^2 + 2(0.5 \text{ kg})(0.1 \text{ m})^2}{5 \times 10^{-2} \text{ kg m}^2 + 2(0.5 \text{ kg})(0.37 \text{ m})^2} \right)$$

$$= (12 \text{ rad s}^{-1}) \left(\frac{6 \times 10^{-2} \text{ kg m}^2}{1.87 \times 10^{-1} \text{ kg m}^2} \right) = \boxed{3.85 \text{ rad s}^{-1}}$$

(b) We use Eq. (10-5a) to calculate the kinetic energy of rotation in each configuration and obtain the difference.

$$\Delta E_{\text{k, rot}} = \tfrac{1}{2}I_1\omega_1^2 - \tfrac{1}{2}I_2\omega_2^2 = \tfrac{1}{2}(I_1\omega_1^2 - I_2\omega_2^2)$$

$$= \tfrac{1}{2}[(6 \times 10^{-2})(12)^2 - (1.87 \times 10^{-1})(3.85)^2] \text{ J} = \boxed{2.93 \text{ J}}$$

Note that work is done by this system in moving the weights from r_1 to r_2.

SUMMARY

1. The *moment of inertia* of a collection of masses relative to an axis of rotation is $I = \Sigma m_i r_1^2$, where r_i is the perpendicular distance from the mass m_i to the axis, and the sum includes all masses in the collection.
2. When the mass of an object is distributed continuously, the moment of inertia can be calculated from the integral $I = \int r^2 \, dm$.
3. The *radius of gyration k* of a body relative to an axis is obtained from its total mass M and its moment of inertia about that axis by solving the equation $I = Mk^2$ for k.
4. The moment of inertia of a body relative to an axis parallel to one that passes through the CM is obtained from the *parallel axis theorem*, $I = I_c + Mb^2$, where I_c is the moment of inertia relative to an axis passing through the CM, M is the total mass of the body, and b is the distance between axes.
5. The *kinetic energy* associated with the rotation of a body about an axis at an angular velocity ω is $E_{k,\,rot} = \frac{1}{2} I \omega^2$, where I is the moment of inertia relative to that axis.
6. The *angular momentum* of a particle that has linear momentum $\mathbf{p}$ and is located at distance $\mathbf{r}$ from the origin is $\mathbf{L} = \mathbf{r} \times \mathbf{p}$.
7. The angular momentum of a rigid body rotating at an angular velocity of $\boldsymbol{\omega}$ about an axis of symmetry that passes through the CM is $\mathbf{L} = I\boldsymbol{\omega}$.
8. Newton's second law for rotational motion states that the net torque acting on a body results in a change in its angular momentum, or $\Sigma \tau = d\mathbf{L}/dt$. If $\mathbf{L}$ can change only in magnitude, not direction, Newton's second law for rotational motion becomes $\Sigma \tau = I\boldsymbol{\alpha}$, where $\boldsymbol{\alpha}$ is the angular acceleration vector. If $\mathbf{L}$ can change only in direction, not magnitude, $\mathbf{L}$ *precesses* about the rotational axis.
9. The *law of conservation of angular momentum* states that if the net external torque acting on a body is zero, the body's angular momentum remains constant in magnitude and direction.

RAISE YOUR GRADES

Can you explain . . . ?

- ☑ how two different disks could have the same mass but different moments of inertia
- ☑ how two different disks could have the same moments of inertia but different masses
- ☑ why a hoop has a greater radius of gyration than a disk when they both have the same mass and radius
- ☑ why a disk has a greater moment of inertia than a sphere when both have the same mass and diameter
- ☑ why an ice skater spins faster when she pulls her arms in toward her body than when she extends them
- ☑ why a gyro top precesses
- ☑ how a gyroscope can be used as a compass
- ☑ why race cars always travel counter-clockwise around race tracks
- ☑ why it is much easier to balance on a moving bicycle than on one that's at rest

SOLVED PROBLEMS

Moment of Inertia

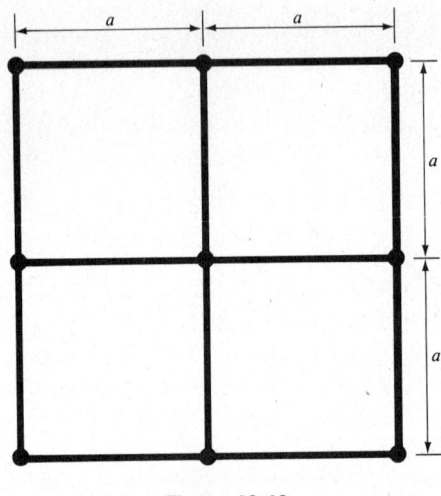

Figure 10-13

PROBLEM 10-1: Calculate the moment of inertia for the structure shown in Figure 10-13 for an axis that passes through its center perpendicular to the plane of the page. Each ball has a mass m, but the connecting rods have negligible mass.

Solution: Use Eq. (10-1). The mass in the center doesn't contribute to the sum because its r value is zero. There are 4 masses located at $r = a$ and 4 located at $r = \sqrt{2}a$ from the axis. Thus

$$I = \sum_i m_i r_i^2 = 4(ma^2) + 4m(\sqrt{2}\,a)^2 = \boxed{12ma^2}$$

PROBLEM 10-2: Determine the moment of inertia of the ring shown in Figure 10-14a relative to the axis passing through its center and perpendicular to the plane of the ring. The ring has mass M.

Solution: Use Eq. (10-2) and consider the ring as a series of hoops. A typical hoop is shown in Figure 10-4b and has a radius r and thickness dr. The moment of inertia of this hoop is $dI = r^2\,dm$, where dm is the mass of the hoop. (See Table 10-1). You can determine dm from σ, the mass per unit area, and dA, the area of the hoop.

(a)

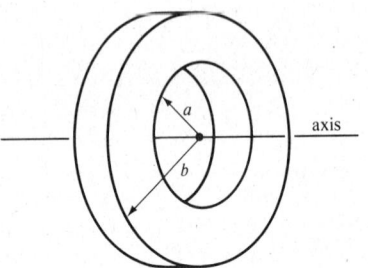

$$\sigma = \frac{M}{\pi b^2 - \pi a^2} = \frac{M}{\pi(b^2 - a^2)}$$

and

$$dA = 2\pi r\,dr$$

Thus

$$dm = \sigma\,dA = \left(\frac{2M}{b^2 - a^2}\right)(r\,dr)$$

(b)

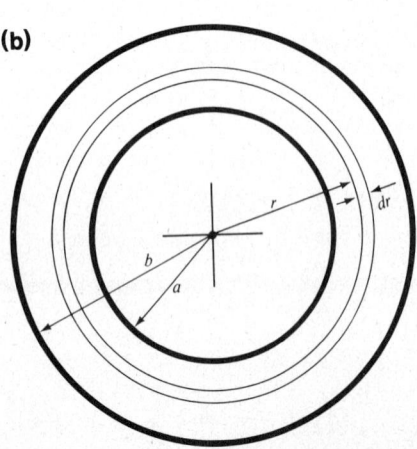

Figure 10-14

and

$$I = \int_a^b r^2\,dm = \int_a^b r^2\left(\frac{2M}{b^2 - a^2}\right)r\,dr$$

$$= \frac{2M}{b^2 - a^2}\int_a^b r^3\,dr$$

$$= \left(\frac{2M}{b^2 - a^2}\right)\left(\frac{1}{4}r^4\Big|_a^b\right)$$

$$= \left(\frac{2M}{b^2 - a^2}\right)\left(\frac{b^4 - a^4}{4}\right)$$

$$= \boxed{\frac{1}{2}M(b^2 + a^2)}$$

Kinetic Energy of Rotation

PROBLEM 10-3: Assume that the earth is a uniform sphere that has a mass $M = 6 \times 10^{24}$ kg and a radius of 6.4×10^6 m. Calculate the kinetic energy of the earth associated with its spin.

Solution: From Table 10-1,

$$I = \tfrac{2}{5}MR^2 = \tfrac{2}{5}(6 \times 10^{24} \text{ kg})(6.4 \times 10^6 \text{ m})^2 = 9.83 \times 10^{37} \text{ kg m}^2$$

and

$$\omega = \left(\frac{2\pi \text{ rad}}{24 \text{ h}}\right)\left(\frac{1 \text{ h}}{3600 \text{ s}}\right) = 7.27 \times 10^{-5} \text{ rad s}^{-1}$$

Thus, from Eq. (10-5a),

$$E_{k,\text{rot}} = \tfrac{1}{2}I\omega^2 = \tfrac{1}{2}(9.83 \times 10^{37} \text{ kg m}^2)(7.27 \times 10^{-5} \text{ rad s}^{-1})^2 = \boxed{2.60 \times 10^{29} \text{ J}}$$

PROBLEM 10-4: (a) What is the minimum value of the moment of inertia of a flywheel that can spin at $\omega = 200$ rad s^{-1} required for the flywheel to store enough energy to propel a 1500-kg car up a 30-m high hill? Neglect any frictional effects. (b) If the flywheel were a disk 1 m in diameter, what would be its mass?

Solution:

(a) Use the principle of conservation of energy: The kinetic energy available in the spinning flywheel will be transformed into the increase in potential energy of the car,

$$E_{k,\text{rot}} = \Delta E_p = mg\,\Delta y,$$

where Δy is the rise due to the hill. So

$$\tfrac{1}{2}I\omega^2 = mg\,\Delta y$$

Solve for I.

$$I = \frac{2mg\,\Delta y}{\omega^2} = \frac{2(1500 \text{ kg})(9.8 \text{ m s}^{-2})(30 \text{ m})}{(200 \text{ rad s}^{-1})^2} = \boxed{22 \text{ kg m}^2}$$

(b) For a disk, $I = \tfrac{1}{2}MR^2$. Solve for M.

$$M = \frac{2I}{R^2} = \frac{2(22 \text{ kg m}^2)}{(0.5 \text{ m})^2} = \boxed{176 \text{ kg}}$$

Angular Momentum

PROBLEM 10-5: Determine the angular momentum of the flywheel in Problem 10-4 when it is spinning at 200 rad s^{-1}.

Solution: Use Eq. (10-7).

$$L = I\omega = (22 \text{ kg m}^2)(200 \text{ rad s}^{-1}) = \boxed{4.4 \times 10^3 \text{ kg m}^2 \text{ s}^{-1}}$$

PROBLEM 10-6: (a) Use the information in Problem 10-3 to calculate the spin angular momentum of the earth. (b) Assume that the earth moves in a circular orbit of radius 1.5×10^{11} m. Calculate its orbital angular momentum.

Solution: For both (a) and (b), use Eq. (10-7).

(a) In this case, $I = 9.83 \times 10^{37}$ kg m^2 and $\omega = 7.27 \times 10^{-5}$ rad s^{-1}. The earth's spin angular momentum is

$$L = I\omega = (9.83 \times 10^{37} \text{ kg m}^2)(7.27 \times 10^{-5} \text{ rad s}^{-1}) = \boxed{7.15 \times 10^{33} \text{ kg m}^2 \text{ s}^{-1}}$$

(b) The moment of inertia of the earth in its orbit is

$$I = MR^2 = (6 \times 10^{24} \text{ kg})(1.5 \times 10^{11} \text{ m})^2 = 1.35 \times 10^{47} \text{ kg m}^2$$

Its angular velocity is

$$\omega = \left(\frac{1 \text{ rev}}{365 \text{ days}}\right)\left(\frac{2\pi \text{ rad}}{\text{rev}}\right)\left(\frac{1 \text{ day}}{8.64 \times 10^4 \text{ s}}\right) = 1.99 \times 10^{-1} \text{ rad s}^{-1}$$

and its orbital angular momentum is

$$L = I\omega = (1.35 \times 10^{47} \text{ kg m}^2)(1.99 \times 10^{-7} \text{ rad s}^{-1}) = \boxed{2.69 \times 10^{40} \text{ kg m}^2\text{s}^{-1}}$$

The earth's orbital angular momentum is 3.76×10^6 times larger than the spin angular momentum.

Torque and Angular Acceleration

PROBLEM 10-7:   (a) How many horsepower are required to bring the flywheel in Problem 10-4 from rest to 200 rad s^{-1} in 2 min? (b) How much torque is required, assuming that the angular acceleration is constant?

Solution:
(a) You obtain the power required from Eq. (5-6).

$$P = \frac{dW}{dt} = \frac{E_{k,\text{rot}}}{\Delta t} = \frac{\frac{1}{2}I\omega^2}{\Delta t} = \frac{\frac{1}{2}(22 \text{ kg m}^2)(200 \text{ rad s}^{-1})^2}{120 \text{ s}} = 3.67 \times 10^3 \text{ W}$$

$$= (3.67 \times 10^3 \text{ W})\left(\frac{1 \text{ hp}}{746 \text{ W}}\right) = \boxed{4.9 \text{ hp}}$$

(a)

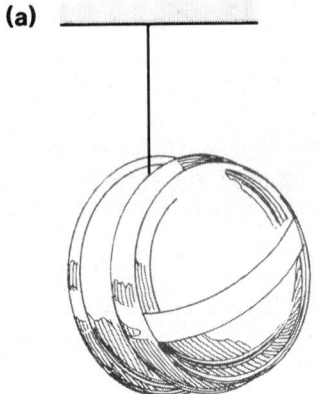

(b) Use Eq. (10-9), assuming that only one torque acts on the flywheel. If α is constant,

$$\alpha = \frac{\omega - \omega_0}{t} = \frac{200 \text{ rad s}^{-1} - 0}{120 \text{ s}} = 1.67 \text{ rad s}^{-2}$$

and

$$\tau = I\alpha = (22 \text{ kg m}^2)(1.67 \text{ rad s}^{-2}) = \boxed{36.7 \text{ N m}}$$

PROBLEM 10-8:   A yo-yo has a moment of inertia about its center of 1.4×10^{-3} kg m^2 and a mass of 400 g. String wrapped around its inner hub (radius of 0.02 m) is attached to a fixed object. Determine the linear acceleration of the yo-yo when it's released from rest. Assume that the supporting string remains vertical.

Solution: A sketch of the yo-yo is shown in Figure 10-15a; the corresponding free-body diagram is shown in Figure 10-15b. It might help if you visualize the string as a vertical track on which the inner hub rolls without sliding. The motion consists of the rotation about the CM coupled to the translation down the string. Because there is no "slipping" the angular acceleration and the linear acceleration are related by $a = \alpha r$.

To obtain the translational component of the motion, apply Newton's second law, assuming that downward is the positive direction.

$$\Sigma F = ma \qquad mg - T = ma$$

(b)

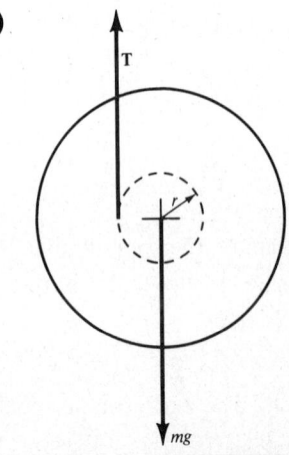

Figure 10-15

Solve for T, the tension in the string, which provides the force for the only torque acting to make the yo-yo spin.

$$T = mg - ma$$

Now apply Eq. (10-9), substitute for T, and solve for a.

$$\Sigma\tau = I\alpha \qquad Tr = I\left(\frac{a}{r}\right)$$

Thus

$$(mg - ma)r = I\left(\frac{a}{r}\right)$$

and

$$a = \frac{mgr^2}{I + mr^2} = \frac{(0.4 \text{ kg})(9.8 \text{ m s}^{-2})(0.02 \text{ m})^2}{1.4 \times 10^3 \text{ kg m}^2 + (0.4 \text{ kg})(0.02 \text{ m})^2} = \boxed{1.005 \text{ m s}^{-2}}$$

Conservation of Angular Momentum

PROBLEM 10-9: Refer to Figure 10-16. Disk 1 has a moment of inertia I_1 of 4.5×10^{-2} kg m^2 and is rotating at 400 rev min^{-1} on frictionless bearings. Disk 2 has a moment of inertia I_2 of 3×10^{-2} kg m^2 and can slide freely on the center spindle. It drops onto disk 1, and after some sliding they both rotate at the same rate. **(a)** Determine this rate. **(b)** Calculate the percentage of the initial kinetic energy lost in this interaction.

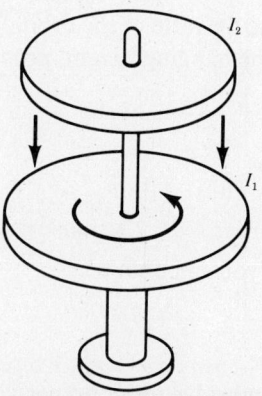

Figure 10-16. Disk 2 slides down the center rod and lands on disk 1, which has been rotating on frictionless bearings.

Solution:

(a) Because no *external* torque acts on this system during the interaction, you can use the principle of conservation of angular momentum. The angular momentum of the system before the interaction is $L_1 = I_1\omega_1$, and

$$\omega_1 = \left(\frac{400 \text{ rev}}{\text{min}}\right)\left(\frac{2\pi \text{ rad}}{\text{rev}}\right)\left(\frac{\text{min}}{60 \text{ s}}\right) = 41.89 \text{ rad s}^{-1}$$

The angular momentum of the system after the interaction is $L_2 = (I_1 + I_2)\omega$, where ω is the angular velocity of the disks after sliding stops. Thus

$$I_1\omega_1 = (I_1 + I_2)\omega$$

and

$$\omega = \omega_1\left(\frac{I_1}{I_1 + I_2}\right) = (41.89 \text{ rad s}^{-1})\left(\frac{4.5 \times 10^{-2} \text{ kg m}^2}{(4.5 + 3) \times 10^{-2} \text{ kg m}^2}\right) = \boxed{25.1 \text{ rad s}^{-1}}$$

(b) The percentage of rotational kinetic energy lost is

$$\left(\frac{\frac{1}{2}I_1\omega_1^2 - \frac{1}{2}(I_1 + I_2)\omega^2}{\frac{1}{2}I_1\omega_1^2}\right)(100\%) = \left[1 - \left(\frac{I_1 + I_2}{I_1}\right)\left(\frac{\omega}{\omega_1}\right)^2\right](100\%)$$

Substituting from **(a)** and simplifying, you can write this expression

$$\text{Percentage of energy lost} = \left(\frac{I_2}{I_1 + I_2}\right)(100\%) = \left(\frac{3 \times 10^{-2} \text{ kg m}^2}{(4.5 + 3) \times 10^{-2} \text{ kg m}^2}\right)(100\%) = \boxed{40\%}$$

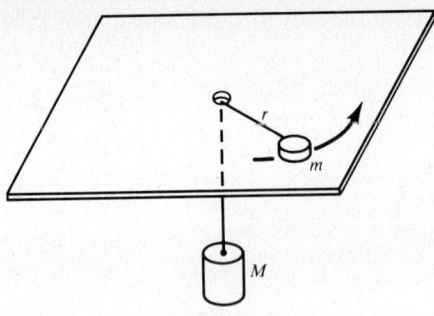

Figure 10-17

PROBLEM 10-10: A puck of mass m is connected by a lightweight cord to a hanging mass M. The cord passes through a hole in a frictionless horizontal surface upon which the puck slides in a circular orbit with angular velocity ω_1. (See Figure 10-17). A second mass M is attached to M and they are allowed to descend slowly so that the new orbit of m is again circular. What is now the angular velocity of the puck?

Solution: The addition of the second mass increases the tension in the cord but no external torque is supplied, so you can apply the principle of conservation of angular momentum, i.e., $L_1 = L_2$, where the subscripts 1 and 2 refer to the initial and final conditions, respectively. From Eq. (10-7), $L = I\omega$ and $I = mr^2$, if you consider the puck to be a point mass. Thus

$$(mr_1^2)\omega_1 = (mr_2^2)\omega_2 \qquad \text{or} \qquad \frac{\omega_2}{\omega_1} = \left(\frac{r_1}{r_2}\right)^2$$

To determine the ratio of the orbital radii you should recognize that the tension in the cord, which is equal to the hanging weight, provides the centripetal force, or

$$F_c = m\frac{v^2}{r} = m\omega^2 r$$

with $v = \omega r$. Thus

$$m\omega_1^2 r_1 = Mg \qquad \text{and} \qquad m\omega_2^2 r_2 = 2Mg$$

from which you can obtain

$$\frac{r_1}{r_2} = \frac{1}{2}\left(\frac{\omega_2}{\omega_1}\right)^2$$

Substitute and solve for ω_2 to get

$$\boxed{\omega_2 = (4)^{1/3}\omega_1 = 1.587\omega_1}$$

Supplementary Exercises

EXERCISE 10-1: What is the moment of inertia of the object in Figure 10-13 when the axis of rotation is the symmetry axis in the plane of the page that divides the object in two?

EXERCISE 10-2: Determine the moment of inertia by integration for the hemisphere of mass M and radius R shown in Figure 10-18. Assume that it is made from a material of uniform density.

EXERCISE 10-3: How much time is required for a 3-hp electric motor to accelerate a flywheel from rest to 10^3 rev min^{-1}? The flywheel has a moment of inertia of 54 kg m^2.

EXERCISE 10-4: A bowling ball with a diameter of 0.26 m and a mass of 8 kg rolls down an incline from a shelf 0.80 m above the floor. How fast is the ball moving when it gets to the floor? Assume that the ball is a solid homogeneous sphere, starts from rest, and rolls without slipping.

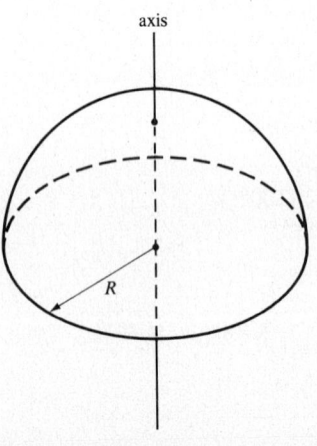

axis

R

Figure 10-18

EXERCISE 10-5: A 250-g ball bearing is released from rest at the point $x = 0.20$ m, $y = 0$ and falls 1.2 m to the floor, as illustrated in Figure 10-19. What is its angular momentum relative to the point of origin at the instant just before it strikes the floor?

EXERCISE 10-6: A meter stick, pivoted at one end, is released from rest in a horizontal position. Calculate its angular momentum at the instant it passes the vertical position, as shown in Figure 10-20. The meter stick has a mass of 350 g.

EXERCISE 10-7: How much torque must an electric motor supply to a flywheel that has a moment of inertia of 1.4 kg m^2 to give the flywheel an angular acceleration of 15 rad s^{-2} when the frictional torque in the bearings is 2 N m?

EXERCISE 10-8: A 2-kg mass is connected to a 3-kg mass by a lightweight cord that passes over a pulley. The pulley has a moment of inertia of 1.2 kg m^2 and a radius of 0.14 m. Determine the linear acceleration of the masses when released from rest, if the cord does not slip on the pulley and there is negligible friction in the pulley bearing.

EXERCISE 10-9: Two blocks of wood, each of mass $M = 800$ g, are mounted on opposite ends of a massless rod so that their centers are 0.12 m from the midpoint of the rod. In Figure 10-21, $r = 0.12$ m. The rod is pivoted about a vertical axis on frictionless bearings. A bullet of mass $m = 20$ g, moving at $v = 240$ m s^{-1}, embeds itself in one of the blocks. Determine the angular velocity of the device if it was initially at rest.

EXERCISE 10-10: The orbit of Halley's comet is shown in Figure 10-22. It is an ellipse with the sun at one focus. The distance from the sun to the comet at perihelion is $r_p = 0.6$ AU and at aphelion is $r_a = 35$ AU, where 1 AU $= 93 \times 10^6$ mi. Determine the ratio of the orbital velocity of the comet at perihelion to that at aphelion. Angular momentum is conserved for periodic solar orbits.

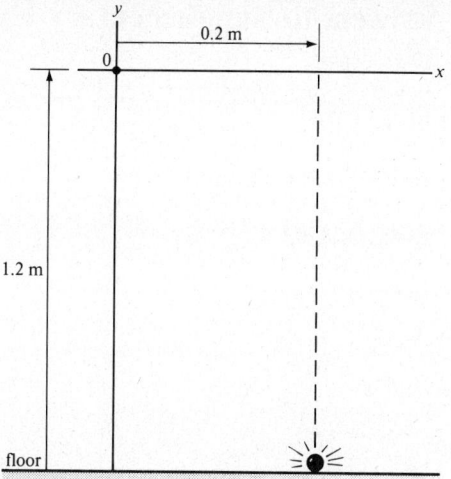

Figure 10-19

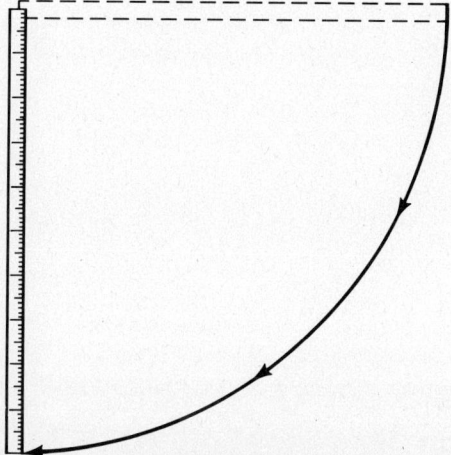

Figure 10-20

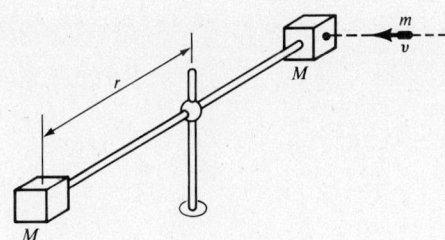

Figure 10-21

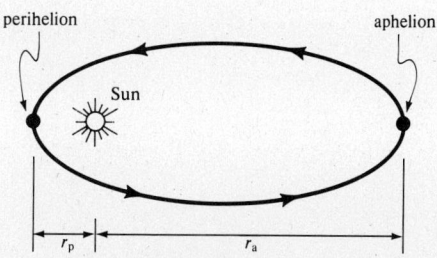

Figure 10-22

Answers to Supplementary Exercises

10-1: $6ma^2$

10-2: $\frac{2}{5}MR^2$

10-3: $132 \text{ s} = 2.2 \text{ min}$

10-4: 3.35 m s^{-1}

10-5: $0.242 \text{ kg m}^2 \text{ s}^{-1}$

10-6: $0.632 \text{ kg m}^2 \text{ s}^{-1}$

10-7: 23 N m

10-8: 0.148 m s^{-2}

10-9: 24.7 rad s^{-1}

10-10: 58.3

11 GRAVITATION

THIS CHAPTER IS ABOUT

☑ **Newton's Law of Universal Gravitation**
☑ **Gravitational Force Due to Nonspherical Bodies**
☑ **The Gravitational Field**
☑ **Gravitational Energy**
☑ **Satellite Motion**

11-1. Newton's Law of Universal Gravitation

When the plague hit London in 1665, Cambridge University closed its doors, and young Isaac Newton went home to Woolsthorpe pondering why the moon moves as it does in its orbit about the earth. It occurred to him the following year that the same force of gravity that causes an apple to fall may be responsible for the motion of the moon, an idea he did not publish until 1687. He concluded that pairs of particles throughout the universe attract each other by a gravitational force that is proportional to the product of their masses and inversely proportional to the square of their separation. This **law of universal gravitation** is stated in the form of an equation as

LAW OF UNIVERSAL GRAVITATION
$$F = G\frac{m_A m_B}{r^2} \qquad (11\text{-}1)$$

where $G = 6.672 \times 10^{-11}\,\text{N}\,\text{m}^2\,\text{kg}^{-2}$ is called the **universal gravitational constant**. Gravity is a mutually attractive force that obeys Newton's third law; that is, the force on m_A due to m_B ($\mathbf{F}_{A,B}$) is equal and opposite to the force on m_B due to m_A ($\mathbf{F}_{B,A}$). (See Figure 11-1.) Thus $\mathbf{F}_{B,A} = -\mathbf{F}_{A,B}$ and the magnitude of both of these is $|\mathbf{F}_{A,B}| = |\mathbf{F}_{B,A}| = F$. The mass that appears in Eq. (11-1), the **gravitational mass**, is equal to the **inertial mass**, the mass we can calculate from Newton's second law; that is, 1 kg of gravitational mass is equal to 1 kg of inertial mass. This equality led Einstein to his **principle of equivalence**: There is no fundamental difference between a gravitational field and an accelerated coordinate system. This principle is the basis for his general theory of relativity.

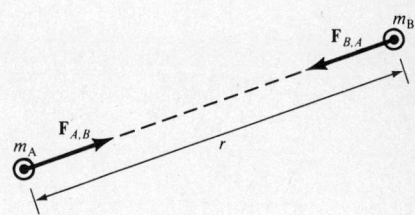

Figure 11-1

EXAMPLE 11-1: Equation (11-1) can be applied to uniform spheres, with r as the distance between their geometric centers. Calculate the gravitational force of attraction between two solid copper spheres that are in contact and have diameters of 0.10 m. The mass per unit volume of copper (its density) is $\rho = 8.92 \times 10^3\,\text{kg}\,\text{m}^{-3}$.

Solution: We use Eq. (11-1), where the mass of each sphere is $m_A = m_B = m$ and $m = \rho V$. The volume of a sphere of radius R is $V = \frac{4}{3}\pi R^3$, and the centers of the spheres in contact are separated by $2R$.

$$F = G\frac{m_A m_B}{r^2} = G\frac{m^2}{(2R)^2} = G\left[\frac{(\rho\frac{4}{3}\pi R^3)^2}{4R^2}\right] = G\rho^2\frac{4}{9}\pi^2 R^4$$

$$= (6.672 \times 10^{-11}\,\text{N}\,\text{m}^2\,\text{kg}^{-2})(8.92 \times 10^3\,\text{kg}\,\text{m}^{-3})^2\left(\frac{4\pi^2}{9}\right)(5 \times 10^{-2}\,\text{m})^4$$

$$= \boxed{1.46 \times 10^{-7}\,\text{N}}$$

Note that this force is negligible compared to the weight of one of the spheres, which is $w = mg = 45.8$ N.

EXAMPLE 11-2: What is the gravitational force of attraction between the earth and the moon? Consider the earth and moon to be spheres, with the mass of the earth $m_e = 6 \times 10^{24}$ kg, the mass of the moon $m_m = 7.4 \times 10^{22}$ kg, and their separation $r = 3.84 \times 10^8$ m.

Solution: Use Eq. (11-1).

$$F = G\frac{m_e m_m}{r^2}$$

$$= (6.672 \times 10^{-11}\,\text{N}\,\text{m}^2\,\text{kg}^{-2})\left[\frac{(6 \times 10^{24}\,\text{kg})(7.4 \times 10^{22}\,\text{kg})}{(3.84 \times 10^8\,\text{m})^2}\right] = \boxed{2.0 \times 10^{20}\,\text{N}}$$

11-2. Gravitational Force Due to Nonspherical Bodies

If a body is not spherical, we cannot calculate the gravitational force due to it from its center, but must integrate the force over the extent of the whole body. The gravitational force on a particle of mass m_p due to an element dm located r from the particle is

$$dF = G\frac{m_p\,dm}{r^2} \qquad (11\text{-}2)$$

The gravitational force due to the entire body is obtained by integration over the extent of the body, i.e., $F = \int dF$.

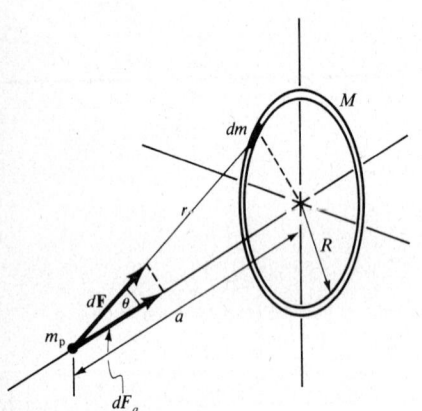

Figure 11-2

EXAMPLE 11-3: Determine the gravitational force of attraction between a particle of mass m_p located on the axis of symmetry at distance a from the center of a ring of mass M and radius R. (See Figure 11-2).

Solution: Use Eq. (11-2). In this case we need to be concerned only with the component of $d\mathbf{F}$ directed along the axis of symmetry; because the particle lies on the ring's axis of symmetry the forces perpendicular to the axis cancel each other.

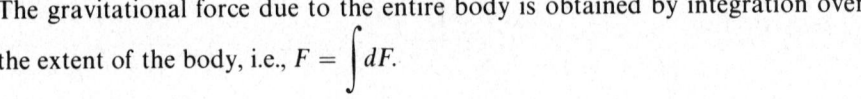

$$dF_a = (dF)(\cos\theta) = \left(G\frac{m_p\,dm}{r^2}\right)\left(\frac{a}{r}\right) = G\left[\frac{m_p a}{(a^2 + R^2)^{3/2}}\right]dm$$

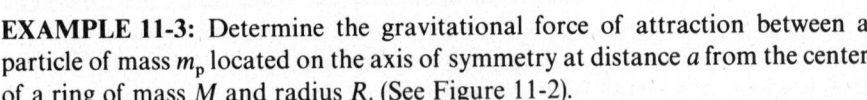

$$F_a = \int dF_a = \left[\frac{Gm_p a}{(a^2 + R^2)^{3/2}}\right]\int dm = \boxed{\frac{Gm_p M a}{(a^2 + R^2)^{3/2}}}$$

Note that $r^3 = (a^2 + R^2)^{3/2}$ and $\int dm = M$.

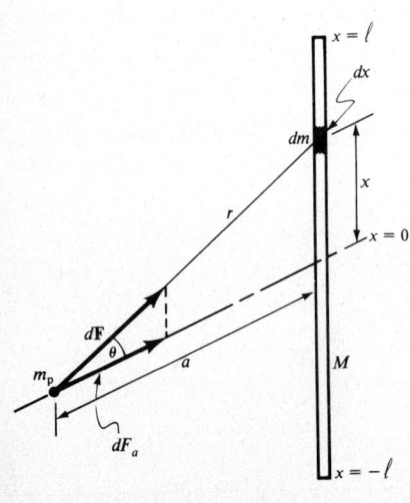

Figure 11-3

EXAMPLE 11-4: Determine the gravitational force of attraction between a particle of mass m_p and a rod of length 2ℓ located at distance a from the center of the rod, which has a mass M. (See Figure 11-3.)

Solution: Use Eq. (11-2). As in Example 11-3, the integral of the component of $d\mathbf{F}$ parallel to the rod is zero. We therefore want the integral of the component of $d\mathbf{F}$ perpendicular to the rod. Let $\lambda = M/2\ell$ be the rod's mass per unit length.

$$dF_a = (dF)(\cos\theta) = \left(G\frac{m_p\,dm}{r^2}\right)\left(\frac{a}{r}\right) = \frac{Gm_p a\lambda\,dx}{(a^2+x^2)^{3/2}}$$

Again, note that $r^3 = (a^2 + x^2)^{3/2}$ and that $dm = \lambda\,dx$. We integrate to obtain F_a. The integral is of a standard form and can be obtained directly or from an integral table.

$$F_a = Gm_p a\lambda \int_{-\ell}^{\ell} \frac{dx}{(a^2+x^2)^{3/2}} = \left(\frac{Gm_p Ma}{2\ell}\right)\left(\frac{x}{a^2(a^2+x^2)^{1/2}}\bigg|_{-\ell}^{\ell}\right) = \boxed{\frac{Gm_p M}{a\sqrt{a^2+\ell^2}}}$$

11-3. The Gravitational Field

The **gravitational field** is the gravitational force per unit mass at a point in space. It is a vector field, is symbolized by $\mathbf{g}$, and has units of acceleration of $\mathrm{m\,s^{-2}}$. For example, in the two-mass system illustrated in Figure 11-1, the gravitational field at the point in space occupied by mass m_A is

GRAVITATIONAL FIELD $$\mathbf{g}_A = \frac{\mathbf{F}_{A,B}}{m_A} \qquad\qquad \textbf{(11-3)}$$

Its magnitude is

$$g_A = \frac{F}{m_A} = G\frac{m_B}{r^2}$$

and is directed toward m_B. If other masses are present, their contributions must be added in a vector sum to obtain the total gravitational field at the point in space.

EXAMPLE 11-5: Determine the gravitational field at a point located at distance a from the center of a rod of mass M and length 2ℓ.

Solution: We use the result of Example 11-4. The magnitude is

$$g_p = \frac{F_a}{m_p} = \boxed{\frac{GM}{a\sqrt{a^2+\ell^2}}}$$

The direction of $\mathbf{g}_p$ is toward the rod, or the same as dF_a in Figure 11-3.

EXAMPLE 11-6: Assume that the earth is a uniform, nonrotating sphere of mass $M = 6 \times 10^{24}$ kg and radius $R = 6.4 \times 10^6$ m. (a) Determine the gravitational field at the earth's surface. (b) Find the height above the earth's surface at which the gravitational field is one-half that at the surface.

Solution: The gravitational field surrounding a nonrotating uniform spherical mass has the same mathematical form as if all its mass were concentrated into a particle located at the geometric center of the sphere.

(a) We use Eq. (11-1), with $m_A = m_p$, the mass of a particle at the earth's surface, $m_B = M$, and $r = R$, so

$$F = G\frac{m_p M}{R^2}$$

In this case F, the gravitational force of attraction between m_p and M, the earth, is the weight of m_p, or $F = w_p = m_p g$. Therefore the gravitational field at the earth's surface is simply the acceleration due to gravity at the earth's surface, i.e.,

$$g = \frac{F}{m_p} = \frac{GM}{R^2} = \frac{(6.672 \times 10^{-11} \text{ N m}^2 \text{ kg}^{-2})(6 \times 10^{24} \text{ kg})}{(6.4 \times 10^6 \text{ m})^2} = 9.8 \text{ m s}^{-2}$$

Its direction is downward, or toward the center of the earth. Note that we can generalize this result to include all planets. The acceleration due to gravity at the surface of a planet is GM/R^2, where M is the mass of the planet and R is its radius, provided that we consider the planet to be a uniform, nonrotating sphere.

(b) The gravitational field at r from the center of the earth is $g_r = (GM)/r^2$. Let $g_r = \frac{1}{2}g$, and solve for r. Use the relationship $GM = gR^2$ from part (a).

$$\frac{GM}{r^2} = \frac{1}{2}g \qquad r^2 = \frac{2GM}{g} = \frac{2gR^2}{g} = 2R^2 \qquad r = \sqrt{2}\,R$$

The height above the earth's surface is

$$h = r - R = \sqrt{2}\,R - R = R(\sqrt{2} - 1) = 0.41R$$
$$= (0.41)(6.4 \times 10^6 \text{ m}) = \boxed{2.62 \times 10^6 \text{ m}}$$

11-4. Gravitational Energy

Gravitational force is conservative; consequently, we can obtain a potential energy function for it by using Eq. (6-2). Let's consider a convenient example: a spherical object of mass M located at the origin and another mass m located at r.

The gravitational force on m is $F = -G\dfrac{Mm}{r^2}$ [Eq. (11-1)]. The negative sign is necessary because $\mathbf{F}$ is directed toward the origin as shown in Figure 11-4. Using Eq. (6-2),

$$\int_1^2 F\,dr = E_{p1} - E_{p2}$$

we get

$$-GMm \int_{r_1}^{r_2} \frac{dr}{r^2} = -GMm\left(-\frac{1}{r}\Big|_{r_1}^{r_2}\right) = -GMm\left(\frac{1}{r_1} - \frac{1}{r_2}\right) = E_{p1} - E_{p2}$$

We commonly take the reference potential energy to be zero when m is infinitely far from M; i.e., as $r_2 \to \infty$, $E_{p2} \to 0$. The subscript 1 is no longer necessary, and we can say that the potential energy of m at distance r from M is

$$E_p(r) = -\frac{GMm}{r} \tag{11-4}$$

This functional relationship is shown in Figure 11-5. The lower limit for r is R, the radius of M.

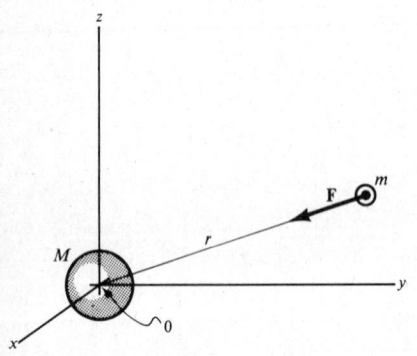

Figure 11-4

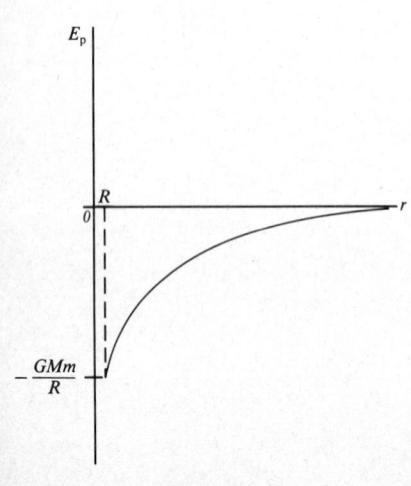

Figure 11-5

EXAMPLE 11-7: What is the potential energy of a 2×10^3 kg satellite in a circular earth orbit at an altitude of 2.3×10^5 m?

Solution: Use Eq. (11-4), but first determine r, the distance of the satellite from the earth's center.

$$r = h + R = 2.3 \times 10^5 \text{ m} + 6.4 \times 10^6 \text{ m} = 6.63 \times 10^6 \text{ m}$$

Now, we can substitute into

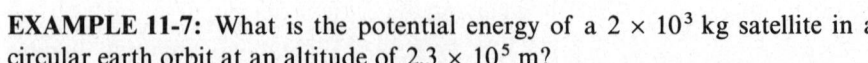

$$E_p(r) = -\frac{GMm}{r}$$

and obtain

$$E_p = -\frac{(6.672 \times 10^{-11} \text{ N m}^2 \text{ kg}^{-2})(6 \times 10^{24} \text{ kg})(2 \times 10^3 \text{ kg})}{6.63 \times 10^6 \text{ m}}$$

$$= \boxed{-1.21 \times 10^{11} \text{ J}}$$

The total energy of a satellite is the sum of its potential and kinetic energies, or $E = E_p + E_k$ [Eq. (6-7)]. For a satellite to escape the gravitational pull of the earth's mass, its total energy must be at least zero. The velocity that accomplishes this condition is called the **escape velocity**, v_e.

EXAMPLE 11-8: Determine the escape velocity for a satellite of mass m in orbit a distance r from a central mass M.

Solution: We set the total energy of the satellite at zero and solve for v_e.

$$E = E_p + E_k \qquad 0 = -\frac{GMm}{r} + \frac{1}{2}mv_e^2 \qquad v_e = \boxed{\sqrt{\frac{2GM}{r}}}$$

11-5. Satellite Motion

Newton's law of universal gravitation provides the basic theory by which we can understand the motion of communications and other types of satellites, moons, planets, binary stars, etc. We can derive Kepler's three laws of planetary (satellite) motion directly from the law of universal gravitation, although Kepler developed the three laws empirically from observational planetary data during the years from about 1600 to 1620. **Kepler's laws of planetary motion**:

(1) The planets move in planar elliptical orbits. The sun is located at one focus of the ellipse and the other focus is empty.
(2) The line that joins the center of the sun to the center of a planet sweeps out equal areas in equal times.
(3) The square of the orbital period is proportional to the cube of the semimajor axis of the planet's orbit.

Figure 11-6 shows the elliptical orbit of a planet. The points marked f indicate the location of the foci on the major axis, which is $2a$ in length. The distance between foci is $2c$ and the *eccentricity* is the ratio $e = c/a$. The eccentricity for the orbit shown is $e = 0.7$, whereas the eccentricity of the earth's orbit about the sun is only 0.017, which means that the orbit is almost circular. All but 2 planetary orbits have eccentricities of less than 0.1. The two exceptions are Mercury, with $e = 0.21$, and Pluto, with $e = 0.25$.

Because the gravitational force acts along the line joining the sun and planet, it cannot exert a torque; consequently, the orbital angular momentum for each planet is constant. Kepler's second law reflects this conservation law. This condition means that $\frac{\Delta A}{\Delta t}$ is constant throughout the orbit, where Δt is the time interval the planet's position vector takes to sweep the area ΔA. For a circular orbit, the semimajor axis is equal to the semiminor axis, and they are equal to the radius of the orbit, a. With this simplification, we can easily derive Kepler's third law for a circular orbit by recognizing that the centripetal force acting on the satellite is provided by the gravitational force. Thus, from Eqs. (4-5) and (11-1),

$$F_{\text{cen}} = F_{\text{grav}} \qquad m\left(\frac{v^2}{a}\right) = G\frac{Mm}{a^2}$$

where m is the planet mass, M is the sun's mass, and v is the orbital velocity. This last quantity is related to the orbital period T by $2\pi a = vT$ (for a circular orbit).

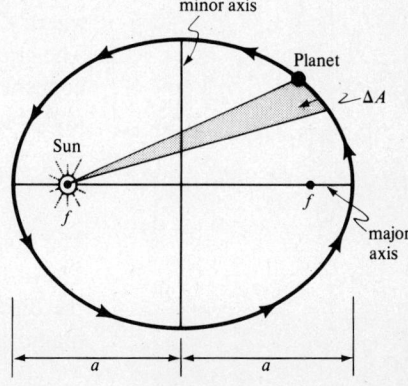

Figure 11-6

So we can substitute $2\pi a/T$ for v and solve for T^2 to get

$$T^2 = \left(\frac{4\pi^2}{GM}\right)a^3 \qquad (11\text{-}5)$$

We can generalize this result to a satellite in an elliptical orbit where a is the semimajor axis and M is the central mass.

EXAMPLE 11-9: One of Jupiter's moons, Io, has an orbital period of 1.77 days and an orbital radius of 4.22×10^8 m. Assuming that Io's orbit is circular, determine the mass of Jupiter.

Solution: We solve Eq. (11-5) for M, first converting the period into seconds, i.e., $T = 1.77$ days $= 1.53 \times 10^5$ s.

$$M = \frac{4\pi^2 a^3}{GT^2} = \frac{4\pi^2(4.22 \times 10^8 \text{ m})^3}{(6.672 \times 10^{-11} \text{ N m}^2 \text{ kg}^{-2})(1.53 \times 10^5 \text{ s})^2} = \boxed{1.9 \times 10^{27} \text{ kg}}$$

EXAMPLE 11-10: A communications satellite is in a geosynchronous orbit above the equator; i.e., its period is 24 h. How far above the earth's surface is it? Assume that the satellite is in a circular orbit. Take the mass and radius of the earth as $M = 6 \times 10^{24}$ kg and $R = 6.4 \times 10^6$ m.

Solution: We first determine the satellite's orbital radius by solving Eq. (11-5) for a.

$$a = \left(\frac{GMT^2}{4\pi^2}\right)^{1/3}$$

$$= \left[\frac{(6.672 \times 10^{-11} \text{ N m}^2 \text{ kg}^{-2})(6 \times 10^{24} \text{ kg})(8.64 \times 10^4 \text{ s})^2}{4\pi^2}\right]^{1/3}$$

$$= 4.23 \times 10^7 \text{ m}$$

We then subtract the radius of the earth to obtain the altitude.

$$h = a - R = 4.23 \times 10^7 \text{ m} - 6.4 \times 10^6 \text{ m} = \boxed{3.59 \times 10^7 \text{ m}}$$

This distance is about 22 000 miles.

SUMMARY

1. Newton's *law of universal gravitation* expresses the gravitational force of attraction between a pair of particles m_A and m_B separated by a distance r as

$$F = G\frac{m_A m_B}{r^2}$$

where G is the *universal gravitational constant* and has the value 6.672×10^{-11} N m^2 kg^{-2}.

2. The gravitational force due to a spherical body is the same as if the body's entire mass were concentrated at the center of the sphere. The gravitational force due to a nonspherical body must be determined by integration over the extent of the body.

$$F = Gm_\text{p} \int \frac{dm}{r^2}$$

The quantity dm is a mass element of the body and is located at distance r from m_p.

3. The *gravitational field* **g** at a point in space is a vector quantity having units of acceleration and the direction of the gravitational force on a particle of mass m_p located at that point. Its magnitude is

$$g_\text{P} = \frac{F}{m_\text{p}}$$

4. The *gravitational potential energy* of a particle of mass m located at distance r from the center of a uniform spherical mass M is

$$E_{\mathrm{p}} = -G\frac{Mm}{r}$$

5. The escape velocity for a satellite in orbit at distance r from the center of a spherical mass M is

$$v_{\mathrm{e}} = \sqrt{\frac{2GM}{r}}$$

6. Kepler's three laws of planetary motion can be derived from Newton's law of universal gravitation. They describe the motion of satellites moving in orbits around a large central mass.

7. The relationship between orbital radius, velocity, and period for a satellite in a circular orbit is $2\pi a = vT$.

8. Kepler's third law for a circular orbit of radius a about a central spherical mass M is

$$T^2 = \left(\frac{4\pi^2}{GM}\right)a^3$$

where T is the orbital period.

RAISE YOUR GRADES

Can you explain . . . ?

☑ why the moon does not crash into the earth despite the large gravitational force acting on it

☑ why an object would weigh less at the equator than at the north pole, even if the earth were perfectly spherical

☑ why the earth has a larger diameter at the equator than it does at the poles

☑ how the period of a simple pendulum would be affected if gravitational mass were greater than inertial mass

☑ why the acceleration due to gravity at the surface of the moon is only about $\frac{1}{6}$ of that at the surface of the earth

☑ why an astronaut is weightless in a spacecraft in orbit around the earth

☑ why a uniform, nonrotating spherical mass can be considered as a particle located at its geometric center

☑ why the direction of a rocket launching a satellite always has an eastward component

SOLVED PROBLEMS

Newton's Law of Universal Gravitation

PROBLEM 11-1: What is the force of gravity on an 80-kg person due to the planet Mars when it is closest to earth? The mass of Mars is 6.5×10^{23} kg and its closest approach to earth is about 78×10^9 m.

Solution: Use Eq. (11-1).

$$F = G\frac{m_A m_B}{r^2} = (6.672 \times 10^{-11}\,\mathrm{N\,m^2\,kg^{-2}})\left[\frac{(80\,\mathrm{kg})(6.5 \times 10^{23}\,\mathrm{kg})}{(78 \times 10^9\,\mathrm{m})^2}\right] = \boxed{5.7 \times 10^{-7}\,\mathrm{N}}$$

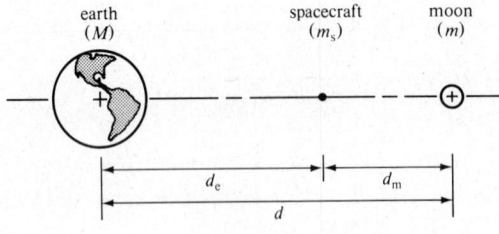

Figure 11-7

PROBLEM 11-2: At what distance from the center of the earth does the gravitational force of the earth cancel the gravitational force of the moon on a spacecraft traveling on a line joining the centers of the earth and moon? The earth's mass $M = 6 \times 10^{24}$ kg, the moon's mass $m = 7.4 \times 10^{22}$ kg, and the distance between their centers $d = 3.84 \times 10^8$ m. (See Figure 11-7.)

Solution: Find the point where the force of the earth on the spacecraft equals the force of the moon on the spacecraft. Let the mass of the spacecraft be m_s, of the earth be M, and of the moon be m; $d = d_e + d_m$. Use Eq. (11-1) for both the earth and the moon.

$$G\frac{M m_s}{d_e^2} = G\frac{m m_s}{d_m^2}$$

Solve for d_e.

$$d_e = d_m \sqrt{\frac{M}{m}} = (d - d_e)\sqrt{\frac{M}{m}} = d\left(\frac{\sqrt{\dfrac{M}{m}}}{1 + \sqrt{\dfrac{M}{m}}}\right) = (3.84 \times 10^8\,\mathrm{m})\left(\frac{\sqrt{\dfrac{6 \times 10^{24}\,\mathrm{kg}}{7.4 \times 10^{22}\,\mathrm{kg}}}}{1 + \sqrt{\dfrac{6 \times 10^{24}\,\mathrm{kg}}{7.4 \times 10^{22}\,\mathrm{kg}}}}\right)$$

$$= (3.84 \times 10^8\,\mathrm{m})\left(\frac{9}{10}\right) = \boxed{3.46 \times 10^8\,\mathrm{m}}$$

Gravitational Force Due to Nonspherical Bodies

PROBLEM 11-3: Calculate the gravitational force of attraction on a particle of mass m_p located along the symmetry axis at distance a from a disk of mass M and radius R.

Solution: Use the results of Example 11-3 and consider that the disk of mass M and radius R is composed of a series of concentric ring elements, each of mass $dm = \sigma 2\pi r\,dr$, where $\sigma = M/(\pi R^2)$ is the mass per unit area of the disk. (See Figure 11-8.) The gravitational force on m_p due to a ring element (from the result of Example 11-3) is

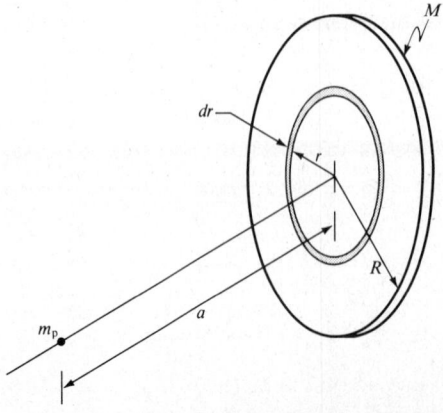

Figure 11-8

$$dF = \frac{G m_p M a}{(a^2 + r^2)^{3/2}} = \frac{G m_p (\sigma 2\pi r\,dr)a}{(a^2 + r^2)^{3/2}}$$

$$F = 2\pi\sigma G m_p a \int_0^R \frac{r\,dr}{(a^2 + r^2)^{3/2}}$$

$$= 2\pi\sigma G m_p a\left(-\frac{1}{\sqrt{a^2 + r^2}}\bigg|_0^R\right)$$

$$= 2\pi\left(\frac{M}{\pi R^2}\right)G m_p a\left(\frac{1}{a} - \frac{1}{\sqrt{a^2 + R^2}}\right)$$

$$\boxed{= \left(\frac{2GMm_p}{R^2}\right)\left(1 - \frac{a}{\sqrt{a^2 + R^2}}\right)}$$

PROBLEM 11-4: Calculate the gravitational force of attraction between a particle of mass m_p and a uniform rod of mass M and length ℓ when the particle is located at distance a from one end of the rod.

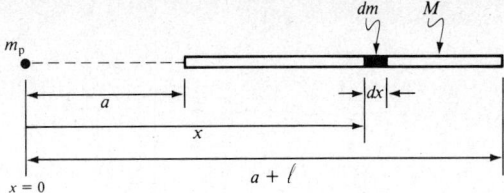

Figure 11-9

Solution: Use Eq. (11-2), with $dm = \lambda\,dx$, where $\lambda = M/\ell$, the mass per unit length of the rod. (See Figure 11-9.)

$$dF = G\frac{m_p\,dm}{r^2} = G\frac{m_p\lambda\,dx}{x^2}$$

$$F = \int dF = Gm_p\lambda\int_a^{\ell+a}\frac{dx}{x^2} = Gm_p\left(\frac{M}{\ell}\right)\left(-\frac{1}{x}\bigg|_a^{\ell+a}\right) = \frac{Gm_pM}{\ell}\left(\frac{1}{a} - \frac{1}{\ell+a}\right) = \boxed{\frac{Gm_pM}{a(\ell+a)}}$$

The Gravitational Field

PROBLEM 11-5: Determine the gravitational field due to the rod at the point occupied by the particle in Problem 11-4.

Solution: Use the definition of gravitational field, Eq. (11-3). The magnitude of the field is

$$g_P = \frac{F}{m_p} = \frac{GMm_p}{m_p a(\ell+a)} = \boxed{\frac{GM}{a(\ell+a)}}$$

The direction of the field is toward the rod.

PROBLEM 11-6: Calculate the gravitational field at the point P in Figure 11-10. The semicircular rod is uniform and has mass M and radius R.

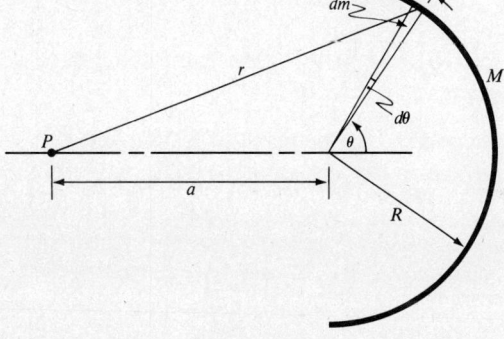

Solution: Use Eq. (11-3). The magnitude of g_P is

$$g_P = \frac{\int G\dfrac{m_p\,dm}{r^2}}{m_p} = G\int\frac{dm}{r^2}$$

Figure 11-10

where

$$dm = \lambda\,ds = \left(\frac{M}{\pi R}\right)(R\,d\theta) \qquad \text{and} \qquad r^2 = a^2 + R^2 + 2aR\cos\theta \qquad \text{(law of cosines)}$$

Thus

$$g_P = \frac{GM}{\pi}\int_{-\pi/2}^{\pi/2}\frac{d\theta}{A + B\cos\theta} = \frac{2GM}{\pi}\int_0^{\pi/2}\frac{d\theta}{A + B\cos\theta}$$

where $A = a^2 + R^2$, and $B = 2aR$. The integral is of a standard form, and you can obtain it from an integral table.

$$\int_0^{\pi/2}\frac{d\theta}{A + B\cos\theta} = \frac{1}{\sqrt{A^2 - B^2}}\tan^{-1}\left(\frac{\sqrt{A^2 - B^2}\sin\theta}{B + A\cos\theta}\right)\bigg|_0^{\pi/2} = \frac{1}{\sqrt{A^2 - B^2}}\tan^{-1}\left(\frac{\sqrt{A^2 - B^2}}{B}\right)$$

$$= \frac{1}{a^2 - R^2}\tan^{-1}\left(\frac{a^2 - R^2}{2aR}\right)$$

Therefore

$$\boxed{g_P = \left(\frac{2GM}{\pi(a^2 - R^2)}\right)\tan^{-1}\left(\frac{a^2 - R^2}{2aR}\right)}$$

Gravitational Energy

PROBLEM 11-7:   How much work must be done to raise a mass m to a height h above the earth's surface? Let the mass of the earth be M and its radius R.

Solution: The work that must be done is equal to the increase in potential energy of the mass. Use Eq. (11-4) for the potential energy.

$$W = E_p(r = R + h) - E_p(r = R)$$

where $E_p(r) = -GMm/r$. Therefore

$$W = -\frac{GMm}{R + h} - \left(-\frac{GMm}{R}\right) = \boxed{GMm\left(\frac{1}{R} - \frac{1}{R + h}\right)}$$

PROBLEM 11-8:   Show that for $h \ll R$ the result of Problem 11-7 becomes $W = mgh$. *Hint:* Use $GM = gR^2$, where g is the acceleration due to gravity at the earth's surface and

$$\frac{1}{1 + x} \cong 1 - x \qquad \text{(for } x \ll 1)$$

Solution: Substitute $gR^2 = GM$ and factor $1/R$, so that

$$W = (gR^2)m\left(\frac{1}{R}\right)\left(1 - \frac{1}{1 + h/R}\right) \cong mgR\left[1 - \left(1 - \frac{h}{R}\right)\right] = \boxed{mgh}$$

Note that this result is the same as Eq. (6-4), or $E_p = mgy$.

Satellite Motion

PROBLEM 11-9:   Determine the escape velocity necessary for a satellite to leave the earth's gravity.

Solution: Use the result of Example (11-8), with $r = R$ (the earth's radius), $R = 6.4 \times 10^8$ m, and $M = 6 \times 10^{24}$ kg.

$$v_e = \sqrt{\frac{2GM}{R}} = \sqrt{\frac{2(6.672 \times 10^{-11}\ \mathrm{N\,m^2\,kg^{-2}})(6 \times 10^{24}\ \mathrm{kg})}{6.4 \times 10^6\ \mathrm{m}}} = \boxed{1.12 \times 10^4\ \mathrm{m\,s^{-1}}}$$

This velocity is slightly more than $25\,000$ mi h^{-1}.

PROBLEM 11-10:   Use Kepler's third law to show that the velocity of a satellite in a circular orbit can be written as $v = \sqrt{GM(a^{-1})}$, where a is the orbit's radius.

Solution: The velocity of a satellite in a circular orbit is

$$v = \frac{2\pi a}{T}$$

Solve for T, substitute into the equation for Kepler's third law, Eq. (11-5), and solve for v.

$$T^2 = \left(\frac{4\pi^2}{GM}\right)a^3 \qquad \left(\frac{2\pi a}{v}\right)^2 = \left(\frac{4\pi^2}{GM}\right)a^3 \qquad v = \boxed{\sqrt{\frac{GM}{a}}}$$

Supplementary Exercises

EXERCISE 11-1:   The radius of the orbit for an electron in the lowest energy state of hydrogen is 5.3×10^{-11} m. The mass of the electron and proton nucleus are $m_e = 9.11 \times 10^{-31}$ kg and $m_p = 1.67 \times 10^{-27}$ kg, respectively. What is the gravitational force on the electron in this orbit?

EXERCISE 11-2: Start with the law of universal gravitation and the average density of the earth (the mass per unit volume, or $\rho = M/V$) and derive an expression for G in terms of ρ, the surface acceleration due to gravity g, and the radius of the earth R.

EXERCISE 11-3: Calculate the gravitational force of attraction between a particle of mass m_p and a disk of "infinite" radius located at distance a from the particle. The disk has a uniform mass per unit area of σ.

EXERCISE 11-4: (a) Calculate the gravitational force of attraction between a particle of mass m_p and a *nonuniform* rod of length ℓ when the particle is located at distance a from one end of the rod. The mass per unit length of the rod is $\lambda = K(x - a)$, where K is a constant that has units of $\mathrm{kg\,m^{-1}}$. Refer again to Figure 11-9. (b) What is the mass of the rod?

EXERCISE 11-5: Assume that the earth is a nonrotating, uniform sphere of mass $M = 6 \times 10^{24}$ kg, with a radius $R = 6.4 \times 10^6$ m. At what height above the earth's surface is the gravitational field 1 percent less than it is at the surface?

EXERCISE 11-6: A spherical, nonrotating planet of radius R has a uniform density (mass per unit volume) of ρ. What is the gravitational field at its surface?

EXERCISE 11-7: Calculate the potential energy of a 10^3-kg satellite in a circular earth orbit of radius 10^7 m. The mass of the earth is $M = 6 \times 10^{24}$ kg.

EXERCISE 11-8: Determine the gravitational potential energy function for satellite of mass m located at distance r from a uniform, spherical, nonrotating mass M if the gravitational force of attraction were $F = G(Mm)/r^3$. Set the reference potential energy for $r \to \infty$ at zero.

EXERCISE 11-9: A satellite in a circular orbit about the earth has a period of 90 min. What is its orbital radius?

EXERCISE 11-10: What is the speed of the satellite in Exercise 11-9?

Answers to Supplementary Exercises

11-1: 3.6×10^{-47} N

11-2: $G = \dfrac{3g}{4\rho R}$

11-3: $F = 2\pi\sigma G m_p$. Note that this force is independent of a!

11-4: (a) $F = G m_p k \left[\ln\!\left(\dfrac{\ell + a}{a}\right) - \dfrac{\ell}{\ell + a} \right]$
(b) $M = \frac{1}{2}k\ell^2$

11-5: $h = 3.22 \times 10^4$ m

11-6: $g = \dfrac{4\pi G R\rho}{3}$

11-7: -4×10^{10} J

11-8: $E_p(r) = -\dfrac{2GMm}{r^2}$

11-9: 6.66×10^6 m

11-10: 7.75×10^3 m s^{-1}

12 SPECIAL RELATIVITY

THIS CHAPTER IS ABOUT

☑ **Einstein's Postulates**
☑ **Length Contraction and Time Dilation**
☑ **Addition of Velocities**
☑ **Relativistic Momentum and Energy**

agreement: **Special relativity** deals with the transformation between coordinate systems that move with *constant* relative velocity. **General relativity**, a subject beyond the scope of this book, deals with transformations between *accelerated* coordinate systems.

12-1. Einstein's Postulates

The concept of relative velocity presented in Section 3-6, often referred to as **Galilean relativity**, is the low-velocity approximation of special relativity. The exact and somewhat more complicated formulas of special relativity must be used when the velocity is about one-half the speed of light, or more. These formulas are based on two postulates published by Albert Einstein in 1905:

(1) All the laws of physics have the same form in any two coordinate systems that move relative to one another with constant velocity.
(2) The speed of light in vacuum has the same value when measured in any two coordinate systems that move relative to one another with constant velocity.

The theory of special relativity has been verified by countless experiments since it was first formulated.

In solving the equations of special relativity, we frequently express the velocity of one coordinate system relative to another as a decimal fraction of the speed of light. For example, $v = 0.6c$, where c is the speed of light (electromagnetic waves) in vacuum. The measured value of c is $299\,792\,458 \pm 1.2 \text{ m s}^{-1}$, but in most calculations we can obtain sufficient accuracy by using $c = 3 \times 10^8 \text{ m s}^{-1}$. For convenience in calculation, we often use two other related quantities,

$$\sqrt{1 - \beta^2} \quad \text{and} \quad \gamma = \frac{1}{\sqrt{1 - \beta^2}}$$

where $\beta = v/c$. Graphs of these quantities are shown in Figure 12-1.

note: Neither of these quantities departs significantly from unity until $v \to 0.6c$ or is greater than $0.6c$.

Let's consider two space–time coordinate systems. By space–time we mean that to describe each point (or event) in a system we need three space coordinates (x, y, and z) to describe the event's location in the three directions of space plus one time coordinate (t) to describe the event's "location" in time. Let's designate one system prime, the other unprime; thus an event has coordinates x, y, z, t in the

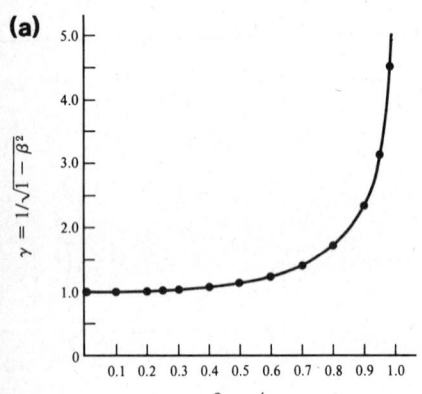

(a)

$\gamma = 1/\sqrt{1 - \beta^2}$

$\beta = v/c$

(b)

$\sqrt{1 - \beta^2}$

$\beta = v/c$

Figure 12-1. Graphs of (a) $\sqrt{1 - \beta^2}$ and (b) $\gamma = \dfrac{1}{\sqrt{1 - \beta^2}}$, as functions of β.

unprime system and x', y', z', t' in the prime system. Suppose that the two systems are parallel, that their origins coincide at time $t = t' = 0$, and that they are moving relative to each other in the x,x' direction. This means that the prime system moves along the positive x axis at velocity v relative to the unprime system and that the unprime system moves along the negative x' axis at velocity $-v$ relative to the prime system (see Figure 12-2). To convert the coordinates of one system to those of the other we use the **Lorentz transformation**:

LORENTZ TRANSFORMATION $x' = \gamma(x - vt)$ **(12-1a)**

$$y' = y \qquad\qquad\qquad \textbf{(12-1b)}$$

$$z' = z \qquad\qquad\qquad \textbf{(12-1c)}$$

$$t' = \gamma\left(t - \frac{vx}{c^2}\right) \qquad \textbf{(12-1d)}$$

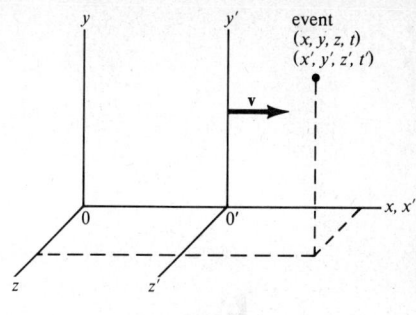

Figure 12-2

12-2. Length Contraction and Time Dilation

The length of a rod measured in a coordinate system in which it is at rest is called its *proper length*. The length of a rod measured in a coordinate system moving at constant velocity parallel to the rod will be shorter than the proper length by the factor $\sqrt{1 - \beta^2}$. This effect is called **Lorentz contraction**. If the proper length of the rod is ℓ_0, its length measured in the moving coordinate system is

LORENTZ CONTRACTION $\ell = \ell_0\sqrt{1 - \beta^2}$ **(12-2)**

Similarly, the time interval between two events at the same location, as measured by a clock at rest relative to that location, is called the *proper time interval* between the events. A clock moving at constant velocity relative to the location of the events will measure an interval longer than the proper time interval by the factor γ. This effect is called **time dilation**. Note that in the coordinate system at rest relative to the moving clock the two events occur at different locations. If T_0 is the proper time interval, the interval measured by the moving clock is T, or

TIME DILATION $T = \gamma T_0 = \dfrac{T_0}{\sqrt{1 - \beta^2}}$ **(12-3)**

EXAMPLE 12-1: Use the Lorentz transformation, Eq. (12-1a), to obtain the Lorentz contraction, Eq. (12-2).

Solution: Let's consider a rod at rest in the prime coordinate system and lying along the x' axis with its end points at x'_A and x'_B. Its proper length ℓ_0 is $x'_B - x'_A$. This rod is moving at velocity v with respect to the unprime system. Its length in the unprime system is $\ell = x_B - x_A$. Using Eq. (12-1a), we obtain

$$\ell_0 = x'_B - x'_A = \gamma(x_B - vt) - \gamma(x_A - vt) = \gamma(x_B - x_A) = \gamma\ell$$

$$\ell = \frac{\ell_0}{\gamma} = \boxed{\ell_0\sqrt{1 - \beta^2}}$$

EXAMPLE 12-2: Use the Lorentz transformation, Eq. (12-1d), to obtain the time dilation, Eq. (12-3).

Solution: Let's consider a clock located at point x' in the prime coordinate system. At that position a clock would measure a proper time interval from an initial instant t'_1 to a final instant t'_2 of $T_0 = t'_2 - t'_1$. We now use Eq. (12-1d) to get

$$T_0 = \gamma\left(t_2 - \frac{vx_2}{c^2}\right) - \gamma\left(t_1 - \frac{vx_1}{c^2}\right) = \gamma(t_2 - t_1) - \left(\frac{\gamma v}{c^2}\right)(x_2 - x_1)$$

This time interval, determined in the unprime system, is $T = (t_2 - t_1)$. The points x_1 and x_2 are the locations of the clock as determined in the unprime system at the initial and final instants. From Eq. (12-1a) we have

$$x' = \gamma(x_2 - vt_2) \quad \text{and} \quad x' = \gamma(x_1 - vt_1)$$

Since the clock doesn't move in the prime system, its initial and final position is x'. We solve the preceding pair of equations for $x_2 - x_1$, which gives us

$$x_2 - x_1 = v(t_2 - t_1) = vT$$

and substitute into the previous expression for T_0 to get

$$T_0 = \gamma T - \frac{\gamma v}{c^2}(vT) = \gamma T\left(1 - \frac{v^2}{c^2}\right) = \gamma T(1 - \beta^2) = T\sqrt{1 - \beta^2}$$

$$T = \frac{T_0}{\sqrt{1 - \beta^2}} = \boxed{\gamma T_0}$$

EXAMPLE 12-3: How fast would a meter stick have to move in a direction parallel to its length so that its measured length would be 1 yd? (1 yd = 36 in; 1 m = 39.4 in.)

Solution: We use Eq. (12-2), with $\ell = 36$ in. and $\ell_0 = 39.4$ in., and solve first for β and then for $v = \beta c$.

$$\ell = \ell_0\sqrt{1 - \beta^2} \qquad \left(\frac{\ell}{\ell_0}\right)^2 = 1 - \beta^2$$

$$\beta = \sqrt{1 - \left(\frac{\ell}{\ell_0}\right)^2} = \sqrt{1 - \left(\frac{36 \text{ in.}}{39.4 \text{ in.}}\right)^2} = 0.406$$

$$v = \boxed{0.406c}$$

EXAMPLE 12-4: A warning light on the control panel of a spacecraft flashes on and off every 0.4 s, as measured by the pilot. The craft passes over a control tower at a speed of $v = 0.6c$. What is the interval between on and off flashes, as measured by the tower's operator?

Solution: The interval measured by the pilot is the proper time interval. We use Eq. (12-3) to calculate T, with $\beta = 0.6$.

$$T = \gamma T_0 = \frac{T_0}{\sqrt{1 - \beta^2}} = \frac{0.4 \text{ s}}{\sqrt{1 - (0.6)^2}} = \boxed{0.5 \text{ s}}$$

12-3. Addition of Velocities

Notice that the Lorentz transformation affects only the component of a coordinate system that is parallel to the system's motion, that is, along the x,x' axes. The components perpendicular to the direction of motion, along the y,y' and z,z' axes, are the same in both the prime and unprime systems. Therefore we can apply Eq. (12-1a) directly to the velocity of objects moving parallel to the x,x' axis. Suppose that an object moves with velocity u, as determined in the unprime coordinate system parallel to the x,x' axes. (See Figure 12-3.) Its velocity in the prime system is

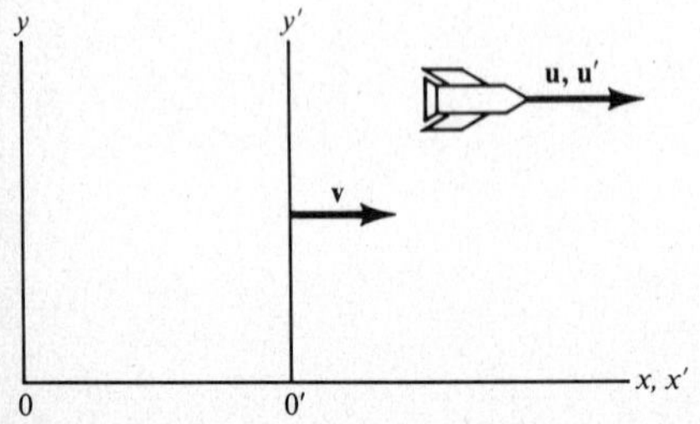

RELATIVISTIC VELOCITY ADDITION $\qquad u' = \dfrac{u - v}{1 - \dfrac{uv}{c^2}}$ **(12-4)**

Equation (12-4) can be derived from Eqs. (12-1a) and (12-1d).

Figure 12-3

EXAMPLE 12-5: The pilot of the spacecraft in Example 12-4 fires a projectile in the direction of motion of the craft. The projectile has a speed 0.4c relative to the craft. What is the speed of the projectile, as determined by the control tower's operator?

Solution: We use Eq. (12-4) and solve for u.

$$u = \frac{u' + v}{1 + \dfrac{u'v}{c^2}} = \frac{0.4c + 0.6c}{1 + \dfrac{(0.4c)(0.6c)}{c^2}} = \boxed{0.806c}$$

Note that Galilean relativity would predict that the projectile has a speed $u = c$.

12-4. Relativistic Momentum and Energy

The **relativistic momentum** of a particle of mass m moving with velocity **u** is

RELATIVISTIC MOMENTUM $\qquad \mathbf{p} = \dfrac{m\mathbf{u}}{\sqrt{1 - \dfrac{u^2}{c^2}}} \qquad$ **(12-5a)**

where the speed of the particle is $u = |\mathbf{u}|$. In some textbooks the **relativistic mass** of a particle of **rest mass** m_0 is defined as

$$\text{Relativistic mass} = \frac{m_0}{\sqrt{1 - \dfrac{u^2}{c^2}}}$$

The relativistic mass is thus a function of the particle's speed. In this book we will use m to represent the rest mass at all times. The magnitude of the relativistic momentum is

$$p = \frac{mu}{\sqrt{1 - \dfrac{u^2}{c^2}}} \qquad \textbf{(12-5b)}$$

The **rest energy** of a particle is mc^2, and its **total energy** is the sum of its kinetic energy and rest energy, i.e.,

TOTAL ENERGY $\qquad E = E_k + mc^2 \qquad$ **(12-6a)**

We can also write the expression for the total energy as

$$E = \frac{mc^2}{\sqrt{1 - \dfrac{u^2}{c^2}}} \qquad \textbf{(12-6b)}$$

and

$$E^2 = p^2 c^2 + (mc^2)^2 \qquad \textbf{(12-6c)}$$

note: The relationship between a particle's total energy, momentum, and rest energy has the form of the Pythagorean theorem. If you visualize the unsquared terms as the hypotenuse and adjacent sides of a right triangle, as shown in Figure 12-4, you can more easily remember the relationship.

The formulas for special relativity are used most frequently for calculations involving subatomic particles. In these calculations the energy of a particle is often measured in MeV, or *mega-electron-volts*, where $1 \text{ eV} = 1.602 \times 10^{-19}$ J and $1 \text{ MeV} = 1.602 \times 10^{-13}$ J. (Also, $1 \text{ GeV} = 10^9$ ev.) Using these units, we can express momentum in units of MeV/c, where $1 \text{ MeV}/c = 5.344 \times 10^{-22} \text{ kg m s}^{-1}$. When you hear or read that a certain particle has an energy of, say, 2.3 MeV, it means that the kinetic energy of the particle is 2.3 MeV.

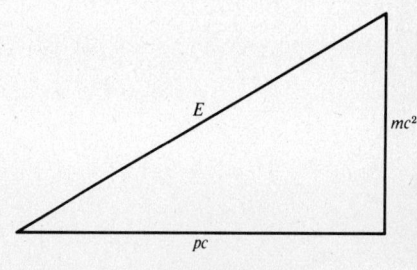

Figure 12-4

EXAMPLE 12-6: Calculate the rest energy in MeV for (**a**) an electron, (**b**) a proton, and (**c**) a neutron. The rest masses of these particles are $m_e = 9.11 \times 10^{-3}$ kg, $m_p = 1.673 \times 10^{-27}$ kg, and $m_n = 1.675 \times 10^{-27}$ kg, respectively. For better accuracy in this problem, use a more precise value for c, 2.998×10^8 m s^{-1}.

Solution: Use the expression for rest energy, mc^2.

(**a**) For an electron,

$$m_e c^2 = (9.11 \times 10^{-31} \text{ kg})(2.998 \times 10^8 \text{ m s}^{-1})^2 = 8.188 \times 10^{-14} \text{ J}$$

$$= (8.188 \times 10^{-14} \text{ J})\left(\frac{1 \text{ MeV}}{1.602 \times 10^{-13} \text{ J}}\right) = \boxed{0.511 \text{ MeV}}$$

(**b**) For a proton,

$$m_p c^2 = (1.673 \times 10^{-27} \text{ kg})(2.998 \times 10^8 \text{ m s}^{-1})^2 = 1.504 \times 10^{-10} \text{ J}$$

$$= (1.504 \times 10^{-10} \text{ J})\left(\frac{1 \text{ MeV}}{1.602 \times 10^{-13} \text{ J}}\right) = \boxed{938 \text{ MeV}}$$

(**c**) For a neutron,

$$m_n c^2 = (1.675 \times 10^{-27} \text{ kg})(2.998 \times 10^8 \text{ m s}^{-1})^2 = 1.505 \times 10^{-10} \text{ J}$$

$$= (1.505 \times 10^{-10} \text{ J})\left(\frac{1 \text{ MeV}}{1.602 \times 10^{-13} \text{ J}}\right) = \boxed{939 \text{ MeV}}$$

EXAMPLE 12-7: What is the momentum of a 260-MeV proton? Express the answer in MeV/c.

Solution: A "260-MeV proton" means that the kinetic energy of the proton is 260 MeV. Use Eq. (12-6a) to obtain E^2 and then substitute the result into Eq. (12-6c) to obtain p.

$$E = E_k + mc^2 \qquad E^2 = E_k^2 + 2E_k mc^2 + (mc^2)^2 \qquad E^2 = p^2 c^2 + (mc^2)^2$$

Thus

$$p^2 c^2 = E_k^2 + 2E_k mc^2$$

$$pc = \sqrt{E_k^2 + 2E_k mc^2} = \sqrt{(260)^2 + 2(260)(938)} \text{ MeV} = 745 \text{ MeV}$$

$$p = \boxed{745 \text{ MeV } c^{-1}}$$

EXAMPLE 12-8: What is the speed of the proton in Example 12-7? Express your answer in terms of c.

Solution: Use Eq. (12-5b) and solve for u/c.

$$p = \frac{mu}{\sqrt{1 - \dfrac{u^2}{c^2}}}$$

$$\frac{p}{mc} = \frac{\dfrac{u}{c}}{\sqrt{1 - \dfrac{u^2}{c^2}}}$$

$$\frac{u}{c} = \frac{\dfrac{p}{mc}}{\sqrt{1 + \dfrac{p^2}{m^2 c^2}}}$$

Next, evaluate p/mc and substitute to obtain the value of u.

$$\frac{p}{mc} = \frac{pc}{mc^2} = \frac{745 \text{ MeV}}{938 \text{ MeV}} = 0.7942$$

$$\frac{u}{c} = \frac{0.7942}{\sqrt{1 + (0.7942)^2}} = 0.622$$

$$u = \boxed{0.622c}$$

EXAMPLE 12-9: A hypothetical particle has a kinetic energy of 200 MeV and a measured momentum of 400 MeV/c. What is the rest energy for this particle?

Solution: Use Eqs. (12-6a) and (12-6c). First, square Eq. (12-6a), then substitute the expression obtained for E^2 into Eq. (12-6c), and solve for mc^2.

$$E = E_k + mc^2 \qquad E^2 = E_k^2 + 2E_k mc^2 + (mc^2)^2 \qquad E^2 = p^2 c^2 + (mc^2)^2$$

Thus

$$E_k^2 + 2E_k mc^2 = p^2 c^2$$

$$mc^2 = \frac{(pc)^2 - E_k^2}{2E_k}$$

Because $p = 400$ MeV/c, $pc = 400$ MeV. Therefore

$$mc^2 = \frac{[(400)^2 - (200)^2](\text{MeV})^2}{2(200 \text{ MeV})} = \boxed{300 \text{ MeV}}$$

SUMMARY

1. The *Lorentz transformation* is used to convert the coordinates of events in space and time between coordinate systems that move relative to each other at constant velocity. When the prime coordinate system moves along the positive unprime x axis at velocity v, with all corresponding axes parallel and the origins coincident at $t = t' = 0$, the Lorentz transformation becomes

$$x' = \gamma(x - vt) \qquad y' = y \qquad z' = z \qquad t' = \gamma\left(t - \frac{vx}{c^2}\right)$$

where

$$\gamma = \frac{1}{\sqrt{1 - \beta^2}} \qquad \text{and} \qquad \beta = \frac{v}{c}$$

2. The *proper length* of a rod ℓ_0 is the length measured in a coordinate system in which the rod is at rest. The length of a rod measured in a coordinate system moving at velocity v parallel to the rod's length can be determined from the *Lorentz contraction*:

$$\ell = \ell_0 \sqrt{1 - \beta^2}$$

3. The *proper time interval* T_0 is the measured interval between events that occur at a fixed point. The time interval measured by a clock moving at velocity v relative to that point is increased and can be determined using the *time dilation* formula:

$$T = \gamma T_0 = \frac{T_0}{\sqrt{1 - \beta^2}}$$

4. If a particle moves with velocity u along the positive x axis in the unprime coordinate system, the direction of its velocity in the prime system is parallel to the x' axis and its magnitude is

$$u' = \frac{u - v}{1 - \frac{uv}{c^2}}$$

5. The magnitude of the *relativistic momentum* of a particle of *rest mass m* moving at speed *u* is

$$p = \frac{mu}{\sqrt{1 - \dfrac{u^2}{c^2}}}$$

6. The *rest energy* of a particle is mc^2, and a particle's *total energy* is

$$E = E_k + mc^2 = \frac{mc^2}{\sqrt{1 - \dfrac{u^2}{c^2}}}$$

where E_k is the kinetic energy of the particle.
7. The relationship between a particle's total energy, momentum, and rest energy is

$$E^2 = p^2c^2 + (mc^2)^2$$

This equation has the form of the Pythagorean theorem and may be more easily remembered that way.
8. Special relativity is used mainly in analyzing the interactions of subatomic particles. Particle kinetic energy is usually specified in MeV or GeV, where 1 MeV = 10^6 eV = 1.602×10^{-13} J, and 1 GeV = 10^9 eV = 1.602×10^{-10} J. Particle momentum is usually specified in MeV/c, where 1 MeV/c = 5.344×10^{-22} kg m s^{-1}.

RAISE YOUR GRADES

Can you explain . . . ?

☑ why special relativity formulas can usually be ignored with regard to the motion of baseballs, bullets, airplanes, planetary motion, etc.

☑ why someone in a rocket ship, rather than an observer on a planet, measures the proper time interval between the rocket's passing two moons of the planet

☑ why special relativity formulas are necessary when $pc \gg mc^2$ and not necessary when $pc \ll mc^2$

☑ the conditions under which a fast-moving circular disk will appear oval

☑ how the spot of light on a cloud, beamed from a searchlight on the ground, could move faster than c

☑ how to identify which of two different values reported for the length of a rod is its proper length, if one of the values is the proper length

☑ why special relativity formulas are necessary for calculations involving subatomic particles

☑ why the kinetic energy of a baseball pitched at 90 km h^{-1} would not be $\frac{1}{2}mv^2$ if the speed of light were 100 km h^{-1}

SOLVED PROBLEMS

Einstein's Postulates

PROBLEM 12-1: Use the Lorentz transformation to show that $(x')^2 - c^2(t')^2 = x^2 - c^2t^2$.

Solution: Begin by squaring Eq. (12-1a) to obtain

$$(x')^2 = \gamma^2(x^2 - 2xvt + v^2t^2)$$

Then use the relationship

$$\gamma^2 = \frac{1}{1 - \dfrac{v^2}{c^2}} = \frac{c^2}{c^2 - v^2}$$

and substitute for γ, so

$$(x')^2 = \frac{c^2x^2 - 2c^2xvt + c^2v^2t^2}{c^2 - v^2}$$

Similarly, rearrange and square Eq. (12-1d) to obtain

$$c^2(t')^2 = c^2\gamma^2\left(t^2 - \frac{2tvx}{c^2} + \frac{v^2x^2}{c^4}\right) = \frac{c^4t^2 - 2c^2tvx + v^2x^2}{c^2 - v^2}$$

Now, combine terms and solve.

$$(x')^2 - c^2(t')^2 = \frac{c^2x^2 - 2c^2xvt + c^2v^2t^2 - c^4t^2 + 2c^2tvx - v^2x^2}{c^2 - v^2}$$

$$= \frac{x^2(c^2 - v^2) - c^2t^2(c^2 - v^2)}{c^2 - v^2} = \boxed{x^2 - c^2t^2}$$

Length Contraction and Time Dilation

PROBLEM 12-2:   How fast must a rocket ship travel so that its measured length is $\frac{1}{2}$ its proper length?

Solution: Use the Lorentz contraction, Eq. (12-2), and solve for $\beta = v/c$, with $\ell/\ell_0 = 0.5$.

$$\ell = \ell_0\sqrt{1 - \beta^2}$$

$$\beta = \sqrt{1 - \left(\frac{\ell}{\ell_0}\right)^2} = \sqrt{1 - (0.5)^2} = 0.866$$

$$v = \boxed{0.866c}$$

PROBLEM 12-3:   Use the Lorentz contraction to determine ℓ/ℓ_0 when $\beta = v/c = 0.999\,999\,99c$.

Solution: The Lorentz contraction is Eq. (12-2), or $\ell = \ell_0\sqrt{1 - \beta^2}$. Note that

$$1 - \beta^2 = (1 + \beta)(1 - \beta) \cong 2(1 - \beta)$$

which is a good approximation when $\beta \cong 1.0$.

$$\frac{\ell}{\ell_0} \cong \sqrt{2(1 - \beta)} = \sqrt{2(1 - 0.999\,999\,99)} = \boxed{1.414 \times 10^{-4}}$$

PROBLEM 12-4:   Astronomers on earth record that it has taken 6 years for a spacecraft to reach a distant planet. The crew of the spacecraft measured the travel time as 1.4 yr. Assuming that constant speed was maintained, how fast was the rocket traveling?

Solution: Use the time dilation formula, Eq. (12-3). The time measured by the crew is the proper time. Solve $\beta = v/c$ for v.

$$T = \gamma T_0 = \frac{T_0}{\sqrt{1 - \beta^2}}$$

$$1 - \beta^2 = \left(\frac{T_0}{T}\right)^2$$

$$\beta = \sqrt{1 - \left(\frac{T_0}{T}\right)^2} = \sqrt{1 - \left(\frac{1.4\ \text{yr}}{6\ \text{yr}}\right)^2} = 0.97$$

$$v = \beta c = \boxed{0.97c}$$

PROBLEM 12-5:   The half-life of a radioactive nucleus in a specimen at rest in the laboratory is 2.7×10^{-5} s. What is its half-life when the nucleus is traveling at $0.98c$ in an accelerator?

Solution: The half-life of a substance is a time interval, so you use the time dilation formula, Eq. (12-3), with $T_0 = 2.7 \times 10^{-5}$ s as the proper time.

$$T = \gamma T_0 = \frac{T_0}{\sqrt{1 - \beta^2}} = \frac{2.7 \times 10^{-5} \text{ s}}{\sqrt{1 - (0.98)^2}} = \boxed{1.36 \times 10^{-4} \text{ s}}$$

Addition of Velocities

Figure 12-5

PROBLEM 12-6:   In a collision type of accelerator, particle A is moving to the right at $0.95c$ and particle B is moving to the left at $0.95c$. The two velocities are measured with respect to the laboratory. How fast is B approaching A in relation to A? (See Figure 12-5.)

Solution: Use the velocity transformation formula, Eq. (12-4). Let the coordinate system in which the laboratory is at rest be the unprime system and the system in which particle A is at rest be the prime system. So, in the unprime system, u is the velocity of particle B and v is the velocity of the prime system, the same as the velocity of A.

$$u' = \frac{u - v}{1 - \dfrac{uv}{c^2}} = \frac{-v_B - v_A}{1 - \dfrac{(-v_B)(v_A)}{c^2}} = -\frac{v_B + v_A}{1 + \dfrac{v_B v_A}{c^2}} = \frac{0.95c + 0.95c}{1 + \dfrac{(0.95c)(0.95c)}{c^2}} = \boxed{-0.999c}$$

PROBLEM 12-7:   A blue spaceship is moving at $0.4c$ relative to earth when it is passed by a red spaceship moving in the same direction. The crew in the blue ship determines that the red ship is moving at $0.3c$ relative to their ship. How fast is the red spaceship moving relative to the earth?

Solution: Use the velocity transformation formula, Eq. (12-4). In this case the blue spaceship should be identified as the prime coordinate system, with a velocity of $v_B = v$. The velocity of the red spaceship relative to the blue ship is $v_{RB} = u'$, and the velocity of the red ship relative to earth is $v_R = u$.

$$u' = \frac{u - v}{1 - \dfrac{uv}{c^2}} \qquad \text{and} \qquad u = \frac{u' + v}{1 + \dfrac{u'v}{c^2}}$$

Thus

$$v_R = \frac{v_{RB} + v_B}{1 + \dfrac{(v_{RB})(v_B)}{c^2}} = \frac{0.3c + 0.4c}{1 + \dfrac{(0.3c)(0.4c)}{c^2}} = \boxed{0.625c}$$

Relativistic Momentum

PROBLEM 12-8:   Calculate the momentum of a 22-MeV electron. The rest energy of an electron is 0.511 MeV. (See Example 12-6.)

Solution: Eliminate the total energy E between Eqs. (12-6a) and (12-6c); solve first for pc, then for p.

$$E = E_k + mc^2 \qquad \text{and} \qquad E^2 = p^2 c^2 + (mc^2)^2$$

$$E^2 = E_k^2 + 2E_k mc^2 + (mc^2)^2 = (pc)^2 + (mc^2)^2$$

$$(pc)^2 = E_k^2\left(1 + \frac{2mc^2}{E_k}\right)$$

$$pc = E_k\sqrt{1 + \frac{2mc^2}{E_k}} \cong E_k\left(1 + \frac{mc^2}{E_k}\right)$$

The approximation is valid because $mc^2/E_k \ll 1$. So

$$pc = (22 \text{ MeV})\left(1 + \frac{0.511 \text{ MeV}}{22 \text{ MeV}}\right) = 22.5 \text{ MeV}$$

and

$$p = 22.5 \text{ MeV}/c$$

PROBLEM 12-9:   What is the velocity of the 22-MeV electron in Problem 12-8?

Solution: Use Eq. (12-5b), divide both sides by mc, and solve for u/c.

$$p = \frac{mu}{\sqrt{1 - \dfrac{u^2}{c^2}}}$$

$$\frac{p}{mc} = \frac{\dfrac{u}{c}}{\sqrt{1 - \dfrac{u^2}{c^2}}}$$

$$\frac{u}{c} = \frac{\dfrac{p}{mc}}{\sqrt{1 + \left(\dfrac{p}{mc}\right)^2}}$$

Now, calculate the value of p/mc, substitute, and solve for u.

$$\frac{p}{mc} = \frac{pc}{mc^2} = \frac{22.5 \text{ McV}}{0.511 \text{ MeV}} = 44.03$$

$$\frac{u}{c} = \frac{44.03}{\sqrt{1 + (44.03)^2}} = 0.99974$$

$$u = \boxed{0.999\,74c}$$

PROBLEM 12-10:   What is the rest mass energy of a 120-MeV particle that has a measured momentum of 200 MeV/c?

Solution: Eliminate the total energy E between Eqs. (12-6a) and (12-6c) and solve for mc^2.

$$E = E_k + mc^2 \quad \text{and} \quad E^2 = p^2c^2 + (mc^2)^2$$

$$E^2 = E_k^2 + 2E_k mc^2 + (mc^2)^2 = p^2c^2 + (mc^2)^2$$

$$mc^2 = \frac{(pc)^2 - E_k^2}{2E_k} = \frac{(200 \text{ MeV})^2 - (120 \text{ MeV})^2}{2(120 \text{ MeV})} = \boxed{107 \text{ MeV}}$$

Supplementary Exercises

EXERCISE 12-1:   Eight microseconds after a spacecraft moving at a speed of $0.9c$ along the positive x axis passes the origin, a flash of light is emitted from the point $x = 2.5 \times 10^3$ m. Determine (**a**) the location where and (**b**) the point in time when an observer on the spacecraft sees the flash.

EXERCISE 12-2:   How fast would a spaceship have to travel for its length as determined by a stationary observer to be only 90 percent of its proper length?

EXERCISE 12-3:   A meter stick moves along the x axis at $0.8c$ and is tilted at an angle of $30°$ with respect to the x' axis, as determined by an observer moving with the stick. What is the angle that a stationary observer would measure?

EXERCISE 12-4:   A space traveler notes that it has taken exactly one year to reach planet X, as measured by her watch. The craft has been traveling at $0.6c$ relative to the earth during this period. How far from earth is planet X, as determined by an observer on earth?

EXERCISE 12-5:   Mesons of certain type have an average half-life of $2.4\ \mu s$ when produced in a reaction in a stationary target in the laboratory. These same mesons are produced by cosmic ray bombardment in the upper atmosphere, where they have an average half-life of $17\ \mu s$. How fast are the mesons produced by cosmic rays traveling?

EXERCISE 12-6:   A tower operator observes spaceship A coming toward him at a speed of $0.6c$ and spaceship B coming toward him from the opposite direction at a speed of $0.8c$. At what speed does the pilot of A measure the approach of B?

EXERCISE 12-7:   A spaceship moving at a speed of $0.6c$ away from earth fires a rocket in the same direction at a speed of $0.7c$, as measured from earth. What is the speed of the rocket relative to the spaceship?

EXERCISE 12-8:   Determine the momentum of a particle of mass m that has kinetic energy three times its rest energy.

EXERCISE 12-9:   What is the speed of the particle in Exercise 12-8?

EXERCISE 12-10:   What is the speed of the particle of Example 12-9?

Answers to Supplementary Exercises

12-1:   (a) $x' = 780$ m
        (b) $t' = 1.15\ \mu s$

12-2: $0.436c$

12-3: $43.9°$

12-4: 7.1×10^{15} m

12-5: $0.99c$

12-6: $v_{BA} = 0.95c$

12-7: $0.17c$

12-8: $\sqrt{15}\ mc$

12-9: $0.968\,25c$

12-10: $0.8c$

13 ELASTICITY OF MATTER

THIS CHAPTER IS ABOUT

- ☑ **Stress and Strain**
- ☑ **Young's Modulus**
- ☑ **Poisson's Ratio**
- ☑ **Shear Modulus**
- ☑ **Bulk Modulus**

13-1. Stress and Strain

When a force is applied to a body to put it in tension, we say that the body is subjected to a **tensile stress**. The force that subjects a body to stress is distributed over an area; for example, in Figure 13-1 the force $F_\perp$ is distributed over the area $A = ab$ of the rod. The bottom end of the rod is firmly attached to a fixed plate. The scale of the drawing is greatly exaggerated for emphasis. Tensile stress is the ratio between tensile force and the cross-sectional area over which the force acts.

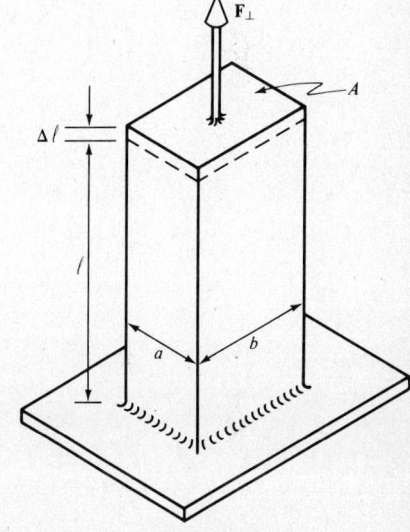

Figure 13-1

TENSILE STRESS $$\text{tensile stress} = \frac{F_\perp}{A} \qquad (13\text{-}1)$$

If the direction of $F_\perp$ were reversed, the rod would be subjected to a **compressional stress**. In SI units, stress is measured in **pascals**: 1 pascal = $1\,\text{Pa} = 1\,\text{N}\,\text{m}^{-2}$. In the British system, stress is measured in lb in.$^{-2}$, or psi; the conversion factor is $1\,\text{Pa} = 1.451 \times 10^{-4}\,\text{lb in.}^{-2}$.

The **tensile strain** in a body is the body's *response* to stress, its change in length $\Delta\ell$, and is a dimensionless quantity.

TENSILE STRAIN $$\text{tensile strain} = \frac{\Delta\ell}{\ell} \qquad (13\text{-}2)$$

Compressional strain is $-\Delta\ell/\ell$. The minus sign indicates that a rod would decrease in length under a compressional stress.

For many materials, notably ductile metals, strain is proportional to stress over a considerable range, beyond which the material begins to flow and ultimately breaks. Figure 13-2a shows a stress–strain diagram for a typical ductile metal subject to a tensile stress. Point p marks the end of the range in which strain is proportional to stress, e indicates the elastic limit (the point at which the material begins to flow), and bp is the breaking point. The material is

(a)

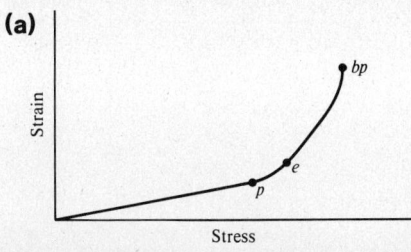

(b)

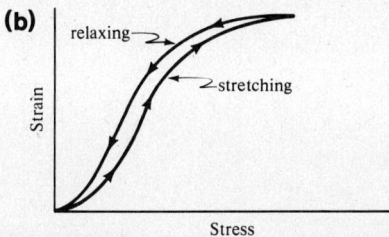

Figure 13-2. (a) Stress–strain diagram for a typical ductile metal. (b) Stress–strain diagram for rubber.

said to be *elastic* in the proportional range. Figure 13-2b shows a stress–strain diagram for rubber.

EXAMPLE 13-1: A 10-kg mass is hung on the end of an 80-cm length of wire 2 mm in diameter. The increase in the length of the wire is measured as 0.3 mm. (**a**) Determine the stress in the wire. (**b**) What is the strain in the wire?

Solution:

(**a**) The wire is subjected to a tensile force equal to the weight of the mass, so

$$F_\perp = mg = (10 \text{ kg})(9.8 \text{ m s}^{-2}) = 98 \text{ N}$$

The cross-sectional area of the wire is

$$A = \pi r^2 = \frac{\pi}{4}d^2 = \frac{\pi}{4}(2 \times 10^{-3} \text{ m})^2 = 3.14 \times 10^{-6} \text{ m}^2$$

Use Eq. (13-1) to calculate the tensile stress.

$$\text{Tensile stress} = \frac{F_\perp}{A}$$

$$= \frac{98 \text{ N}}{3.14 \times 10^{-6} \text{ m}^2} = \boxed{3.12 \times 10^7 \text{ N m}^2 = 3.12 \times 10^7 \text{ Pa}}$$

(**b**) Use Eq. (13-2) to determine the tensile strain.

$$\text{Tensile strain} = \frac{\Delta\ell}{\ell} = \frac{0.3 \times 10^{-3} \text{ m}}{80 \times 10^{-2} \text{ m}} = \boxed{3.75 \times 10^{-4}}$$

Note that we could also write the strain as $3.75 \times 10^{-2} \%$.

13-2. Young's Modulus

As indicated in Figure 13-2a, for a ductile metal, strain is proportional to stress over a considerable range. We generally express this proportional relationship between stress and strain in the form

$$\text{Stress} = (\text{Elastic modulus}) \times (\text{Strain})$$

Because strain is a dimensionless quantity, the elastic modulus has units of stress. In the case of a tensile (or compressional) stress, the modulus is **Young's modulus**, which is usually represented by Y (sometimes by E).

YOUNG'S MODULUS $$\underset{\substack{\text{Tensile} \\ \text{stress}}}{\frac{F_\perp}{A}} = Y \underset{\substack{\text{Tensile} \\ \text{strain}}}{\frac{\Delta\ell}{\ell}}$$ (13-3)

TABLE 13-1: Elastic Moduli of Selected Materials

Material	Y Young's modulus 10^{10} Pa	S Shear modulus 10^{10} Pa	B Bulk modulus 10^{10} Pa	σ Poisson's ratio
Aluminium	6.9	3.0	7.2	0.16
Brass	9.1	3.7	6.1	0.26
Copper	11	4.4	14	0.32
Cast iron	9.2	3.8	7	0.27
Lead	1.7	0.56	0.8	0.43
Steel	22	8	16	0.19
Glass	6	2.4	4.6	0.19

Table 13-1 lists Young's modulus, the shear and bulk moduli, and Poisson's ratio for a few selected materials.

EXAMPLE 13-2: A copper rod 80 cm long has a square cross-section with sides of 2 mm. It is clamped at its upper end, and a 12-kg mass is attached to its lower end. By how much does the rod elongate?

Solution: We use Eq. (13-3), with $Y = 11 \times 10^{10}$ Pa (from Table 13-1). $F_\perp$ is the weight of the mass, or $F_\perp = mg$. Solve for $\Delta\ell$.

$$\frac{F_\perp}{A} = Y\frac{\Delta\ell}{\ell}$$

$$\Delta\ell = \frac{\ell(mg)}{YA} = \frac{(0.8 \text{ m})(12 \text{ kg})(9.8 \text{ m s}^{-2})}{(11 \times 10^{10} \text{ Pa})(2 \times 10^{-3} \text{ m})^2} = \boxed{2.14 \times 10^{-4} \text{ m} = 0.214 \text{ mm}}$$

13-3. Poisson's Ratio

When a rod is subjected to a tensile stress, it decreases in width as well as increases in length. The fractional decrease in its width is proportional to the tensile strain if the rod is composed of a substance that is *isotropic*, i.e., has the same structure in all directions. (This excludes objects, such as crystals, that have a definite internal structure.) We refer to the fractional decrease in width as the **transverse strain**. (See Figure 13-3). The proportionality constant in the relationship between tensile and transverse strains is called **Poisson's ratio** and is represented by σ.

$$\frac{\Delta b}{b} = -\sigma\frac{\Delta\ell}{\ell} \tag{13-4}$$

The negative sign indicates a decrease in the rod's width. Conversely, a compressional stress will cause the width to increase.

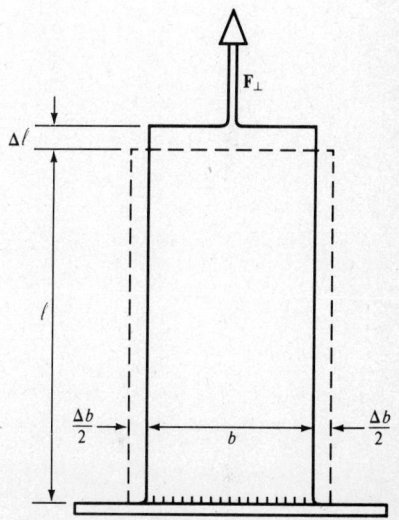

Figure 13-3

EXAMPLE 13-3: What is the decrease in width of the rod in Example 13-2?

Solution: We use Eq. (13-4) and solve for Δb, obtaining the value of Poisson's ratio from Table 13-1.

$$\frac{\Delta b}{b} = -\sigma\frac{\Delta\ell}{\ell}$$

$$\Delta b = -\frac{\sigma b \Delta\ell}{\ell} = -\frac{(0.32)(2 \times 10^{-3} \text{ m})(2.14 \times 10^{-4} \text{ m})}{0.8 \text{ m}} = \boxed{-1.71 \times 10^{-7} \text{ m}}$$

13-4. Shear Modulus

Shear stress can be described in two ways: linear and torsional. Linear shear stress is shown in Figure 13-4. The shearing force $F_\parallel$ is defined as distributed over the top surface area A, so **linear shear stress** is

LINEAR SHEAR STRESS $\qquad\qquad \dfrac{F_\parallel}{A} \qquad\qquad$ (13-5)

Linear shear strain is

LINEAR SHEAR STRAIN $\qquad\qquad \dfrac{s}{\ell} = \phi \qquad\qquad$ (13-6)

where ϕ is measured in radians. The **shear modulus** S relates the proportionality

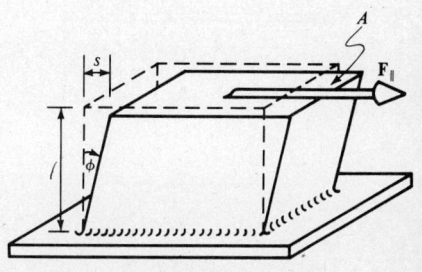

Figure 13-4

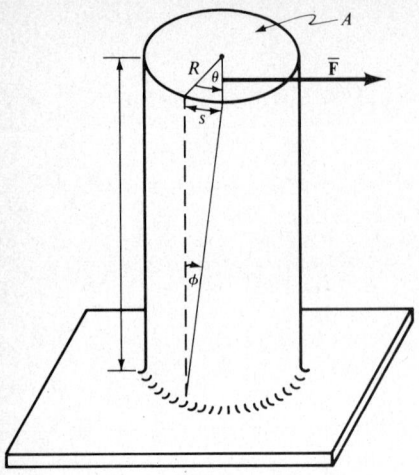

Figure 13-5

between linear shear stress and strain as

$$\frac{F_\parallel}{A} = S\left(\frac{s}{\ell}\right) = S\phi \tag{13-7}$$

$$\underset{\text{Shear}}{\underbrace{}} \qquad \underset{\text{Shear}}{\underbrace{}}$$
$$\text{stress} \qquad \text{strain}$$

The torsional case is illustrated in Figure 13-5. The lower end of the rod of length ℓ and circular cross-section is fixed. The average force $\bar{F}$ applied at the point $R/2$ is distributed over the area $A = \pi R^2$ and results in a torque $\tau = (R/2)\bar{F}$. Thus **torsional shear stress** can be thought of as

TORSIONAL SHEAR STRESS

$$\frac{\bar{F}}{A} = \frac{\dfrac{2\pi}{R}}{\pi R^2} = \frac{2\tau}{\pi R^3}$$

The **torsional shear strain** is $\phi = s/\ell$ and is proportional to the shear stress according to

$$\frac{2\tau}{\pi R^3} = S\left(\frac{s}{\ell}\right) = S\phi \tag{13-8}$$

The angle through which the rod is twisted is $\theta = s/R$ and is related to the strain by $\theta R = \phi \ell$. This permits us to rearrange Eq. (13-8) in terms of θ, i.e.,

$$\theta = \frac{2\tau\ell}{\pi R^4 S} \tag{13-9}$$

Keep in mind that θ is measured in radians.

EXAMPLE 13-4: How much torque is required to twist a steel rod 12 cm long and 2 cm in diameter through an angle of $5°$?

Solution: We use Eq. (13-9) and solve for τ, but first we have to convert $\theta = 5°$ into radians.

$$\theta = 5°\left(\frac{\pi \text{ rad}}{180°}\right) = \frac{\pi}{36} \text{ rad}$$

Substituting, we get

$$\tau = \frac{\theta \pi R^4 S}{2\ell} = \frac{(\pi/36)(\pi)(10^{-2} \text{ m})^4(8 \times 10^{10} \text{ Pa})}{2(12 \times 10^{-2} \text{ m})} = \boxed{914 \text{ N m}}$$

13-5. Bulk Modulus

If a block of material of volume V is immersed in a fluid and subjected to a *hydrostatic pressure p* (equal to force per unit area; Chapter 14), it undergoes a decrease in volume ΔV, as shown in Figure 13-6. The pressure is the stress in this case, and the relative change in volume is the strain. Stress and strain are related by

$$p = -B\left(\frac{\Delta V}{V}\right) \tag{13-10}$$

where B is the **bulk modulus** and the negative sign indicates that the volume decreases. The *compressibility* of a substance is the reciprocal of its bulk modulus, or $k = 1/B$. If we know Young's modulus and Poisson's ratio, we can estimate the

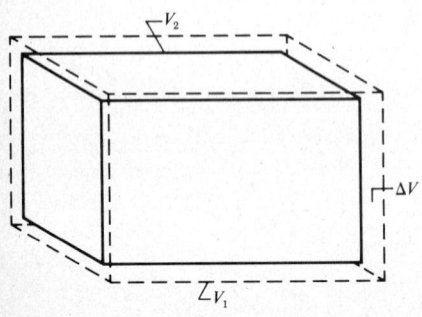

Figure 13-6

shear and bulk moduli accurately by using

$$S = \frac{Y}{2(1 + \sigma)} \qquad \text{(13-11a)}$$

and

$$B = \frac{Y}{3(1 - 2\sigma)} \qquad \text{(13-11b)}$$

EXAMPLE 13-5: How much hydrostatic pressure is needed to compress a sample of lead by 0.1 percent of its original volume?

Solution: Use Eq. (13-10). The bulk modulus of lead is $B = 0.8 \times 10^{10}$ Pa (Table 13-1). A 0.1 percent compression in volume means that the volumetric strain is $\Delta V/V = -0.001 = -10^{-3}$, so

$$p = -B\frac{\Delta V}{V} = -(0.8 \times 10^{10} \text{ Pa})(-10^{-3}) = \boxed{8 \times 10^6 \text{ Pa}}$$

SUMMARY

1. *Young's modulus* expresses the proportional relationship between tensile stress and strain, i.e.,

$$\frac{F_\perp}{A} = Y\left(\frac{\Delta \ell}{\ell}\right)$$

This relationship is also valid for compressional stress.

2. *Poisson's ratio* relates transverse strain to tensile strain by

$$\frac{\Delta b}{b} = -\sigma \frac{\Delta \ell}{\ell}$$

3. The *shear modulus* relates shear stress to shear strain by

$$\frac{F_\parallel}{A} = S\phi$$

4. The angle through which a rod of length ℓ and circular cross-section of radius R is twisted by an applied torque τ is

$$\theta = \frac{2\tau\ell}{\pi R^4 S}$$

5. Hydrostatic pressure, the force per unit area exerted by a fluid ($p = F/A$), produces stress on an immersed volume of a substance. The *bulk modulus* relates the hydrostatic stress to the volumetric strain by

$$p = -B\frac{\Delta V}{V}$$

6. The shear and bulk moduli can be estimated from Young's modulus and Poisson's ratio by

$$S = \frac{Y}{2(1 + \sigma)} \qquad \text{and} \qquad B = \frac{Y}{3(1 - 2\sigma)}$$

RAISE YOUR GRADES

Can you explain . . . ?

☑ why steel rods are embedded in structural concrete
☑ the relationship between Young's modulus for steel and the spring constant for a steel spring
☑ why Young's modulus for steel is about twice that for copper
☑ why it would be incorrect to identify a Young's modulus for rubber
☑ why a given specimen of wood does not have a unique value of Young's modulus
☑ why a shear modulus can be identified for a solid substance but not for a liquid
☑ why the bulk modulus of ice is greater (by about four times) than that of water
☑ why a piece of metal gets hot along the bend as it is bent back and forth

SOLVED PROBLEMS

Stress and Strain

PROBLEM 13-1: A 2000-kg van is parked on the street. Each tire is in contact with the street over an area of $2.5 \times 10^{-2} \text{ m}^2$. Calculate the resulting stress on the street surface. Assume that the weight supported by each tire is equal.

Solution: The weight supported by each tire is $w = \frac{1}{4}Mg = (1/4)(2000 \text{ kg})(9.8 \text{ m s}^{-2}) = 4.9 \times 10^3 \text{ N}$. Use Eq. (13-1) to calculate the compressional stress.

$$\text{Compressional stress} = \frac{w}{A} = \frac{4.9 \times 10^3 \text{ N}}{2.5 \times 10^{-2} \text{ m}^2} = \boxed{1.96 \times 10^5 \text{ Pa}}$$

PROBLEM 13-2: If the tensile strain in a 1-m glass rod is 0.2 percent, how much has the rod's length increased?

Solution: Use Eq. (13-2), which gives you the fractional increase in the length of a rod of length ℓ. The percent increase in length is 100 times the tensile strain.

$$\text{Percent increase in length} = (\text{Tensile strain})(100\%) = \left(\frac{\Delta\ell}{\ell}\right)(100\%)$$

$$0.2\% = \left(\frac{\Delta\ell}{\ell}\right)(100\%) \qquad \Delta\ell = \frac{0.2\%\,\ell}{100\%} = \frac{(0.2)(1 \text{ m})}{100} = \boxed{2 \times 10^{-3} \text{ m, or 2 mm}}$$

Young's Modulus

PROBLEM 13-3: A 50-cm length of wire, 2 mm in diameter, is clamped at its upper end, and a 20-kg mass is attached to its lower end. The wire stretches 0.4 mm. What is Young's modulus for the material?

Solution: Use Eq. (13-3) and solve for Y. The stretching force in this case is the weight of the 20-kg mass.

$$\frac{F_\perp}{A} = Y\frac{\Delta\ell}{\ell}$$

$$Y = \frac{F_\perp \ell}{\frac{\pi}{4}d^2\,\Delta\ell} = \frac{4Mg\ell}{\pi d^2 \Delta\ell} = \frac{4(20 \text{ kg})(9.8 \text{ m s}^{-2})(0.5 \text{ m})}{\pi(2 \times 10^{-3} \text{ m})^2(0.4 \times 10^{-3} \text{ m})} = \boxed{7.8 \times 10^{10} \text{ Pa}}$$

PROBLEM 13-4: The breaking point for aluminum is 1.4×10^8 Pa. How much does an aluminum rod 1 m long and 2 cm in diameter elongate when it is subjected to a stress equal to 1 percent of its breaking point?

Solution: Use Eq. (13-3) and solve for $\Delta\ell$. Recall that the breaking point is the tensile stress in the material when it ruptures. Obtain the value of Young's modulus from Table 13-1.

$$\frac{F_\perp}{A} = Y\left(\frac{\Delta\ell}{\ell}\right)$$

$$\Delta\ell = \left(\frac{F_\perp}{A}\right)\left(\frac{\ell}{Y}\right) = (0.01)(1.4 \times 10^8 \text{ Pa})\left(\frac{1 \text{ m}}{6.9 \times 10^{10} \text{ Pa}}\right) = \boxed{2.03 \times 10^{-5} \text{ m}}$$

Poisson's Ratio

PROBLEM 13-5: How much does the diameter of the rod in Problem 13-4 decrease?

Solution: Use Eq. (13-4) and solve for Δb, where b represents the diameter of the rod. Obtain the value of Poisson's ratio from Table 13-1.

$$\frac{\Delta b}{b} = -\sigma\frac{\Delta\ell}{\ell}$$

$$\Delta b = -\frac{\sigma\,\Delta\ell\,b}{\ell} = -\frac{(0.16)(2.03 \times 10^{-5} \text{ m})(2 \times 10^{-2} \text{ m})}{1 \text{ m}} = \boxed{-6.5 \times 10^{-8} \text{ m}}$$

The negative sign indicates that the change in diameter is a decrease.

Shear Modulus

PROBLEM 13-6: One end of a rod 50 cm in length and 2 cm in diameter is clamped, and a torque of 150 N m is applied to the other end, twisting it 10°. What is the shear modulus for this material?

Solution: Use Eq. (13-9) and solve for S. Keep in mind that θ must be in radians, so

$$\theta = (10°)\left(\frac{\pi \text{ rad}}{180°}\right) = \frac{\pi}{18} \text{ rad} = \frac{2\tau\ell}{\pi R^4 S}$$

and

$$S = \frac{2\tau\ell}{\pi R^4\theta} = \frac{2(150 \text{ N m})(0.5 \text{ m})}{\pi(10^{-2} \text{ m})^4(\pi/18)} = \boxed{2.7 \times 10^{10} \text{ Pa}}$$

PROBLEM 13-7: An electric motor drives a pump by means of a steel shaft 20 cm long and 2 cm in diameter. The motor supplies 12 hp to the shaft as it spins at 200 rad s^{-1}. Find the angle through which the shaft is twisted.

Solution: Use Eq. (13-9), but first determine how much torque is delivered by the motor. Equation (5-7) states that the power transmitted by the force **F** is $P = \mathbf{F} \cdot \mathbf{v}$. For a shaft of circular cross-section and radius R, $v = \omega r$ so you can also write Eq. (5-7) as

$$P = \mathbf{F} \cdot \mathbf{v} = F\omega R = \tau\omega$$

Thus the transmitted torque is $\tau = P/\omega$, or

$$\tau = \frac{(20 \text{ hp})(746 \text{ W hp}^{-1})}{200 \text{ rad s}^{-1}} = 74.6 \text{ N m}$$

and

$$\theta = \frac{2\tau\ell}{\pi R^4 S} = \frac{2(74.6 \text{ N m})(0.2 \text{ m})}{\pi(10^{-2} \text{ m})(8 \times 10^{10} \text{ Pa})} = \boxed{1.19 \times 10^{-2} \text{ rad, or } 0.68°}$$

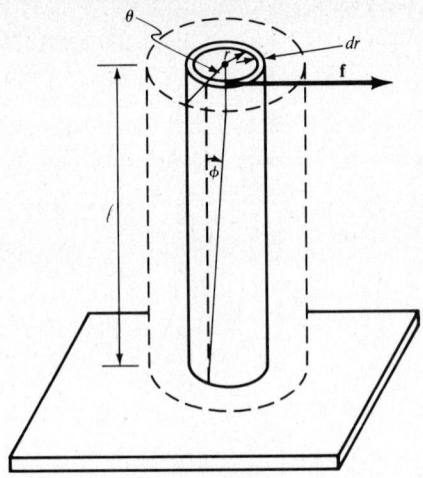

Figure 13-7

PROBLEM 13-8:   Derive Eq. (13-9) by considering a solid cylinder to be made up of a succession of hollow cylinders of radius r and thickness dr. A force f and, consequently, a torque rf is applied to each hollow cylinder. (See Figure 13-7.) Integrate from $r = 0$ to $r = R$ to obtain the total torque τ.

Solution: The shear stress on the hollow cylinder is f/dA, where $dA = 2\pi r\, dr$. The shear strain is ϕ and, as before, $\theta r = \phi\ell$. The relationship between shear stress and strain is

$$\frac{f}{dA} = S\phi \qquad \frac{f}{2\pi r\, dr} = S\phi$$

Solve for f.

$$f = S\frac{\theta r}{\ell} 2\pi r\, dr = \frac{2\pi S\theta}{\ell} r^2\, dr$$

The torque on this hollow cylinder is

$$rf = \frac{2\pi S\theta}{\ell} r^3\, dr$$

You can now determine the total torque on the solid cylinder by integrating, or

$$\tau = \frac{2\pi S\theta}{\ell} \int_0^R r^3\, dr = \left(\frac{2\pi S\theta}{\ell}\right)\left(\frac{R^4}{4}\right)$$

Solve for θ to obtain Eq. (13-9).

$$\boxed{\theta = \frac{2\tau\ell}{\pi R^4 S}}$$

Bulk Modulus

PROBLEM 13-9:   A hydraulic jack contains 0.25 m^3 of oil. Calculate the decrease in volume when the pressure inside the jack is 3.6×10^7 Pa. The bulk modulus for this oil is 50×10^8 Pa.

Solution: Use Eq. (13-10) and solve for ΔV.

$$p = -B\frac{\Delta V}{V}$$

$$\Delta V = -\frac{pV}{B} = -\frac{(3.6 \times 10^7 \text{ Pa})(0.25 \text{ m}^3)}{50 \times 10^8 \text{ Pa}} = \boxed{-1.8 \times 10^{-3} \text{ m}^3}$$

The negative sign indicates that the volume decreases.

PROBLEM 13-10: Poisson's ratio for many substances is approximately 0.3. (See Table 13-1.) *Estimate* the shear and bulk moduli for brass, for which Young's modulus is 9.1×10^{10} Pa and $\sigma \cong 0.3$.

Solution: Use Eqs. (13-11a) and (13-11b), with $\sigma = 0.3$

$$S = \frac{Y}{2(1 + \sigma)} = \frac{9.1 \times 10^{10} \text{ Pa}}{2(1 + 0.3)} = \boxed{3.5 \times 10^{10} \text{ Pa}}$$

From Table 13-1, $S = 3.7 \times 10^{10}$ Pa. The calculated value isn't a bad estimate.

$$B = \frac{Y}{3(1 - 2\sigma)} = \frac{9.1 \times 10^{10} \text{ Pa}}{3[1 - 2(0.3)]} = \boxed{7.6 \times 10^{10} \text{ Pa}}$$

From Table 13-1, $B = 6.1 \times 10^{10}$ Pa. The calculated value is also a reasonable estimate.

Supplementary Exercises

EXERCISE 13-1: The elastic limit for mild steel is 2.1×10^8 Pa. How many kilograms of mass will a mild steel wire 1 mm in diameter support at this stress?

EXERCISE 13-2: A steel rod 2 ft in length is stretched 0.002 in. Calculate the tensile strain in the rod.

EXERCISE 13-3: What is the tensile stress in the rod in Exercise 13-2?

EXERCISE 13-4: A rod 30 cm in length is subjected to a tensile stress of 4×10^8 Pa. It elongates 2 mm. What is Young's modulus for the material in the rod?

EXERCISE 13-5: A rod 14 cm in length and 1 cm in diameter is used to support a 150-kg weight. Young's modulus for the material in the rod is 4×10^{10} Pa. How much is the rod compressed?

EXERCISE 13-6: How much torque is necessary to twist one end of a cast-iron bolt 3 cm in diameter and 20 cm long through an angle of 10°? Assume that the other end of the bolt is fixed.

EXERCISE 13-7: Derive an expression for the angle through which a hollow circular cylinder of radius R and wall thickness t will twist when subjected to a torque τ. Assume that the cylinder has a length of ℓ and a shear modulus of S.

EXERCISE 13-8: A hollow steel shaft is used to transmit 500 hp from a diesel engine to a pump 2 m from the engine. The diameter of the shaft is 10 cm, and the wall thickness is 1 cm. Through what angle is the shaft twisted if it rotates at 20 rad s^{-1}?

EXERCISE 13-9: The bulk modulus for brass is 6.1×10^{10} Pa, whereas the bulk modulus for copper is 1.4×10^{10} Pa. Which material is more compressible?

EXERCISE 13-10: A certain material has values of Young's modulus and shear modulus of 8×10^{10} Pa and 3.2×10^{10} Pa, respectively. Estimate Poisson's ratio for this material.

Answers to Supplementary Exercises

13-1: 16.8 kg

13-2: 8.33×10^{-5}

13-3: 1.83×10^7 Pa

13-4: 6×10^{10} Pa

13-5: 2.06×10^{-4} m, or 0.206 mm

13-6: 8.44×10^4 N m

13-7: $\theta = \dfrac{\tau \ell}{2\pi S R^3 t}$

13-8: 3.4°

13-9: Brass

13-10: 0.25

14 FLUID STATICS

THIS CHAPTER IS ABOUT

☑ **Density and Specific Gravity**
☑ **Static Pressure and Pascal's Principle**
☑ **Archimedes' Principle**
☑ **Surface Tension**

14-1. Density and Specific Gravity

A. Density

The **density** of a substance is its mass per unit volume and is usually represented by ρ (rho):

DENSITY
$$\rho = \frac{m}{V} \tag{14-1}$$

where m is the mass of a specimen and V is its volume. Eq. (14-1) defines the average density, which of course is the same as the density of any part of the specimen for uniform substances. The density of a substance can be considered uniform unless otherwise specified. In the SI system m and V are in kilograms and cubic meters, respectively, so that ρ has units of $\mathrm{kg\,m^{-3}}$. Many books, however, continue to specify density in grams per cubic centimeter, $\mathrm{g\,cm^{-3}}$. It's easy to show that $1\ \mathrm{g\,cm^{-3}} = 10^3\ \mathrm{kg\,m^{-3}}$.

The quantity defined by Eq. (14-1) is sometimes referred to as the **mass density**, as contrasted with the **weight density** usually used in the British system of units. We obtain the weight density D from the mass density by multiplying by g:

WEIGHT DENSITY
$$D = \rho g = \frac{mg}{V} = \frac{w}{V} \tag{14-2}$$

In the British system D has units of pounds per cubic foot, $\mathrm{lb\,ft^{-3}}$. Table 14-1 lists the density of several substances.

Temperature and external pressure affect the density of substances somewhat. Generally an increase in external pressure causes a decrease in volume. (Thermal expansion will be considered in Chapter 16; compressibility was reviewed in Chapter 13). When the external pressure is one atmosphere ($1\ \mathrm{atm} = 1.013 \times 10^5\ \mathrm{Pa}$), the density of pure water reaches its maximum at $4°C$. Under these conditions the density of water is $1.000\ \mathrm{g\,cm^{-3}} = 10^3\ \mathrm{kg\,m^{-3}} = 62.43\ \mathrm{lb\,ft^{-3}}$.

B. Specific gravity

The **specific gravity**, SG, of a substance is the ratio of its density to the density of water at $4°C$.

SPECIFIC GRAVITY
$$\mathrm{SG} = \frac{\rho_{\text{substance}}}{\rho_{\text{water at }4°C}} \tag{14-3}$$

TABLE 14-1: Density of Selected Materials

Material	ρ (10^3 kg m^{-3})	D (lb ft^{-3})
Solids		
Aluminum	2.7	168.5
Brass	8.6	540
Copper	8.89	555
Glass	2.6	160
Gold	19.3	1205
Ice	0.91	57.2
Iron	7.9	493
Lead	11.4	712
Platinum	21.4	1330
Steel	7.83	489
Liquids		
Ethyl alcohol	0.79	49.4
Gasoline	0.69	42
Mercury	13.6	849
Sea water	1.03	64

EXAMPLE 14-1: What is the volume in m^3 of 8.7 kg of a substance that has a specific gravity of 6.3?

Solution: From Eq. (14-3) the density of this substance is

$$\rho = (SG)(\rho_{\text{water}}) = 6.3 \times 10^3 \text{ kg m}^{-3}$$

Solve Eq. (14-1) for the volume.

$$\rho = \frac{m}{V} \qquad V = \frac{m}{\rho} = \frac{8.7 \text{ kg}}{6.3 \times 10^3 \text{ kg m}^{-3}} = \boxed{1.38 \times 10^{-3} \text{ m}^3}$$

EXAMPLE 14-2: A block of stone has dimensions of 16 in. × 8 in. × 4 in. and weighs 55 lb. **(a)** What is its weight density? **(b)** Determine the specific gravity of the stone.

Solution:

(a) Use Eq. (14-2), but first determine the volume of the stone in cubic feet.

$$V = 16 \text{ in.} \times 8 \text{ in.} \times 4 \text{ in.} = 512 \text{ in.}^3$$

One cubic foot is 12 in. × 12 in. × 12 in. = 1728 in.^{3} Thus

$$V = 512 \text{ in.}^3 \times \frac{\text{ft}^3}{1728 \text{ in.}^3} = 0.2963 \text{ ft}^3$$

$$D = \frac{w}{V} = \frac{55 \text{ lb}}{0.2963 \text{ ft}^3} = \boxed{185.6 \text{ lb ft}^{-3}}$$

(b) Modify Eq. (14-3) for weight density:

$$SG = \frac{\rho_{\text{substance}}}{\rho_{\text{water}}} = \frac{\rho_{\text{substance}} g}{\rho_{\text{water}} g} = \frac{D_{\text{substance}}}{D_{\text{water}}} = \frac{185.6 \text{ lb ft}^{-3}}{62.43 \text{ lb ft}^{-3}} = \boxed{2.97}$$

14-2. Static Pressure and Pascal's Principle

A. Static pressure

Pressure is the force per unit area acting on a surface:

PRESSURE $$P = \frac{F}{A}$$ (14-4)

and has the same units as stress: pascals in the SI, $1 \text{ Pa} = 1 \text{ N m}^{-2}$, and pounds per square inch (psi) in the British system. Pressure is also commonly measured in atmospheres and other units:

$$1 \text{ atm} = 1.01325 \times 10^5 \text{ Pa} = 14.7 \text{ lb in.}^{-2}$$

$$1 \text{ atm} = 760 \text{ mm of Hg} = 29.9 \text{ in. of Hg}$$

$$1 \text{ mm of Hg} = 1 \text{ torr}$$

$$1 \text{ bar} = 10^5 \text{ Pa}$$

Think of a tank containing a fluid (Fig. 14-1). The pressure at any point beneath the surface of the fluid is due in part to the weight of the fluid and in part to the weight of the atmosphere or other gas above the fluid. The *total* pressure, called the **hydrostatic pressure** or the **absolute pressure**, at a depth h below the surface of a fluid of density ρ is

ABSOLUTE PRESSURE $$P = P_0 + \rho g h$$ (14-5)

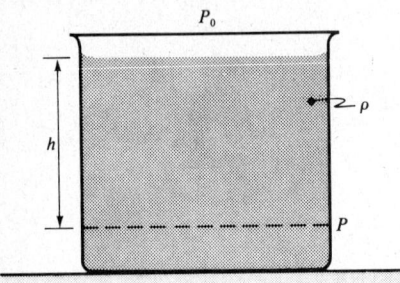

Figure 14-1

P_0 is the pressure above the fluid; $\rho g h$, called the **gauge pressure**, is the pressure due to the fluid alone.

B. Pascal's principle

Pascal's principle states that a change in the pressure applied to a fluid, that is, a change in P_0, is transmitted uniformly throughout the fluid and to the walls of the container.

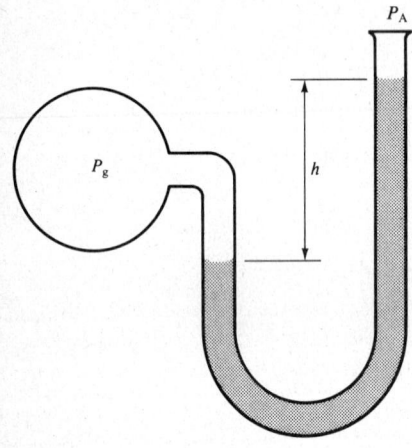

Figure 14-2. Mercury manometer.

EXAMPLE 14-3: Figure 14-2 depicts a **manometer**, a U-shaped tube containing mercury. One end of the tube is open to the atmosphere. Compute the pressure of the gas in the container on the left when the difference in the height of the columns of mercury is 22 cm. Assume that atmospheric pressure $P_A = 1.013 \times 10^5$ Pa.

Solution: By Pascal's principle, the pressure at the interface between the confined gas and the mercury is equal to the pressure at this same level in the right-hand column. You can calculate the pressure at this level from Eq. (14-5) by using P_A for P_0. From Table 14-1, the density of mercury is $13.6 \times 10^3 \text{ kg m}^{-3}$.

$$P_g = P_A + \rho g h = 1.013 \times 10^5 \text{ Pa} + (13.6 \times 10^3 \text{ kg m}^{-3})(9.8 \text{ m s}^{-2})(0.22 \text{ m})$$
$$= 1.013 \times 10^5 \text{ Pa} + 2.93 \times 10^4 \text{ Pa} = \boxed{1.306 \times 10^5 \text{ Pa}}$$

Notice that the gauge pressure of the gas is 2.93×10^4 Pa.

EXAMPLE 14-4: An automobile lift system (Fig. 14-3) consists of a smooth-fitting piston of circular cross-section with a diameter of 20 cm. The piston and platform have a total mass of 200 kg. What must be the gauge pressure of the compressed air below the piston to support a 1200-kg car?

Solution: From Eq. (14-5), gauge pressure equals total pressure minus atmospheric pressure. Atmospheric pressure acts equally on all sides of the piston, so the gauge pressure of the compressed air alone provides the upward force. To support the car, the upward force must equal the weight of the car plus piston

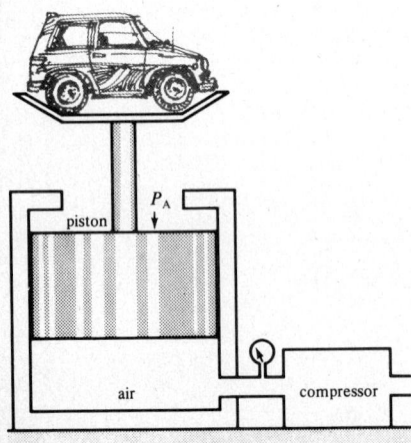

Figure 14-3. Automobile lift system.

plus platform. Use the definition of pressure, $P = F/A$, to express the upward force in terms of the pressure, then solve for P.

$$F_{up} = F_{down}$$

$$AP = \tfrac{1}{4}\pi d^2 P_{gauge} = mg$$

$$P_{gauge} = \frac{4mg}{d^2} = \frac{4(1200 + 200)\text{kg}(9.8 \text{ m s}^{-1})}{\pi(0.2 \text{ m})^2} = \boxed{4.37 \times 10^5 \text{ Pa}}$$

This is a gauge pressure of about 63 lb in.$^{-2}$

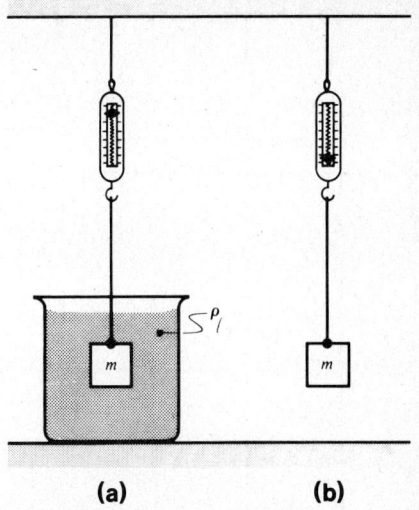

14-3. Archimedes' Principle

Archimedes' principle states that a body partly or completely submerged in a fluid experiences an upward, buoyant force equal to the weight of the fluid displaced by the body. This explains why a body weighs less when immersed in a fluid than when it is weighed outside of the fluid (Figure 14.4). The difference in the measured weights is the buoyant force.

Figure 14-4. Scale indicates **(a)** w_i (or m_i) when object is immersed in the liquid, **(b)** w (or m) when object is weighed in air.

EXAMPLE 14-5: Determine the density of an object that weighs 56 N when weighed in air and 47 N when immersed in a liquid that has a specific gravity of 1.2.

Solution: Let w_i be the apparent weight of the object when immersed in the fluid, w the actual weight of the object, V the volume of the object, ρ the density of the object, ρ_ℓ the density of the liquid, and F_B the buoyant force. First solve the definition of buoyant force, $F_B = w - w_i$, for w_i, then divide both sides of the equation by w. According to Archimedes' principle, $F_B = m_\ell g = \rho_\ell V g$ where m_ℓ is the mass of the liquid displaced by the object. Also, since $\rho = m/V = (w/g)/V$, $w = \rho V g$.

$$w_i = w - F_B \qquad \frac{w_i}{w} = 1 - \frac{F_B}{w} = 1 - \frac{\rho_\ell V g}{\rho V g} = 1 - \frac{\rho_\ell}{\rho}$$

Then calculate ρ_ℓ as ρ_{water} SG, and solve for ρ:

$$\rho = \frac{\rho_\ell}{1 - \dfrac{w_i}{w}} = \frac{1.2 \times 10^3 \text{ kg m}^{-3}}{1 - \dfrac{47 \text{ N}}{56 \text{ N}}} = \boxed{7.47 \times 10^3 \text{ kg m}^{-3}}$$

Notice that $w_i/w = m_i/m$, where m_i is the "apparent mass" of the immersed object in case you find the "weight" of the object by reading a scale or balance in grams or kilograms.

EXAMPLE 14-6: What is the buoyant force due to the air acting on the object of Example 14-5 when it is weighed in air? The density of air ρ_a is 1.3 kg m^{-3}.

Solution: By using the definition of density, $\rho = m/V$, you can express the volume of air displaced by the object as

$$V = \frac{m}{\rho} = \frac{w}{g\rho}$$

The buoyant force F_B is the weight of this volume of air. Use your result from Example 14-5 for ρ.

$$F_B = w_a = \rho_a V g = \rho_a \left(\frac{w}{g\rho}\right) g = w \frac{\rho_a}{\rho} = 56 \text{ N} \left(\frac{1.3 \text{ kg m}^{-3}}{7.47 \times 10^3 \text{ kg m}^{-3}}\right) = 9.7 \times 10^{-3} \text{ N}$$

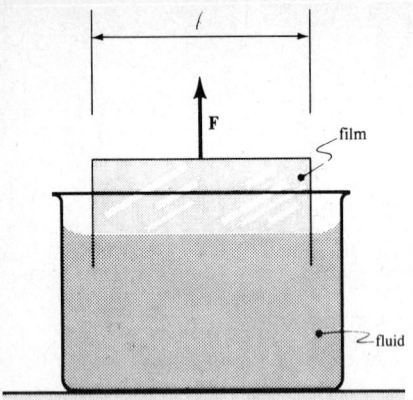

Figure 14-5

14-4. Surface Tension

Suppose we sink a wire frame into a fluid, such as a soap solution, then draw it out (Figure 14-5). The fluid forms a film over the frame, and the film exerts a force on the frame. We can calculate this force by

**FORCE OF
SURFACE TENSION**
$$F = 2\ell S \tag{14-6}$$

where S is the **surface tension** of the fluid and has units of $\mathrm{N\,m^{-1}}$ in the SI. The factor of 2 appears because the film has two surfaces, a front and back. Surface tension, like density and specific gravity, is a characteristic of each type of substance.

EXAMPLE 14-7: Determine the surface tension of the liquid that forms a film on a wire frame 6 cm in length if the net force lifting the frame is $8.2 \times 10^{-3}\,\mathrm{N}$ (you should neglect the weight of the frame).

Solution: Solve Eq. (14-6) for S.

$$F = 2\ell S \qquad S = \frac{F}{2\ell} = \frac{8.2 \times 10^{-3}\,\mathrm{N}}{2(6 \times 10^{-2}\,\mathrm{m})} = \boxed{6.83 \times 10^{-2}\,\mathrm{N\,m^{-1}}}$$

Surface tension also governs the height h to which a fluid of density ρ will rise in a narrow tube (a **capillary tube**) of radius r:

**CAPILLARY
RISE**
$$h = \frac{2S\cos\theta}{\rho g r} \tag{14-7}$$

where θ is the angle of contact between the surface of the fluid and the capillary wall (Figure 14-6a). Equation (14-7) applies both to fluids that "wet" (adhere) to the surfaces of their containers and to those that do not wet surfaces (Figure 14-6b). The level of a nonwetting fluid in a capillary tube is *depressed* below the level of the surrounding fluid.

(a)

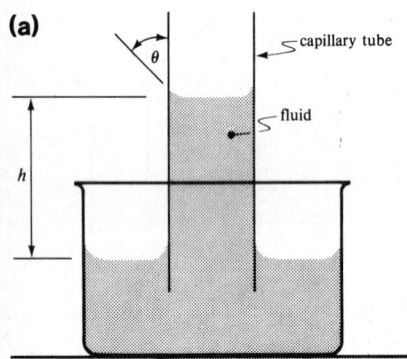

(b)

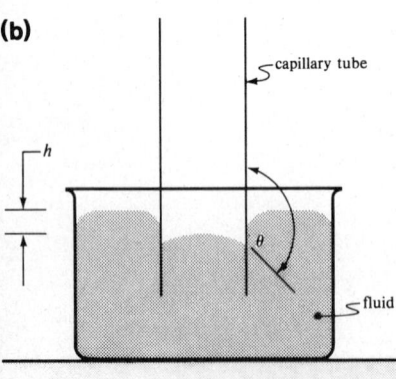

Figure 14-6. (a) Capillary rise of wetting fluid. **(b)** Capillary depression of nonwetting fluid.

EXAMPLE 14-8: Water has a surface tension of about $73 \times 10^{-3}\,\mathrm{N\,m^{-1}}$. The contact angle for water and glass is about $25°$. How high will water rise in a glass capillary that has a diameter of 0.5 mm?

Solution: Use Eq. (14-7) with $r = \frac{1}{2}(0.5\,\mathrm{mm}) = 2.5 \times 10^{-4}\,\mathrm{m}$.

$$h = \frac{2S\cos\theta}{\rho g r}$$

$$= \frac{2(73 \times 10^{-3}\,\mathrm{N\,m^{-1}})\cos 25°}{(10^3\,\mathrm{kg\,m^{-3}})(9.8\,\mathrm{m\,s^{-2}})(2.5 \times 10^{-4}\,\mathrm{m})} = \boxed{5.4 \times 10^{-2}\,\mathrm{m} = 5.4\,\mathrm{cm}}$$

The net pressure (gauge pressure) inside a droplet of fluid of radius r is

**PRESSURE WITHIN
A DROPLET**
$$P = \frac{2S}{r} \tag{14-8a}$$

A hollow bubble of fluid has two surfaces (inside and outside), so the net pressure inside a bubble of radius r is twice the pressure inside a droplet of the same size of the same liquid.

**PRESSURE WITHIN
A BUBBLE**
$$P = \frac{4S}{r} \tag{14-8b}$$

EXAMPLE 14-9: What is the net pressure inside a water droplet of diameter 2×10^{-6} m?

Solution: Use Eq. (14-8a).

$$P = \frac{2S}{r} = \frac{2(73 \times 10^{-3}\,\mathrm{N\,m^{-1}})}{10^{-6}\,\mathrm{m}} = \boxed{1.46 \times 10^{5}\,\mathrm{Pa}}$$

SUMMARY

1. The mass and weight *densities* of an object are, respectively,

$$\rho = \frac{m}{V} \quad \text{and} \quad D = \frac{w}{V}$$

where V is the volume of the object. Mass density is commonly used in the metric system; weight density is used in the British system.

2. The *specific gravity* of a substance is the ratio of the density of the substance to the density of water: $\rho_{\text{substance}}/\rho_{\text{water}}$.

3. The pressure acting on a surface is the force per unit area: $P = F/A$.

4. The *hydrostatic pressure* at a depth h below the surface of a fluid is $P = P_0 + \rho gh$.

5. *Pascal's principle* states that a change in the pressure applied to a fluid is transmitted uniformly throughout the fluid.

6. *Archimedes' principle* states that the buoyant force experienced by a body partly or completely submerged in a fluid is equal to the weight of the fluid displaced by the body.

7. The net force exerted by a film of fluid stretched over a wire frame of length ℓ is $F = 2\ell S$ where S is the *surface tension* of the fluid.

8. In a capillary of radius r a fluid of density ρ will rise or be depressed a distance h according to $h = 2S\cos\theta/\rho gr$, depending on whether the fluid does or does not wet the surface of the capillary walls.

9. The net pressure inside a droplet of liquid of radius r is $P = 2S/r$; inside a bubble, $P = 4S/r$.

RAISE YOUR GRADES

Can you explain . . . ?

☑ why a boat floats higher in the ocean than in a lake

☑ whether the pressure at a given depth will increase, decrease, or remain the same if a liquid becomes compressible

☑ what supports the column of mercury in a barometer

☑ why a person is not crushed when the average surface area of a human body is about 1.5 m² and atmospheric pressure is about 10^5 Pa = 10^5 N/m²

☑ why pressure in a fluid depends only on depth beneath the surface and not on the shape of the container

☑ why a container of water weighs more when a lead weight suspended by a string is immersed in it than when the lead weight is removed

☑ how to tell if an irregularly shaped object is made of solid gold or lead with a thin coating of gold without scratching the surface of the object

☑ why a smaller soap bubble will blow up a larger soap bubble when the bubbles are connected by a hollow tube

☑ why a dam is built thicker near the base than near the top

SOLVED PROBLEMS

Density and Specific Gravity

PROBLEM 14-1:   What is the volume of 4 kg of aluminum?

Solution: Use Eq. (14-1) and solve for V. Find the density of aluminum from Table 14-1.

$$\rho = \frac{m}{V} \qquad V = \frac{m}{\rho} = \frac{4\text{ kg}}{2.7 \times 10^3\text{ kg m}^{-3}} = \boxed{1.48 \times 10^{-3}\text{ m}^3 = 1.48 \times 10^3\text{ cm}^3}$$

PROBLEM 14-2:   The specific gravity of a certain mix of concrete is 3. What is the weight of a cubic yard of this concrete? (A cubic yard of material is often simply called a "yard" by construction engineers.)

Solution: Use Eq. (14-2) and solve for w. One cubic yard is $V = 3\text{ ft} \times 3\text{ ft} \times 3\text{ ft} = 27\text{ ft}^3$.

$$D = \frac{w}{V}$$

$$w = DV = (SG)D_{\text{water}}V = (3)(62.43\text{ lbs ft}^{-3})(27\text{ ft}^3) = \boxed{5.06 \times 10^3\text{ lb} = 2.53\text{ tons}}$$

Static Pressure and Pascal's Principle

PROBLEM 14-3:   What is the pressure in the ocean at a depth of 33 m (about 100 ft)?

Solution: Use Eq. (14-5) with $P_0 = P_A$, atmospheric pressure. Find the density of sea water from Table 14-1.

$$P = P_0 + \rho gh = 1.013 \times 10^5\text{ Pa} + (1.03 \times 10^3\text{ kg m}^{-3})(9.8\text{ m s}^{-2})(33\text{ m})$$

$$= 1.013 \times 10^5\text{ Pa} + 3.331 \times 10^5\text{ Pa} = \boxed{4.34 \times 10^5\text{ Pa}}$$

PROBLEM 14-4:   Determine the total force on the dam shown in Figure 14-7. The water behind the dam is fresh (i.e., pure) water. The wall is 25 m wide and 12 m high.

Solution: Solve Eq. (14-4), $P = F/A$, to find the force on the dam's area A: $F = PA = Pwh$. The pressure is not constant over the height of the dam, but varies according to depth y, by $P = P_0 + \rho gy$. Atmospheric pressure, $P_A = P_0$, acts on both sides of the dam, so its net effect is zero, and you can substitute ρgy for P. To find F, integrate ρgyw over the dam's height h:

$$F = \int dF = \int (\rho gy)w\,dy = \rho gw \int_0^h y\,dy = \tfrac{1}{2}\rho gwh^2$$

$$= \tfrac{1}{2}(10^3\text{ kg m}^{-3})(9.8\text{ m s}^{-2})(25\text{ m})(12\text{ m})^2 = \boxed{1.76 \times 10^7\text{ N}}$$

PROBLEM 14-5:   The flask shown in Figure 14-8 has a volume of 1 L $(= 10^3\text{ cm}^3)$. One liter of water has a mass of 1 kg and thus weighs 9.8 N. Calculate the force on the bottom of the flask when filled with water.

Solution: Use Eq. (14-4) to compute the force and Eq. (14-5) for the pressure. The net effect of atmospheric pressure is zero, since atmospheric pressure acts on all sides of the flask. So $F = PA$, where A is the area of the base of the flask and P is the gauge pressure.

$$F = (\rho gh)A = (\rho gh)(\pi r^2) = (10^3\text{ kg m}^{-3})(9.8\text{ m s}^{-2})[(2.77 + 0.05)\text{m}]\pi(0.05\text{ m})^2 = \boxed{217\text{ N}}$$

Notice the apparent paradox: The force on the bottom of the container is considerably greater

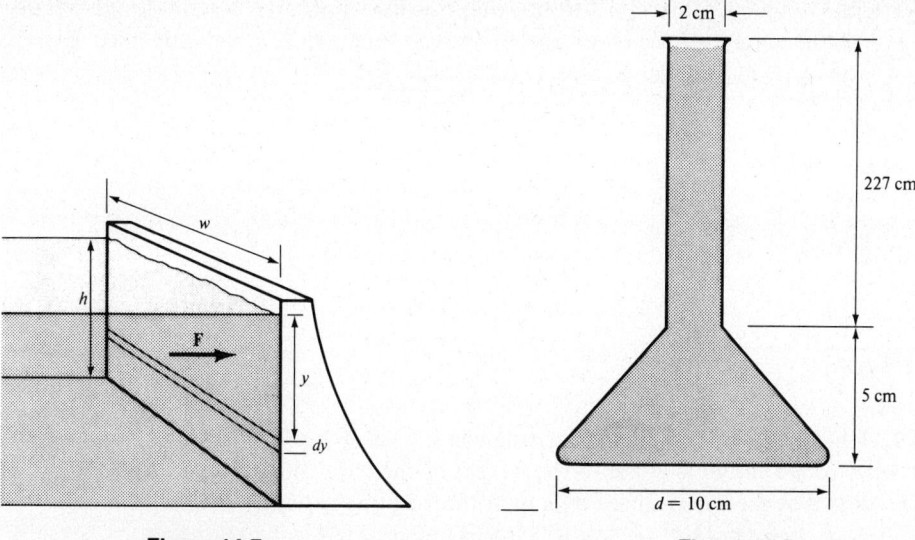

Figure 14-7   **Figure 14-8**

than the weight of the liquid producing it! What is the origin of this "extra" force? The paradox is resolved by considering the constraining force that the sloping walls of the container exert on the liquid.

Archimedes' Principle

PROBLEM 14-6: An object "weighs" 147 g in air and 104 g when immersed in a light oil that has a specific gravity of 0.85. What is the specific gravity of the object?

Solution: From Example 14-5 you know that

$$\frac{w_i}{w} = 1 - \frac{\rho_\ell}{\rho}$$

Because specific gravity is just density divided by a constant, you can substitute SG for ρ in the equation.

$$\frac{w_i}{w} = 1 - \frac{\rho_\ell}{\rho} = 1 - \frac{SG_{liquid}}{SG_{object}}$$

Now solve for the specific gravity of the object.

$$SG_{object} = \frac{SG_{liquid}}{1 - \dfrac{w_i}{w}} = \frac{0.85}{1 - \dfrac{104\ g}{147\ g}} = \boxed{2.9}$$

PROBLEM 14-7: A solid cork ball 8 cm in diameter is immersed in water and held down by a light thread attached to the bottom of the tank. Compute the tension in the thread. The SG of cork is 0.3.

Solution: Use Archimedes' principle. The tension T in the thread equals the buoyant force on the ball minus the ball's weight. The buoyant force is $F_B = w_{water} = \rho_{water} V g$. The weight of the ball is $m_{ball} V g = \rho_{cork} V g = \rho_{water} SG_{cork} V g$. So

$$T = F_B - w_{ball} = \rho_{water} V g - \rho_{water} SG_{cork} V g$$
$$= V g \rho_{water}(1 - SG_{cork}) = \tfrac{4}{3} \pi r^3 g \rho_{water}(1 - SG_{cork})$$
$$= \tfrac{4}{3}\pi(0.04\ m)^3(9.8\ m\,s^{-1})(1 \times 10^3\ kg\,m^{-3})(1 - 0.3) = \boxed{1.84\ N}$$

PROBLEM 14-8: As a function of specific gravity, what fraction of the volume of a block of wood is immersed when the block floats on water?

Solution: Use Archimedes' principle. In this case, the buoyant force is equal to the weight of the wood. Let V = the volume of the block and V_i = the volume of the wood immersed. Express F_B in terms of V_i and w_{wood} in terms of V, then find the ratio V_i/V.

$$F_B = w$$

$$\rho_{water}V_i g = mg = \rho_{wood}Vg = (SG)\rho_{water}Vg$$

$$V_i = (SG)V$$

$$\frac{V_i}{V} = \boxed{SG}$$

Surface Tension

PROBLEM 14-9:   A piece of wire 6 cm long has a mass of 1 g. The net force needed to pull this piece of wire from a liquid is equal to the weight of the wire. What is the surface tension of the liquid? Assume that the wire is lifted by a light thread attached at its midpoint.

Solution: Solve Eq. (14-6) for S where F is the weight of the wire.

$$F = 2\ell S$$

$$S = \frac{F}{2\ell} = \frac{mg}{2\ell} = \frac{(10^{-3}\ kg)(9.8\ m\ s^{-2})}{2(6 \times 10^{-2}\ m)} = \boxed{8.2 \times 10^{-2}\ N\,m^{-1}}$$

PROBLEM 14-10:   The liquid in Problem 14-9 rises to a height of 4.8 cm in a capillary tube that has a diameter of 0.5 mm. The specific gravity of this liquid is 1.1. Find the contact angle between the liquid surface and the tube wall.

Solution: Use Eq. (14-7) and solve for θ.

$$h = \frac{2S\cos\theta}{\rho g r}$$

$$\theta = \arccos\left(\frac{h\rho g r}{2S}\right) = \cos^{-1}\left[\frac{(4.8 \times 10^{-2}\ m)(1.1 \times 10^3\ kg\,m^{-3})(9.8\ m\,s^{-2})(2.5 \times 10^{-4}\ m)}{2(8.2 \times 10^{-2}\ N\,m^{-1})}\right]$$

$$= \boxed{37.9°}$$

Supplementary Exercises

EXERCISE 14-1:   One pound (450 g) of gold was used to cover the "flame" on the Statue of Liberty. (a) How much gold (in cm^3) was used? (b) What area (in m^2) would this much gold cover if its thickness were 0.0001 mm?

EXERCISE 14-2:   What is the weight of air in a room that has dimensions 3 m × 4 m × 5 m? The specific gravity of air is 1.3×10^{-3}.

EXERCISE 14-3:   The specific gravity of the dark fluid in the U-tube in Figure 14-9 is 6. What is the specific gravity of the light fluid if $h_1 = 8$ cm and $h_2 = 3$ cm?

EXERCISE 14-4:   Determine the pressure inside the closed container in Figure 14-10. P_0 is atmospheric pressure, the fluid in the U-tube is mercury, and h is 12 cm.

EXERCISE 14-5:   From Eq. (14-5) you can easily obtain $dP = \rho g\,dh$. Using this relationship, find the pressure at a depth h in a fluid in which the density is directly proportional to the pressure: $\rho = kP$ where k is the proportionality constant. Assume P_0 is the pressure above the fluid.

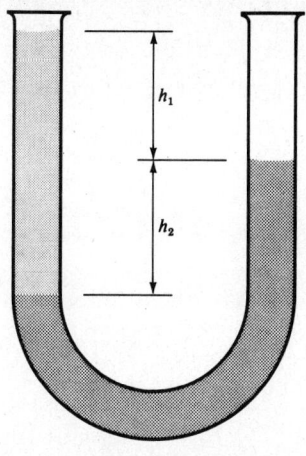

Figure 14-9   **Figure 14-10**

EXERCISE 14-6:   What is the density of an object that weighs 1.27 stone in air but only 0.98 stone when submerged in water? (The "stone" is a British unit of weight.)

EXERCISE 14-7:   An object is known to be made of lead and gold. It "weighs" 2.500 kg in air and 2.252 kg when immersed in an oil that has a specific gravity of 1.5. What percentage of the object is gold?

EXERCISE 14-8:   What is the wall thickness of a lead balloon 20 cm in diameter that floats in water with half its volume submerged?

EXERCISE 14-9:   What is the net pressure inside a soap bubble 4 cm in diameter? The surface tension of soap film is $25 \times 10^{-3} \, \mathrm{N \, m^{-1}}$.

EXERCISE 14-10:   How far below the surrounding surface will the level of mercury be depressed in a glass capillary 2 mm in diameter? The surface tension of mercury is $0.513 \, \mathrm{N \, m^{-1}}$ and the contact angle of mercury with glass is 140°.

Answers to Supplementary Exercises

14-1: (a) $23.3 \, \mathrm{cm^3}$ (b) $233 \, \mathrm{m^2}$

14-2: 78 kg

14-3: 1.64

14-4: $8.53 \times 10^4 \, \mathrm{Pa}$

14-5: $P = P_0 e^{kgh}$

14-6: $4.38 \times 10^3 \, \mathrm{kg \, m^{-3}}$

14-7: 60% by weight

14-8: 1.48 mm

14-9: 5 Pa

14-10: 5.9 mm

15 FLUID DYNAMICS

THIS CHAPTER IS ABOUT

☑ Equation of Continuity
☑ Bernoulli's Equation
☑ Viscosity and Poiseuille's Equation
☑ Stokes' Laws
☑ Reynolds Number

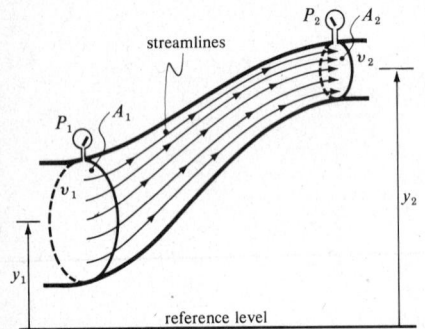

Figure 15-1. Flow of an incompressible fluid in a pipe.

15-1. Equation of Continuity

When an incompressible fluid flows through a pipe, the law of conservation of mass requires that the amount of fluid that flows into the pipe must equal the quantity that flows out. This results in the fact that the product of the pipe's cross-sectional area and the flow velocity, assumed uniform over the area, is constant at all points in the pipe (Figure 15-1). We can express this by the **equation of continuity** for any two points 1 and 2 in the pipe:

EQUATION OF CONTINUITY $\qquad A_1 v_1 = A_2 v_2$ $\qquad$ (15-1)

EXAMPLE 15-1: Water enters the large end of a hose nozzle 6 cm in diameter at a speed of 2.1 m s^{-1}. With what speed does it exit the small end, 2.2 cm in diameter?

Solution: Use the equation of continuity and solve for v_2.

$$A_1 v_1 = A_2 v_2$$

$$v_2 = v_1 \left(\frac{A_1}{A_2}\right) = v_1 \left(\frac{\pi r_1^2}{\pi r_2^2}\right) = v_1 \left(\frac{r_1}{r_2}\right)^2 = (2.1 \text{ m s}^{-1})\left(\frac{3 \text{ cm}}{1.1 \text{ cm}}\right)^2 = \boxed{15.6 \text{ m s}^{-1}}$$

15-2. Bernoulli's Equation

If the fluid is incompressible, if its flow is *laminar* (nonturbulent), and if there are no viscous (frictional) forces on the fluid, the energy of the fluid is constant at each point in the pipe as well. **Bernoulli's equation** relates the kinetic and potential energy of a fluid to the absolute pressure at any two points along a streamline in a pipe.

BERNOULLI'S EQUATION $\qquad P_1 + \rho g y_1 + \frac{1}{2}\rho v_1^2 = P_2 + \rho g y_2 + \frac{1}{2}\rho v_2^2$ $\qquad$ (15-2)

EXAMPLE 15-2: What is the gauge pressure at the hose end of the nozzle in Example 15-1 if the nozzle is held horizontally?

Solution: Use Bernoulli's equation, keeping in mind that the pressures in it are absolute pressures. Absolute pressure equals gauge pressure plus atmospheric pressure, so we can express P_1 as $P_{1(gauge)} + P_A$. When the water leaves the nozzle, atmospheric pressure is the only pressure acting on it, so $P_2 = P_A$. Because the nozzle is horizontal, $y_1 = y_2$ and $y_2 - y_1 = 0$.

$$P_1 + \rho g y_1 + \tfrac{1}{2}\rho v_1^2 = P_2 + \rho g y_2 + \tfrac{1}{2}\rho v_2^2$$

$$P_{1(gauge)} + P_A + \rho g y_1 + \tfrac{1}{2}\rho v_1^2 = P_A + \rho g y_2 + \tfrac{1}{2}\rho v_2^2$$

$$P_{1(gauge)} = \rho g(y_2 - y_1) + \tfrac{1}{2}\rho(v_2^2 - v_1^2) = \tfrac{1}{2}\rho(v_2^2 - v_1^2)$$

Substitute for v_1 and v_2 the values from Example 15-1.

$$P_{1(gauge)} = \tfrac{1}{2}(10^3\ \mathrm{kg\,m^{-3}})[(15.6)^2 - (2.1)^2]\ \mathrm{m^2\,s^{-2}} = \boxed{1.19 \times 10^5\ \mathrm{Pa}}$$

Notice that this result is about 18 lb in.$^{-2}$.

EXAMPLE 15-3: The air pressure above the water in a large tank is maintained at 2 atmospheres (absolute). A length of pipe runs from the base of the tank to a valve on the ground. The distance from the water's surface to the ground is 4 m. At what velocity does the water flow from the pipe when the valve is open? Assume that the tank is sufficiently large so that the velocity of upper surface of the water can be neglected (see Figure 15-2).

Solution: Use Bernoulli's equation with 1 representing conditions at the upper water surface and 2 representing conditions at the ground. Solve for v_2.

$$P_1 + \rho g y_1 + \tfrac{1}{2}\rho v_1^2 = P_2 + \rho g y_2 + \tfrac{1}{2}\rho v_2^2$$

$$v_2^2 = \frac{2(P_1 - P_2)}{\rho} + 2g(y_1 - y_2) + v_1^2$$

Let $y_1 - y_2 = h = 4$ m, $P_1 - P_2 = (2 - 1)$ atm $= 1.013 \times 10^5$ Pa, and $v_1 \cong 0$.

$$v_2 = \sqrt{\frac{2(P_1 - P_2)}{\rho} + 2gh} = \sqrt{\frac{2(1.013 \times 10^5\ \mathrm{Pa})}{10^3\ \mathrm{kg\,m^{-3}}} + 2(9.8\ \mathrm{m\,s^{-2}})(4\ \mathrm{m})}$$

$$= \boxed{16.8\ \mathrm{m\,s^{-1}}}$$

Notice that if the upper part of the tank were open to the atmosphere, $P_1 = P_2 = P_A$ so that $v_2 = \sqrt{2gh}$, the same result as for any object freely falling a distance h.

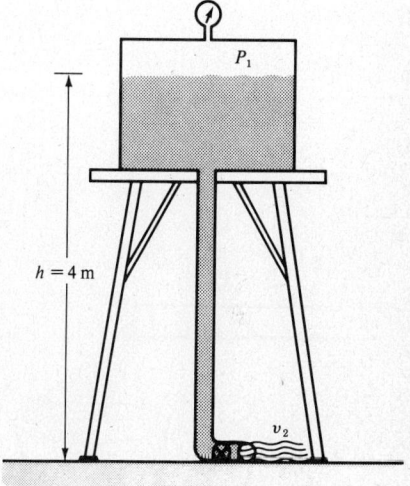

Figure 15-2

15-3. Viscosity and Poiseuille's Equation

Figure 15-3 shows a slab of area A sliding uniformly on a film of thickness h at constant speed v. The force $F_\parallel$ does not accelerate the slab because it is balanced by the viscous forces in the fluid. We assume that the fluid in contact with each surface is at rest with respect to that surface, so that a uniform velocity gradient is present in the fluid, v/h. The fluid therefore is subject to a shear stress $F_\parallel/A$ (see Sec. 13-4) and the rate of change of shear strain in it is equal to the velocity gradient. The **coefficient of viscosity** η relates these quantities:

COEFFICIENT OF VISCOSITY

$$\underbrace{\frac{F_\parallel}{A}}_{\substack{\text{shear}\\\text{stress}}} = \underbrace{\eta\left(\frac{v}{h}\right)}_{\substack{\text{rate of change}\\\text{of shear strain}}} \tag{15-3}$$

The unit of η in the SI is the pascal · second: 1 Pa s $= 1\ \mathrm{N\,m^{-2}\,s}$. The unit in common usage, however, is the **poise**: 10 poise $= 10^3$ centipoise $= 1$ Pa s. The value of η for any liquid depends strongly on temperature, so be cautious when you use values in handbooks. Be sure to note the temperature at which the reported measurement was made. Equation (15-3) describes an idealization that not all fluids obey. Fluids that obey Eq. (15-3) are called **newtonian fluids**.

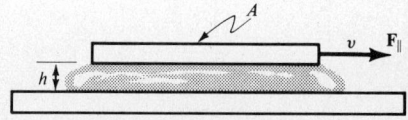

Figure 15-3

EXAMPLE 15-4: How much force is needed to pull a sheet of aluminum measuring 20 cm × 12 cm at a speed of 30 cm s^{-1} over a flat surface covered with oil 0.2 mm thick? The oil has a viscosity of 8 poise.

Solution: Use Eq. (15-3) and solve for $F_{||}$. The coefficient of viscosity $\eta = 8$ poise $= 0.8$ Pa s.

$$\frac{F_{||}}{A} = \eta\left(\frac{v}{h}\right) \qquad F_{||} = \frac{A\eta v}{h} = \frac{[(0.2)(0.12)\text{ m}^2](0.8\text{ Pa s})(0.3\text{ m s}^{-1})}{(0.2 \times 10^{-3}\text{ m})} = \boxed{28.8\text{ N}}$$

When a newtonian fluid flows in a pipe of circular cross-section, the flow velocity is greatest at the center and decreases to zero at the inner wall. The quantity of fluid that flows per second (the number of cubic meters per second) in a pipe of inner radius r is proportional to the pressure gradient dP/dx in the pipe and is given by **Poiseuille's equation**:

POISEUILLE'S EQUATION
$$\frac{\Delta V}{\Delta t} = \frac{\pi r^4}{8\eta}\left(\frac{P_1 - P_2}{x}\right) \qquad (15\text{-}4)$$

where $(P_1 - P_2)/x$ is the uniform pressure gradient, as shown in Figure 15-4.

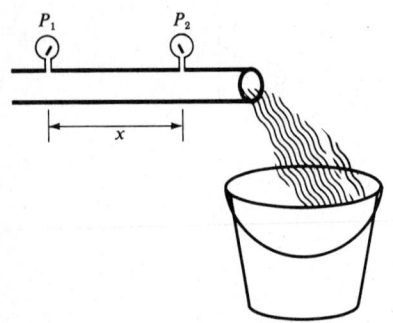

P_1 P_2

x

Figure 15-4. Flow of a newtonian fluid in a pipe.

EXAMPLE 15-5: Determine the flow rate for the oil in Example 15-4 that flows in a pipe with an inside diameter of 4 cm if the pressure gradient along the pipe is 3×10^4 Pa m^{-1}.

Solution: Use Eq. (15-4)

$$\frac{\Delta V}{\Delta t} = \frac{\pi r^4}{8\eta}\left(\frac{P_1 - P_2}{x}\right)$$

$$= \frac{\pi(2 \times 10^{-2}\text{ m})^4}{8(0.8\text{ Pa s})}(3 \times 10^4\text{ Pa m}^{-1}) = \boxed{2.36 \times 10^{-3}\text{ m}^3\text{ s}^{-1}}$$

Notice that this result can also be expressed as 2.36 L s^{-1}.

15-4. Stokes' Laws

When a sphere of radius r moves in a newtonian fluid at a constant velocity v relative to the fluid, it experiences a retarding force, given by Stokes' law.

STOKES' LAW $\qquad\qquad F = 6\pi\eta r v \qquad\qquad (15\text{-}5)$

If the sphere has density ρ, the fluid has density ρ_ℓ, and the sphere is falling vertically in the fluid, the sphere's constant downward velocity is

$$v = \frac{2r^2 g(\rho - \rho_\ell)}{9\eta} \qquad (15\text{-}6)$$

Equation (15-6) is sometimes called Stokes' law too. It is derived from Eq. (15-5) by considering the buoyant force on the sphere.

EXAMPLE 15-6: Use Stokes' law, Eq. (15-5), and Archimedes' principle to derive Eq. (15-6).

Solution: If a sphere is falling at a constant velocity (the *terminal velocity*) in the fluid, the buoyant force plus the retarding force must just balance the weight of the sphere.

$$F_B + F = mg$$

$$\rho_1 V g + 6\pi\eta r v = \rho V g$$

Solve for v:

$$v = \frac{Vg(\rho - \rho_\ell)}{6\pi\eta r} = \frac{(\frac{4}{3}\pi r^3)g(\rho - \rho_\ell)}{6\pi\eta r} = \boxed{\frac{2r^2 g(\rho - \rho_\ell)}{9\eta}}$$

EXAMPLE 15-7: A steel ball bearing 3 mm in diameter is released in a cylinder of glycerin. Calculate the constant velocity with which this bearing descends. The specific gravities of steel and glycerin are 7.83 and 1.26, respectively. The coefficient of viscosity of glycerin is 8.3 poise.

Solution: Use the terminal velocity form of Stokes' law, Eq. (15-6). The coefficient of viscosity in SI units is 0.83 Pa s.

$$v = \frac{2r^2 g(\rho - \rho_\ell)}{9\eta}$$

$$= \frac{2(1.5 \times 10^{-3}\ \mathrm{m})^2 (9.8\ \mathrm{m\,s^{-2}})(7.8 - 1.26) \times 10^3\ \mathrm{kg\,m^{-3}}}{9(0.83\ \mathrm{Pa\,s})}$$

$$= \boxed{3.86 \times 10^{-2}\ \mathrm{m\,s^{-1}}}$$

15-5. Reynolds Number

The Reynolds number $\mathcal{R}$ is a dimensionless quantity that can be used to estimate whether the flow of a fluid through a pipe or around an object is **laminar** (smooth) or **turbulent**. Turbulent flow is characterized by swirls and vortices in the fluid that greatly increase the resistance to flow. For flow in a circular pipe of radius r or around a sphere of radius r the Reynolds number is

REYNOLDS NUMBER

$$\mathcal{R} = \frac{2r\,\rho_\ell v}{\eta} \tag{15-7}$$

When $\mathcal{R}$ is below about 2000 the flow is laminar; when it is above 3000 the flow is turbulent. For fluids not flowing through a circular pipe or around a sphere, $2r$ must be replaced by another quantity characteristic of the size of the object.

EXAMPLE 15-8: The oil in Example 15-5 has specific gravity of 0.9. **(a)** Decide if the flow in the pipe is laminar or turbulent. **(b)** At what flow velocity could you be sure the flow would be turbulent?

Solution:

(a) For an incompressible fluid, the volumetric flow rate in a pipe of cross-sectional area A is $\Delta V/\Delta t = A\,\Delta x/\Delta t = Av$, when v, the flow velocity, is constant across the area. For the pipe in Example 15-5,

$$v = \frac{\Delta V/\Delta t}{A} = \frac{\Delta V/\Delta t}{\pi r^2} = \frac{2.36 \times 10^{-3}\ \mathrm{m^3\,s^{-1}}}{\pi(2 \times 10^{-2}\ \mathrm{m})^2} = 1.88\ \mathrm{m\,s^{-1}}$$

Now use Eq. (15-7):

$$\mathcal{R} = \frac{2r\,\rho_\ell v}{\eta} = \frac{2(2 \times 10^{-2}\ \mathrm{m})(0.9 \times 10^3\ \mathrm{kg\,m^{-3}})(1.88\ \mathrm{m\,s^{-1}})}{0.8\ \mathrm{Pa\,s}} = 84.6$$

$\mathcal{R} \ll 2000$, so clearly the flow is laminar.

(b) Set $\mathcal{R} = 3000$ and solve for v:

$$v = \frac{\mathcal{R}\eta}{2r\,\rho_\ell} = \frac{(3 \times 10^3)(0.8\ \mathrm{Pa\,s})}{2(2 \times 10^{-2}\ \mathrm{m})(0.9 \times 10^3\ \mathrm{kg\,m^{-3}})} = \boxed{66.7\ \mathrm{m\,s^{-1}}}$$

SUMMARY

1. The *equation of continuity* relates the cross-sectional area and fluid flow velocity at two different points in a pipe: $A_1v_1 = A_2v_2$.
2. *Bernoulli's equation* relates the absolute pressure, height, and flow velocity at two different points in a pipe through which an ideal fluid flows:

$$P_1 + \rho gy_1 + \tfrac{1}{2}\rho v_1^2 = P_2 + \rho gy_2 + \tfrac{1}{2}\rho v_2^2$$

3. The *coefficient of viscosity* is the proportionality constant between the shear stress and the rate of change of the shear strain for a newtonian fluid:

$$\frac{F_{\|}}{A} = \eta\left(\frac{v}{h}\right)$$

4. *Poiseuille's equation* relates the quantity of fluid flow per second to the pressure gradient in a pipe of radius r:

$$\frac{\Delta V}{\Delta t} = \frac{\pi r^4}{8\eta}\left(\frac{P_1 - P_2}{x}\right)$$

5. Stokes' law describes the retarding force acting on a sphere of radius r moving with relative velocity v in a fluid: $F = 6\pi\eta\, rv$
6. The terminal velocity of a sphere of radius r and density ρ falling in a fluid of density ρ_1 is

$$v = \frac{2r^2 g(\rho - \rho_\ell)}{9\eta}$$

RAISE YOUR GRADES

Can you explain . . . ?

☑ what modification is necessary to the equation of continuity if the fluid is compressible

☑ why it is necessary to use absolute pressure in Bernoulli's equation

☑ a major difference between $F_{\|}$ in Eq. (15-3) and the force of sliding friction between surfaces

☑ how to use Poiseuille's equation to show that the flow velocity depends on the square of pipe radius

☑ the relationship between Stokes' law, Eq. (15-5), and Eq. (4-6), $f = bv$ (see Sec. 4-7)

☑ how Stokes' law can be used to determine the viscosity of a fluid

☑ how to write an expression for the Reynolds number in terms of the *kinematic viscosity* $v = \eta/\rho_\ell$

SOLVED PROBLEMS

Equation of Continuity

PROBLEM 15-1:   A large pipe, 10 cm in diameter, carries water flowing at $5.4 \times 10^{-2}\ \mathrm{m\,s^{-1}}$ into a smaller pipe, 3 cm in diameter (Figure 15-5). **(a)** What is the flow velocity in the smaller pipe? **(b)** At what rate (in $\mathrm{L\,s^{-1}}$) does water flow out of the smaller pipe?

Solution:

(a) Use the equation of continuity, Eq. (15-1), and solve for v_2. Remember that $A = \pi r^2 = (\pi/4)d^2$, where d = diameter.

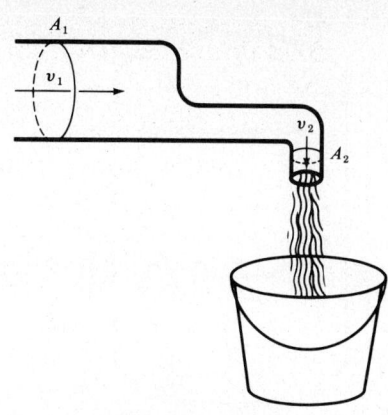

$$A_1 v_1 = A_2 v_2$$

$$v_2 = v_1\left(\frac{A_1}{A_2}\right) = v_1\frac{(\pi/4)d_1^2}{(\pi/4)d_2^2} = v_1\left(\frac{d_1}{d_2}\right)^2$$

$$= (5.4 \times 10^{-2}\ \mathrm{m\,s^{-1}})\left(\frac{10\ \mathrm{cm}}{3\ \mathrm{cm}}\right)^2 = \boxed{0.6\ \mathrm{m\,s^{-1}}}$$

(b) Recall from Example 15-8 that $\Delta V/\Delta t = Av$ for an incompressible fluid. Thus

$$\frac{\Delta V}{\Delta t} = A_2 v_2 = \left(\frac{\pi}{4}\right)d_2^2 v_2 = \frac{\pi}{4}(3 \times 10^{-2}\ \mathrm{m})^2(0.6\ \mathrm{m\,s^{-1}})$$

$$= \boxed{4.24 \times 10^{-4}\ \mathrm{m^3\,s^{-1}} = 0.424\ \mathrm{L\,s^{-1}}}$$

Figure 15-5

Bernoulli's Equation

PROBLEM 15-2: The specific gravity of the liquid flowing in the pipe shown in Figure 15-6 is 1.2. The inside diameter of the pipe narrows from 10 to 6 cm, and the difference in pressure $P_1 - P_2 = 2 \times 10^5$ Pa. What is the flow rate in the pipe in liters per second?

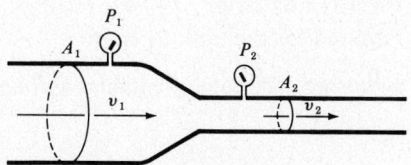

Figure 15-6

Solution: Use the equation of continuity, Eq. (15-1), to express v_2 in terms of v_1 and the pipe diameters.

$$A_1 v_1 = A_2 v_2 \qquad v_2 = v_1\left(\frac{A_1}{A_2}\right) = v_1\frac{(\pi/4)d_1^2}{(\pi/4)d_2^2} = v_1\left(\frac{d_1}{d_2}\right)^2$$

Substitute this result into Bernoulli's equation and solve for v_1. Because $y_1 = y_2$, the terms $\rho g y_1$ and $\rho g y_2$ cancel out.

$$P_1 + \frac{1}{2}\rho v_1^2 = P_2 + \frac{1}{2}\rho v_2^2 = P_2 + \frac{1}{2}\rho v_1^2\left(\frac{d_1}{d_2}\right)^4$$

$$\frac{1}{2}\rho v_1^2\left[\left(\frac{d_1}{d_2}\right)^4 - 1\right] = P_1 - P_2$$

$$v_1 = \sqrt{\frac{2(P_1 - P_2)}{\rho\left[\left(\frac{d_1}{d_2}\right)^4 - 1\right]}} = \sqrt{\frac{2(2 \times 10^5\ \mathrm{Pa})}{(1.2 \times 10^3\ \mathrm{kg\,m^{-3}})\left[\left(\frac{10\ \mathrm{cm}}{6\ \mathrm{cm}}\right)^4 - 1\right]}} = 7.045\ \mathrm{m\,s^{-1}}$$

Use this result to determine the flow rate:

$$\frac{\Delta V}{\Delta t} = A_1 v_1 = \frac{\pi}{4}d_1^2 v_1 = \frac{\pi}{4}(0.1\ \mathrm{m})^2(7.045\ \mathrm{m\,s^{-1}}) = \boxed{5.53 \times 10^{-2}\ \mathrm{m^3\,s^{-1}} = 55.3\ \mathrm{L\,s^{-1}}}$$

PROBLEM 15-3: A right circular cylindrical tank, open at the top, is filled with water to a height h. A small pipe with an inside diameter of d_2 fits into the wall of the tank at its base (Figure 15-7). Use Bernoulli's equation to show that the exit velocity of the water is proportional to the square root of the remaining height of the water.

Solution: Bernoulli's equation, Eq. (15-2), is

$$P_1 + \rho g y_1 + \tfrac{1}{2}\rho v_1^2 = P_2 + \rho g y_2 + \tfrac{1}{2}\rho v_2^2$$

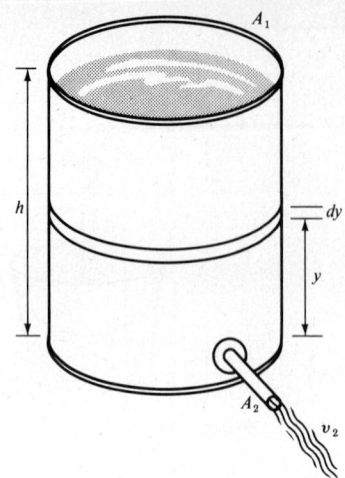

Figure 15-7

Use the equation of continuity to express v_1 in terms of v_2:

$$A_1 v_1 = A_2 v_2 \qquad v_1 = v_2 \left(\frac{A_2}{A_1} \right)$$

Substitute this result into Bernoulli's equation and note that because the tank is open, $P_1 = P_2 = P_A$. Therefore

$$\rho g y_1 + \frac{1}{2} \rho v_2^2 \left(\frac{A_2}{A_1} \right)^2 = \rho g y_2 + \frac{1}{2} \rho v_2^2$$

Solve for v_2 with $y_2 = 0$ and $y_1 - y_2 = y$.

$$v_2 = \sqrt{\frac{2gy}{1 - \left(\frac{A_2}{A_1} \right)^2}}$$

Notice that this result could be written as

$$v_2 = K \sqrt{y} \qquad \text{where} \qquad K = \sqrt{\frac{2g}{1 - \left(\frac{A_2}{A_1} \right)^2}}$$

PROBLEM 15-4: Use the result of Problem 15-3 to obtain an expression for the amount of time required for the tank to empty.

Solution: The rate at which water leaves the tank is

$$\frac{dV}{dt} = A_2 v_2 = A_2 K \sqrt{y}$$

From Figure (15-7) you can see that $dV = A_1 \, dy$. Thus

$$\frac{A_1 \, dy}{dt} = A_2 K \sqrt{y} \qquad \frac{dy}{\sqrt{y}} = \frac{A_2 K}{A_1} dt$$

$$\int_h^0 \frac{dy}{\sqrt{y}} = \frac{A_2 K}{A_1} \int_0^t dt \qquad 2\sqrt{y} \Big|_h^0 = -2\sqrt{h} = \frac{A_2 K}{A_1} t$$

$$\boxed{t = -\frac{2 A_1 \sqrt{h}}{A_2 K}}$$

Notice that the negative sign could be dropped. It indicates only that the tank is being emptied.

Viscosity and Poiseuille's Equation

PROBLEM 15-5: A force of 9 N is required to pull a thin sheet of metal 8 cm × 14 cm over a film of oil at a constant speed of 20 cm s^{-1}. What is the viscosity of the oil if its thickness is 0.3 mm?

Solution: Use Eq. (15-3) and solve for η.

$$\frac{F_\parallel}{A} = \eta \left(\frac{v}{h} \right) \qquad \eta = \frac{F_\parallel h}{Av} = \frac{(9 \text{ N})(3 \times 10^{-4} \text{ m})}{(8 \times 10^{-2} \text{ m})(14 \times 10^{-2} \text{ m})(0.2 \text{ m s}^{-1})} = \boxed{1.2 \text{ Pa s} = 12 \text{ poise}}$$

PROBLEM 15-6: Determine the pressure gradient necessary for a pipe 4 cm in diameter to carry a flow of 2 L s^{-1} of the oil in Problem 15-5.

Solution: Solve Poiseuille's equation, Eq. (15-4), for the pressure gradient.

$$\frac{\Delta V}{\Delta t} = \frac{\pi r^4}{8\eta} \left(\frac{P_1 - P_2}{x} \right)$$

$$\frac{P_1 - P_2}{x} = \frac{\Delta V}{\Delta t} \left(\frac{8\eta}{\pi r^4} \right) = (2 \times 10^{-3} \text{ m}^3 \text{ s}^{-1}) \left[\frac{8(1.2 \text{ Pa s})}{\pi (2 \times 10^{-2} \text{ m})^4} \right] = \boxed{3.82 \times 10^4 \text{ Pa m}^{-1}}$$

PROBLEM 15-7: How much fluid flows through a pipe per second when the pressure gradient in the pipe is 3×10^5 Pa m^{-1}, the inside diameter of the pipe is 8 cm, and the fluid has a viscosity of 0.3 poise?

Solution: Use Poiseuille's equation, Eq. (15-4).

$$\frac{\Delta V}{\Delta t} = \frac{\pi r^4}{8\eta}\left(\frac{P_1 - P_2}{x}\right)$$

$$= \frac{\pi (0.04 \text{ m})^4}{8(0.03 \text{ Pa s})}(3 \times 10^5 \text{ Pa m}^{-1}) = \boxed{10.05 \text{ m}^3 \text{ s}^{-1} = 1.005 \times 10^4 \text{ L s}^{-1}}$$

Stokes' Laws

PROBLEM 15-8: Steel BBs 2 mm in diameter are released in a tall column of heavy machine oil that has a specific gravity of 0.95. They descend at a constant speed of 2.4 cm s^{-1}. The specific gravity of steel is 7.8. What is the viscosity of the oil?

Solution: Use Eq. (15-6) and solve for η.

$$v = \frac{2r^2 g(\rho - \rho_\ell)}{9\eta}$$

$$\eta = \frac{2r^2 g(\rho - \rho_\ell)}{9v} = \frac{2(10^{-3} \text{ m})^2 (9.8 \text{ m s}^{-2})(7.8 - 0.95) \times 10^3 \text{ kg m}^{-3}}{9(2.4 \times 10^{-2} \text{ m s}^{-1})}$$

$$= \boxed{0.62 \text{ Pa s} = 6.2 \text{ poise}}$$

PROBLEM 15-9: (a) What is the magnitude of the retarding force on each BB in Problem 15-8? (b) How does the retarding force compare with the weight of a single BB?

Solution:

(a) Use Stokes' law, Eq. (15-5).

$$F = 6\pi\eta r v = 6\pi(0.62 \text{ Pa s})(10^{-3} \text{ m})(2.4 \times 10^{-2} \text{ m s}^{-1}) = \boxed{2.80 \times 10^{-4} \text{ N}}$$

(b) Find the weight of a single BB from $w = mg$ with $m = \rho V = \rho(\frac{4}{3}\pi r^3)$.

$$w = \rho(\tfrac{4}{3}\pi r^3)g = (7.8 \times 10^3 \text{ kg m}^{-3})[\tfrac{4}{3}\pi(10^{-3} \text{ m})^3](9.8 \text{ m s}^{-2}) = \boxed{3.20 \times 10^{-4} \text{ N}}$$

The ratio is $\dfrac{F}{w} = \dfrac{2.80 \times 10^{-4} \text{ N}}{3.20 \times 10^{-4} \text{ N}} = 0.875$

Reynolds Number

PROBLEM 15-10: Calculate the Reynolds number for the steel BBs in Problem 15-8.

Solution: Use Eq. (15-7).

$$\mathcal{R} = \frac{2r\,\rho_\ell v}{\eta}$$

$$= \frac{2(10^{-3} \text{ m})(0.95 \times 10^3 \text{ kg m}^{-3})}{0.62 \text{ Pa s}} = \boxed{3.06}$$

This low value indicates that the flow around the BBs is laminar.

Supplementary Exercises

EXERCISE 15-1:   The exit nozzle for a vertical fountain is 4 cm in diameter. The water leaves with a velocity of 15 m s^{-1} directed straight up. The nozzle is fed from a pipe whose inside diameter is 8 cm. (**a**) What is the velocity of the water in the large pipe? (**b**) What is the flow rate out the nozzle?

EXERCISE 15-2:   (**a**) Determine the necessary gauge pressure in the large pipe in Exercise 15-1 so that the water leaves the nozzle at 15 m s^{-1}. (**b**) How high is the jet? (**c**) What minimum power (in hp) must a pump have to operate this fountain?

EXERCISE 15-3:   The tank in Figure 15-7 has an inside diameter of 80 cm and the exit pipe at the base has an inside diameter of 6 cm. What is the flow velocity of the water leaving the tank when the level in the tank is 60 cm above the base?

EXERCISE 15-4:   Notice that in Problem 15-3 $K \cong \sqrt{2g}$ if $A_1 \gg A_2$. Use this approximation to derive an equation for the time to empty the tank from a level h.

EXERCISE 15-5:   How long will it take to empty the tank in Exercise 15-3?

EXERCISE 15-6:   At what angle should a plane be inclined so that a 1.5-kg block of metal with an area 6 cm × 12 cm slides down on a film of oil 0.2 mm thick at a constant speed of 0.5 m s^{-1}? The viscosity of the oil is 3 poise.

EXERCISE 15-7:   What is the viscosity of a fluid that flows at the rate of 4 L min^{-1} through a pipe with an inside diameter of 2 cm when the pressure gradient is 10^5 Pa m^{-1}?

EXERCISE 15-8:   Determine the velocity with which an air bubble 1 cm in diameter will rise in glycerin. The viscosity and density of glycerin are 15 poise and 1.26×10^3 kg m^{-3}, respectively. The density of air is 1.3 kg m^{-3}.

EXERCISE 15-9:   Through a microscope, a student observes a tiny droplet falling at 0.2 mm s^{-1}. What is the mass of the droplet if its specific gravity is 0.95? The viscosity and density of air are 181×10^{-6} poise and 1.3 kg m^{-3}, respectively.

EXERCISE 15-10:   A raindrop of mass 4×10^{-6} kg reaches a terminal velocity of 6 m s^{-1}. Find the Reynolds number for raindrops in air.

Answers to Supplementary Exercises

15-1: (**a**) 3.75 m s^{-1} (**b**) 18.8 L s^{-1}

15-2: (**a**) 1.05×10^5 Pa   (**b**) 11.5 m
(**c**) 2.84 hp

15-3: 3.44 m s^{-1}

15-4: $t = \left(\dfrac{A_1}{A_2}\right)\sqrt{\dfrac{2h}{g}}$

15-5: 62.2 s

15-6: 21.5°

15-7: 5.9 Pa s

15-8: 4.57×10^{-2} m s^{-1}

15-9: 9.2×10^{-15} kg

15-10: 849

EXAM 2 (Chapters 9 to 15)

1. Locate the center of mass of a uniform rod bent into the shape of a quarter circle of radius R. **[Ch. 9]**

2. A uniform rod 1.8 m long, weighing 80 N, is supported from the ceiling at one end by a frictionless pivot. The other end of the rod is held by a cord that makes an angle of $\phi = 40°$ with the ceiling. The rod makes an angle of $\theta = 35°$ with the ceiling. A 120-N weight hangs from the free end of the rod. (see Fig. E-4).

    (a) Find the x and y components of the reaction force at the pivot **[Ch. 9]**
    (b) Find the tension in the supporting cord. **[Ch. 9]**

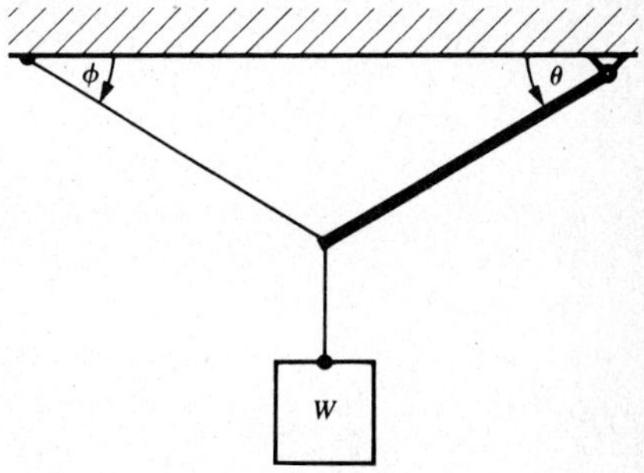

Figure E-4

3. Calculate the moment of inertia of a hoop of mass M and radius R when the hoop is rotated about a diameter. [*Hint:* $\int_0^{2\pi} \cos^2 \theta \, d\theta = \pi$]

4. A disk mounted on a motor shaft has a moment of inertia of 0.12 kg m^2.

    (a) If the disk starts from rest, how much work is done by the motor to spin the disk at 160 rad s^{-1}? **[Ch. 10]**
    (b) It takes the motor 3 s to attain this speed. How much torque does the motor supply? **[Ch. 10]**

5. A puck on a horizontal frictionless surface slides in a circle around a small hole in the surface because it is held by a light cord that passes through the hole. The puck has a mass of 300 g and the length of string from puck to hole is 40 cm.

    (a) Find the tension in the string if the angular velocity of the puck is 12 rad s^{-1}. **[Ch. 4]**
    (b) The string is drawn down through the hole until the distance from puck to hole is reduced to 20 cm. What is the angular velocity of the puck? **[Ch. 10]**
    (c) How much work was done in pulling the string through the hole? **[Ch. 11]**

6. Derive an expression for the universal gravitational constant in terms of g, R (the radius of the earth), and ρ (the density of the earth). **[Ch. 11]**

7. Derive an expression for the mass of the sun in terms of r, T, and G, where r and T are the radius and period, respectively, of the earth's orbit around the sun. **[Ch. 11]**

8. The pilot of a rocket states that his ship is 4 m long and that he is traveling at $0.55c$ relative to your position.

    (a) What length do you measure for his ship? **[Ch. 12]**
    (b) How long does it take for the ship to pass your position? **[Ch. 12]**

9. What is the kinetic energy of a proton that has a momentum of $2 \text{ GeV } c^{-1}$? The rest energy of a proton is 938 MeV and $1 \text{ GeV} = 10^3 \text{ MeV}$. **[Ch. 12]**

10. A piece of wire 1.2 m in length and 2 mm in diameter elongates 0.56 mm when a load of 14 kg is attached to its lower end. Its upper end is securely fixed.

    (a) What is the tensile stress in the wire? **[Ch. 13]**
    (b) What is Young's modulus for the material of the wire? **[Ch. 13]**
    (c) Assume that Poisson's ratio for the material is 0.3. What is the wire's shear modulus? **[Ch. 13]**

11. A solid cylindrical steel rod 2 cm in diameter is tightly clamped in a vise so that a 12-cm length protrudes above the vise jaws. How much torque applied to the upper end of the rod will cause it to twist through 6°? The shear modulus of steel is $8 \times 10^{10} \text{ Pa}$. **[Ch. 13]**

12. A swimmer dives to a depth of 12 m in the Pacific Ocean. The specific gravity of sea water is 1.03. Take atmospheric pressure to be 10^5 Pa.

    (a) What is the gauge pressure at this depth? **[Ch. 14]**
    (b) What is the absolute pressure at this depth? **[Ch. 14]**

13. What is the apparent weight of a 2-kg block of material that has a specific gravity of 2.6 when the block is immersed in water? **[Ch. 14]**

14. Use the principle of conservation of mass to derive the equation of continuity for an incompressible ideal fluid flowing at velocity v_1 through a pipe of cross-sectional area A_1 into a pipe of cross-sectional area A_2, in which the flow velocity is v_2 (see Fig. E-5).   **[Ch. 15]**

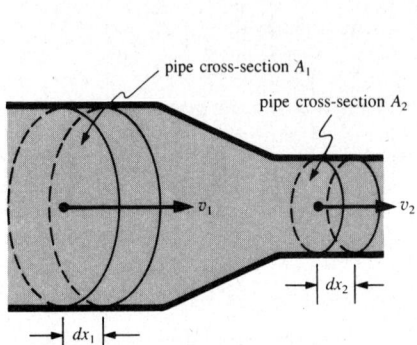

Figure E-5   **Figure E-6**

15. Figure E-6 shows water in a cylindrical tank that has an inner diameter d_1 and is open to the atmosphere. The outlet pipe has an inner diameter d_2. Derive an expression for the flow velocity of the water out of the lower pipe as a function of h, d_1, and d_2. Assume that water is an ideal fluid.   **[Ch. 15]**

Solutions to Exam 2

1. Define a coordinate system so that the ends of the rods are on the axes and are equidistant from the origin (see Fig. E-7). The CM is at some point along the rod's axis of symmetry, which in this system is the line $y = x$. Therefore $y_{CM} = x_{CM}$. The length of the rod, a quarter circumference, is $\pi R/2$, so if the rod's mass is M, its mass per unit length is $\mu = 2M/\pi R$. Use

$$x_{CM} = \frac{\int x\, dm}{M}$$

with $dm = \mu\, ds = \mu r\, d\theta$ and $x = r\cos\theta$

$$x_{CM} = \frac{1}{M}\int_0^{\pi/2} R\cos\theta\, \mu R\, d\theta = \frac{\mu R^2}{M}\int_0^{\pi/2}\cos\theta\, d\theta = \left(\frac{2M}{\pi R}\right)\left(\frac{R^2}{M}\right)\left(\sin\theta\Big|_0^{\pi/2}\right) = \boxed{\frac{2R}{\pi}}$$

2. Figure E-8 is a free-body diagram for the rod where $\phi = 40°, \theta = 35°, \alpha = 90° - \theta = 55°$, and $\gamma = \phi + \theta = 75°$, $W = 120$ N, $w = 80$ N, $L = 1.8$ m, and T is the tension in the cord. Apply the equations for both translational and rotational equilibrium.

$$\Sigma F_x = R_x - T\cos\phi = 0 \qquad\qquad\text{(i)}$$

$$\Sigma F_y = R_y + T\sin\phi - W - w = 0 \qquad\qquad\text{(ii)}$$

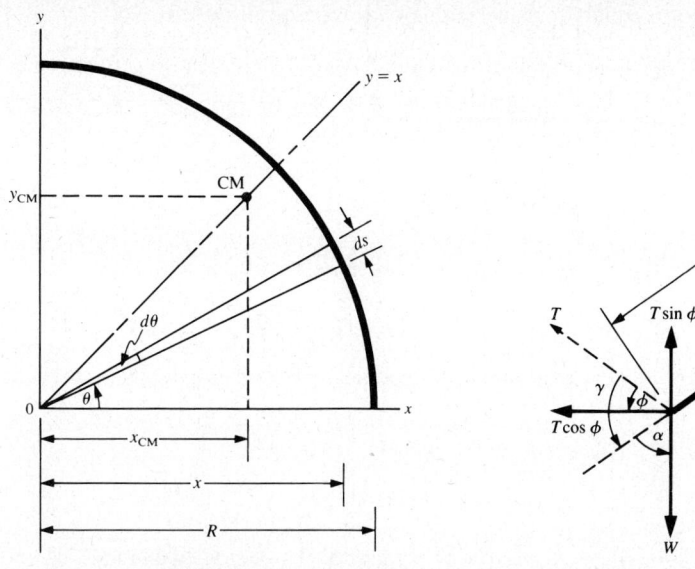

Figure E-7   **Figure E-8**

For rotational equilibrium, sum the torques about the pivot with the ccw direction positive.

$$\Sigma\tau = \left(\frac{L}{2}\right)w\sin\alpha + LW\sin\alpha - LT\sin\gamma = 0 \qquad \text{(iii)}$$

To find R_x and R_y you must know T, so answer part (b) of the question first.

(b) Solve the expression for rotational equilibrium, eq. (iii), for T.

$$T = \left(W + \frac{w}{2}\right)\left(\frac{\sin\alpha}{\sin\gamma}\right) = \left(120\text{ N} + \frac{80\text{ N}}{2}\right)\left(\frac{\sin 55°}{\sin 75°}\right) = \boxed{136\text{ N}}$$

(a) Solve eq. (i) for R_x and eq. (ii) for R_y.

$$R_x = T\cos\phi = (136\text{ N})(\cos 40°) = \boxed{104\text{ N}}$$

$$R_y = W + w - T\sin\phi = 120\text{ N} + 80\text{ N} - (136\text{ N})(\sin 40°) = \boxed{113\text{ N}}$$

3. Use the equation for the moment of inertia of a body whose mass is continuously distributed, $I = \int r^2\,dm$, with $dm = \mu R\,d\theta$, $\mu = M/2\pi R$, and $r = R\cos\theta$ (see Fig. E-9).

$$I = \int_0^{2\pi} (R\cos\theta)^2(\mu R\,d\theta) = \mu R^3 \int_0^{2\pi} \cos^2\theta\,d\theta = \frac{M}{2\pi R}R^3\pi = \boxed{\frac{1}{2}MR^2}$$

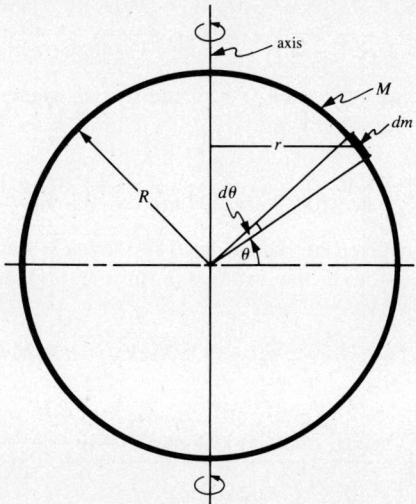

Figure E-9

4. (a) The work done by the motor is equal to the increase in kinetic energy of the disk. So

$$W = \Delta E_k = E_{k\,final} = \tfrac{1}{2}I\omega^2 = \tfrac{1}{2}(0.12\ \text{kg m}^2)(160\ \text{rad s}^{-1})^2 = \boxed{1.54 \times 10^3\ \text{J}}$$

(b)

$$\tau = I\alpha = I\frac{\Delta\omega}{\Delta t} = (0.12\ \text{kg m}^2)\left(\frac{160\ \text{rad s}^{-1}}{3\ \text{s}}\right) = 6.4\ \text{N m}$$

5. (a) The tension in the string provides the centripetal force.

$$T = F_c = m\frac{v^2}{r} = m\omega^2 r \qquad \text{because} \qquad v = \omega r$$

$$T = (0.3\ \text{kg})(12\ \text{rad s}^{-1})^2(0.4\ \text{m}) = \boxed{17.3\ \text{N}}$$

(b) Use the principle of conservation of angular momentum because there are no external torques acting on the system.

$$I_1\omega_1 = I_2\omega_2 \qquad \text{where} \qquad I = mr^2$$

Solve for ω_2.

$$\omega_2 = \omega_1\left(\frac{I_1}{I_2}\right) = \omega_1\left(\frac{mr_1^2}{mr_2^2}\right) = \omega_1\left(\frac{r_1}{r_2}\right)^2 = 12\ \text{rad s}^{-1}\left(\frac{40\ \text{cm}}{20\ \text{cm}}\right)^2 = \boxed{48\ \text{rad s}^{-1}}$$

(c) The work done is equal to the increase in the puck's kinetic energy.

$$W = E_{k2} - E_{k1} = \tfrac{1}{2}I_2\omega_2^2 - \tfrac{1}{2}I_1\omega_1^2 = \tfrac{1}{2}mr_2^2\omega_2^2 - \tfrac{1}{2}mr_1^2\omega_1^2$$

$$= \frac{m}{2}[(r_2\omega_2)^2 - (r_1\omega_1)^2] = \left(\frac{0.3\ \text{kg}}{2}\right)\{[(0.2\ \text{m})(48\ \text{rad s}^{-1})]^2 - [(0.4\ \text{m})(12\ \text{rad s}^{-1})]^2\} = \boxed{10.4\ \text{J}}$$

6. At the surface of the earth $g = GM/R^2$ where M is the mass of the earth. The density of the earth is

$$\rho = \frac{M}{V} = \frac{M}{\frac{4}{3}\pi R^3} = \frac{3M}{4\pi R^3} \qquad \text{so} \qquad M = \frac{4\pi\rho R^3}{3}$$

Substitute this into the equation for g and solve for G.

$$G = \frac{gR^2}{M} = \frac{3gR^2}{4\pi\rho R^3} = \boxed{\frac{3g}{4\pi\rho R}}$$

7. The gravitational force provides the centripetal force so

$$G\frac{M_E M_S}{r^2} = M_E\frac{v^2}{r} = \frac{M_E}{r}\left(\frac{2\pi r}{T}\right)^2$$

Solve for M_S.

$$M_S = \left(\frac{r^2}{GM_E}\right)\left(\frac{M_E}{r}\right)\left(\frac{2\pi r}{T}\right)^2 = \boxed{\frac{4\pi^2 r^3}{GT^2}}$$

8. (a)

$$L = L_0\sqrt{1 - (u/c)^2} = (4\ \text{m})\sqrt{1 - (0.55c/c)^2} = \boxed{3.34\ \text{m}}$$

(b) Because the ship travels at constant speed $u = 0.55c$ use $L = ut$, where L is the observed length of the ship and t is the time required for the ship to pass.

$$t = \frac{L}{u} = \frac{3.34\ \text{m}}{(0.55)(3 \times 10^8\ \text{m s}^{-1})} = \boxed{2.02 \times 10^{-8}\ \text{s}}$$

9. The total energy of the proton is given by both $E = \sqrt{(pc)^2 + (mc^2)^2}$ and $E = E_k + mc^2$. Solve the second equation for E_k and substitute for E the expression in the first equation. The rest energy of the proton is mc^2 and $pc = 2 \times 10^3$ MeV.

$$E_k = \sqrt{(pc)^2 + (mc^2)^2} - mc^2 = \sqrt{(2 \times 10^3\ \text{MeV})^2 + (938\ \text{MeV})^2} - 938\ \text{MeV} = \boxed{1.27 \times 10^3\ \text{MeV} = 1.27\ \text{GeV}}$$

10. (a) The tensile stress is

$$\frac{F}{A} = \frac{mg}{\frac{\pi}{4}d^2} = \frac{(14\ \text{kg})(9.8\ \text{m s}^{-2})}{\left(\frac{\pi}{4}\right)(2 \times 10^{-3}\ \text{m})^2} = \boxed{4.37 \times 10^7\ \text{Pa}}$$

(b) Use $\dfrac{F}{A} = Y\dfrac{\Delta\ell}{\ell}$ and solve for Y.

$$Y = \frac{F/A}{\Delta\ell/\ell} = \frac{4.37 \times 10^7 \text{ Pa}}{(0.56 \times 10^{-3} \text{ m})/(1.2 \text{ m})} = \boxed{9.36 \times 10^{10} \text{ Pa}}$$

(c) The shear modulus can be estimated from Young's modulus and Poisson's ratio by

$$S = \frac{Y}{2(1+\sigma)} = \frac{9.36 \times 10^{10} \text{ Pa}}{2(1+0.3)} = \boxed{3.60 \times 10^{10} \text{ Pa}}$$

11. Use $\theta = 2\tau\ell/\pi R^4 S$, solve for τ, and remember to express θ in radians.

$$\tau = \frac{\theta\pi R^4 S}{2\ell} = \frac{(6°)(\pi \text{ rad}/180°)(\pi)(1 \times 10^{-2} \text{ m})^4(8 \times 10^{10} \text{ Pa})}{2(12 \times 10^{-2} \text{ m})} = \boxed{1.1 \times 10^3 \text{ N m}}$$

12. (a) The gauge pressure is ρgh. The density of sea water is its specific gravity times the density of pure water, 10^3 kg m^{-3}.

$$P_{\text{gauge}} = (SG)\rho_{\text{water}}gh = (1.03 \times 10^3 \text{ kg m}^{-3})(9.8 \text{ m s}^{-2})(12 \text{ m}) = \boxed{1.21 \times 10^5 \text{ Pa}}$$

(b)
$$P = P_0 + P_{\text{gauge}} = (10^5 + 1.21 \times 10^5) \text{ Pa} = \boxed{2.21 \times 10^5 \text{ Pa}}$$

13. Use Archimedes' principle in the form $\dfrac{w_i}{w} = 1 - \dfrac{\rho_\ell}{\rho}$ and solve for w_i. In this problem the liquid is water, so

$$\frac{\rho_\ell}{\rho} = \frac{\rho_{\text{water}}}{(SG)(\rho_{\text{water}})} = \frac{1}{SG}$$

Thus

$$w_i = w\left(1 - \frac{1}{SG}\right) = (2 \text{ kg})(9.8 \text{ m s}^{-2})\left(1 - \frac{1}{2.6}\right) = \boxed{12.1 \text{ N}}$$

14. Because the fluid is incompressible the mass entering per unit time is equal to the mass leaving per unit time:

$$\frac{dm_1}{dt} = \frac{dm_2}{dt}$$

$$dm_1 = \rho\, dV_1 = \rho A_1\, dx_1 \quad \text{and} \quad dm_2 = \rho\, dV_2 = \rho A_2\, dx_2$$

where ρ is the fluid's density and dV is a volume element. Substitute for dm_1 and dm_2, cancel out ρ, and note that $dx/dt = v$.

$$\frac{\rho A_1\, dx_1}{dt} = \frac{\rho A_2\, dx_2}{dt} \qquad A_1 v_1 = A_2 v_2$$

15. Use Bernoulli's equation and the equation of continuity. Designate variables that apply to conditions in the tank by subscript 1 and variables that apply to conditions in the outlet pipe by subscript 2.

$$P_1 + \rho g y_1 + \tfrac{1}{2}\rho v_1^2 = P_2 + \rho g y_2 + \tfrac{1}{2}\rho v_2^2$$

Because the tank is open $P_1 = P_2 = P_A$. Also, $y_1 - y_2 = h$.

$$\rho g y_1 + \tfrac{1}{2}\rho v_1^2 = \rho g y_2 + \tfrac{1}{2}\rho v_2^2$$

$$v_1^2 - v_2^2 = 2g(y_1 - y_2) = 2gh$$

Solve the equation of continuity for v_1, substitute, and solve for v_2.

$$A_1 v_1 = A_2 v_2 \qquad v_1 = v_2\left(\frac{A_2}{A_1}\right) = v_2\frac{(\pi/4)d_2^2}{(\pi/4)d_1^2} = v_2\left(\frac{d_2}{d_1}\right)^2$$

$$\left[v_2\left(\frac{d_2}{d_1}\right)^2\right]^2 - v_2^2 = 2gh \qquad v_2^2\left[1 - \left(\frac{d_2}{d_1}\right)^4\right] = 2gh \qquad v_2 = \sqrt{\frac{2gh}{1 - \left(\frac{d_2}{d_1}\right)^4}}$$

16 TEMPERATURE AND THERMAL EXPANSION

THIS CHAPTER IS ABOUT

- ☑ **The Celsius and Fahrenheit Temperature Scales**
- ☑ **The Kelvin and Rankine Scales**
- ☑ **Thermal Expansion**
- ☑ **Thermal Stress**

The **temperature** of an object is a measure of the average kinetic energy of the atoms or molecules that constitute the object.

16-1. The Celsius and Fahrenheit Temperature Scales

The **Celsius temperature scale** is defined so there are 100 Celsius degrees between the *ice point*, the temperature of melting ice, and the *steam point*, the temperature of boiling water, when both are measured at a pressure of one atmosphere, 1.013×10^5 Pa. The zero point of this scale, $0°C$, is located at the ice point. Because of experimental difficulties in measuring the melting and boiling points of water precisely, an international agreement has defined the Celsius scale with respect to the **triple point** of water, the condition under which the solid (ice), liquid, and gaseous phases can coexist in equilibrium: $0.01°C$ at a pressure of 610 Pa.

The **Fahrenheit temperature scale**, commonly used in the United States, has 180 Fahrenheit degrees between the ice point, $32°F$, and the steam point, $212°F$. Thus an interval of $100 C°$ equals an interval of $180 F°$. We describe particular temperatures by statements like "The ice point is at 32 degrees Fahrenheit ($32°F$)," but describe ranges or intervals of temperature by statements like "There are 180 Fahrenheit degrees ($180 F°$) between the ice and steam points." To convert between the Celsius and Fahrenheit scales,

CELSIUS/FAHRENHEIT TEMPERATURE CONVERSION

$$T_C = \frac{5}{9}(T_F - 32°F) \tag{16-1a}$$

$$T_F = \frac{9}{5}T_C + 32°F \tag{16-1b}$$

EXAMPLE 16-1: Normal human body temperature is $98.6°F$. What is this temperature in degrees Celsius?

Solution: Use Eq. (16-1a)

$$T_C = \frac{5}{9}(T_F - 32°F)$$

$$= \frac{5}{9}(98.6 - 32) = \boxed{37.0°C}$$

16-2. The Kelvin and Rankine Scales

As the temperature of an object decreases, the kinetic energy of the atoms constituting the object decreases. There is a minimum energy that atoms must have; a decrease below this energy is not possible. This minimum possible energy defines the minimum possible temperature: the **absolute zero**. We call temperature scales whose zero point equals absolute zero **absolute scales**. Two absolute scales are in use, the **Kelvin scale** in science and the **Rankine scale** in engineering, although the trend is toward exclusive use of the Kelvin scale. The Kelvin scale is also an absolute scale in the sense that an object at 100 K is twice as hot as one at 50 K (100°C is *not* twice as hot as 50°C). A kelvin is an absolute, not relative, unit so the word and symbol for degree are not used with kelvins.

The Kelvin scale is defined so that the triple point of water is at a temperature of 273.16 K, read as "273.16 kelvins." Thus the ice and steam point temperatures are at 273.15 K and 373.15 K, respectively, because one kelvin equals one Celsius degree. For most calculations, 273 K for the ice point is sufficiently accurate. To convert between the Celsius and Kelvin scales, use

KELVIN/CELSIUS	$T_K = T_C + 273.15$	**(16-2a)**
TEMPERATURE CONVERSION	$T_C = T_K - 273.15$	**(16-2b)**

Notice that absolute zero, 0 K, is −273.15°C.

EXAMPLE 16-2: Determine the Kelvin temperature of human body temperature, 37°C.

Solution: Use Eq. (16-2a):

$$T_K = T_C + 273.15 = 37 + 273.15 = 310.15 \text{ K} \cong 310 \text{ K}$$

The Rankine scale is the absolute Fahrenheit scale: one Rankine degree is equal to one Fahrenheit degree, and 0°R = 0 K. Therefore, to convert from the Kelvin to the Rankine scale, use

KELVIN/RANKINE		
TEMPERATURE CONVERSION	$T_R = \dfrac{9}{5} T_K$	**(16-3)**

Figure 16-1 illustrates the relationship among the four scales.

EXAMPLE 16-3: Derive a conversion equation between Fahrenheit degrees and Rankine degrees.

Solution: Use Eqs. (16-3), (16-2a), and (16-1a)

$$T_R = \frac{9}{5} T_K = \frac{9}{5}(T_C + 273.15) = \frac{9}{5} T_C + 491.67$$

$$T_R = \frac{9}{5}\left[\frac{5}{9}(T_F - 32)\right] + 491.67 = T_F - 32 + 491.67 = \boxed{T_F + 459.67}$$

Notice that the Fahrenheit temperature of absolute zero, 0°R, is −459.67°F. This result can be rounded to −460°F for most calculations.

16-3. Thermal Expansion

When the temperature of a rod of length ℓ is increased by an amount ΔT, the length of the rod increases by an amount $\Delta \ell$ given by

THERMAL LINEAR		
EXPANSION	$\Delta \ell = \alpha \ell \, \Delta T$	**(16-4a)**

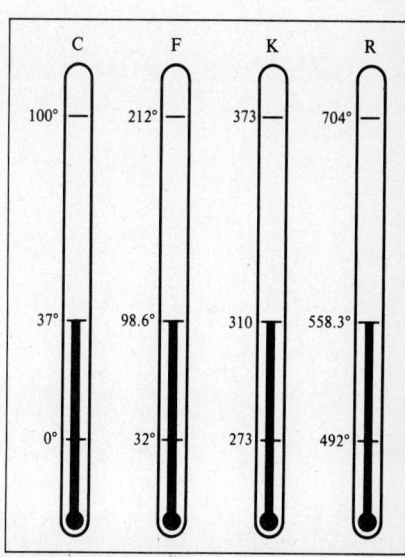

Figure 16-1. Four identical thermometers with different scales measuring the same temperature.

The quantity α is characteristic of a substance and is called the **coefficient of linear expansion**, which has units of $(C°)^{-1}$.

The thermal increase in area A of a sheet of material is

**THERMAL AREA
   EXPANSION**
$$\Delta A = 2\alpha A\,\Delta T \qquad\qquad (16\text{-}4\text{b})$$

The increase in volume of a substance is given by

**THERMAL VOLUME
   EXPANSION**
$$\Delta V = \beta V\,\Delta T \qquad\qquad (16\text{-}4\text{c})$$

The quantity β is called the **coefficient of volumetric expansion** and has units of $(C°)^{-1}$. For solids, $\beta = 3\alpha$. Equations (16-4a), (16-4b), and (16-4c) are valid when the temperature change is not too great. They may be used also to calculate decreases in length, area, and volume caused by a drop in temperature. Table 16-1 lists values of α and β for some materials.

EXAMPLE 16-4: During a sunny day, the temperature of a steel rail increases from 20°C to 94°C. Find the increase in the length of the rail if it is initially 12 m long.

Solution: Use Eq. (16-4a) with $\alpha = 1.2 \times 10^{-5}\,(C°)^{-1}$ from Table 16-1.

$$\Delta\ell = \alpha\ell\,\Delta T$$
$$= [1.2 \times 10^{-5}\,(C°)^{-1}](12\text{ m})(94 - 20)C° = \boxed{1.07 \times 10^{-2}\text{ m} = 1.07\text{ cm}}$$

EXAMPLE 16-5: An aluminum container has a capacity of 1 L and is filled to the brim with carbon tetrachloride at 20°C. How much liquid overflows when the temperature is raised to 25°C?

TABLE 16-1: Expansion Coefficients of Selected Materials

Material	α $10^{-5}(C°)^{-1}$	β $10^{-5}(C°)^{-1}$
Solids		
Aluminum	2.4	7.2
Brass	1.9	5.7
Copper	1.7	5.1
Glass (average)	0.6	1.8
Steel	1.2	3.6
Invar (an iron–nickel alloy)	0.1	0.3
Liquids		
Methyl alcohol	—	120
Gasoline	—	110
Carbon tetrachloride	—	124
Glycerin	—	51
Mercury	—	18
Turpentine	—	94
Water (at 20°C)	—	21
Water (at 50°C)	—	60

Solution: Use Eq. (16-4c) with $\beta_{Al} = 7.2 \times 10^{-5}$ (C°)$^{-1}$ for aluminum and $\beta_{CCl_4} = 124 \times 10^{-5}$ (C°)$^{-1}$ for carbon tetrachloride. Because β_{CCl_4} is greater than β_{Al}, the volume increase of the liquid is greater than that of the solid. Let $V_s =$ the volume of the spill. Then

$$V_s = \Delta V_{CCl_4} - \Delta V_{Al} = \beta_{CCl_4} V \Delta T - \beta_{Al} V \Delta T = (\beta_{CCl_4} - \beta_{Al}) V \Delta T$$

$$= (124 - 7.2) \times 10^{-5} \text{ (C°)}^{-1} (10^3 \text{ cm}^3)(25 - 20)\text{C°} = \boxed{5.8 \text{ cm}^3}$$

16-4. Thermal Stress

Suppose a rod of length ℓ is held fixed at both ends so that it can't expand when its temperature is increased by ΔT. Its fractional increase in length would have been $\Delta \ell / \ell = \alpha \Delta T$ according to Eq. (14-4a), but because it can't expand it experiences compressional stress. According to Eq. (13-3), the compressional stress in the rod is

THERMAL COMPRESSIONAL STRESS

$$\frac{F}{A} = Y\frac{\Delta \ell}{\ell} = \underbrace{Y\alpha \Delta T}_{\text{thermally induced compressional stress}} \qquad (16\text{-}5)$$

where Y is Young's modulus for the material. Equation (16-5) applies also to the thermally induced *tensile* stress that results from a temperature *decrease*.

EXAMPLE 16-6: Assume that the cross-sectional area of the steel rail in Example 16-4 is 100 cm^2 = 10^{-2} m^2. Calculate the compressional force the rail would experience at 94°C if there were no expansion gaps between rails at 20°C. $Y = 22 \times 10^{10}$ Pa for steel.

Solution: Use Eq. (16-5) and solve for F:

$$\frac{F}{A} = Y\alpha \Delta T$$

$$F = AY\alpha \Delta T$$

$$= (10^{-2} \text{ m}^2)(22 \times 10^{10} \text{ Pa})(1.2 \times 10^{-5})(\text{C°})^{-1}(94 - 20)\text{C°} = \boxed{1.95 \times 10^6 \text{ N}}$$

In British units, this is nearly 220 tons.

SUMMARY

1. Convert between the Celsius and Fahrenheit temperature scales with

$$T_C = \frac{5}{9}(T_F - 32°F) \qquad \text{and} \qquad T_F = \frac{9}{5}T_C + 32°F$$

2. Convert between the Celsius and Kelvin temperature scales with $T_K = T_C + 273.15 \cong T_C + 273$

3. Convert between the Fahrenheit and Rankine temperature scales with $T_R = T_F + 459.67 \cong T_F + 460$

4. Increases in length, area, and volume due to an increase in temperature ΔT are

$$\Delta \ell = \alpha \ell \, \Delta T$$

$$\Delta A = 2\alpha A \, \Delta T$$

$$\Delta V = 3\alpha V \Delta T = \beta V \Delta T$$

5. The thermally induced stress in a rod due to a change in temperature ΔT is

$$\frac{F}{A} = Y\alpha \, \Delta T$$

RAISE YOUR GRADES

Can you explain . . . ?

☑ why colored alcohol rather than mercury is used as the fluid in some thermometers
☑ why mercury or alcohol thermometers cannot be used to measure very low temperatures
☑ what advantage the Kelvin scale has over the Celsius scale
☑ why it makes no sense to say an object at 40°C is twice as hot as one at 20°C
☑ why a hole in a piece of sheet metal gets larger rather than smaller when the temperature is raised
☑ why expansion containers are necessary on automobile cooling systems
☑ why there is no need for a coefficient of linear expansion for liquids
☑ why one end of a highway bridge rests on rollers if the other end is fixed

SOLVED PROBLEMS

The Celsius and Fahrenheit Temperature Scales

PROBLEM 16-1:   What are the Celsius temperatures corresponding to (a) 0°F and (b) 72°F?

Solution: Use Eq. (16-1a).

(a) $T_C = \dfrac{5}{9}(T_F - 32°F) = \dfrac{5}{9}(0 - 32) = \boxed{-17.8°C}$

(b) $T_C = \dfrac{5}{9}(72 - 32) = \boxed{22.2°C}$

PROBLEM 16-2:   What are the Fahrenheit temperatures corresponding to (a) 50°C and (b) 900°C?

Solution: Use Eq. (16-1b), $T_F = \frac{9}{5}T_C + 32°F$.

(a) $T_F = \dfrac{9}{5}(50) + 32 = \boxed{122°F}$

(b) $T_F = \dfrac{9}{5}(900) + 32 = \boxed{1652°F}$

PROBLEM 16-3:   At what temperature are the Celsius and Fahrenheit scales numerically equal?

Solution: Let this "unknown" temperature be T and use Eq. (16-1a). Solve for T:

$$T_C = \frac{5}{9}(T_F - 32°F)$$

$$T = \frac{5}{9}(T - 32) \qquad 9T = 5T - 160 \qquad 4T = -160$$

$$T = \boxed{-40} \qquad \text{So } -40°C = -40°F$$

The Kelvin and Rankine Scales

PROBLEM 16-4: What Kelvin temperature corresponds to $-18°C$?

Solution: Use Eq. (16-2a):

$$T_K = T_C + 273.15 = -18 + 273.15 \cong -18 + 273 = \boxed{255 \text{ K}}$$

PROBLEM 16-5: What Rankine temperature corresponds to $100°F$?

Solution: Use the result of Example 16-3:

$$T_R = T_F + 459.67 \cong T_F + 460 = 100 + 460 = \boxed{560°R}$$

Thermal Expansion

PROBLEM 16-6: A carpenter's steel measuring tape is exactly 100 ft long at $40°F$. What is its length at $80°F$?

Solution: Use Eq. (16-4a) to calculate the increase in length. From Table 16-1 obtain $\alpha = 1.2 \times 10^{-5}$ $(C°)^{-1}$ for steel. It is easy to show, from Eq. (16-1a), that $\Delta T_C = \frac{5}{9}\Delta T_F$. Thus $\Delta T_C = \frac{5}{9}(80 - 40) = 22.22$ $C°$.

$$\Delta\ell = \alpha\ell\,\Delta T = (1.2 \times 10^{-5})(C°)^{-1}(100 \text{ ft})(22.22 \text{ } C°) = \boxed{2.67 \times 10^{-2} \text{ ft} = 0.32 \text{ in.}}$$

Notice that any system of length units may be used in Eqs. (16-1).

PROBLEM 16-7: A hole 8 cm in diameter is bored through a sheet of aluminum when the temperature is $20°C$. **(a)** What is the diameter of the hole when the temperature is raised to $60°C$? **(b)** Calculate the percentage increase in the area of the hole.

Solution: To aid your intuition regarding this problem, consider the sheet of aluminum shown in Figure 16-2 that has two half-circular holes cut in such a way that a nearly complete hole 8 cm in diameter is produced. The remaining strip of aluminum is as thin as you please and its length is the diameter of the hole.

(a) Use Eq. (16-4a) to calculate the increase in length of this strip.

$$\Delta\ell = \alpha\ell\,\Delta T = (2.4 \times 10^{-5})(C°)^{-1}(8 \text{ cm})(60 - 20)(C°) = 7.68 \times 10^{-3} \text{ cm}$$

Figure 16-2

Therefore the diameter of the hole at $60°C$ is

$$(8 + 7.68 \times 10^{-3}) \text{ cm} = \boxed{8.00768 \text{ cm}}$$

(b) Use Eq. (16-4b) and solve for the fractional increase in area $\Delta A/A$.

$$\Delta A = 2\alpha A\,\Delta T$$

$$\frac{\Delta A}{A} = 2\alpha\,\Delta T = 2(2.4 \times 10^{-5})(C°)^{-1}(40 C°) = 1.92 \times 10^{-3}$$

To express this increase as a percentage, multiply by 100%:

$$1.92 \times 10^{-3} \times 100\% = \boxed{0.192\%}$$

PROBLEM 16-8: Estimate the amount of water that overflows a 500-ml glass flask completely filled at $20°C$ as the temperature rises to $50°C$.

Solution: Use the expression for spill volume that you derived in Example 16-5. Because the coefficient of volumetric expansion for water is 21×10^{-5} $(C°)^{-1}$ at $20°C$ and 60×10^{-5} $(C°)^{-1}$ at $50°C$ (see Table 16-1), assume it has an "average value" of 40×10^{-5} $(C°)^{-1}$ for lack of more detailed information.

$$V_s = (\beta_w - \beta_g)V\,\Delta T = (40 - 1.8) \times 10^{-5} \text{ } (C°)^{-1}(500 \text{ ml})(50 - 20)C° = \boxed{5.7 \text{ ml}}$$

Thermal Stress

PROBLEM 16-9:   A brass rod, 1.6 m long and 3 cm in diameter, is wedged horizontally between two stone buildings on a day when the air temperature is −18°C. How much force does the rod exert on the buildings on a summer day when the temperature is 30°C? Young's modulus for brass is 9.1×10^{10} Pa.

Solution: Use Eq. (16-5) and solve for *F*. In this case, because the temperature increases, the stress in the rod is compressional.

$$\frac{F}{A} = Y\alpha\,\Delta T$$

$$F = AY\alpha\,\Delta T = [\pi(1.5 \times 10^{-2}\text{ m})^2](9.1 \times 10^{10}\text{ Pa})(1.9 \times 10^{-5})(\text{C}°)^{-1}[30 - (-18)]\text{C}°$$

$$= \boxed{5.87 \times 10^4\text{ N}}$$

This is about 6.6 tons force in British units.

Supplementary Exercises

EXERCISE 16-1:   What Celsius temperatures correspond to 90°F and 1000°F?

EXERCISE 16-2:   What Fahrenheit temperatures correspond to −190°C and 212°C?

EXERCISE 16-3:   What Kelvin temperatures correspond to 20°C and −190°C?

EXERCISE 16-4:   What Rankine temperatures correspond to −10°F and 72°F?

EXERCISE 16-5:   Derive a formula to convert Fahrenheit temperatures directly to the Kelvin scale.

EXERCISE 16-6:   What Kelvin temperature corresponds to −10°F?

EXERCISE 16-7:   A length of #10 copper wire is used to form a circular ring with an inside diameter of 15 cm at 20°C. What is the ring's inside diameter when the temperature is 65°C?

EXERCISE 16-8:   Derive a formula for the length of a rod that has undergone a temperature rise ΔT if its original length is ℓ and its coefficient of linear expansion is α.

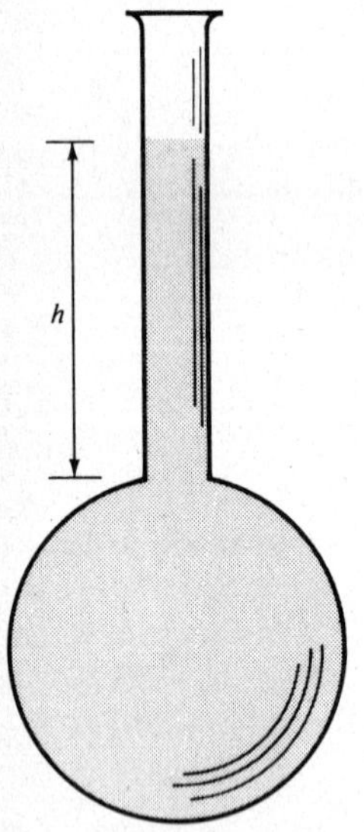

EXERCISE 16-9:   A spherical glass flask has an inside diameter of 6 cm and is fitted with a neck that has an inside diameter of 5 mm (see Figure 16-3). The spherical part of the flask is filled with carbon tetrachloride at 20°C. How far up the neck does the liquid rise when the temperature reaches 65°C? Assume that the volume of the flask remains constant.

EXERCISE 16-10:   How far up the neck of the tube does the liquid in Exercise 16-9 rise when the expansion of the glass flask is accounted for?

Figure 16-3

Answers to Supplementary Exercises

16-1:   32.2°C; 538°C

16-2:   −310°F; 320°F

16-3:   293 K; 83 K

16-4:   450°R; 537°R

16-5: $T_K = \frac{5}{9}T_F + 255.37$

16-6:   249.8 K

16-7:   15.011 cm

16-8: $\ell' = \ell(1 + \alpha\Delta T)$

16-9:   8.03 cm

16-10:   7.91 cm

17 HEAT AND HEAT TRANSFER

THIS CHAPTER IS ABOUT

☑ **Quantity of Heat**
☑ **Heat Capacity**
☑ **Change of Phase**
☑ **Heat Transfer**

17-1. Quantity of Heat

Heat, or **thermal energy**, is the kinetic energy of atoms and molecules, and it obeys the law of conservation of energy. Mechanical kinetic energy is readily transformed into thermal energy (heat) when frictional forces are present. The transformation of heat into *organized* mechanical energy cannot be accomplished with perfect efficiency (see Chapter 20).

The unit of heat historically used is the **calorie** (cal). One calorie is the quantity of heat required to raise the temperature of one gram of water from 14.5°C to 15.5°C. The nutritional unit of food energy is the Calorie (spelled with a capital C):

$$1 \text{ Calorie} = 10^3 \text{ cal} = 1 \text{ kcal}$$

Because heat is a form of energy, we can express it in the same units as the other forms of energy. In SI units,

DEFINITION OF THE CALORIE
$$1 \text{ cal} = 4.186 \text{ J} \tag{17-1}$$

Equation (17-1) is referred to as the *mechanical equivalent of heat* in some texts.

In the British system, the quantity of heat is the **British thermal unit**, the BTU. One BTU is the quantity of heat needed to raise the temperature of one pound of water from 63°F to 64°F.

DEFINITION OF THE BTU
$$1 \text{ BTU} = 252 \text{ cal} \tag{17-2}$$

EXAMPLE 17-1: How many joules are there in one BTU?

Solution: Combine Eqs. (17-1) and (17-2):

$$1 \text{ BTU} = (252 \text{ cal})\left(\frac{4.186 \text{ J}}{\text{cal}}\right) = \boxed{1055 \text{ J}}$$

17-2. Heat Capacity

When a quantity of heat Q is transferred to a mass m of a substance, it causes a rise in temperature ΔT. Within a limited range, the rise in temperature is proportional to Q according to:

HEAT CAPACITY
$$Q = mc\,\Delta T \tag{17-3a}$$

TABLE 17-1: Specific Heat Capacity of Selected Materials

Material	$kJ\,kg^{-1}(C°)^{-1}$	$kcal\,kg^{-1}(C°)^{-1}$ $cal\,g^{-1}(C°)^{-1}$
Solids		
Aluminum	0.900	0.215
Brass	0.385	0.092
Copper	0.385	0.092
Glass (average)	0.674	0.161
Steel	0.448	0.107
Lead	0.128	0.0306
Zinc	0.389	0.093
Ice	2.04	0.487
Liquids		
Water	4.186	1.000
Methyl alcohol	2.52	0.602
Turpentine	1.72	0.411

where the proportionality constant c is called the **specific heat capacity** of the substance or sometimes just the *specific heat*.

The units of c depend upon the units used for Q, m, and ΔT. Table 17-1 lists values of c for a few solids and liquids. The specific heats of gases are discussed in Chapter 18. If c varies as a function of T, Eq. (17-3a) must be modified to

$$Q = m \int c(T)\,dT \qquad \textbf{(17-3b)}$$

EXAMPLE 17-2: A styrofoam cup contains $300\ cm^3$ of water at $22°C$. A 200-g block of copper at $90°C$ is dropped into the water. Find the temperature to which the water rises. Assume that styrofoam has a negligible specific heat.

Solution: To solve this problem we use the *method of mixtures*, which is in essence the principle of the conservation of energy: the heat lost by the hot copper is equal to the heat gained by the water. Both water and copper come to the same final temperature T_f. Let T_{Cu} and T_w be the initial temperatures of the copper and water, respectively. Use Eq. (17-3a):

$$\text{heat lost by copper} = \text{heat gained by water}$$

$$m_{Cu}c_{Cu}\,\Delta T_{Cu} = m_w c_w\,\Delta T_w$$

$$m_{Cu}c_{Cu}(T_{Cu} - T_f) = m_w c_w(T_f - T_w)$$

Now solve for T_f. Notice that in *this* problem you may use any consistent system of units for m and c. Use the values of c_{Cu} and c_w from Table (17-1).

$$T_f = \frac{m_{Cu}c_{Cu}T_{Cu} + m_w c_w T_w}{m_{Cu}c_{Cu} + m_w c_w}$$

$$= \frac{(0.2\ kg)[0.385\ kJ\,kg^{-1}(C°)^{-1}](90°C) + (0.3\ kg)[4.186\ kJ\,kg^{-1}(C°)^{-1}](22°C)}{(0.2\ kg)[0.385\ kJ\,kg^{-1}(C°)^{-1}] + (0.3\ kg)[4.186\ kJ\,kg^{-1}(C°)^{-1}]}$$

$$= \boxed{25.9°C}$$

EXAMPLE 17-3: A 250-g aluminum cup contains 300 ml of water at 22°C. The cup is completely insulated from its surroundings. A 200-g chunk of a metal at 80°C is dropped into the water, whose temperature then rises to 27°C. What is the specific heat capacity of this metal?

Solution: Use the method of mixtures but, in this case, account for the heat absorbed by the aluminum cup, which undergoes the same rise in temperature as the water.

$$\text{heat lost by metal} = \text{heat gained by water and cup}$$

$$m_m c_m \Delta T_m = (m_w c_w + m_{Al} c_{Al}) \Delta T_w$$

Now solve for c_m:

$$c_m = \frac{(m_w c_w + m_{Al} c_{Al}) \Delta T_w}{m_w \Delta T_m}$$

$$= \frac{[(0.3)(4.186) + (0.25)(0.9)](27 - 22)\,\text{kJ}}{[(0.2)(80 - 27)]\,\text{kg}\,(\text{C}°)} = \boxed{0.698\ \text{kJ}\,\text{kg}^{-1}(\text{C}°)^{-1}}$$

EXAMPLE 17-4: The specific heat capacity of a certain substance depends on the Kelvin temperature according to $c(T) = BT^2$ where $B = 1.2 \times 10^{-6}\ \text{kJ}\,\text{kg}^{-1}\,\text{K}^{-3}$ between 100 K and 150 K. How much heat is required to raise the temperature of 300 g of this substance from 100 K to 150 K?

Solution: Use Eq. (17-3b).

$$Q = m \int c(T)\,dT = m \int_{T_i}^{T_f} BT^2\,dT$$

$$= mB \frac{T^3}{3}\Big|_{T_i}^{T_f} = \frac{mB}{3}(T_f^3 - T_i^3)$$

$$= \frac{(0.3\ \text{kg})(1.2 \times 10^{-6}\ \text{kJ}\,\text{kg}^{-1}\,\text{K}^{-3})}{3}[(150\ \text{K})^3 - (100\ \text{K})^3] = \boxed{0.285\ \text{kJ}}$$

17-3. Change of Phase

Many substances can exist in three phases—solid, liquid, and gas—determined by temperature and pressure. Converting solid to liquid or liquid to gas requires heat, a quantity of heat *in addition* to the quantity needed to increase temperature. For example, heating 1 g of ice from $-1°\text{C}$ to $+1°\text{C}$ requires 2.04 J to heat the ice from $1°\text{C}$ to $0°\text{C}$, 4.186 J to heat liquid water from $0°\text{C}$ to $1°\text{C}$, *plus* 335 J to convert the ice from solid to liquid (that is, to melt it). The quantity of heat required to melt a mass m of a substance at its normal melting temperature (the melting point at one atmosphere ambient pressure) is

LATENT HEAT OF FUSION
$$Q = mL_f \tag{17-4a}$$

where L_f is called **latent heat of fusion** and has units of $\text{kJ}\,\text{kg}^{-1}$. (In some books L_f is called simply the heat of fusion.)

The quantity of heat required to vaporize a mass m of a substance at its normal boiling temperature is

LATENT HEAT OF VAPORIZATION
$$Q = mL_v \tag{17-4b}$$

where L_v is called the **latent heat of vaporization**. Table 17-2 lists the latent heats for a few substances.

TABLE 17-2: Latent Heats of Selected Materials

Material	Normal melting temperature (°C)	L_f (kJ kg^{-1})	Normal boiling temperature (°C)	L_v (kJ kg^{-1})
Copper	1083	134	2567	5069
Gold	1064	64	3080	1578
Lead	327	24	1740	871
Mercury	− 39	12	357	272
Water	0	335	100	2256
Nitrogen	− 210	26	− 196	201
Helium	− 271	2.1	− 269	21

EXAMPLE 17-5: A styrofoam cup holds 300 g of water at a temperature of 22°C. Then 50 g of ice at its melting temperature of 0°C is added. What is the final temperature of the contents of the cup? Assume that the specific heat capacity of the cup is negligible.

Solution: It is perhaps best to solve problems involving phase changes or possible phase changes in steps.

(1) Compute the decrease in temperature of the water produced by the melting ice. Let $m_w = 0.3$ kg, the mass of the water, and $m_i = 0.05$ kg, the mass of the ice. Assume all the ice melts. (Sometimes this assumption will prove to be false—see Example 17-6.) Use conservation of energy:

$$\text{heat lost by water} = \text{heat necessary to melt ice}$$

$$m_w c_w \Delta T_w = m_i L_f$$

Solve for ΔT_w:

$$\Delta T_w = \frac{m_i L_f}{m_w c_w}$$

$$= \frac{(0.05 \text{ kg})(335 \text{ kJ kg}^{-1})}{(0.3 \text{ kg})[4.186 \text{ kJ kg}^{-1} (\text{C}°)^{-1}]} = 13.33 \text{ C}°$$

So the transfer of heat from the water to the melting ice lowers the water temperature to $T_w = 22 - 13.33 = 8.66°$C. Notice that if ΔT_w had been greater than 22°C we would have concluded that the final temperature of the water was 0°C but not all the ice had melted.

(2) We now have 300 g of water at 8.66°C and 50 g of water (the melted ice) at 0°C. Use the method of mixtures to obtain the final temperature. Let m_i represent the mass of melted ice at 0°C and T_f the final temperature.

$$\text{heat lost by water} = \text{heat gained by melted ice}$$

$$m_w c_w \Delta T_w = m_i c_w \Delta T_i$$

$$m_w(T_w - T_f) = m_i(T_f - 0°\text{C}) = m_i T_f$$

Solve for T_f:

$$T_f = T_w\left(\frac{m_w}{m_w + m_i}\right)$$

$$= 8.66°\text{C}\left(\frac{300 \text{ g}}{(300 + 50) \text{ g}}\right) = \boxed{7.42°\text{C}}$$

EXAMPLE 17-6: Now suppose we add 150 g of ice at its melting temperature to a styrofoam cup holding 300 g of water at 22°C. (**a**) What is the final temperature of the contents of the cup? (**b**) How much ice melts?

Solution:

(**a**) Proceed as in Example (17-5). The decrease in water temperature would be:

$$\Delta T_w = \frac{m_i L_f}{m_w c_w} = \frac{(0.15 \text{ kg})(335 \text{ kJ kg}^{-1})}{(0.3 \text{ kg})[4.186 \text{ kJ kg}^{-1}(\text{C}°)^{-1}]} = \boxed{40 \text{ C}°}$$

This result seems to imply that the temperature of the water decreased to $T_w = 22 - 40 = -18°C$. Since the final temperature cannot be lower than the lowest initial temperature, we infer that only part of the ice melted and the temperature is 0°C, the temperature of a water–ice mixture.

(**b**) Let m_x = the mass of ice that melts. Use conservation of energy.

$$\text{heat lost by water} = \text{heat to melt } m_x \text{ ice}$$

$$m_w c_w \Delta T_w = m_x L_f$$

Solve for m_x with $\Delta T_w = T_w - T_f = 22°C - 0°C = 22°C$.

$$m_x = \frac{m_w c_w \Delta T_w}{L_f} = \frac{(0.3 \text{ kg})[4.186 \text{ kJ kg}^{-1}(\text{C}°)^{-1}](22\text{C}°)}{335 \text{ kJ kg}^{-1}} = \boxed{0.082 \text{ kg}}$$

The contents of the cup are $(300 + 82)\text{ g} = 382\text{ g}$ of water and $(150 - 82)\text{ g} = 68$ g of ice at 0°C.

17-4. Heat Transfer

There are three principal ways in which heat is transferred: *conduction, convection,* and *radiation.* In a given situation all three may participate to various degrees.

A. Conduction

- **Conduction** is heat flow through a substance by means of intermolecular collisions transferring kinetic energy from one molecule to another.

Figure 17-1 shows a uniform flow of heat from a hot reservoir at temperature $T_1 = T_H$ through a substance of thickness $\Delta x = x_2 - x_1$ and cross-sectional area A to a cold reservoir at temperature $T_2 = T_C$. Let's assume that no heat flows out the sides the substance. In a time interval Δt a quantity of heat ΔQ flows so that the rate of flow (the **heat current**) is

HEAT CONDUCTION
$$\frac{\Delta Q}{\Delta t} = KA\frac{T_H - T_C}{\Delta x} \qquad \text{(17-5a)}$$

where K, called the **thermal conductivity**, is characteristic of the substance. K has units of $W\,m^{-1}(\text{C}°)^{-1}$ in the SI system. In the British system, ΔQ is measured in BTU, Δt in hours, A in square feet, T_H and T_C in °F, and Δx in inches, so that K has units of $\text{BTU in.}\,h^{-1}\,ft^{-2}(F°)^{-1}$. Table 17-3 lists the thermal conductivity of a few materials. In terms of the *temperature gradient*, $\Delta T/\Delta x = (T_2 - T_1)/(x_2 - x_1) = (T_C - T_H)/\Delta x$, Eq. (17-5a) can be written as

$$\frac{\Delta Q}{\Delta t} = -KA\frac{\Delta T}{\Delta x} \qquad \text{(17-5b)}$$

In the building trades, thermal insulating materials are usually specified by their *R value*, designated by R and defined by

R VALUE OF THERMAL INSULATION
$$R = \frac{\Delta x}{K} \qquad \text{(17-6)}$$

where Δx is measured in inches and K is specified in British units.

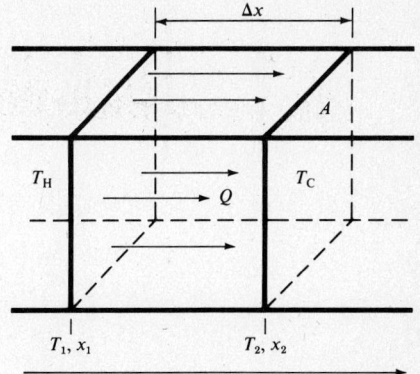

Figure 17-1. Uniform flow of heat from a hot reservoir at temperature T_H through a slab of material with a thickness Δx and cross sectional area A to a cold reservoir at temperature T_C.

TABLE 17-3: Thermal Conductivities of Selected
Materials

Material	$W\,m^{-1}(C^\circ)^{-1}$	$BTU\,in.\,h^{-1}\,ft^{-2}(F^\circ)^{-1}$
Aluminum	224	1530
Copper	390	2660
Silver	410	2800
Steel	48	328
Concrete	~1	~7
Wood	~0.14	~1
Glass	0.8	5.5
Fiberglass	0.041	0.28
Styrofoam	0.02	0.14
Air	0.025	0.17

EXAMPLE 17-7: The thermal conductivity of fiberglass is

$$K = 0.28\ BTU\,in.\,h^{-1}\,ft^{-2}(F^\circ)^{-1}$$

What is the R value of $3\frac{1}{2}$ inches of fiberglass insulation?

Solution: Use Eq. (17-6) with just the numerical values of Δx in inches and K in British units.

$$R = \frac{\Delta x}{K} = \frac{3.5}{0.28} = \boxed{12.5}$$

EXAMPLE 17-8: Estimate the rate at which heat flows through a wood panel 2 cm thick that measures 1 m × 2 m and separates a region where the temperature is 22°C from a region where the temperature is −2°C.

Solution: Use Eq. (17-5a) with $A = 1\ m \times 2\ m = 2\ m^2$. From Table 17-3, $K = 0.14\ W\,m^{-1}(C^\circ)^{-1}$.

$$\frac{\Delta Q}{\Delta t} = KA\frac{T_H - T_C}{\Delta x} = [0.14\ W\,m^{-1}(C^\circ)^{-1}](2m^2)\frac{[22 - (-2)]\,C^\circ}{2 \times 10^{-2}\ m} = \boxed{336\ W}$$

EXAMPLE 17-9: (a) How much heat flows through a wall measuring 8 ft × 10 ft that has an R value of 10 when the temperature is 68°F on one side and 10°F on the other side? (b) Convert your answer to watts.

Solution:

(a) Use Eqs. (17-5a) and (17-6) with A = 8 ft × 10 ft = 80 ft².

$$\frac{\Delta Q}{\Delta t} = KA\frac{T_H - T_C}{\Delta x} = \frac{A}{R}(T_H - T_C)$$

$$= \frac{80\ ft^2}{10\ BTU\,h^{-1}\,ft^2(F^\circ)^{-1}}(68°F - 10°F) = \boxed{464\ BTU\,h^{-1}}$$

(b) $\dfrac{\Delta Q}{\Delta t} = \left(\dfrac{464\ BTU}{h}\right)\left(\dfrac{1055\ J}{BTU}\right)\left(\dfrac{h}{3600\ s}\right) = 136\ W$

B. Convection

- **Convection** is heat transfer due to a flowing fluid.

Newton's law of cooling can be used to estimate the heat flow from a hot object at temperature T_H through a surface area A of the object in contact with a moving fluid at temperature T_F:

NEWTON'S LAW OF COOLING $$\frac{\Delta Q}{\Delta t} = hA(T_H - T_F) = hA\,\Delta T \tag{17-7}$$

where h is called the **coefficient of convection** and has units of $W\,m^{-2}(C°)^{-1}$. Only approximate values of h are available because it depends strongly on the natures of the surface and the fluid and the flow rate.

EXAMPLE 17-10: A fan blows air at a temperature of 20°C at an aluminum panel 20 cm × 30 cm, keeping the temperature of the panel at 55°C. How much heat is flowing through the panel? Use $300\ W\,m^{-2}(C°)^{-1}$ as your value of h.

Solution: Use Eq. (17-7) with $A = (0.2)(0.3)\ m^2 = 6 \times 10^{-2}\ m^2$.

$$\frac{\Delta Q}{\Delta t} = hA(T_H - T_F)$$
$$= [300\ W\,m^{-2}(C°)^{-1}](6 \times 10^{-2}\ m^2)(55°C - 20°C) = \boxed{630\ W}$$

C. Radiation

- **Radiation** is heat transfer by electromagnetic waves.

The rate at which a body with a surface area A at an *absolute* temperature T radiates heat is given by the **Stefan–Boltzmann law**:

STEFAN–BOLTZMANN LAW $$\frac{\Delta Q}{\Delta t} = Ae\sigma T^4 \tag{17-8}$$

where $\sigma = 5.67 \times 10^{-8}\ W\,m^{-2}\,K^{-4}$, called the **Stefan–Boltzmann constant**, and e is the body's **emissivity**. The emissivity may range from 0, for a perfectly reflecting surface, to 1, for a perfectly absorbing surface (called a **blackbody**). The *net* rate of heat radiation by a body at absolute temperature T_H when the body is enclosed in a chamber with inner walls at absolute temperature T_C is

$$\frac{\Delta Q}{\Delta t} = Ae\sigma(T_H^4 - T_C^4) \tag{17-9}$$

EXAMPLE 17-11: The emissivity of a copper sphere is 0.36. Its diameter is 20 cm and its temperature is maintained at 45°C by means of a small electric heater inside. How much heat flows through the surface of the sphere when it is kept in a room that has a temperature of 20°C?

Solution: Use Eq. (17-9) with $A = 4\pi r^2 = 4\pi(0.1\ m)^2 = 0.1257\ m^2$. Remember to express temperatures in kelvins: $T_H = 45 + 273 = 328\ K$ and $T_C = 20 + 273 = 293\ K$.

$$\frac{\Delta Q}{\Delta t} = Ae\sigma(T_H^4 - T_C^4)$$
$$= (0.1257\ m^2)(0.36)(5.67 \times 10^{-8}\ W\,m^2\,K^{-4})[(328)^4 - (293)^4]\ K^4 = \boxed{10.8\ W}$$

SUMMARY

1. The *mechanical equivalent of heat* can be written as 1 calorie = 4.186 J or 1 BTU = 1055 J.
2. The quantity of heat needed to raise by ΔT the temperature of a mass m of a substance that has a *specific heat capacity* c is $Q = mc\,\Delta T$. When the specific heat varies significantly with temperature you must use

$$Q = m \int c(T)\,dT$$

3. A substance's normal melting and boiling temperatures are its melting and boiling points when the ambient pressure is one atmosphere.
4. The quantity of heat needed to melt a mass m of a substance is $Q = mL_f$ at the substance's normal melting temperature, where L_f is the substance's *latent heat of fusion*. The quantity of heat needed to vaporize a mass m of a substance is $Q = mL_v$ at the substance's normal boiling temperature, where L_v is the substance's *latent heat of vaporization*.
5. The flow of heat through a cross-sectional area A of a material that has a *thermal conductivity* K and thickness Δx is

$$\frac{\Delta Q}{\Delta t} = KA\frac{T_H - T_C}{\Delta x}$$

where T_H and T_C are, respectively, the temperatures on the hot and cold faces of the material.
6. The R value of a thermal insulating material is $R = \Delta x/K$ where Δx is the thickness of the material in inches and K is the thermal conductivity, specified in British units. The R value is usually reported as a pure number although it actually has units of h (F°) ft² BTU⁻¹.
7. *Newton's law of cooling* gives the heat flowing through a surface of area A at temperature T_H when a fluid at temperature T_F is moving past it:

$$\frac{\Delta Q}{\Delta t} = hA(T_H - T_F)$$

where h is the coefficient of convection appropriate for the particular situation.
8. The *Stefan–Boltzmann law* gives the heat radiated from a surface of area A at an absolute temperature T:

$$\frac{\Delta Q}{\Delta t} = Ae\sigma T^4$$

where e is the surface's emissivity and σ is the Stefan–Boltzmann constant.

RAISE YOUR GRADES

Can you explain . . . ?

☑ why a cool piece of aluminum feels cooler than a piece of wood at the same temperature
☑ why temperature changes are smaller in a city located near a large body of water than in a similar city located far from the water
☑ why, in cold weather, water freezes more quickly on a bridge surface than on a road
☑ why storm windows consisting of two sheets of glass separated by an air space insulate better than the same thickness of solid glass
☑ why you feel warmer in a dark shirt than in a white one when you are outdoors on a sunny day
☑ why a Thermos flask is made of glass
☑ why the space between the walls of a Thermos flask is evacuated
☑ why the glass surfaces in a Thermos flask are silvered.

SOLVED PROBLEMS

Quantity of Heat

PROBLEM 17-1: How many BTU h^{-1} are equivalent to one horsepower?

Solution: 1 hp = 746 W = 746 J s^{-1} (Sec. 5-4). From Example 17-1, BTU = 1055 J. Thus

$$1 \text{ hp} = \left(\frac{746 \text{ J}}{\text{s}}\right)\left(\frac{1 \text{ BTU}}{1055 \text{ J}}\right)\left(\frac{3600 \text{ s}}{\text{h}}\right) = \boxed{2546 \frac{\text{BTU}}{\text{h}}}$$

Heat Capacity

PROBLEM 17-2: Eight hundred grams of a metal at 95°C is dropped into a styrofoam cup containing 400 g of water at 20°C. The water temperature rises to 25°C. What is the specific heat capacity of the metal? Assume that the specific heat capacity of the styrofoam is negligible.

Solution: Use Eq. (17-3a) and the method of mixtures.

$$\text{heat lost by metal} = \text{heat gained by water}$$

$$m_\text{m} c_\text{m} \Delta T_\text{m} = m_\text{w} c_\text{w} \Delta T_\text{w}$$

Solve for c_m:

$$c_\text{m} = c_\text{w} \frac{m_\text{w} \Delta T_\text{w}}{m_\text{m} \Delta T_\text{m}} = [4.186 \text{ kJ kg}^{-1}(\text{C}°)^{-1}]\left[\frac{(400 \text{ g})(25-20)\text{ C}°}{(800 \text{ g})(95-25)\text{ C}°}\right] = \boxed{0.1495 \text{ kJ kg}^{-1}(\text{C}°)^{-1}}$$

PROBLEM 17-3: A fluid whose specific gravity is 0.94 flows through a pipe at the rate of 3.5 cm^3 s^{-1}. Electrical energy supplied to a heater inside the pipe at the rate of 50 W heats the fluid from 19°C to 24°C (Figure 17-2). What is the specific heat capacity of the fluid?

Solution: You need to modify Eq. (17-3a) to express the rate of heating, $\Delta Q/\Delta t$, as a function of the flow rate, $\Delta V/\Delta t$. First, divide both sides of Eq. (17-3a) by Δt. Then use the definitions of density ($\rho = m/V$) and specific gravity ($\text{SG} = \rho/\rho_\text{water}$) to express the mass Δm in terms of SG and ρ_water: $\Delta m = \text{SG}\rho_\text{water}\Delta V$. Finally, solve for the specific heat capacity c.

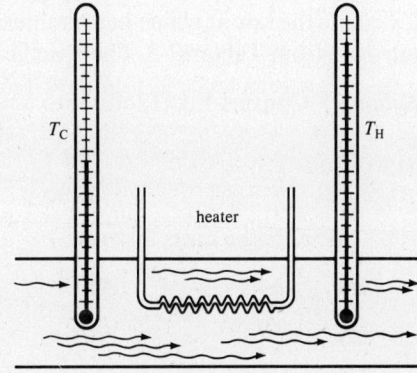

Figure 17-2

$$\Delta Q = \Delta m c \, \Delta T \qquad \frac{\Delta Q}{\Delta t} = \frac{\Delta m}{\Delta t} c \, \Delta T \qquad \frac{\Delta Q}{\Delta t} = \frac{\text{SG}\rho_\text{water}}{\Delta t}\frac{\Delta V}{} c \, \Delta T$$

$$c = \frac{\Delta Q/\Delta t}{(\Delta V/\Delta t)\text{SG}\rho_\text{water}\Delta T}$$

$$= \frac{50 \text{ W}}{(3.5 \times 10^{-6} \text{ m}^3 \text{s}^{-1})(0.94)(10^3 \text{ kg m}^{-3})(24-19)\,°\text{C}} = \boxed{3.04 \text{ kJ kg}^{-1}(\text{C}°)^{-1}}$$

Change of Phase

PROBLEM 17-4: One hundred grams of ice taken from a freezer at −17°C is added to a styrofoam cup holding 400 g of water at 20°C. Describe the contents of the cup after equilibrium is established. Assume that the specific heat of the cup is negligible. Obtain the specific heat capacities for ice and water from Table 17-1 and the latent heats from Table 17-2.

Solution: To get a feeling for the magnitudes of quantities involved in this problem let's first determine **(a)** how much heat is required to raise the temperature of the ice from $-17°C$ to $0°C$ and **(b)** how much heat is needed to decrease the temperature of the water from $20°C$ to $0°C$. Use Eq. (17-3a).

(a)
$$Q = mc\,\Delta T = (0.1\text{ kg})[2.04\text{ kJ kg}^{-1}(\text{C}°)^{-1}](17\text{C}°) = 3.468\text{ kJ}$$

(b)
$$Q = mc\,\Delta T = (0.4\text{ kg})[4.186\text{ kJ kg}^{-1}(\text{C}°)^{-1}](20\text{C}°) = 33.488\text{ kJ}$$

Because $33.488 > 3.468$, sufficient heat is available in the water to raise the temperature of the ice to $0°C$. Next determine how much heat is required to melt all the ice. Use Eq. (17-4a).

$$Q = mL_f = (0.1\text{ kg})(335\text{ kJ kg}^{-1}) = 33.5\text{ kJ}$$

Because $(33.488 - 3.468) < 33.5$, there is not sufficient heat remaining in the water to melt all the ice. Use the conservation of energy principle to calculate the mass m_x of ice that melts. It should be clear now that the contents of the cup are water and ice at $0°C$.

$$\text{heat lost by water} = \text{heat to raise temperature of ice to melting point}$$
$$+ \text{ heat to melt } m_x \text{ kilograms of ice}$$

$$33.488\text{ kJ} = 3.468\text{ kJ} + m_x L_f$$

$$m_x = \frac{(33.488 - 3.468)\text{ kJ}}{L_f} = \frac{30.020\text{ kJ}}{335\text{ kJ kg}^{-1}} = 0.09\text{ kg} = 90\text{ g}$$

Therefore the contents of the cup are $(400 + 90)\text{ g} = 490$ g of water and $(100 - 90)\text{ g} = 10$ g of ice at $0°C$.

PROBLEM 17-5: Half a liter of liquid nitrogen (at its normal boiling temperature) must be added every 30 min to a **dewar** (a vacuum flask for extremely cold liquids) in a laboratory experiment to replenish the nitrogen that boils away because of heat leaking into the dewar. Calculate the rate at which heat is entering the dewar. Find the latent heat of vaporization of liquid nitrogen from Table 17-2. The specific gravity of liquid nitrogen is 0.81.

Solution: Convert Eq. (17-4b) into a rate formula, as you did with Eq. (17-3a) in Problem 17-2.

$$\frac{\Delta Q}{\Delta t} = \frac{\Delta m}{\Delta t} L_v$$

Next calculate the mass loss rate:

$$\frac{\Delta m}{\Delta t} = \rho \frac{\Delta V}{\Delta t} = SG\rho_{\text{water}} \frac{\Delta V}{\Delta t} = (0.81)(10^3\text{ kg m}^{-3})\left(\frac{0.5 \times 10^{-3}\text{ m}^3}{1800\text{ s}}\right) = 2.25 \times 10^{-4}\text{ kg s}^{-1}$$

Therefore

$$\frac{\Delta Q}{\Delta t} = (2.25 \times 10^{-4}\text{ kg s}^{-1})(201\text{ kJ kg}^{-1}) = \boxed{4.52 \times 10^{-2}\text{ kW} = 45.2\text{ W}}$$

PROBLEM 17-6: How much ice at $0°C$ will it take to cool 300 g of molten copper from its melting temperature of $1083°C$ to $0°C$?

Solution: Let m, L, and c represent the mass, latent heat of fusion, and specific heat of copper, and let m_i and L_i be the mass and latent heat of fusion of ice. Use conservation of energy:

$$\text{heat lost by copper} = \text{heat to melt ice}$$

$$mL + mc\,\Delta T = m_i L_i$$

Solve for m_i:

$$m_i = m\left(\frac{L + c\,\Delta T}{L_i}\right) = (0.3\text{ kg})\left[\frac{134\text{ kJ kg}^{-1} + [0.385\text{ kJ kg}^{-1}(\text{C}°)^{-1}](1083 - 0)\text{C}°}{335\text{ kJ kg}^{-1}}\right] = \boxed{1.64\text{ kg}}$$

Heat Transfer

PROBLEM 17-7: A block of steel 1 cm thick is put in contact with a block of copper 6 cm thick but of the same cross-sectional area as the steel. The steel is in contact with a hot reservoir at 100°C and the copper is in contact with a cold reservoir at 0°C. The two blocks are covered with a thermal insulator so that negligible heat escapes from their surface (Figure 17-3). What is the equilibrium temperature at the steel–copper interface?

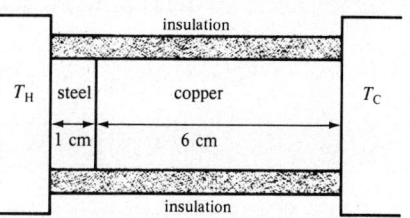

Figure 17-3.

Solution: Let the interface temperature be T and the cross-sectional area of the metals be A. In the steady state you can conclude that the rate of heat flow through the steel must be the same as the rate through the copper. Use the equation for heat conduction, Eq. (17-5a).

$$\left(\frac{\Delta Q}{\Delta t}\right)_{\text{steel}} = \left(\frac{\Delta Q}{\Delta t}\right)_{\text{copper}}$$

$$K_s A \frac{T_H - T}{\Delta x_s} = K_{Cu} A \frac{T - T_C}{\Delta x_{Cu}}$$

Now solve for T:

$$\frac{K_s}{\Delta x_s} T_H + \frac{K_{Cu}}{\Delta x_{Cu}} T_C = \left(\frac{K_s}{\Delta x_s} + \frac{K_{Cu}}{\Delta x_{Cu}}\right) T$$

$$T = \frac{\dfrac{K_s}{\Delta x_s} T_H + \dfrac{K_{Cu}}{\Delta x_{Cu}} T_C}{\dfrac{K_s}{\Delta x_s} + \dfrac{K_{Cu}}{\Delta x_{Cu}}}$$

Omitting the units for clarity,

$$T = \frac{\left(\dfrac{48}{1}\right)(100) + \left(\dfrac{390}{6}\right)(0)}{\left(\dfrac{48}{1}\right) + \left(\dfrac{390}{6}\right)} = \boxed{42.5°C}$$

PROBLEM 17-8: A new construction panel measuring 4 ft × 8 ft × 4 in. is said to have an R value of 30. **(a)** What is the thermal conductivity of this panel? **(b)** What is the thermal conductivity in SI units?

Solution:

(a) Use the definition Eq. (17-6) and solve for K. Remember to specify Δx in inches.

$$R = \frac{\Delta x}{K}$$

$$K = \frac{\Delta x}{R} = \frac{4}{30} = \boxed{0.133 \text{ BTU in. h}^{-1} \text{ ft}^{-2} (\text{F}°)^{-1}}$$

(b) Let K_{SI} represent K in SI units, let K_B represent K in British units. K_B has units of BTU in. h^{-1} ft^{-2} (F°)$^{-1}$, so convert this to SI units:

$$\frac{\text{BTU in.}}{\text{h ft}^2 (\text{F}°)} = \left(\frac{\text{BTU in.}}{\text{h ft}^2 (\text{F}°)}\right)\left(\frac{\text{ft}^2}{144 \text{ in.}^2}\right)\left(\frac{\text{in.}}{2.54 \times 10^{-2} \text{ m}}\right)\left(\frac{1055 \text{ J}}{\text{BTU}}\right)\left(\frac{\text{h}}{3600 \text{ s}}\right)\left(\frac{9(\text{F}°)}{5(\text{C}°)}\right)$$

$$= 0.1442 \text{ W m}^{-1} (\text{C}°)^{-1}$$

Divide the right side of this equation by the left side to get the unity conversion factor:

$$1 = \frac{0.1442 \text{ W m}^{-1} (\text{C}^\circ)^{-1}}{\text{BTU in. h}^{-1} \text{ ft}^{-2} (\text{F}^\circ)^{-1}}$$

$$K_{SI} = K_B = (0.133 \text{ BTU in. h}^{-1} \text{ ft}^{-2} (\text{F}^\circ)^{-1}) \left(\frac{0.1442 \text{ W m}^{-1} (\text{C}^\circ)^{-1}}{\text{BTU in. h}^{-1} \text{ ft}^{-2} (\text{F}^\circ)^{-1}} \right)$$

$$= \boxed{1.92 \times 10^{-2} \text{ W m}^{-1} (\text{C}^\circ)^{-1}}$$

PROBLEM 17-9: The skin temperature of a wet person emerging from a swimming pool is only 27°C. If she exposes 1.2 m^2 of skin and the air temperature is 22°C, estimate her rate of heat loss in a light breeze for which $h = 200$ W m^{-2} (C$^\circ$)$^{-1}$.

Solution: Use Eq. (17-1).

$$\frac{\Delta Q}{\Delta t} = hA(T_H - T_F) = [200 \text{ W m}^{-2} (\text{C}^\circ)^{-1}](1.2 \text{ m}^2)(27°\text{C} - 22°\text{C}) = \boxed{1200 \text{ W}}$$

PROBLEM 17-10: Dry skin temperature is about 30°C. Estimate the net radiant heat loss from the swimmer from Problem 17-9 when she is dry. Assume her skin radiates as a blackbody.

Solution: Use Eq. (17-9), but first convert the skin and ambient temperatures to the Kelvin scale. The emissivity of a blackbody is 1.

$$T_H = 30 + 273 = 303 \text{ K} \qquad T_C = 22 + 273 = 295 \text{ K}$$

$$\frac{\Delta Q}{\Delta t} = Ae\sigma[T_H^4 - T_C^4] = (1.2 \text{ m}^2)(1)(5.67 \times 10^{-8} \text{ W m}^2 \text{ K}^{-4})[(303)^4 - (295)^4] \text{ K}^4 = \boxed{58 \text{ W}}$$

Supplementary Exercises

EXERCISE 17-1: An "average" person takes in 2500 Calories in a 24-hour period. Assuming no net gain or loss of body weight, what is the average person's rate of heat loss?

EXERCISE 17-2: A 250-g chunk of glass is dropped into a styrofoam cup containing 400 g of water initially at 22°C. The water's temperature rises to 25°C when thermal equilibrium is established. What was the initial temperature of the glass? (*Hint:* Use Table 17-1.)

EXERCISE 17-3: An insulated copper can contains 400 g of methyl alcohol at 16°C. The mass of the can is 300 g. Determine the final temperature when 500 g of turpentine at 90°C is mixed in.

EXERCISE 17-4: How much liquid helium will be vaporized in the process of cooling a 160-g copper device from the temperature liquid nitrogen, $-196°$C, to the temperature of liquid helium, $-269°$C? Use 0.385 kJ kg^{-1} (C$^\circ$)$^{-1}$ as the specific heat capacity of copper.

EXERCISE 17-5: How long will a 600-W heater immersed in 250 g of water at its boiling temperature take to boil the water away?

EXERCISE 17-6: A styrofoam cup contains 400 g of water at 22°C. Describe the contents of the cup after 1.5 kg of ice at a temperature of $-15°$C has been added.

EXERCISE 17-7: What is the rate of heat flow through the block of steel in Problem 17-7 if the cross section of the blocks measures 3 cm × 3 cm?

EXERCISE 17-8: (a) What is the R value for a pane of glass 2 ft wide, 3 ft high, and $\frac{1}{4}$ inch thick? (b) What is the rate of heat flow through this pane on a day when the indoor temperature is 68°F and the outdoor temperature is 20°F?

EXERCISE 17-9:   Estimate the surface area of an aluminum heat sink needed to cool a power transistor that produces heat at the rate of 6 W. The temperature of the transistor should not exceed 30°C when the ambient air temperature is 22°C. Assume $h = 10 \text{ W m}^{-2} (\text{C}°)^{-1}$.

EXERCISE 17-10:   Estimate the effective surface area of the filament in a 150-W light bulb that operates at a temperature of 1500°C when the ambient temperature is 22°C. Assume the filament has an emissivity of 0.8.

Answers to Supplementary Exercises

17-1:   121 W

17-2:   54.8°C

17-3:   49.8°C

17-4:   214 g, which corresponds to 1.46 L of liquid helium.

17-5:   940 s = 15.6 min

17-6:   373 g of water and 1.527 kg of ice at 0°C.

17-7:   248 W

17-8:   (a) 4.55×10^{-2} (b) 6.34×10^{3} BTU/h $= 1.86 \text{ kW}$

17-9: $7.5 \times 10^{-2} \text{ m}^2$

17-10: $3.35 \times 10^{-4} \text{ m}^2$

18 IDEAL GAS AND KINETIC THEORY

THIS CHAPTER IS ABOUT

- ☑ **Equation of State**
- ☑ **Pressure of an Ideal Gas**
- ☑ **Density of an Ideal Gas**
- ☑ **Specific Heats of Gases**
- ☑ **Equipartition of Energy**

18-1. Equation of State

Early experiments on confined gases revealed two basic relationships:

BOYLE'S LAW PV = a constant for a fixed mass and temperature   **(18-1)**

CHARLES' LAW V/T = a constant for a fixed mass and pressure   **(18-2)**

where P = the absolute pressure, V = the volume, and T = the absolute temperature of the gas. Combining these two equations gives us the **equation of state of an ideal gas:**

EQUATION OF STATE OF AN IDEAL GAS $\dfrac{PV}{T}$ = a constant for a fixed mass   **(18-3a)**

This equation, commonly called the **ideal gas law**, can be written in two other forms:

IDEAL GAS LAW

$$PV = nRT \qquad \textbf{(18-3b)}$$

$$PV = NkT \qquad \textbf{(18-3c)}$$

In Eqs. (18-3b) and (18-3c) n, the number of **moles** (mol) of the gas, is the mass of the gas in grams divided by the molecular mass (commonly called the "molecular weight"), R is the **universal gas constant**, $8.314\ \mathrm{J\,mol^{-1}\,K^{-1}}$, N is the number of molecules in the gas, and k is **Boltzmann's constant**, $1.38 \times 10^{-23}\ \mathrm{J\,K^{-1}}$. Clearly, $nR \times NK$, so if $n = 1$ mol,

$$N = \frac{(1\ \mathrm{mol})(8.314\ \mathrm{J\,mol^{-1}\,K^{-1}})}{1.38 \times 10^{-23}\ \mathrm{J\,K^{-1}}} = 6.022 \times 10^{-23}$$

which is **Avogadro's number** N_A, the number of molecules in a mole of any substance.

The behavior of real gases conforms closely to the ideal gas law, so long as the pressure is not too great (not much more that 10 to 20 atmospheres) and the temperature is not too low (not much less that 200 K). To understand a gas, it is often convenient to think of confining it in a cylinder equipped with a movable piston, a pressure gauge, a thermometer, and a supply tank, as shown in Figure 18-1.

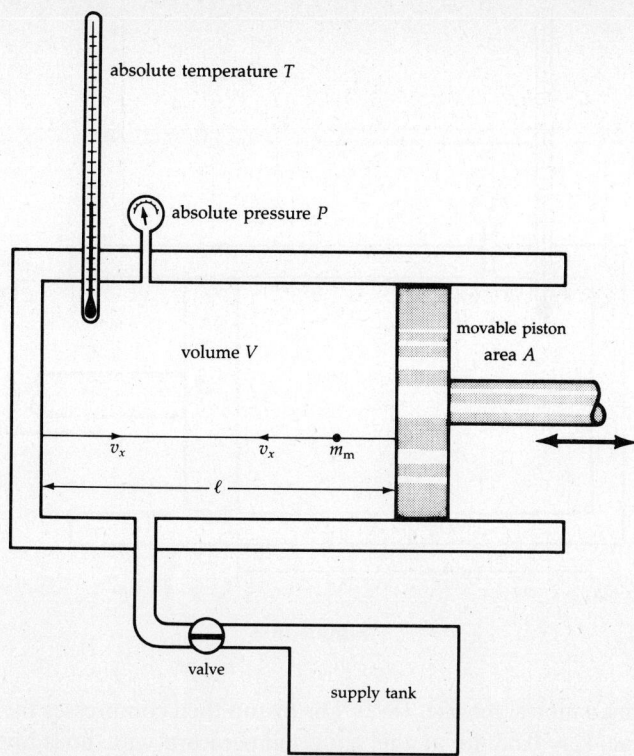

Figure 18-1. Cylinder with movable piston to control the volume. Thermometer and pressure gauge indicate temperature and pressure.

EXAMPLE 18-1: What is the mass m of nitrogen gas in a 60-L tank at a gauge pressure of 4 atm and a temperature of 22°C?

Solution: Use the ideal gas law, Eq. (18-3b), to obtain the number of moles of the gas and from that the mass. But first convert the given information into appropriate units:

$$V = 60 \text{ L} = 60 \times 10^3 \text{ cm}^3 = 60 \times 10^{-3} \text{ m}^3$$

$$P = (4 + 1) \text{ atm} = 5 \text{ atm} \times 1.013 \times 10^5 \text{ Pa atm}^{-1} = 5.065 \times 10^5 \text{ Pa}$$

$$T = 22 + 273 = 295 \text{ K}$$

Then solve Eq. (18-3b) for n:

$$PV = nRT$$

$$n = \frac{PV}{RT} = \frac{(5.065 \times 10^5 \text{ Pa})(60 \times 10^{-3} \text{ m}^3)}{(8.314 \text{ J mol}^{-1} \text{ K}^{-1})(295 \text{ K})} = 12.39 \text{ mol}$$

Nitrogen gas exists as diatomic molecules. From the periodic table, its molecular mass (molecular "weight") is $M = 2 \times 14 = 28 \text{ g mol}^{-1}$. Thus

$$m = nM = (12.39 \text{ mol})(28 \text{ g mol}^{-1}) = \boxed{347 \text{ g}}$$

EXAMPLE 18-2: What volume of air at 1 atm pressure must be added to a 0.65-m³ tank containing air at 1 atm pressure to increase the absolute pressure to 12 atm? Assume that the initial and final temperatures are the same.

Solution: One way to analyze this problem is to think of all the air as being contained in one large cylinder with an initial volume $V_i = V_f + V_x$ at temperature

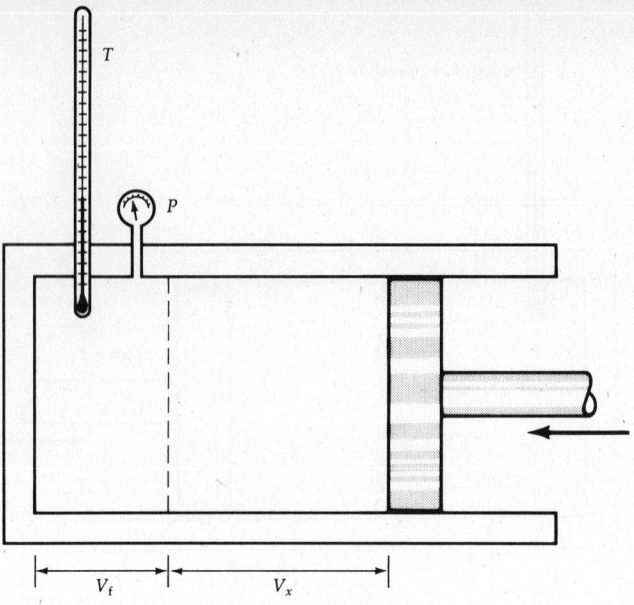

Figure 18-2

T containing *n* moles (see Fig. 18-2). The piston then compresses the air into the final volume $V_f = 0.65 \text{ m}^3$ at the same temperature with no addition or loss of air. In this situation, nRT is a constant, so by the ideal gas law, Eq. (18-3b), PV equals the same constant. Therefore, $P_i V_i = P_f V_f$. Now solve for V_x.

$$P_i(V_f + V_x) = P_f V_f$$

$$P_i V_x = P_f V_f - P_i V_f$$

$$V_x = V_f\left(\frac{P_f}{P_i} - 1\right) = (0.65 \text{ m}^3)\left(\frac{12 \text{ atm}}{1 \text{ atm}} - 1\right) = \boxed{7.15 \text{ m}^3}$$

EXAMPLE 18-3: Assume that the compression of the air in Example 18-2 is done so quickly that the temperature of the air rises from 22°C to 300°C before it begins to cool. Determine the pressure of the air at 300°C.

Solution: Use the ideal gas law, Eq. (18-3b), with $n = $ constant.

$$PV = nRT$$

$$\frac{P_i V_i}{T_i} = \frac{P_f V_f}{T_f} = nR$$

Solve for P_f:

$$P_f = P_i\left(\frac{V_i}{V_f}\right)\left(\frac{T_f}{T_i}\right) = (1 \text{ atm})\left(\frac{(7.15 + 0.65)\,\text{m}^3}{0.65}\right)\left(\frac{(300 + 273)\,\text{K}}{(22 + 273)\,\text{K}}\right) = 23.3 \text{ atm}$$

18-2. Pressure of an Ideal Gas

The pressure of a confined gas is due to the elastic collisions of the gas molecules with the container's walls, so it depends on the velocity of the molecules when they strike the walls. But the molecules do not all move at the same velocity, so to relate pressure to molecular velocity, we must use a measure of the *average* velocity. Because the kinetic energy of any particular molecule is $E_k = \frac{1}{2}m_m v^2$ (where m_m is the mass of one molecule), we could, in principle, find the average squared velocity (*not* the average velocity squared) from

$$\overline{v^2} = \frac{N_1 v_1^2 + N_2 v_2^2 + N_3 v_3^2 + \cdots}{N_1 + N_2 + N_3 + \cdots} = \frac{\Sigma(N_i v_i^2)}{\Sigma N_i} = \frac{\Sigma(N_i v_i^2)}{N} \qquad \textbf{(18-4)}$$

where N_1 = the number of molecules having velocity v_1, N_2 = the number of molecules having velocity v_2, etc., and $N = N_1 + N_2 + N_3 + \cdots = \Sigma N_i$. The total kinetic energy of the gas, also called the **internal energy** of the gas, is

$$E_{k\,tot} = N_1(\tfrac{1}{2}m_m v_1^2) + N_2(\tfrac{1}{2}m_m v_2^2) + N_3(\tfrac{1}{2}m_m v_3^2) + \cdots = \tfrac{1}{2}m_m \Sigma(N_i v_i^2)$$

From Eq. (18-4), $\Sigma(N_i v_i^2) = N\overline{v^2}$, so

INTERNAL ENERGY OF A GAS
$$E_{k\,tot} = N(\tfrac{1}{2}m_m \overline{v^2}) \qquad \textbf{(18-5)}$$

If we define the **root-mean-square speed**, v_{rms}, of a molecule as $v_{rms} = \sqrt{\overline{v^2}}$, we can express Eq. (18-5) as

$$E_{k\,tot} = N(\tfrac{1}{2}m_m v_{rms}^2) \qquad \textbf{(18-6)}$$

Refer again to Figure 18-1. If the component of a molecule's velocity perpendicular to the face of the piston is v_x, the molecule's change in momentum during the collision is $2m_m v_x$, which is equal to the impulse given by the molecule to the piston (see Sec. 8-1). So if v_x is the average perpendicular molecular velocity, the impulse on the wall during each collision is $F_{av}\Delta t = 2m_m v_x$. The average time between collisions between a molecule with perpendicular velocity v_x and the piston is the time the molecule takes to travel to the opposite wall, rebound, and return: $\Delta t = 2\ell/v_x$, where ℓ is the distance between the piston and the opposite wall. Therefore the average force on the piston due to this molecule is the impulse per collision divided by the time between collisions:

$$F_{av} = \frac{2m_m v_x}{\Delta t} = \frac{2m_m v_x}{2\ell/v_x} = \frac{m_m v_x^2}{\ell}$$

In three dimensions, the relationship between the velocity of a molecule and its three components is $v^2 = v_x^2 + v_y^2 + v_z^2$. On the average, the three velocity components are equal: $v_x^2 = v_y^2 = v_z^2$, so $v_x^2 = \tfrac{1}{3}\overline{v^2}$ and we can write the average force due to the average molecule as

$$F_{av} = \frac{m_m \overline{v^2}}{3\ell}$$

The average force due to all the gas molecules is NF_{av}, and therefore the pressure on the piston face is

$$P = N\frac{F_{av}}{A} = N\frac{m_m \overline{v^2}}{3\ell A} = \frac{2}{3}\frac{N}{V}\left(\frac{1}{2}m_m \overline{v^2}\right)$$

where $\ell A = V$, the volume of the cylinder. Multiply by V and use Eqs. (18-6) and (18-3c) to get

$$PV = \tfrac{2}{3}N(\tfrac{1}{2}m_m \overline{v^2}) = \tfrac{2}{3}N(\tfrac{1}{2}m_m v_{rms}^2) = \tfrac{2}{3}E_{k\,tot} = NkT \qquad \textbf{(18-7)}$$

So we can express the average kinetic energy of a molecule as

$$\tfrac{1}{2}m_m v_{rms}^2 = \tfrac{3}{2}kT \qquad \textbf{(18-8)}$$

Temperature is thus identified as a measure of the average kinetic energy of the molecules in a volume. (Therefore it makes no sense to speak of "the temperature of a molecule.") Solve Eq. (18-8) for v_{rms}:

ROOT-MEAN-SQUARE SPEED OF A MOLECULE
$$v_{rms} = \sqrt{\frac{3kT}{m_m}} \qquad \textbf{(18-9)}$$

EXAMPLE 18-4: Because air consists essentially of 80% nitrogen ($M = 2 \times 14 = 28$ g mol^{-1}) and 20% oxygen ($M = 2 \times 16 = 32$ g mol^{-1}) it can be thought of as having a molecular "weight" of approximately 29 g mol^{-1}. (a) Determine the mass of one "molecule of air" and (b) find its root-mean-square speed at 22°C.

Solution

(a) One mole of air, 29 g, contains N_A molecules. Thus the mass of a molecule of air is

$$m_m = \frac{M}{N_A} = \frac{29 \text{ g mol}^{-1}}{6.022 \times 10^{23} \text{ molecules mol}^{-1}}$$

$$= \boxed{4.82 \times 10^{-23} \text{ g} = 4.82 \times 10^{-26} \text{ kg}}$$

(b) Use Eq. (18-9).

$$v_{rms} = \sqrt{\frac{3kT}{m_m}} = \sqrt{\frac{3(1.38 \times 10^{-23} \text{ J K}^{-1})(22 + 273) \text{ K}}{4.82 \times 10^{-26} \text{ kg}}} = \boxed{503 \text{ m s}^{-1}}$$

EXAMPLE 18-5: (a) Find the number of molecules in a cubic meter of air at a pressure of 1 atm and temperature of 22°C. (b) What is the average kinetic energy of a molecule of air under these conditions?

Solution:

(a) Use Eq. (18-3c) and solve for $\mathcal{N} = N/V$, the number of molecules per unit volume.

$$PV = NkT$$

$$\mathcal{N} = \frac{N}{V} = \frac{P}{kT} = \frac{1.013 \times 10^5 \text{ Pa}}{(1.38 \times 10^{-23} \text{ J K}^{-1})(22 + 273) \text{ K}}$$

$$= \boxed{2.49 \times 10^{25} \text{ m}^{-3} = 2.49 \times 10^{19} \text{ cm}^{-3}}$$

(b) Solve Eq. (18-7) for $E_k = \frac{1}{2} m_m v_{rms}^2$.

$$P = \tfrac{2}{3} \mathcal{N} (\tfrac{1}{2} m_m v_{rms}^2)$$

$$\frac{1}{2} m_m v_{rms}^2 = \frac{3P}{2\mathcal{N}} = \frac{3(1.013 \times 10^5 \text{ Pa})}{2(2.49 \times 10^{25} \text{ m}^{-3})} = \boxed{6.11 \times 10^{-21} \text{ J}}$$

Another way is to use Eq. (18-8):

$$\tfrac{1}{2} m_m v_{rms}^2 = \tfrac{3}{2} kT$$

$$= \tfrac{3}{2}(1.38 \times 10^{-23} \text{ J K}^{-1})(22 + 273) \text{ K} = \boxed{6.11 \times 10^{-21} \text{ J}}$$

18-3. Density of an Ideal Gas

The density of an ideal gas is

$$\rho = \frac{PM}{RT} \tag{18-10}$$

where M is the molecular mass specified in kilograms per mole, to yield ρ in SI units, kg m^{-3}.

EXAMPLE 18-6: Determine the density of air at a pressure of 1 atm and a temperature of 22°C.

Solution: Use Eq. (18-10) with $M = 29\,\text{g mol}^{-1} = 29 \times 10^{-3}\,\text{kg mol}^{-1}$.

$$\rho = \frac{PM}{RT} = \frac{(1.013 \times 10^5\,\text{Pa})(29 \times 10^{-3}\,\text{kg mol}^{-1})}{(8.314\,\text{J mol K}^{-1})(22 + 273)\,\text{K}} = \boxed{1.20\,\text{kg m}^{-3}}$$

EXAMPLE 18-7: Start with the ideal gas law, Eq. (18-3b), and derive Eq. (18-10).

Solution: Solve Eq. (18-3b) for n, the number of moles.

$$PV = nRT \qquad n = \frac{PV}{RT}$$

The mass of a quantity of gas is $m = nM$. Thus

$$m = \left(\frac{PV}{RT}\right)M = \left(\frac{PM}{RT}\right)V$$

Since density is mass per unit volume,

$$\rho = \frac{m}{V} = \boxed{\frac{PM}{RT}}$$

18-4. Specific Heats of Gases

Consider a fixed number of moles of a gas confined to a cylinder like that shown in Figure 18-1. If the *volume V* is held constant, a quantity of heat supplied to the gas produces a rise in temperature (plus a rise in pressure) when the volume is held fixed, according to

SPECIFIC HEAT AT CONSTANT VOLUME $\qquad Q = nC_v\Delta T \qquad\qquad$ **(18-11)**

where C_v is called the **molar specific heat capacity at constant volume** and has units of $\text{J mol}^{-1}\,\text{K}^{-1}$.

  If the *pressure P* is held constant (by moving the piston back to increase the volume), a quantity of heat supplied to the gas produces a rise in temperature, according to

SPECIFIC HEAT AT CONSTANT PRESSURE $\qquad Q = nC_p\Delta T \qquad\qquad$ **(18-12)**

where C_p is called the **molar specific heat capacity at constant pressure** and has units of $\text{J mol}^{-1}\,\text{K}^{-1}$. The ratio of these two quantities is designated by γ:

$$\gamma = \frac{C_p}{C_v} \qquad\qquad \textbf{(18-13)}$$

For a monatomic ideal gas, $C_p = \frac{5}{2}R$ and $C_v = \frac{3}{2}R$, so that their difference is $C_p - C_v = R$. The molar specific heats and γ for a few gases are listed in Table 18-1.

EXAMPLE 18-8: **(a)** How much heat is required to raise the temperature of $0.6\,\text{m}^3$ of nitrogen at a pressure of 3 atm from 20°C to 26°C? **(b)** What is the pressure of the gas at 26°C?

TABLE 18-1: Molar Specific Heat Capacities of Selected Gases*

Gas	C_p (J mol^{-1} K^{-1})	C_v (J mol^{-1} K^{-1})	$C_p - C_v$ (J mol^{-1} K^{-1})	$\gamma = \dfrac{C_p}{C_v}$
Monatomic				
Helium	20.8	12.5	8.3	1.66
Argon	20.9	12.5	8.4	1.67
Diatomic				
Nitrogen	29.1	20.8	8.3	1.40
Oxygen	29.4	21.1	8.3	1.40
Carbon monoxide	29.2	20.9	8.3	1.40
Triatomic				
Carbon dioxide	36.9	28.4	8.5	1.30
Hydrogen sulfide	34.6	26.0	8.6	1.33

* Tabulated values are for a pressure of 1 atmosphere and temperature near 22°C.

Solution:

(a) The volume is held constant, so use Eq. (18-7). But first use Eq. (18-3b) to obtain the number of moles of nitrogen in the tank with $P = 3(1.013 \times 10^5$ Pa$) = 3.039 \times 10^5$ Pa. Find C_v for nitrogen from Table 18-1.

$$PV = nRT$$

$$n = \frac{PV}{RT} = \frac{(3.039 \times 10^5 \text{ Pa})(0.6 \text{ m}^3)}{(8.314 \text{ J mol}^{-1} \text{ K}^{-1})(20 + 273) \text{ K}} = 74.85 \text{ mol}$$

$$Q = nC_v \Delta T = (74.85 \text{ mol})(20.8 \text{ J mol}^{-1} \text{ K}^{-1})(26 - 20)(\text{C}°)$$
$$= \boxed{9.34 \times 10^3 \text{ J} = 9.34 \text{ kJ}}$$

Notice that because a temperature *interval* is involved the temperature units cancel: $(\text{C}°)(\text{K}^{-1}) = 1$.

(b) Use the ideal gas law to find P.

$$PV = nRT$$

$$P = \frac{nRT}{V} = \frac{(74.85 \text{ mol})(8.314 \text{ J mol}^{-1} \text{ K}^{-1})(26 + 273) \text{ K}}{0.6 \text{ m}^3}$$

$$= \boxed{3.10 \times 10^5 \text{ Pa} = 3.06 \text{ atm}}$$

EXAMPLE 18-9: Suppose that the nitrogen in Example 18-8 is contained in a cylinder like that of Fig. 18-1. (a) Find the temperature rise if the same quantity of heat is added but the piston is pulled back so as to keep the pressure constant. (b) Determine the new volume the gas occupies.

Solution:

(a) Solve Eq. (18-12) for ΔT.

$$Q = nC_p \Delta T$$

$$\Delta T = \frac{Q}{nC_p} = \frac{9.34 \times 10^3 \text{ J}}{(74.85 \text{ mol})(29.1 \text{ J mol}^{-1} \text{ K}^{-1})} = \boxed{4.29 \text{ C}°}$$

Because the initial temperature was 20°C the final temperature is 24.29°C.

(b) Use the ideal gas law, Eq. (18-3b), to obtain the volume.

$$PV = nRT \qquad \frac{V}{T} = \frac{nR}{P} = \text{constant}$$

Thus

$$\frac{V_i}{T_i} = \frac{V_f}{T_f} \qquad V_f = V_i\left(\frac{T_f}{T_i}\right) = (0.6 \text{ m}^3)\left(\frac{(24.29 + 273)\,\text{K}}{(20 + 273)\,\text{K}}\right) = \boxed{0.609 \text{ m}^3}$$

EXAMPLE 18-10: Compute C_p, C_v, and γ for a monatomic ideal gas.

Solution:

$$C_p = \tfrac{5}{2}R = \tfrac{5}{2}(8.314 \text{ J mol}^{-1}\text{K}^{-1}) = \boxed{20.78 \text{ J mol}^{-1}\text{K}^{-1}}$$

$$C_v = \tfrac{3}{2}R = \tfrac{3}{2}(8.314 \text{ J mol}^{-1}\text{K}^{-1}) = \boxed{12.47 \text{ J mol}^{-1}\text{K}^{-1}}$$

$$\gamma = \frac{C_p}{C_v} = \frac{\tfrac{5}{2}R}{\tfrac{3}{2}R} = \frac{5}{3} = \boxed{1.67}$$

Notice how good the agreement is between these values and those for helium and argon in Table 18-1.

18-5. Equipartition of Energy

The **principle of the equipartition of energy** states that

- On the average, a kinetic energy of $\tfrac{1}{2}kT$ is associated with each **degree of freedom** that a molecule has.

Each degree of freedom represents one way in which a molecule may possess energy. *Monatomic* gases have *three* degrees of freedom, one for each component of translational velocity, v_x, v_y, and v_z. Therefore, $E_{k\,tot} = 3 \times N \times \tfrac{1}{2}kT = \tfrac{3}{2}NkT = \tfrac{3}{2}nRT$, consistent with Eqs. (18-6) and (18-8). For an ideal monatomic gas, $C_v = \tfrac{3}{2}R$ (see Sec. 19-4), and for any ideal gas, $C_p - C_v = R$ (Sec. 19-5), so $C_p = C_v + R$. Therefore,

$$C_p = \frac{3}{2}R + R = \frac{5}{2}R \qquad \text{and} \qquad \gamma = \frac{C_p}{C_v} = \frac{\tfrac{5}{2}R}{\tfrac{3}{2}R} = \frac{5}{3} = 1.67$$

Diatomic gases have *five* degrees of freedom: three associated with translation and two with rotation (see Fig. 18-3). Thus $E_{k\,tot} = 5 \times N \times \tfrac{1}{2}kT = \tfrac{5}{2}NkT = \tfrac{5}{2}nRT$ and $C_v = \tfrac{5}{2}R$, therefore

$$C_p = C_v + R = \tfrac{7}{2}R$$

Triatomic and *polyatomic* gases have *six* degrees of freedom: three associated with translation and three with rotation about all three axes. Thus $E_{k\,tot} = N \times 6 \times \tfrac{1}{2}kT = \tfrac{6}{2}NkT = \tfrac{6}{2}nRT$ and $C_v = \tfrac{6}{2}R = 3R$, therefore

$$C_p = C_v + R = 4R$$

Yet another mode of motion is possible for molecules of two or more atoms: vibration. At room temperature, however, gas molecules do not possess any energy in this mode. But at very high temperatures, vibration becomes important and, in the case of diatomic molecules, adds two more degrees of freedom, one each for kinetic and potential energy of rotation. At low temperatures, the rotational degrees of freedom drop out, so that polyatomic molecules can possess energy among the three components of translational motion only.

EXAMPLE 18-11: Compute C_p, C_v, and γ for a diatomic molecule with rotational degrees of freedom. Compare to values for nitrogen, oxygen, and carbon monoxide in Table 18-1.

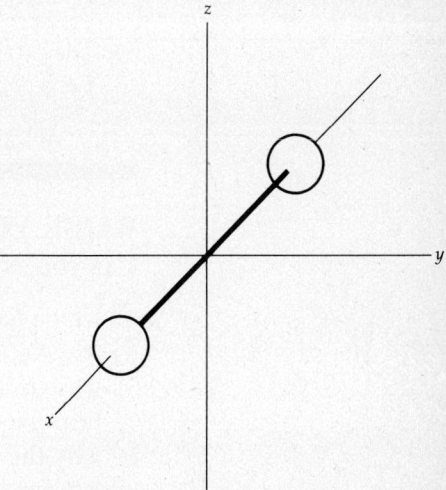

Figure 18-3. The diatomic molecule lies along the *x* axis. It is "free" to rotate only about the *y* and *z* axes in the sense that rotation about the *x* axis cannot be initiated by a collision with another molecule.

Solution: For a diatomic gas, neglecting vibrational modes,

$$C_p = \tfrac{7}{2}R = \tfrac{7}{2}(8.314\,\mathrm{J\,mol^{-1}\,K^{-1}}) = \boxed{29.1\,\mathrm{J\,mol^{-1}\,K^{-1}}}$$

$$C_v = \tfrac{5}{2}R = \tfrac{5}{2}(8.314\,\mathrm{J\,mol^{-1}\,K^{-1}}) = \boxed{20.8\,\mathrm{J\,mol^{-1}\,K^{-1}}}$$

$$\gamma = \frac{C_p}{C_v} = \frac{\tfrac{7}{2}R}{\tfrac{5}{2}R} = \frac{7}{5} = \boxed{1.40}$$

SUMMARY

1. The *equation of state of an ideal gas*, or *ideal gas law*, has two common forms

$$PV = nRT \qquad \text{or} \qquad PV = NkT$$

where P and T are absolute pressure and absolute temperature, respectively, n is the number of moles of the gas, R is the universal gas constant, N is the number of molecules, and k is Boltzmann's constant.

2. $R = N_A k$, where N_A is *Avogadro's number.*

3. The *total kinetic energy* of a gas consisting of N molecules, also called the *internal energy* of the gas, is

$$E_{k\,tot} = N(\tfrac{1}{2}m_m v^2_{rms}) = \tfrac{3}{2}PV = \tfrac{3}{2}NkT$$

where m_m is the mass of a molecule and v_{rms} is the root-mean-square speed of a molecule:

$$v_{rms} = \sqrt{\frac{3kT}{m_m}}$$

4. The density of an ideal gas that has a molecular mass M (also called the molecular weight) is

$$\rho = \frac{PM}{RT}$$

5. The relationships between the quantity of heat transferred to a gas and its corresponding rise in temperature are

$$\text{volume constant:} \qquad Q = nC_v \Delta T$$

$$\text{pressure constant:} \qquad Q = nC_p \Delta T$$

where C_v and C_p are the *molar specific heat capacities* at constant volume and constant pressure, respectively.

6. The ratio of the molar specific heats is $\gamma = C_p/C_v$

7. Each *degree of freedom* of a molecule has an associated kinetic energy of $\tfrac{1}{2}kT$.

RAISE YOUR GRADES

Can you explain . . . ?

☑ why a fixed volume of gas has a higher pressure at an elevated temperature
☑ why an isolated molecule does not have a temperature
☑ why decreasing the volume of a gas causes its pressure to rise if the temperature is held fixed
☑ why the ideal gas law is invalid for real gases at very low temperatures
☑ what happens to molecular motion at the absolute zero of temperature
☑ how to determine experimentally the Celsius value of absolute zero by measuring the pressure of a fixed volume of gas at several temperatures
☑ why C_p is greater than C_v
☑ why $C_v = \tfrac{3}{2}R$ for an ideal gas

SOLVED PROBLEMS

Equation of State

PROBLEM 18-1: What is the mass of helium (assume it to be an ideal gas) contained in a tank that has a volume of 0.1 m^3 and a gauge pressure of 2000 psi at room temperature, 22°C?

Solution: First convert the temperature to kelvins and the pressure to pascals. Then use the ideal gas law, Eq. (18-3b), to find the number of moles; finally, compute the mass.

$$T = 22 + 273 = 275 \text{ K}$$

The *gauge* pressure is:

$$P = 2000 \text{ lbs in.}^{-2} \times \frac{1 \text{ atm}}{14.7 \text{ lb in.}^{-2}} \times \frac{1.013 \times 10^5 \text{ Pa}}{\text{atm}} = 1.378 \times 10^7 \text{ Pa}$$

The *absolute* pressure is $1.378 \times 10^7 \text{ Pa} + 1.013 \times 10^5 \text{ Pa} = 1.388 \times 10^7 \text{ Pa}$

$$PV = nRT \qquad n = \frac{PV}{RT} = \frac{(1.388 \times 10^7 \text{ Pa})(0.1 \text{ m}^3)}{(8.314 \text{ J mol}^{-1} \text{ K}^{-1})(293 \text{ K})} = 570 \text{ mol}$$

The molecular weight of helium is $M = 4 \text{ g mol}^{-1}$. Therefore

$$m = nM = (570 \text{ mol})(4 \text{ g mol}^{-1}) = \boxed{2.28 \times 10^3 \text{ g} = 2.28 \text{ kg}}$$

PROBLEM 18-2: In distant space the temperature is about 3 K and there are about five hydrogen molecules in a cubic centimeter. What pressure does this correspond to?

Solution: Use the ideal gas law, Eq. (18-3c), and solve for *P*. Remember, $1 \text{ cm}^3 = 10^{-6} \text{ m}^3$.

$$PV = NkT$$

$$P = \frac{NkT}{V} = \frac{5(1.38 \times 10^{-23} \text{ J K}^{-1})(3 \text{ K})}{10^{-6} \text{ m}^3} = \boxed{2.07 \times 10^{-16} \text{ Pa}}$$

PROBLEM 18-3: A cylinder like that shown in Fig. 18-1 has an initial volume of 0.3 m^3 and contains air at an absolute pressure of 2 atm. Assume the air behaves like an ideal gas as it is heated from 22°C to 260°C while the volume increases to 1.2 m^3. What is the final pressure?

Solution: Use the ideal gas law in the form of Eq. (18-3a) and solve for P_2, the final pressure.

$$\frac{P_1 V_1}{T_1} = \frac{P_2 V_2}{T_2} = \text{constant}$$

$$P_2 = P_1 \left(\frac{T_2}{T_1}\right)\left(\frac{V_1}{V_2}\right) = (2 \text{ atm})\left(\frac{(260 + 273) \text{ K}}{(22 + 273) \text{ K}}\right)\left(\frac{0.3 \text{ m}^3}{1.2 \text{ m}^3}\right) = \boxed{0.903 \text{ atm}}$$

PROBLEM 18-4: The gauge pressure in an automobile tire is 40 psi. What volume of air at atmospheric pressure must be released to reduce the pressure to 32 psi? Assume that the volume of the tire remains constant at 0.05 m^3 and that the temperature is 22°C.

Solution: First change the 40 psi = 40 lb in^{-2} to SI units. The gauge pressure is

$$P = 40 \text{ lb in.}^{-2} \times \frac{1 \text{ atm}}{14.7 \text{ lb in.}^{-2}} \times \frac{1.013 \times 10^5 \text{ Pa}}{\text{atm}} = 2.756 \times 10^5 \text{ Pa}$$

The *absolute* pressure therefore is

$$P = 2.756 \times 10^5 \text{ Pa} + 1.013 \times 10^5 \text{ Pa} = 3.769 \times 10^5 \text{ Pa}$$

Then use the ideal gas law, Eq. (18-3b), to calculate the initial number of moles of air in the tire.

$$n_1 = \frac{P_1 V_1}{RT_1} = \frac{(3.769 \times 10^5 \text{ Pa})(0.05 \text{ m}^3)}{(8.314 \text{ J mol}^{-1} \text{ K}^{-1})(22 + 273) \text{ K}} = 7.68 \text{ mol}$$

Because V and T are constant we can write

$$\frac{P_1}{n_1} = \frac{P_2}{n_2} = \frac{RT}{V} = \text{constant}$$

Now solve for n_2, the number of moles left in the tire at the final pressure

$$n_2 = n_1\left(\frac{P_2}{P_1}\right) = (7.68 \text{ mol})\left(\frac{(32 + 14.7) \text{ lb in.}^{-2}}{(40 + 14.7) \text{ lb in.}^{-2}}\right) = 6.56 \text{ mol}$$

Notice that you must use absolute pressures. The pressures can be left in British units because only the ratio of absolute pressures is needed. The number of moles that must be released is therefore $(7.68 - 6.56) \text{ mol} = 1.12 \text{ mol}$. Use the ideal gas law to compute the volume of this much gas at atmospheric pressure.

$$V = \frac{nRT}{P}$$

$$= \frac{(1.12 \text{ mol})(8.314 \text{ J mol}^{-1} \text{ K}^{-1})(295 \text{ K})}{1.013 \times 10^5 \text{ Pa}} = \boxed{2.71 \times 10^{-2} \text{ m}^3}$$

Pressure of an Ideal Gas

PROBLEM 18-5: A laboratory vacuum system is pumped to a pressure of 0.2 torr. If the remaining gas in the system is air at 22°C, find the number of molecules in a cubic centimeter.

Solution: Use Eq. (18-3c) and solve for $\mathcal{N} = N/V$, but first convert the pressure to SI units.

$$1 \text{ torr} = 1 \text{ mm of mercury} = \frac{1}{760} \text{ atm} = \frac{1.013 \times 10^5 \text{ Pa}}{760} = 1.333 \times 10^2 \text{ Pa}$$

$$P = 0.2 \text{ torr} = (0.2 \text{ torr})(1.333 \times 10^2 \text{ Pa torr}^{-1}) = 26.66 \text{ Pa}.$$

$$PV = NkT$$

$$\mathcal{N} = \frac{N}{V} = \frac{P}{kT} = \frac{26.66 \text{ Pa}}{(1.38 \times 10^{-23} \text{ J K}^{-1})(22 + 273) \text{ K}}$$

$$= \boxed{6.55 \times 10^{21} \text{ molecules m}^{-3} = 6.55 \times 10^{15} \text{ molecules cm}^{-3}}$$

PROBLEM 18-6: In Problem 11-9 you calculated the escape velocity from the earth's surface to be $1.12 \times 10^4 \text{ m s}^{-1}$. Find the Celsius temperature at which a helium atom would have an rms speed equal to this escape velocity.

Solution: First calculate the mass of a helium atom. Helium has a molecular mass of 4 g mol^{-1}. Therefore

$$m_m = \frac{M}{N_A} = \frac{4 \text{ g mol}^{-1}}{6.022 \times 10^{23} \text{ mol}^{-1}} = \frac{4 \times 10^{-3} \text{ kg}}{6.022 \times 10^{23}} = 6.64 \times 10^{-27} \text{ kg}$$

Now solve Eq. (18-8) for T and convert it to Celsius degrees.

$$\tfrac{1}{2}m_m v_{rms}^2 = \tfrac{3}{2}kT$$

$$T = \frac{m_m v_{rms}^2}{3k} = \frac{(6.64 \times 10^{-27} \text{ kg})(1.12 \times 10^4 \text{ m s}^{-1})^2}{3(1.38 \times 10^{-23} \text{ J K}^{-1})} = 2.01 \times 10^4 \text{ K}$$

The Celsius temperature is: $2.01 \times 10^4 - 273 = \boxed{1.98 \times 10^4 \text{ °C}}$

Density of an Ideal Gas

PROBLEM 18-7: (a) Determine the volume of helium needed to support a balloon and its load, total mass 100 kg, in air at 22°C and a pressure of one atmosphere. Ignore the mass of helium. (b) Calculate the mass of helium needed.

Solution:

(a) Use Archimedes' principle. In this case the buoyant force is equal to the weight of the load $w = mg$.

$$F_B = w \qquad \rho_\ell V g = mg$$

In this case the "liquid" is air and its density is, from Eq. (18-10), $\rho_a = PM/RT = 1.20 \text{ kg m}^{-3}$ (from Example 18-6). Therefore

$$V = \frac{m}{\rho_a} = \frac{100 \text{ kg}}{1.20 \text{ kg m}^{-3}} = \boxed{83.3 \text{ m}^3}$$

(b) Use the ideal gas law and find the number of moles, then use $m = nM$ to get the mass of helium.

$$PV = nRT$$

$$n = \frac{PV}{RT} = \frac{(1.013 \times 10^5 \text{ Pa})(83.3 \text{ m}^3)}{(8.314 \text{ J mol}^{-1} \text{ K}^{-1})(22 + 273) \text{ K}} = 3.44 \times 10^3 \text{ mol}$$

$$m = nM = (3.44 \times 10^3 \text{ mol})(4 \text{ g mol}^{-1}) = 1.38 \times 10^4 \text{ g} = \boxed{13.8 \text{ kg}}$$

PROBLEM 18-8: The specific gravity of liquid nitrogen at its boiling temperature of $-196°C$ at atmospheric pressure is 0.81. Assuming that nitrogen obeys the ideal gas law at temperatures this low, calculate the pressure necessary to produce this density.

Solution: Use Eq. (18-10) and solve for P. The molecular mass of nitrogen is $2 \times 14 = 28 \text{ g mol}^{-1} = 28 \times 10^{-3} \text{ kg mol}^{-1}$.

$$\rho = \frac{PM}{RT}$$

$$P = \frac{\rho RT}{M} = \frac{(0.81 \times 10^3 \text{ kg m}^{-3})(8.314 \text{ J mol}^{-1} \text{K}^{-1})(-196 + 273) \text{ K}}{28 \times 10^{-3} \text{ kg mol}^{-1}}$$

$$= \boxed{1.85 \times 10^7 \text{ Pa} = 183 \text{ atm}}$$

Specific Heats of Gases

PROBLEM 18-9: Two moles of a gas in a tank of fixed volume is heated from 20°C to 30°C by the transfer of 400 J of heat into the gas. Is the gas monatomic, diatomic, triatomic, or ...?

Solution: Use Eq. (18-11) and solve for C_v. Compare this value with values in Table 18-1 and make a judgment.

$$Q = nC_v \Delta T$$

$$C_v = \frac{Q}{n \Delta T} = \frac{400 \text{ J}}{(2 \text{ mol})(30 - 20) \text{ K}} = 20 \text{ J mol}^{-1} \text{K}^{-1}$$

Inspection of Table 18-1 indicates that diatomic gases have specific heats at constant volume of 20 to 21 $\text{J mol}^{-1} \text{K}^{-1}$. Therefore this gas is most likely diatomic.

PROBLEM 18-10: Find the temperature rise that would result from the transfer of 400 J of heat into the 2 moles of gas of Problem 18-9 if the gas were contained in a cylinder equipped with a piston that moved so as to keep the pressure constant.

Solution: Use Eq. (18-12) and solve for ΔT. From Table 18-1, $\gamma = 1.40$ for diatomic gases. According to Eq. (18-13), $\gamma = C_p/C_v$, thus $C_p = \gamma C_v$.

$$Q = nC_p\Delta T$$

$$\Delta T = \frac{Q}{nC_p} = \frac{Q}{n\gamma C_v} = \frac{400 \text{ J}}{(2 \text{ mol})(1.40)(20 \text{ J mol}^{-1}\text{K}^{-1})} = \boxed{7.14 \text{ K}}$$

Notice that this could also be written as $\Delta T = 7.14 \text{ C}°$. If the initial temperature was 20°C the final temperature would be 27.14°C.

Equipartition of Energy

PROBLEM 18-11: Compute C_p, C_v, and γ for a triatomic or polyatomic molecule with rotational degrees of freedom. Compare to values for carbon dioxide and hydrogen sulfide in Table 18-1.

Solution: For a tri- or polyatomic gas, neglecting vibrational modes,

$$C_p = 4R = 4(8.314 \text{ J mol}^{-1}\text{K}^{-1}) = \boxed{33.3 \text{ J mol}^{-1}\text{K}^{-1}}$$

$$C_v = 3R = 3(8.314 \text{ J mol}^{-1}\text{K}^{-1}) = \boxed{24.9 \text{ J mol}^{-1}\text{K}^{-1}}$$

$$\gamma = \frac{C_p}{C_v} = \frac{4R}{3R} = \boxed{1.33}$$

The concordance between this simple theory and experimental results is not too bad but not as good as for diatomic gases. Evidently a more detailed theory is necessary to explain more accurately the thermal properties of gases composed of complex molecules.

Supplementary Exercises

EXERCISE 18-1: A tank contains 12 mol of argon, 20 mol of nitrogen, and 30 mol of carbon dioxide at a temperature of 22°C. The volume of the tank is 0.4 m³. What is the absolute pressure of these gases?

EXERCISE 18-2: A 0.3-m³ tank contains air at a gauge pressure of 3 atm and temperature of 22°C. How many moles of air must be added to raise the gauge pressure to 4 atm at this temperature?

EXERCISE 18-3: A 0.2-m³ tank contains 800 g of methane, CH_4, at 32°C. What is the absolute pressure of this gas?

EXERCISE 18-4: A cylinder like that of Fig. 18-1 contains 4 ft³ of argon at an absolute pressure of 80 psi and a temperature of 75°F. What is the absolute pressure if the argon is compressed to a volume of 0.8 ft³ at a temperature of 180°F?

EXERCISE 18-5: At a temperature of 40°C iodine has a vapor pressure of 1 mm of mercury. How many iodine atoms are there in a cubic centimeter under these conditions?

EXERCISE 18-6: Start with Eq. (18-7) and show that $P = \frac{1}{3}\rho v_{\text{rms}}^2$

EXERCISE 18-7: The volume of a tank containing 2 kg of a gas is 2.4 m³. The pressure in the tank is 1.3 atm. What is the rms speed of molecules of this gas?

EXERCISE 18-8: What is the total kinetic energy of 3 mol of an ideal gas at a temperature of 200°C?

EXERCISE 18-9: How much heat must be added to a tank containing 6 mol of hydrogen sulfide if its temperature is to rise from 20°C to 32°C?

EXERCISE 18-10: A cylinder like that shown in Fig. 18-1 contains 4 mol of carbon dioxide. If the initial temperature is 22°C, compute the amount of heat that must be added to keep the pressure constant when the volume is doubled.

Answers to Supplementary Exercises

18-1: 3.80×10^5 Pa $= 3.75$ atm

18-2: 12.4 mol

18-3: 6.34×10^5 Pa $= 6.26$ atm

18-4: 479 psi $= 32.6$ atm

18-5: 3.08×10^{16} atoms cm^{-3}

18-7: 689 m s^{-1}

18-8: 1.77×10^4 J

18-9: 1.87×10^3 J

18-10: 435×10^4 J

19 THE FIRST LAW OF THERMODYNAMICS

THIS CHAPTER IS ABOUT

☑ **Work and Internal Energy**

☑ **The First Law of Thermodynamics**

☑ **Isothermal Processes**

☑ **Isovolumic Processes**

☑ **Isobaric Processes**

☑ **Adiabatic Processes**

19-1. Work and Internal Energy

Thermodynamics is a branch of physics and engineering that deals mainly with the conversion of thermal energy into useful work. To see the application of thermodynamic principles, think of a monatomic ideal gas confined to a cylinder equipped with a movable frictionless piston, as shown in Figure 19-1. The cylinder is thermally insulated except in a small region where heat may be transferred. If the gas expands and the piston moves to the right, work is done *by the gas*. If the piston moves to the left, compressing the gas, work is done *on the gas*. Work done by the gas is positive; work done on the gas is negative. The work done by the gas in forcing the piston to the right a distance dx is $dW = F\,dx = PA\,dx = P\,dV$. The total work done in expanding from an initial volume V_1 to

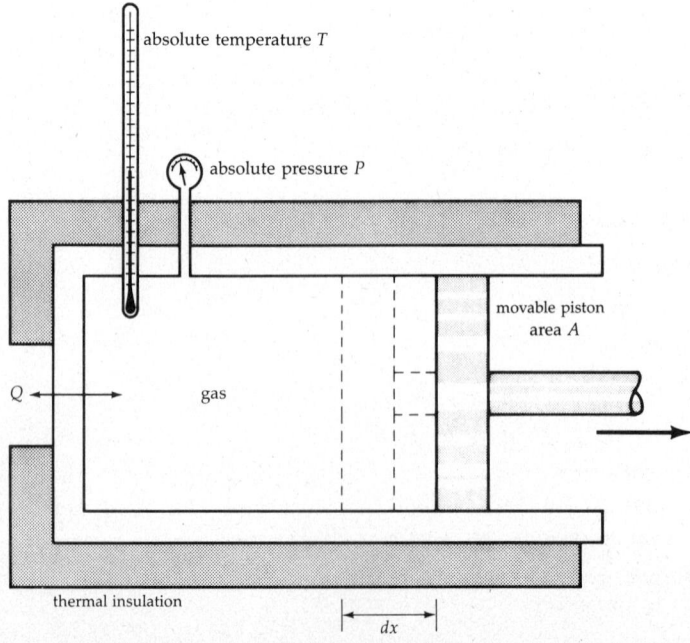

Figure 19-1. Thermally insulated cylinder with a movable piston.

236

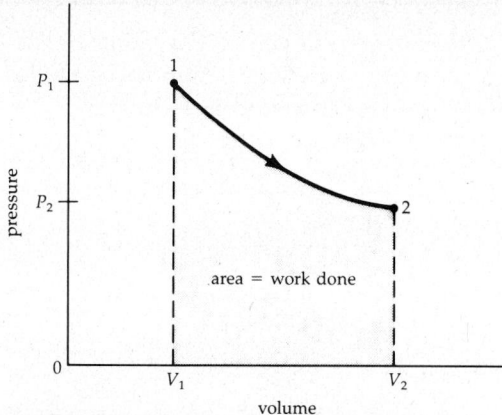

Figure 19-2. The work done by the gas in expanding from V_1 to V_2 is equal to the area under the curve on a *PV* diagram.

a final volume V_2 is

WORK DONE BY AN IDEAL GAS

$$W = \int_{V_1}^{V_2} P \, dV \qquad (19\text{-}1)$$

This integral is represented by the area under the curve on a pressure–volume diagram, as shown in Figure 19-2. Clearly, if there is no change in volume during a process, no work is done either by or on the gas.

The *internal energy* of a monatomic ideal gas, its total kinetic energy, was defined in Eq. (18-7) as

$$E_{k\,tot} = N(\tfrac{1}{2}m_m v_{rms}^2) = \tfrac{3}{2}PV = \tfrac{3}{2}NkT = \tfrac{3}{2}nRT$$

A change in the internal energy is thus

$$\Delta E_{k\,tot} = \tfrac{3}{2}\Delta(PV) \qquad (19\text{-}2a)$$

or

$$\Delta E_{k\,tot} = \tfrac{3}{2}Nk\,\Delta T \qquad (19\text{-}2b)$$

or

$$\Delta E_{k\,tot} = \tfrac{3}{2}nR\,\Delta T \qquad (19\text{-}2c)$$

19-2. The First Law of Thermodynamics

The **first law of thermodynamics** is a statement of the law of conservation of energy. It is

- When heat (thermal energy) is transferred to a system, the internal energy of the system increases and/or the system does work.

We can express this as

FIRST LAW OF THERMODYNAMICS

$$Q = \Delta E_{k\,tot} + W \qquad (19\text{-}3)$$

19-3. Isothermal Processes

During an **isothermal process**, the temperature of the system remains constant: $\Delta T = 0$. Equations (19-2b) and (19-2c) indicate that if $\Delta T = 0$ there is no change in the internal energy; therefore Eq. (19-3) becomes

FIRST LAW FOR AN ISOTHERMAL PROCESS

$$Q = W \qquad (19\text{-}4)$$

The system does an amount of work exactly equal to the amount of heat transferred to it.

EXAMPLE 19-1: Calculate the amount of work done by the system shown in Figure 19-1 when a quantity Q of heat is transferred to the gas at constant temperature T and the volume expands from V_1 to V_2.

Solution: Solve the ideal gas law, $PV = nRT$, for P and substitute into Eq. (19-1). In an isothermal process, T is a constant, so it may be factored out of the integrand with the other constants.

$$P = \frac{nRT}{V}$$

$$W = Q = \int_{V_1}^{V_2} P\,dV = \int_{V_1}^{V_2} \frac{nRT}{V}\,dV = nRT \int_{V_1}^{V_2} \frac{dV}{V}$$

WORK DONE DURING AN ISOTHERMAL PROCESS $= \boxed{nRT\ln\left(\frac{V_2}{V_1}\right)}$

Notice that this result can be written also as $W = Q = NkT\ln(V_2/V_1)$.

EXAMPLE 19-2: Heat is transferred to 0.12 mol of an ideal gas in a cylinder as the piston moves, increasing the volume from 2.25×10^{-4} m³ to 2.5×10^{-3} m³. The temperature remains fixed at 240°C. (a) How much work is done by the gas? (b) How much heat is transferred?

Solution:

(a) Use the result of Example (19-1).

$$W = nRT\ln\left(\frac{V_2}{V_1}\right)$$

$$= (0.12 \text{ mol})(8.314 \text{ J mol}^{-1}\text{ K}^{-1})(240 + 273) \text{ K } \ln\left(\frac{2.5 \times 10^{-3} \text{ m}^3}{2.25 \times 10^{-4} \text{ m}^3}\right)$$

$$= \boxed{1.23 \times 10^3 \text{ J}}$$

(b) Because this is an isothermal process, $Q = W$ (Eq. 19-4). Thus the amount of heat added is equal to the work done by the gas, $Q = 1.23 \times 10^3$ J.

19-4. Isovolumic Processes

During an **isovolumic process**, the volume of the gas remains constant: $\Delta V = 0$. In some books this process is referred to as an **isochoric process**. Because the volume does not change, the piston remains fixed, and no work can be done by or on the gas. $W = 0$, so in this case Eq. (19-3) reduces to

FIRST LAW FOR AN ISOVOLUMIC PROCESS $Q = \Delta E_{k\,tot}$ (19-5)

All the heat transferred to the system goes into increasing the system's internal energy. Consequently the temperature increases according to Eq. (18-11), $Q = nC_v\Delta T$. Thus

$$nC_v\Delta T = \Delta E_{k\,tot} \tag{19-6}$$

Conversely, heat transferred out of the system as the volume is held fixed reduces the temperature and internal energy. Comparison of Eq. (19-6) with Eq. (19-2c) confirms the fact that for a monatomic ideal gas $C_v = \frac{3}{2}R$.

19-5. Isobaric Processes

During an **isobaric process**, the pressure of the gas remains constant: $\Delta P = 0$. When heat is transferred to the gas in an isobaric process, the piston must move out, increasing the volume to keep the pressure constant. The expanding gas does work, according to Eq. (19-1):

WORK DONE DURING AN ISOBARIC PROCESS
$$W = \int_{V_1}^{V_2} P\,dV = P\int_{V_1}^{V_2} dV \tag{19-7}$$
$$= P(V_2 - V_1) = P\,\Delta V$$

Because this process is not isothermal, there is a rise in temperature, i.e., a rise in the internal energy. If we substitute $nC_p\,\Delta T$ for Q (Eq. 18-12), $nC_v\,\Delta T$ for $E_{k\,tot}$ (Eq. 19-6), and $P\,\Delta V$ for W (Eq. 19-7), Eq. (19-3) becomes

FIRST LAW FOR AN ISOBARIC PROCESS
$$nC_p\,\Delta T = nC_v\,\Delta T + P\,\Delta V \tag{19-8}$$

We can rewrite this as

$$P\,\Delta V = n(C_p - C_v)\,\Delta T$$

From the ideal gas law, $P\,\Delta V = nR\,\Delta T$, so we can conclude that for an ideal gas $C_p - C_v = R$.

19-6. Adiabatic Processes

During an **adiabatic process**, no heat is transferred into or out of the system: $Q = 0$. Equation (19-3) then becomes

FIRST LAW FOR AN ADIABATIC PROCESS
$$0 = \Delta E_{k\,tot} + W \tag{19-9}$$

Work done by the system results in an equal decrease in its internal energy.

For an adiabatic process of an ideal gas it is also true that

$$PV^\gamma = \text{constant} \tag{19-10a}$$

$$TV^{\gamma-1} = \text{constant} \tag{19-10b}$$

where $\gamma = C_p/C_v$.

EXAMPLE 19-3: Derive Eq. (19-10a) from Eq. (19-9) for an ideal gas.

Solution: An infinitesimal change in the work done by the gas is $dW = P\,dV$ (Eq. 19-1) and results in an infinitesimal drop in the internal energy of $dE_{k\,tot} = nC_v\,dT$ (Eq. 19-6). Therefore Eq. (19-9) can be written in differential form as

$$0 = nC_v\,dT + P\,dV \qquad dT = -\frac{P\,dV}{nC_v}$$

The differential form of the ideal gas law is

$$d(PV) = d(nRT)$$

$$P\,dV + V\,dP = nR\,dT$$

Substitute dT from above and collect like terms.

$$P\,dV + V\,dP = nR\left(-\frac{P\,dV}{nC_v}\right) = -\frac{R}{C_v}P\,dV$$

$$\left(\frac{R}{C_v} + 1\right)P\,dV + V\,dP = 0$$

Because $R = C_p - C_v$, $\left(\dfrac{R}{C_v} + 1\right) = \dfrac{C_p}{C_v} = \gamma$. Divide by PV:

$$\gamma \frac{dV}{V} + \frac{dP}{P} = 0$$

Integrate to get

$$\gamma \int \frac{dV}{V} + \int \frac{dP}{P} = 0$$

$\gamma \ln V + \ln P = \text{constant}$, or $\boxed{PV^\gamma = \text{constant}}$

EXAMPLE 19-4: In a certain engine the compression ratio is 12 to 1, which means that after the compression stroke the volume of gas is $\frac{1}{12}$ of its original volume. Assume the initial fuel–air mixture is at a temperature of 35°C, a pressure of 1 atm, and that $\gamma = 1.40$. Assume also that the compression is adiabatic. Compute the (**a**) final pressure and (**b**) temperature of the mixture.

Solution:

(**a**) Use Eq. (19-10a) and solve for P_2.

$$P_1 V_1^\gamma = P_2 V_2^\gamma$$

$$P_2 = P_1 \left(\frac{V_1}{V_2}\right)^\gamma = (1 \text{ atm})\left(\frac{V_1}{\frac{1}{12}V_1}\right)^{1.4} = (1 \text{ atm})(12)^{1.4} = \boxed{32.4 \text{ atm}}$$

Notice that this result can also be written as

$$P_2 = (32.4 \text{ atm})(1.013 \times 10^5 \text{ Pa atm}^{-1}) = 3.28 \times 10^6 \text{ Pa}$$

(**b**) Use Eq. (19-10b) and solve for T_2. Temperatures must be in absolute degrees.

$$T_1 V_1^{\gamma - 1} = T_2 V_2^{\gamma - 1}$$

$$T_2 = T_1 \left(\frac{V_1}{V_2}\right)^{\gamma - 1} = (35 + 273)\,\text{K} \times \left(\frac{V_1}{\frac{1}{12}V_1}\right)^{1.4 - 1} = (308 \text{ K})(12)^{0.4}$$

$$= \boxed{832 \text{ K} = 559°\text{C}}$$

Notice that this temperature may be sufficiently high to ignite the fuel–air mixture and cause the engine to "diesel."

SUMMARY

1. The work done by an expanding gas, as in moving a piston, for example, is

$$W = \int_{V_1}^{V_2} P \, dV$$

Work done *by* the system is positive, work done *on* the system is negative.

2. The *internal energy* of a monatomic ideal gas can be written as

$$E_{k\,\text{tot}} = \tfrac{3}{2}PV = \tfrac{3}{2}NkT = \tfrac{3}{2}nRT$$

3. The *first law of thermodynamics* states that thermal energy transferred to a system increases its internal energy and/or results in work done by the system:

$$Q = \Delta E_{k\,\text{tot}} + W$$

4. The first law for an *isothermal process* becomes $Q = W$. The work done by a system during an isothermal process is

$$W = nRT \ln\left(\frac{V_2}{V_1}\right)$$

5. The first law for an *isovolumic process* becomes $Q = \Delta E_{k\,tot} = nC_v\,\Delta T$. During an isovolumic process no work is done by or on the system.
6. The first law for an *isobaric process* becomes $nC_p\,\Delta T = nC_v\,\Delta T + P\,\Delta V$. The work done by a system during an isobaric process is $W = P\,\Delta V$.
7. The first law for an *adiabatic process* becomes $0 = \Delta E_{k\,tot} + W$. During an adiabatic process on an ideal gas, $PV^\gamma = $ constant and $TV^{\gamma-1} = $ constant, where $\gamma = C_p/C_v$.

RAISE YOUR GRADES

Can you explain . . . ?

☑ why the internal energy of an ideal gas depends only on temperature
☑ the difference between heat and temperature
☑ why the temperature of adiabatically compressed gas is greater than before the gas was compressed
☑ why C_v increases with temperature for diatomic molecules
☑ why C_p is greater than C_v for any gas
☑ the nature of the kinetic energy associated with the molecules of a solid substance at temperature T
☑ why it is meaningless to consider temperatures below absolute zero
☑ why it is not possible to convert all of the internal energy of a real gas into work

SOLVED PROBLEMS

The following problems and exercises are not grouped by chapter section but instead integrate the concepts presented in all sections. Problems 19-1 to 19-7 deal with the system described in Problem 19-1.

PROBLEM 19-1: There is 0.15 mol of a monatomic ideal gas contained in a cylinder with a movable piston like that shown in Figure 19-1. The gas undergoes the cycle shown in the pressure–volume diagram in Figure 19-3. The pressures and volumes are $P_1 = 2 \times 10^5$ Pa, $P_2 = 5 \times 10^5$ Pa, $V_1 = 0.2$ L $= 0.2 \times 10^{-3}$ m^3, and $V_2 = 2$ L $= 2 \times 10^{-3}$ m^3. **(a)** Identify process A. **(b)** How much work is done by the gas? **(c)** What is the change in temperature of the gas?

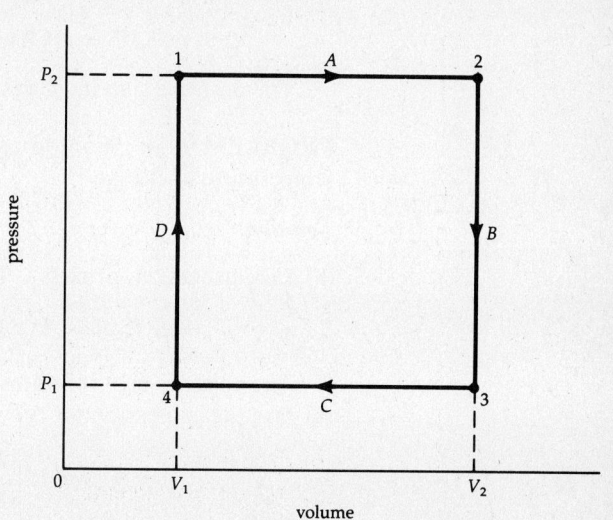

Figure 19-3

Solution:

(a) Process A is an isobaric (constant pressure) expansion from V_1 to V_2.

(b) For an isobaric process, the work done by the gas is

$$W = \int_{V_1}^{V_2} P\,dV = P\,\Delta V = P(V_2 - V_1)$$

$$W_A = P_2(V_2 - V_1)$$
$$= (5 \times 10^5 \text{ Pa})(2 \times 10^{-3} - 0.2 \times 10^{-3})\text{ m}^3$$
$$= \boxed{900 \text{ J}}$$

(c) Use the first law in the form of Eq. (19-8) and solve for ΔT.

$$nC_p \Delta T = nC_v \Delta T + P \Delta V$$

$$\Delta T = \frac{P \Delta V}{n(C_p - C_v)} = \frac{W}{nR} = \frac{900 \text{ J}}{(0.15 \text{ mol})(8.314 \text{ J mol}^{-1} \text{K}^{-1})} = \boxed{722 \text{ K} = 449 \text{ C}^\circ}$$

PROBLEM 19-2: (a) What are the initial and final temperatures of the gas in Problem 19-1? (b) How much heat was transferred during process A?

Solution:

(a) Use the ideal gas law and solve for T_1 and T_2.

$$PV = nRT$$

$$T_1 = \frac{P_2 V_1}{nR} = \frac{(5 \times 10^5 \text{ Pa})(0.2 \times 10^{-3} \text{ m}^3)}{(0.15 \text{ mol})(8.314 \text{ J mol}^{-1} \text{K}^{-1})} = \boxed{80.2 \text{ K} = 193^\circ\text{C}}$$

$$T_2 = \frac{P_2 V_2}{nR} = \frac{(5 \times 10^5 \text{ Pa})(2 \times 10^{-3} \text{ m}^3)}{(0.15 \text{ mol})(8.314 \text{ J mol}^{-1} \text{K}^{-1})} = \boxed{802 \text{ K} = 529^\circ\text{C}}$$

Notice that $T_2 = T_1 + \Delta T = 80.2 \text{ K} + 722 \text{ K} = 802 \text{ K}$.

(b) For an isobaric process, $Q = nC_p \Delta T$ (Eq. 18-12), and $C_p = \frac{5}{2}R$ for a monatomic ideal gas.

$$Q = nC_p \Delta T = n(\tfrac{5}{2}R)\Delta T = (0.15 \text{ mol})(\tfrac{5}{2})(8.314 \text{ J mol}^{-1} \text{K}^{-1})(722 \text{ K}) = \boxed{2.25 \times 10^3 \text{ J}}$$

PROBLEM 19-3: (a) Identify process B. How much work is done during process B? (b) Find the change in temperature of the gas. (c) Determine the quantity of heat transferred.

Solution:

(a) Process B is isovolumic. No work is done either on or by the gas.
(b) Use the ideal gas law to determine the temperature at point 3. You found the temperature at point 2 in Problem 19-2a.

$$\frac{P_2 V_2}{T_2}\bigg|_{\text{point 2}} = \frac{P_1 V_2}{T_3}\bigg|_{\text{point 3}}$$

$$T_3 = T_2\left(\frac{P_1}{P_2}\right) = (802 \text{ K})\left(\frac{2 \times 10^5 \text{ Pa}}{5 \times 10^5 \text{ Pa}}\right) = 321 \text{ K} = 47.8^\circ\text{C}$$

$$\Delta T = T_3 - T_2 = 321 \text{ K} - 802 \text{ K} = \boxed{-481 \text{ K} = -481 \text{ C}^\circ}$$

(c) For an isovolumic process the quantity of heat transferred is $Q = nC_v \Delta T$ (Eqs. 19-5 and 19-6). For a monatomic ideal gas $C_v = \frac{3}{2}R$.

$$Q = nC_v \Delta T = n(\tfrac{3}{2}R)\Delta T = (0.15 \text{ mol})(\tfrac{3}{2})(8.314 \text{ J mol}^{-1} \text{K}^{-1})(-481 \text{ K}) = \boxed{-900 \text{ J}}$$

The negative sign indicates that this quantity of heat is transferred *out* of the system.

PROBLEM 19-4: For process C, (a) How much work is done? (b) What is the change in temperature of the gas?

Solution:

(a) Like process A, process C is isobaric. Use Eq. (19-7), $W = P \Delta V$.

$$W = P_1(V_1 - V_2) = (2 \times 10^5 \text{ Pa})(0.2 \times 10^{-3} \text{ m}^3 - 2 \times 10^{-3} \text{ m}^3) = \boxed{-360 \text{ J}}$$

The negative sign indicates that work is done *on* the gas in compressing it from V_2 to V_1.

(b) Use the expression for ΔT that you found in Problem 19-1c.

$$\Delta T = \frac{W}{nR} = \frac{-360 \text{ J}}{(0.15 \text{ mol})(8.314 \text{ J mol}^{-1} \text{K}^{-1})} = \boxed{-289 \text{ K}}$$

The negative sign indicates that the temperature decreases.

PROBLEM 19-5: **(a)** What is the final temperature for process C? **(b)** How much heat is transferred during process C?

Solution:

(a) Use the ideal gas law or use the results of Problems 19-4b and 19-3a.

$$T_4 = T_3 + \Delta T = 321 \text{ K} - 289 \text{ K} = \boxed{32 \text{ K} = -241°\text{C}}$$

(b) Because process C is isobaric $Q = nC_p \Delta T$ with $C_p = \frac{5}{2}R$.

$$Q = (0.15 \text{ mol})(\tfrac{5}{2})(8.314 \text{ J mol}^{-1}\text{K}^{-1})(-289 \text{ K}) = \boxed{-901 \text{ J}}$$

The negative sign indicates that this quantity of heat is transferred out of the system.

PROBLEM 19-6: For process D, **(a)** How much work is done? **(b)** How much heat is transferred?

Solution:

(a) Like process B, process D is isovolumic, therefore no work is done either by or on the gas.
(b) The quantity of heat transferred is $Q = nC_v \Delta T$ with $C_v = \frac{3}{2}R$. The change in temperature is

$$\Delta T = T_1 - T_4 = 80.2 \text{ K} - 32 \text{ K} = 48.2 \text{ K} = 48.2 \text{ C}°$$

Therefore

$$Q = n(\tfrac{3}{2}R)\Delta T = (0.15 \text{ mol})(\tfrac{3}{2})(8.314 \text{ J mol}^{-1}\text{K}^{-1})(48.2 \text{ K}) = \boxed{90.2 \text{ J}}$$

PROBLEM 19-7: **(a)** Find the net work done for the complete cycle. **(b)** Calculate the net heat transferred in for the complete cycle.

Solution:

(a) Work is done during processes A and C only, so

$$W_{net} = W_A + W_C = 900 \text{ J} - 360 \text{ J} = \boxed{540 \text{ J}}$$

Notice that the net work done per cycle is equal to the area inside the cycle on the pressure–volume diagram. This is true *in general*, regardless of the shape of the diagram. In this case

$$W_{net} = \Delta P \Delta V = (P_2 - P_1)(V_2 - V_1)$$
$$= (5 \times 10^5 - 2 \times 10^5)\text{Pa} \times (2 \times 10^{-3} - 0.2 \times 10^{-3})\text{m}^3 = 540 \text{ J}$$

(b) $\qquad Q_{net} = Q_A + Q_B + Q_C + Q_D = 2.25 \times 10^3 \text{ J} - 900 \text{ J} - 901 \text{ J} + 90.2 \text{ J} = \boxed{539 \text{ J}}$

Notice that the principle of the conservation of energy, the first law of thermodynamics, applies: $W_{net} = Q_{net}$ (to within round-off error). The net work done per cycle is equal to the net heat transferred into the system per cycle.

We can symbolize the heat transferred into or out of the system during each process by arrows pointing into or out of our diagram, as shown in Figure 19-4. In processes A and D heat is transferred into the system, but in B and C heat is transferred out. Notice also that if the cycle shown in Figures 19-3 and 19-4 were traversed in the opposite direction, from point 1 to 4 to 3 to 2 and back to 1, net work of 540 J would have been done *on* the system and a net quantity of heat of 540 J would have been transferred *out*.

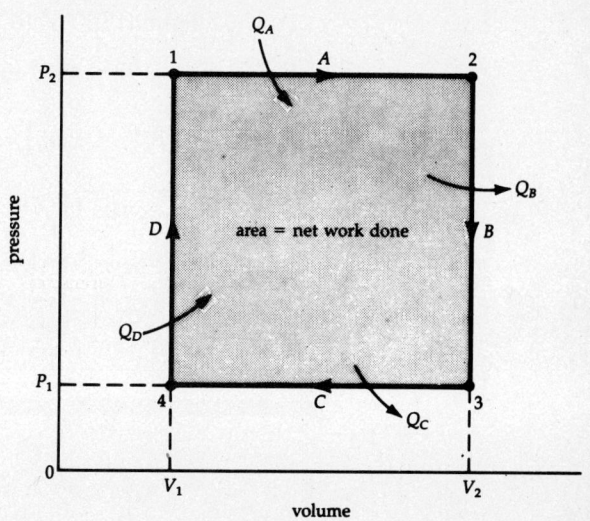

Figure 19-4

PROBLEM 19-8: How much work will the system indicated in Figure 19-3 do in proceeding isothermally from point 1 to volume $V_2 = 2 \times 10^{-3}$ m?

Solution: Recall from Problem 19-1 that $n = 0.15$ mol, $P_2 = 5 \times 10^5$ Pa, $V_1 = 0.2 \times 10^{-3}$ m^3, and $T_1 = 80.2$ K. From Example 19-1, the work done during an isothermal process is $W = nRT \ln(V_2/V_1)$, thus

$$W = (0.15 \text{ mol})(8.314 \text{ J mol}^{-1}\text{K}^{-1})(80.2 \text{ K}) \ln\left(\frac{2 \times 10^{-3} \text{ m}^3}{0.2 \times 10^{-3} \text{ m}^3}\right) = \boxed{230 \text{ J}}$$

Notice that the quantity of heat transferred into the system during this process is equal to W (Eq. 19-4): $Q = W = 230$ J. The temperature would remain at 80.2 K and the pressure at this point would be

$$P = \frac{nRT_1}{V_2} = \frac{(0.15 \text{ mol})(8.314 \text{ J mol}^{-1}\text{K}^{-1})(80.2 \text{ K})}{2 \times 10^{-3} \text{ m}^3} = 0.5 \times 10^5 \text{ Pa}$$

which is considerably less than $P_1 = 2 \times 10^5$ Pa, the pressure at point 3 in Figure 19-3.

PROBLEM 19-9: Derive an expression for the work done by a gas during an adiabatic expansion.

Solution: Use Eq. (19-10a), $PV^\gamma = $ constant. Let the constant be B and the initial and the final states be 1 and 2, respectively. Thus $P_1 V_1^\gamma = P_2 V_2^\gamma = B$. Equation (19-1) states that

$$W = \int_{V_1}^{V_2} P\,dV \qquad \text{where} \qquad P = \frac{B}{V^\gamma} \qquad \text{so}$$

$$W = B \int_{V_1}^{V_2} \frac{dV}{V^\gamma} = B\left(\frac{1}{1-\gamma} V^{1-\gamma}\bigg|_{V_1}^{V_2}\right)$$

$$= \frac{1}{1-\gamma}(P_2 V_2^\gamma V_2^{1-\gamma} - P_1 V_1^\gamma V_1^{1-\gamma})$$

WORK DONE DURING AN ADIABATIC PROCESS
$$= \boxed{\frac{1}{1-\gamma}(P_2 V_2 - P_1 V_1)}$$

PROBLEM 19-10: How much work is done *on* the fuel–air mixture in the adiabatic compression of Example 19-4 if the initial volume is $V_1 = 3.5 \times 10^{-3}$ m^3?

Solution: Use the results from Problem 19-9 and Example 19-4.

$$P_1 = 1 \text{ atm} = 1.013 \times 10^5 \text{ Pa} \qquad V_1 = 3.5 \times 10^{-3} \text{ m}^3 \qquad \gamma = 1.4$$

$$P_2 = (32.4 \text{ atm})(1.013 \times 10^5 \text{ Pa atm}^{-1}) = 3.28 \times 10^6 \text{ Pa}$$

$$V_2 = \tfrac{1}{12}V_1 = 2.92 \times 10^{-4} \text{ m}^3$$

$$W = \frac{1}{1-\gamma}(P_2 V_2 - P_1 V_1) = \frac{1}{1-1.4}[(3.28 \times 10^6)(2.92 \times 10^{-4}) - (1.013 \times 10^5)(3.5 \times 10^{-3})] \text{ J}$$

$$= \boxed{-1.51 \times 10^3 \text{ J}}$$

Notice that the negative sign reminds us that work is done *on* the gas during this compression.

Supplementary Exercises

Exercises 19-1 to 19-10 deal with the system shown in Figure 19-5. At point 1, $P_3 = 10 \times 10^5$ Pa, $V_1 = 2 \times 10^{-3}$ m^3, and $T_1 = 600$ K; $\gamma = \frac{5}{3}$, i.e., the gas is a monatomic ideal gas.

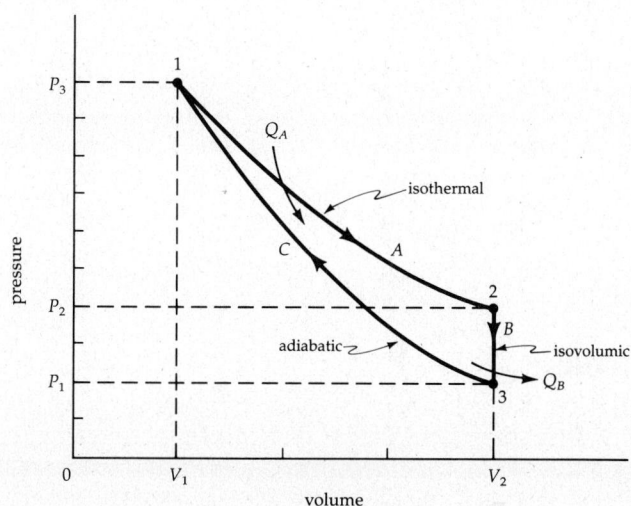

Figure 19-5

EXERCISE 19-1: How many moles of gas are contained in the system?

EXERCISE 19-2: Process A is an isothermal expansion to $V_2 = 5 \times 10^{-3}$ m^3. Determine the pressure P_2.

EXERCISE 19-3: How much heat is transferred into the system during process A?

EXERCISE 19-4: How much work is done by the expanding gas?

EXERCISE 19-5: Process B is isovolumic from P_2 to $P_1 = 2.17 \times 10^5$ Pa. Find the temperature at point 3.

EXERCISE 19-6: How much heat is transferred out of the system during process B?

EXERCISE 19-7: Prove that process C is adiabatic.

EXERCISE 19-8: How much work is done *on* the system during process C?

EXERCISE 19-9: What is the net work done by the system?

EXERCISE 19-10: What is the net quantity of heat transferred into the system?

Answers to Supplementary Exercises

19-1: 0.401 mol

19-2: 4×10^5 Pa

19-3: 1.83×10^3 J

19-4: 1.83×10^3 J

19-5: 326 K

19-6: 1.37×10^3 J

19-7: $P_1 V_2^\gamma$ must equal $P_3 V_1^\gamma$.
$P_1 V_2^\gamma = 31.7; P_3 V_1^\gamma = 31.7$

19-8: 1.37×10^3 J

19-9: 460 J

19-10: 460 J

20 THE SECOND LAW OF THERMODYNAMICS AND ENTROPY

THIS CHAPTER IS ABOUT

☑ **Heat Engines**
☑ **Efficiency of Heat Engines**
☑ **Refrigerators**
☑ **The Carnot Cycle**
☑ **The Second Law of Thermodynamics**
☑ **Entropy**

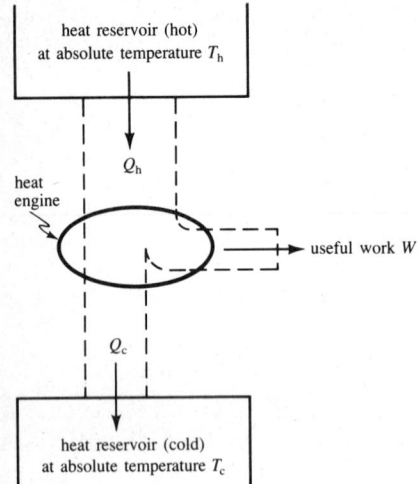

Figure 20-1. Flow of heat (thermal energy) and useful output work through a heat engine.

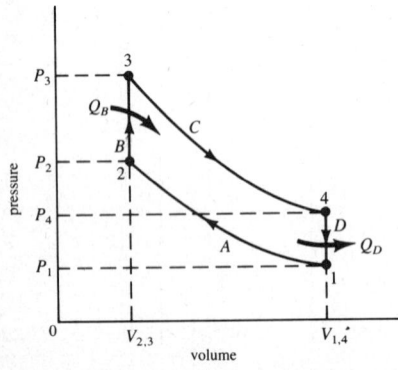

Figure 20-2. The Otto cycle of an internal combustion engine during the compression and power strokes.

20-1. Heat Engines

A **heat engine** is a device that converts thermal energy, i.e., heat, into useful work. A heat engine extracts a quantity of heat Q_h from a hot reservoir at absolute temperature T_h, converts some of this energy into useful work W, and transfers the remainder Q_c to a cold reservoir at absolute temperature T_c (Figure 20-1). The efficiency at which heat engines convert thermal energy into mechanical energy is generally considerably less than 100%. The first law of thermodynamics requires that a heat engine, like all other systems, conserve energy. Thus

$$Q_h = W + Q_c \tag{20-1}$$

The substance, called the **working substance**, that brings the heat Q_h from the surroundings to the engine and carries the heat Q_c from the engine to the surroundings is normally a gas or mixture of gases.

A. The Otto cycle

The internal combustion engine of an automobile is an example of a heat engine. The **Otto cycle** is an idealized description of the operation of such an engine in which frictional and other losses are neglected. Figure 20-2 is a pressure–volume diagram for the Otto cycle during the compression and power strokes. Starting at point 1, process A is an adiabatic compression of the fuel–air mixture. At point 2, a spark plug ignites the mixture, and the burning fuel releases a quantity of heat Q_B. This quickly raises the pressure to P_3 and an adiabatic expansion takes place, process C. During process D, heat Q_D is exhausted; that is, it is transferred from the engine to the surroundings, which function as the cold reservoir. In the standard four-cycle gasoline engine the exhaust and intake strokes follow, releasing the waste products of combustion, bringing in fresh fuel and air, and returning the engine to point 1 on the PV diagram.

B. The Diesel cycle

The **Diesel cycle** is an idealized description of the operation of a diesel internal combustion engine in which frictional and other losses are neglected. Figure 20-3 is a PV diagram for the Diesel cycle during the compression and power strokes. Starting at point 1, process A is an adiabatic compression of air, raising the pressure and temperature. At point 2 fuel is injected and combustion occurs, producing the quantity of heat Q_B as the volume increases slightly during process *B*. Process *C* is an adiabatic expansion to point 4, during which Q_D is exhausted. Process *D* is an isovolumic reduction in pressure. The standard four-cycle diesel engine also has exhaust and intake strokes that return the engine to point 1.

20-2. Efficiency of Heat Engines

The theoretical efficiency of a heat engine is defined as the ratio of the output work to the input heat.

EFFICIENCY OF A HEAT ENGINE
$$e = \frac{W}{Q_h} = \frac{Q_h - Q_c}{Q_h} = 1 - \frac{Q_c}{Q_h} \qquad \textbf{(20-2)}$$

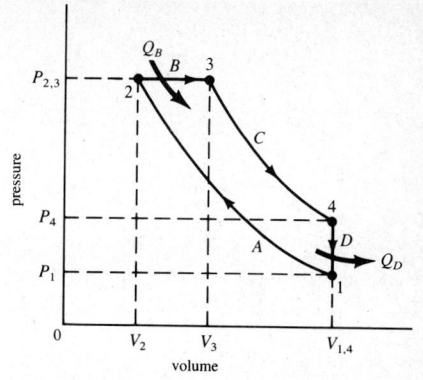

Figure 20-3. The Diesel cycle during the compression and power strokes.

The second step above follows from Eq. (20-1), $W = Q_h - Q_c$. When frictional and other losses are accounted for, the efficiency of a real engine is always less than that predicted by Eq. (20-2). The theoretical efficiency of the Otto cycle during the compression and power strokes is

THEORETICAL EFFICIENCY OF THE OTTO CYCLE
$$e_O = 1 - \frac{Q_D}{Q_B} \qquad \textbf{(20-3a)}$$

This can also be written as

$$e_O = 1 - \frac{1}{r^{\gamma - 1}} \qquad \textbf{(20-3b)}$$

where $\gamma = C_p/C_v$ and $r = V_{1,4}/V_{2,3}$, the engine's **compression ratio**. The compression ratio of a typical gasoline engine is about 8. When the intake and exhaust strokes are considered the practical theoretical efficiency of a four-cycle engine is exactly one-half the value calculated from Eq. (20-3b).

EXAMPLE 20-1: Find the theoretical efficiency of a gasoline engine that has a compression ratio of 8. Assume that $\gamma = 1.4$ for an oxygen–nitrogen–fuel mixture.

Solution: Use Eq. (20-3b).

$$e_O = 1 - \frac{1}{r^{\gamma - 1}} = 1 - \frac{1}{8^{(1.4 - 1)}} = 1 - \frac{1}{8^{0.4}} = \boxed{0.565}$$

This result is more commonly expressed as a percentage: 56.5%. The practical theoretical efficiency is about 28%.

EXAMPLE 20-2: Derive Eq. (20-3b) from Eq. (20-3a).

Solution: First express the ratio of output heat to input heat Q_D/Q_B in terms of temperatures. Because processes *D* and *B* are isovolumic you can use Eq. (18-11), $Q = nC_v\Delta T$. Thus

$$\frac{Q_D}{Q_B} = \frac{nC_v(T_4 - T_1)}{nC_v(T_3 - T_2)}$$

Processes *A* and *C* are adiabatic so you can apply Eq. (19-10b): $TV^{\gamma - 1} =$ constant. So for process *C*, $T_3V_{2,3}^{\gamma-1} = T_4V_{1,4}^{\gamma-1}$ and

$$T_3 = T_4\left(\frac{V_{1,4}}{V_{2,3}}\right)^{\gamma - 1} = T_4r^{\gamma - 1}$$

For process *A*, $T_1V_{1,4}^{\gamma-1} = T_2V_{2,3}^{\gamma-1}$ and

$$T_2 = T_1\left(\frac{V_{1,4}}{V_{2,3}}\right)^{\gamma - 1} = T_1r^{\gamma - 1}$$

Now express the difference $T_3 - T_2$ in terms of T_4 and T_1:

$$T_3 - T_2 = T_4 r^{\gamma-1} - T_1 r^{\gamma-1} = (T_4 - T_1)r^{\gamma-1}$$

Thus

$$\frac{Q_D}{Q_B} = \frac{T_4 - T_1}{T_3 - T_2} = \frac{T_4 - T_1}{(T_4 - T_1)r^{\gamma-1}} = \frac{1}{r^{\gamma-1}}$$

and therefore

$$e_O = 1 - \frac{Q_D}{Q_B} = \boxed{1 - \frac{1}{r^{\gamma-1}}}$$

The theoretical efficiency of the Diesel cycle during the compression and power strokes is

THEORETICAL EFFICIENCY OF THE DIESEL CYCLE
$$e_D = 1 - \frac{Q_D}{Q_B} = 1 - \frac{(V_{1,4}/V_3)^{-\gamma} - (V_{1,4}/V_2)^{-\gamma}}{\gamma[(V_{1,4}/V_3)^{-1} - (V_{1,4}/V_2)^{-1}]} \qquad (20\text{-}4)$$

where $V_{1,4}/V_2$ is called the **compression ratio** and $V_{1,4}/V_3$ is called the **expansion ratio**. The compression ratio of real diesel engines ranges from 15 to 20 and the expansion ratio is about 5. The practical theoretical efficiency of a four-cycle diesel engine is one-half the value determined from Eq. (20-4).

EXAMPLE 20-3: Find the theoretical efficiency of a diesel engine that has a compression ratio of 14 and an expansion ratio of 5. Let $\gamma = 1.4$.

Solution: Use Eq. (20-4).

$$e_D = 1 - \frac{(V_{1,4}/V_3)^{-\gamma} - (V_{1,4}/V_2)^{-\gamma}}{\gamma[(V_{1,4}/V_3)^{-1} - (V_{1,4}/V_2)^{-1}]}$$

$$= 1 - \frac{(5)^{-1.4} - (15)^{-1.4}}{1.4[(5)^{-1} - (15)^{-1}]}$$

$$= 1 - \frac{8.249 \times 10^{-2}}{1.4(0.1333)} = \boxed{0.558}$$

This result could be reported as 55.8%.

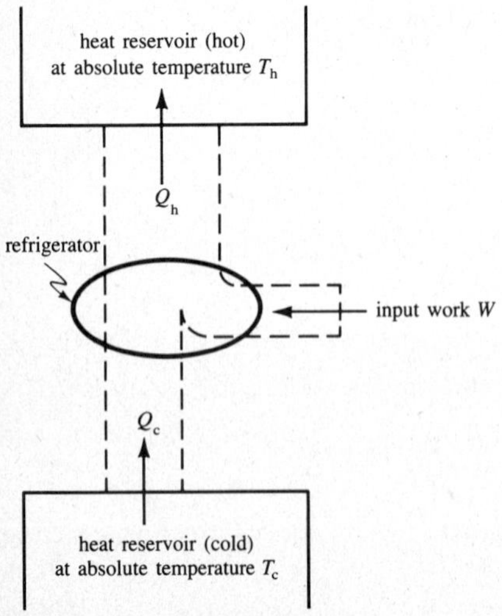

20-3. Refrigerators

In a theoretical sense a refrigerator can be thought of as a heat engine run in reverse. A refrigerator is a device that uses a quantity of input energy (input work) to transfer a quantity of heat Q_c from a cold reservoir at absolute temperature T_c to a hot reservoir at absolute temperature T_h. During this process the refrigerator transforms the input work into heat and transfers it to the hot reservoir along with Q_c so that the quantity of heat transferred to the hot reservoir is $Q_h = Q_c + W$ (see Fig. 20-4). The **coefficient of performance** of a refrigerator is the ratio of heat extracted from the cold reservoir to the amount of work required to extract that amount of heat.

COEFFICIENT OF PERFORMANCE
$$\eta = \frac{Q_c}{W} = \frac{Q_c}{Q_h - Q_c} \qquad (20\text{-}5)$$

The coefficient of performance is at most 6 or 7 and more often 2 or 3 for real refrigerators.

Figure 20-4. Flow of heat and input work through a refrigerator.

heat reservoir (hot) at absolute temperature T_h

Q_h

refrigerator

input work W

Q_c

heat reservoir (cold) at absolute temperature T_c

EXAMPLE 20-4: A small air conditioner has a coefficient of performance of 4 and can transfer $20\,000$ BTU h^{-1} from a room. What input power is required to operate this unit?

Solution: Equation (20-5) can be easily modified to solve this problem by noting that

$$\eta = \frac{\text{the } \textit{rate} \text{ at which a refrigerator transfers heat from the cold reservoir}}{\text{the } \textit{rate} \text{ at which a refrigerator uses energy}}$$

$$\eta = \frac{dQ_c/dt}{dW/dt} = \frac{dQ_c/dt}{P}$$

$$P = \frac{dQ_c/dt}{\eta}$$

Convert $20\,000$ BTU h^{-1} into SI units (watts) by using 1 BTU $= 1055$ J.

$$\frac{20,000 \text{ BTU}}{h} = \left(\frac{20,000 \text{ BTU}}{h}\right)\left(\frac{1055 \text{ J}}{\text{BTU}}\right)\left(\frac{h}{3600 \text{ s}}\right) = 5.861 \times 10^3 \text{ W}$$

Therefore

$$P = \frac{5.861 \times 10^3 \text{ W}}{4} = \boxed{1.47 \times 10^3 \text{ W} = 1.47 \text{ kW}}$$

20-4. The Carnot Cycle

Figure 20-5 shows a thermodynamic cycle in which processes A and C are isothermal and processes B and D are adiabatic. Such a cycle is called a **Carnot cycle** and a theoretical engine that operates around a Carnot cycle is called a **Carnot engine**. No such real engine can be constructed, but we can show that no other type of engine, real or theoretical, operating in a cycle with the same pair of heat reservoirs, can have an efficiency greater than that of a Carnot engine. The efficiency of a Carnot engine is called the **Carnot efficiency** and is

CARNOT EFFICIENCY $$e_C = 1 - \frac{T_c}{T_h} \qquad (20\text{-}6)$$

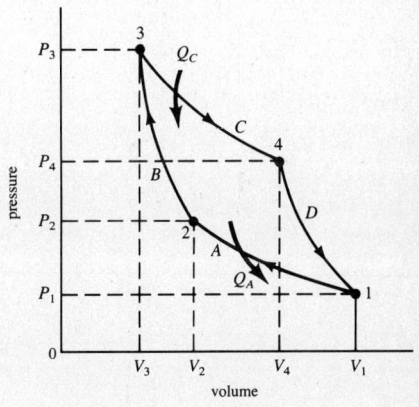

Figure 20-5. The Carnot cycle.

EXAMPLE 20-5: Start with Eq. (20-2) and derive Eq. (20-6).

Solution: To derive Eq. (20-6) we must show that

$$\frac{T_c}{T_h} = \frac{Q_c}{Q_h}$$

In the Carnot cycle shown in Figure 20-5 process C is the "hot" isothermal process, so $T_h = T_C$ and $Q_h = Q_C$. Process A is the "cold" isothermal process, so $T_c = T_A$ and $Q_c = Q_A$. Therefore we want to show that

$$\frac{T_A}{T_C} = \frac{Q_A}{Q_C}$$

The heat transfer in an isothermal process is the same as the work done in the process (Eq. 19-4) and is equal to $Q = nRT \ln(V_2/V_1)$ (Example 19-1). If we consider Q_A a positive quantity,

$$\frac{Q_A}{Q_C} = \frac{n\cancel{R}T_A \ln(V_1/V_2)}{n\cancel{R}T_C \ln(V_4/V_3)} = \frac{T_A}{T_C} \qquad \text{if} \qquad \ln\left(\frac{V_1}{V_2}\right) = \ln\left(\frac{V_4}{V_3}\right)$$

Now use Eq. (19-9b), $TV^{\gamma-1} = $ constant, for the adiabatic processes D and B to find a relationship between V_1/V_2 and V_4/V_3.

$$\text{Process } D: \quad T_A V_1^{\gamma-1} = T_C V_4^{\gamma-1}$$
$$\text{Process } B: \quad T_A V_2^{\gamma-1} = T_C V_3^{\gamma-1}$$

Divide the first equation by the second to get

$$\frac{T_A V_1^{\gamma-1}}{T_A V_2^{\gamma-1}} = \frac{T_C V_4^{\gamma-1}}{T_C V_3^{\gamma-1}}$$

Therefore

$$\left(\frac{V_1}{V_2}\right)^{\gamma-1} = \left(\frac{V_4}{V_3}\right)^{\gamma-1} \qquad \text{so} \qquad \frac{V_1}{V_2} = \frac{V_4}{V_3} \qquad \text{and} \qquad \ln\left(\frac{V_1}{V_2}\right) = \ln\left(\frac{V_4}{V_3}\right)$$

The Carnot cycle is reversible and the theoretical device described by the reverse of the cycle in Figure 20-5 is called a **Carnot refrigerator**. The transfers of heat in a Carnot refrigerator are also in the directions opposite those in a Carnot engine. The coefficient of performance of a Carnot refrigerator is

CARNOT COEFFICIENT OF PERFORMANCE
$$\eta_C = \frac{T_c}{T_h - T_c} \qquad\qquad \textbf{(20-7)}$$

No real refrigerator, operating in a cycle between the same two heat reservoirs, has a coefficient of performance greater than that of a Carnot refrigerator.

EXAMPLE 20-6: Begin with Eq. (20-5) and verify Eq. (20-7).

Solution: Divide both numerator and denominator by Q_h. In a Carnot cycle you can substitute T_c/T_h for Q_c/Q_h.

$$\eta = \frac{Q_c}{Q_h - Q_c} = \frac{Q_c/Q_h}{1 - Q_c/Q_h} = \frac{T_c/T_h}{1 - T_c/T_h}$$

Multiply the numerator and denominator by T_h.

$$\eta_C = \frac{T_c}{T_h - T_c}$$

EXAMPLE 20-7: Calculate the coefficient of performance of a Carnot refrigerator operating between the two temperatures of a household refrigerator, an inside temperature of $38°F$ and an outside temperature equal to room temperature, $72°F$.

Solution: Convert the temperatures into absolute units and use Eq. (20-7).

$$T_c = 38°F = 3.33°C = (3.33 + 273)\text{K} = 276.33 \text{ K}$$
$$T_h = 72°F = 22.22°C = (22.22 + 273)\text{K} = 295.22 \text{ K}$$
$$\eta_C = \frac{T_c}{T_h - T_c} = \frac{276.33 \text{ K}}{(295.22 - 276.33)\text{K}} = \boxed{14.6}$$

20-5. The Second Law of Thermodynamics

The fundamental result of our study of engines is the **second law of thermodynamics:** The efficiency at which real engines convert heat to mechanical energy is always less than 100%. There are three equivalent ways in which the second law is commonly stated. The law asserts what *cannot* be done.

The Kelvin statement:

- It is *not* possible for a real process to extract a quantity of heat from a reservoir, convert it completely into useful work, and produce no other effect.

The Clausius statement:

- It is *not* possible for a real process to have as its only result the transfer of a quantity of heat from a cooler to a hotter reservoir.

The Carnot cycle statement:

- It is *not* possible to produce an engine, operating in a cycle between the same two heat reservoirs, that has an efficiency greater than that of a Carnot engine.

20-6. Entropy

Entropy, commonly denoted by S, is a useful thermodynamic state variable, much like P, V, and T, that we can use to describe the state of a system. Its nature, however, is somewhat more abstract.

First consider a *system*, its *surroundings*, and *everywhere else* (Fig. 20-6). For a concrete example of a system, think of an ideal gas confined to a cylinder equipped with a movable piston (Fig. 19-1). Some kind of *process* then takes the system from one state (state 1) to another (state 2). We know that a change has occurred if there is a change in the system's state variables (P, V, T, S,...). This process must also produce a change in the system's surroundings. For example, if a quantity of heat is transferred to the system, that amount of heat must be transferred out of the surroundings.

- A **reversible process** is one in which the system and its surroundings can be restored to their original states without producing any change anywhere else. An **irreversible process** does not meet these rigorous conditions.

The concept of a reversible process is an idealization; all real, natural processes are irreversible to some degree. Some processes deviate from the reversible ideal more than others. An example of a reversible process would be the expansion of the gas in the cylinder in Figure 19-1 if the piston moved without friction. The gas does an amount of work, say W_R, in going from state 1 to state 2. If this amount of work is done on the gas by a force pushing the piston back to its original position the system will be restored to its original state. If, on the other hand, friction is present, as is the case for all *real* piston–cylinder systems, the gas will do an amount of mechanical work W_l to move the piston plus an amount of work W_f due to the frictional force. W_f is converted into heat (thermal energy) in the process. To restore this system to its original state will require an amount of work W_l from the surroundings plus W_f plus an *additional* W_f needed to move the piston against friction during the compression. This additional W_f must come from somewhere other than the surroundings in order to restore the system *and* its surroundings to their original states. Hence this real process is irreversible.

The *change* in entropy during a process is a measure of the process' *reversibility*, because entropy is defined for a reversible process so that when the entropy is integrated over the process from the initial state (state 1) to the final state (state 2) and then back to the initial state, the result is zero:

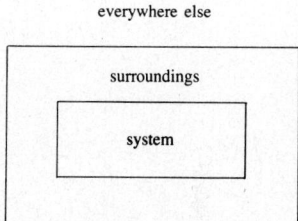

everywhere else

surroundings

system

Figure 20-6

ENTROPY CHANGE AROUND A REVERSIBLE CYCLE

$$\oint dS = 0$$

The integration from the final state back to the original state need not be along the same path, but this reverse process must be reversible too. The change in entropy of a system can be calculated *by integration* only for a reversible process:

ENTROPY CHANGE DURING A REVERSIBLE PROCESS

$$\Delta S = S_2 - S_1 = \int_1^2 dS$$

To clarify, let's consider a system consisting of an ideal gas, as shown in Figure 19-1, to which a quantity of heat is added in a reversible, but unspecified, process. According to the first law of thermodynamics, Eq. (19-3), $Q = E_{k\,tot} + W$. We can express this in differential form as

$$dQ = dE_{k\,tot} + dW = nC_v\,dT + P\,dV$$

(You may wish to review Example 19-3). For an ideal gas in equilibrium $PV = nRT$, so

$$dQ = nC_v\,dT + nRT\frac{dV}{V}$$

To calculate $Q = \int dQ$ we must integrate, but because of the form of the second term on the right, we can integrate only if we identify the process explicitly by its volume–temperature relationship. But if we divide the equation by T we get a new equation that *can* be integrated, that is *independent* of any process, and that has a solution that depends only on the initial and final states!

$$\frac{dQ}{T} = nC_v\frac{dT}{T} + nR\frac{dV}{V}$$

The quantity on the left is, by definition, the differential form of the entropy: $dS = dQ/T$ for any reversible process. We define the change in entropy for a reversible process in a system going from state 1 to state 2 as

$$\Delta S = S_2 - S_1 = \int_1^2 \frac{dQ}{T} \tag{20-8}$$

where dQ is an infinitesimal transfer of heat at temperature T during an intermediate step in the reversible process. For an ideal gas the change in entropy, if C_v is constant, is

$$\Delta S = S_2 - S_1 = \int_1^2 \frac{dQ}{T} = nC_v\int_{T_1}^{T_2}\frac{dT}{T} + nR\int_{V_1}^{V_2}\frac{dV}{V}$$

$$= nC_v\ln\!\left(\frac{T_2}{T_1}\right) + nR\ln\!\left(\frac{V_2}{V_1}\right)$$

If C_v is not constant, it is a function of temperature only, not of the process, so our differential equation for entropy can still be integrated. Entropy has units of joules per kelvin, $J\,K^{-1}$, in the SI system.

Notice that, like potential energy, only a *change* in entropy is defined. In the example of the ideal gas, if the reference entropy is $S_r = S_1$, then the entropy with respect to this is $S = S_2$:

$$S = nC_v\ln\!\left(\frac{T}{T_r}\right) + nR\ln\!\left(\frac{V}{V_r}\right) + S_r$$

for an ideal gas with C_v constant.

The usefulness of the concept of entropy comes from the fact that if the internal constraints on a system in equilibrium are removed, the system changes to a final equilibrium state in which entropy is maximized. Textbooks on thermodynamics cover the details of calculations of this type. Notice that for a reversible adiabatic process Q is zero and consequently so is ΔS. Because the entropy is constant during this process it is called an **isentropic process**.

EXAMPLE 20-8: What is the change in entropy of 2 kg of water heated in a reversible process from 20°C to 50°C? Assume the specific heat of water is constant in this temperature range.

Solution: Use Eq. (17-3a), $Q = mc \Delta T$, for an infinitesimal transfer of heat: $dQ = mc\, dT$. Then use Eq. (20-8) to compute the change in entropy for the water:

$$\Delta S = \int_1^2 \frac{dQ}{T} = mc \int_{T_1}^{T_2} \frac{dT}{T} = mc \left(\ln T \Big|_{T_1}^{T_2} \right) = mc \ln\left(\frac{T_2}{T_1}\right)$$

$$= (2 \text{ kg})(4.186 \times 10^3 \text{ J kg}^{-1}\text{ K}^{-1}) \ln\left[\frac{(50 + 273)\text{K}}{(20 + 273)\text{K}}\right] = \boxed{816 \text{ J K}^{-1}}$$

Notice that the temperature must be specified in absolute units.

SUMMARY

1. *Heat engines* are devices that extract some of the thermal energy from a heat reservoir and convert it into useful work.
2. The *efficiency* of a heat engine is the ratio of the useful work output from the engine to the quantity of thermal energy extracted from the heat reservoir: $e = W/Q_h$. It is often expressed as a percentage.
3. The *coefficient of performance* for a refrigerator is the ratio of the quantity of heat extracted from the cold reservoir to the work needed to remove that heat: $\eta = Q_c/W$.
4. The *Carnot cycle* consists of two isothermal and two adiabatic processes. A *Carnot engine* is described by a clockwise traversal of the cycle on its *PV* diagram. A *Carnot refrigerator* is described by a counter-clockwise traversal.
5. The *second law of thermodynamics* states that heat cannot be converted to mechanical work with 100% efficiency.
6. The Carnot cycle statement of the second law of thermodynamics defines the theoretical upper limit for the efficiency of a heat engine operating between a pair of heat reservoirs.
7. The *change in entropy* for a reversible process is

$$\Delta S = \int_1^2 \frac{dQ}{T}$$

where 1 and 2 represent the initial and final states of the system. Entropy has units of joules per kelvin in the SI system.

RAISE YOUR GRADES

Can you explain . . . ?

☑ one advantage of a large compression ratio for a gasoline engine
☑ one disadvantage of a large compression ratio for a gasoline engine
☑ one advantage of raising the temperature of the input steam to the turbine at a power plant
☑ one disadvantage of raising the temperature of the input steam to the turbine at a power plant
☑ why the temperature in a kitchen will rise when the door to the refrigerator is left open
☑ why the Carnot efficiency can never be 100%
☑ how the Kelvin statement of the second law of thermodynamics applies to heat engines
☑ how the Clausius statement of the second law of thermodynamics applies to refrigerators
☑ why the net entropy change in a Carnot cycle is zero

SOLVED PROBLEMS

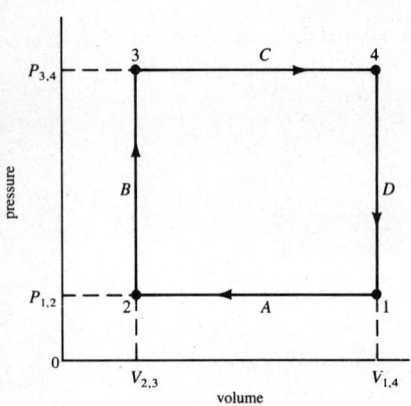

Figure 20-7

Heat Engines

PROBLEM 20-1:   Calculate the net work done per cycle by the heat engine operating around the cycle shown in Figure 20-7. $P_{1,2} = 1 \times 10^5$ Pa, $P_{3,4} = 2.5 \times 10^5$ Pa, $V_{2,3} = 4 \times 10^{-4}$ m^3, and $V_{1,4} = 3 \times 10^{-3}$ m^3.

Solution: Processes A and C are isobaric. Work is done by the expanding gas in process C and on the gas during the compression process A. No work is done during the isovolumic processes B and D. Use Eq. (19-1), $W = \int P \, dV$.

$$W = W_C + W_A = \int_3^4 P_{3,4} \, dV + \int_1^2 P_{1,2} \, dV$$

$$= P_{3,4}(V_{1,4} - V_{2,3}) + P_{1,2}(V_{2,3} - V_{1,4}) = (P_{3,4} - P_{1,2})(V_{1,4} - V_{2,3})$$

$$= [(2.5 - 1) \times 10^5 \text{ Pa}][(30 - 4) \times 10^{-4} \text{ m}^3] = \boxed{390 \text{ J}}$$

Efficiency of Heat Engines

PROBLEM 20-2:   A two-cycle gasoline engine operating on the Otto cycle extracts 20 kJ of heat each cycle and does work with an efficiency of 30%. Assume $\gamma = 1.4$. **(a)** How much work is done each cycle? **(b)** How much heat is transferred to the cold reservoir each cycle? **(c)** What is the compression ratio for this engine?

Solution:

(a) Use Eq. (20-2) and solve for W.

$$e = \frac{W}{Q_h} \qquad W = eQ_h = (0.3)(20 \text{ kJ}) = \boxed{6.0 \text{ kJ}}$$

(b) Use Eq. (20-1) and solve for Q_c.

$$Q_h = W + Q_c$$

$$Q_c = Q_h - W = 20 \text{ kJ} - 6 \text{ kJ} = \boxed{14 \text{ kJ}}$$

(c) Equate Eq. (20-3a) with Eq. (20-3b) and solve for r. Q_D is the heat exhausted to the cold reservoir, Q_c, and Q_B is the heat extracted from the hot reservoir, Q_h (see Fig. 20-2).

$$1 - \frac{Q_D}{Q_B} = 1 - \frac{1}{r^{\gamma - 1}}$$

$$r^{\gamma - 1} = \frac{Q_B}{Q_D} = \frac{Q_h}{Q_c}$$

$$r = \left(\frac{Q_h}{Q_c}\right)^{1/(\gamma - 1)} = \left(\frac{20 \text{ kJ}}{14 \text{ kJ}}\right)^{1/(1.4 - 1)} = (1.43)^{2.5} = \boxed{2.44}$$

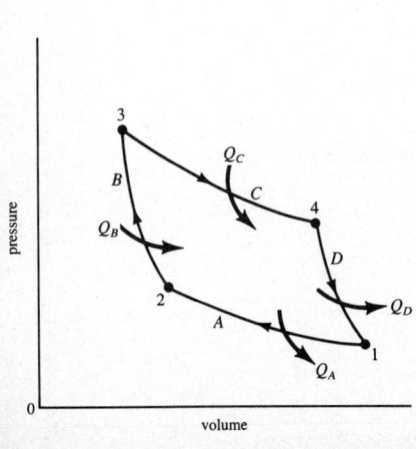

Figure 20-8

PROBLEM 20-3:   Figure 20-8 shows the PV diagram for a certain heat engine. The heat transfers shown in the figure are $Q_A = 20$ kJ, $Q_B = 10$ kJ, $Q_C = 30$ kJ, and $Q_D = 8$ kJ. **(a)** How much work is done by this engine during the cycle shown? **(b)** What is the efficiency of this engine?

Solution: Heat is transferred into the engine during processes B and C and out of the engine during processes A and D.

(a) Use Eq. (20-1) where $Q_h = Q_B + Q_C$ and $Q_c = Q_A + Q_D$. Solve for W.

$$Q_h = W + Q_c$$

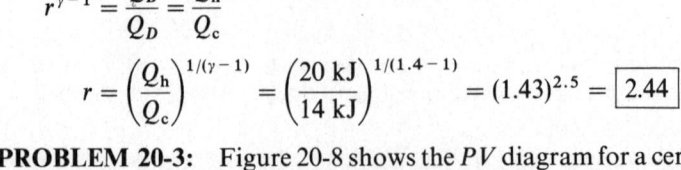

(b) Use Eq. (20-2).

$$e = \frac{W}{Q_h} = \frac{12 \text{ kJ}}{(10 + 30)\text{kJ}} = \boxed{0.3 = 30\%}$$

PROBLEM 20-4: A certain diesel engine has a theoretical efficiency of 60% and a compression ratio of 12. Find the engine's expansion ratio. Assume $\gamma = 1.4$.

Solution: To make the algebra a bit easier let the expansion ratio be represented by E and the compression ratio by C. With these substitutions, Eq. (20-4) is

$$e_D = 1 - \frac{E^{-\gamma} - C^{-\gamma}}{\gamma[E^{-1} - C^{-1}]} = 1 - e_D$$

An attempt to solve this equation quickly reveals the fact that an analytical solution is not possible. In the real world of physics, not all problems have simple mathematical solutions! We must try repeated approximations with a calculator or microcomputer to find a satisfactory estimate for E. Rearrange this equation in the form $f(E) = 0$.

$$\frac{E^{-\gamma} - C^{-\gamma}}{E^{-1} - C^{-1}} + \gamma(e_D - 1) = 0$$

Substitute the numerical values given in the problem to get:

$$f(E) = \frac{E^{-1.4} - 12^{-1.4}}{E^{-1} - 12^{-1}} + (1.4)(0.6 - 1) = 0$$

or

$$f(E) = \frac{E^{-1.4} - 3.084 \times 10^{-2}}{E^{-1} - 8.333 \times 10^{-2}} - 0.56 = 0$$

Now systematically substitute different values for E until you find the one for which $f(E)$ is zero or a sufficiently close approximation. If we try values of E from 7.0 to 9.5 in steps of 0.5 our results are

E	$f(E)$	E	$f(E)$
7.0	2.383×10^{-2}	8.5	-0.221×10^{-2}
7.5	1.424×10^{-2}	9.0	-0.935×10^{-2}
8.0	0.561×10^{-2}	9.5	-1.588×10^{-2}

Clearly E has a value between 8.0 and 8.5. If we improve our approximation by using ever smaller increments between our trial values of E, our result, to four significant figures, is $E \cong \boxed{8.353}$.

Refrigerators

PROBLEM 20-5: Heat flows into a small freezer at a rate of 250 W when its inside temperature is $-18°C$. The average input power needed to maintain this temperature is 75 W. What is this refrigerator's coefficient of performance?

Solution: Use Eq. (20-5) in the differential form of Example 20-4:

$$\eta = \frac{Q_c}{W} = \frac{dQ_c/dt}{P} = \frac{250 \text{ W}}{75 \text{ W}} = \boxed{3.33}$$

The Carnot Cycle

PROBLEM 20-6: Calculate the hot reservoir temperature for a Carnot engine that has an efficiency of 60% when the temperature of the waste discharged by the engine to the environment is 27°C.

Solution: Apply Eq. (20-6), and be sure to express temperatures in kelvins. Solve for T_h.

$$e_C = 1 - \frac{T_c}{T_h}$$

$$T_h = \frac{T_c}{1 - e_C} = \frac{(27 + 273)\text{K}}{1 - 0.6} = 750\text{ K}$$

The Celsius temperature is $750 - 273 = \boxed{477°\text{C}}$.

PROBLEM 20-7: The Carnot engine of Problem 20-6 is doing work at the rate of 1 hp. (a) At what rate does it extract heat from the hot reservoir? (b) At what rate does it transfer heat to the environment?

Solution:

(a) Differentiate the numerator and denominator of Eq. (20-2) with respect to t.

$$e = \frac{W}{Q_h} = \frac{dW/dt}{dQ_h/dt} = \frac{P}{dQ_h/dt}$$

where P is the power output of the engine. Solve for dQ_h/dt, the rate at which heat is extracted from the hot reservoir. In SI units, $P = 1$ hp $= 746$ W.

$$\frac{dQ_h}{dt} = \frac{P}{e} = \frac{746\text{ W}}{0.6} = \boxed{1.24 \times 10^3\text{ W} = 1.24\text{ kW}}$$

(b) In Example 20-5 we showed that for a Carnot engine $T_c/T_h = Q_c/Q_h$, which can also be written as

$$\frac{T_c}{T_h} = \frac{dQ_c/dt}{dQ_h/dt}$$

Solve for dQ_c/dt.

$$\frac{dQ_c}{dt} = \left(\frac{dQ_h}{dt}\right)\left(\frac{T_c}{T_h}\right) = (1.24\text{ kW})\left(\frac{300\text{ K}}{750\text{ K}}\right) = \boxed{497\text{ W}}$$

PROBLEM 20-8: The heat input during the isothermal process C of the Carnot cycle shown in Figure 20-5 is $Q_h = Q_c = 5 \times 10^3$ J at a temperature $T_3 = T_4 = 500$ K. Assume that the working substance is a diatomic ideal gas with $\gamma = 1.4$. Some of the conditions of the cycle are $P_3 = 8 \times 10^5$ Pa, $V_3 = 0.01$ m^3, and $P_1 = 1.5 \times 10^5$ Pa. (a) Compute V_4. (b) Compute V_1. (c) Find $T_1 = T_2$. (d) Obtain $Q_A = Q_c$. (e) What is the Carnot efficiency?

Solution:

(a) Use the expression for heat transferred in an isothermal process (Example 19-1) and solve for V_4.

$$Q_C = nRT_3 \ln\left(\frac{V_4}{V_3}\right)$$

Before proceeding with that calculation, use the ideal gas law to substitute for nRT_3: $P_3V_3 = nRT_3$

$$Q_C = P_3V_3 \ln\left(\frac{V_4}{V_3}\right)$$

$$\ln\left(\frac{V_4}{V_3}\right) = \frac{Q_C}{P_3V_3} = \frac{5 \times 10^3\text{ J}}{(8 \times 10^5\text{ Pa})(0.01\text{ m}^3)} = 0.625$$

$$V_4 = V_3e^{0.625} = (0.01\text{ m}^3)e^{0.625} = \boxed{1.868 \times 10^{-2}\text{ m}^3}$$

Also, $P_4V_4 = P_3V_3$, so

$$P_4 = P_3\left(\frac{V_3}{V_4}\right) = (8 \times 10^5\text{ Pa})\left(\frac{1 \times 10^{-2}\text{ m}^3}{1.868 \times 10^{-2}\text{ m}^3}\right) = 4.282 \times 10^5\text{ Pa}$$

(b) Process D is adiabatic, so use Eq. (19-9a) to obtain V_1.

$$P_1 V_1^\gamma = P_4 V_4^\gamma$$

$$V_1 = V_4 \left(\frac{P_4}{P_1}\right)^{1/\gamma} = (1.868 \times 10^{-2}\ \text{m}^3)\left(\frac{4.282 \times 10^5\ \text{Pa}}{1.5 \times 10^5\ \text{Pa}}\right)^{1/1.4} = \boxed{3.952 \times 10^{-2}\ \text{m}^3}$$

(c) Use the ideal gas law to compute $T_1 = T_2$:

$$\frac{P_1 V_1}{T_1} = \frac{P_3 V_3}{T_3}$$

$$T_1 = T_3 \left(\frac{P_1 V_1}{P_3 V_3}\right) = (500\ \text{K})\frac{(1.5 \times 10^5\ \text{Pa})(3.952 \times 10^{-2}\ \text{m}^3)}{(8 \times 10^5\ \text{Pa})(0.01\ \text{m}^3)} = \boxed{370\ \text{K}}$$

(d) In Example 20-5 we showed that for a Carnot engine

$$\frac{T_c}{T_h} = \frac{Q_c}{Q_h}$$

Solve for Q_c and substitute the appropriate values.

$$Q_c = Q_h \left(\frac{T_c}{T_h}\right) = Q_A = Q_c \left(\frac{T_1}{T_3}\right) = (5 \times 10^3\ \text{J})\left(\frac{370\ \text{K}}{500\ \text{K}}\right) = \boxed{3.7 \times 10^3\ \text{J}}$$

(e) Use Eq. (20-6):

$$e_C = 1 - \frac{T_c}{T_h} = 1 - \frac{T_1}{T_3} = 1 - \frac{370\ \text{K}}{500\ \text{K}} = \boxed{0.26 = 26\%}$$

Entropy

PROBLEM 20-9: Compute the increase in entropy when one kilogram of water boils into steam. The latent heat of vaporization of water is 2256 kJ kg^{-1} at 100°C.

Solution: Use Eq. (20-8). The heat absorbed by boiling water is $Q = mL_v$, Eq. (17-4b). Because the temperature is constant in this process, $T = 100°C$, we can factor it out of the integrand. Remember that T must be in absolute degrees.

$$\Delta S = \int_1^2 \frac{dQ}{T} = \frac{1}{T} \int_1^2 dQ = \frac{1}{T}(Q_2 - Q_1) = \frac{Q}{T} = \frac{mL_v}{T} = \frac{(1\ \text{kg})(2256\ \text{kJ kg}^{-1})}{(100 + 273)\text{K}} = \boxed{6.05 \times 10^3\ \text{J K}^{-1}}$$

PROBLEM 20-10: Draw a diagram showing the relationship between temperature and entropy for the Carnot cycle illustrated in Figure 20-5.

Solution: The Carnot cycle consists of two isothermal processes and two adiabatic processes. Temperature is constant during an isothermal process. Because dQ is zero during an adiabatic process, the change in entropy is also zero, according to Eq. (20-8). Therefore entropy is constant during an adiabatic process. Let the temperatures (in kelvins) during processes A and C in Figure 20-5 be T_A and T_C; notice that $T_A < T_C$. Let the entropies during processes B and D be S_B and S_D. Use Eq. (20-8) during the isothermal process C:

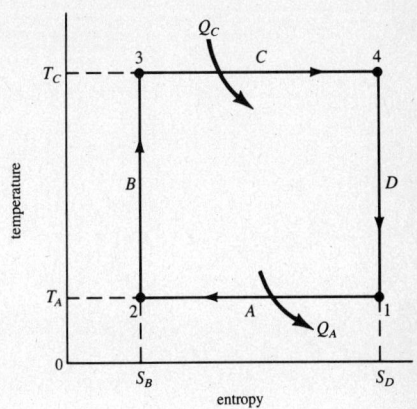

$$\Delta S = \int_3^4 \frac{dQ}{T} = \frac{1}{T_C} \int_3^4 dQ = \frac{Q_C}{T_C} = S_D - S_B$$

Thus $S_D > S_B$ because Q_C is positive. Combining all these relationships gives us Figure 20-9 as our final temperature–entropy diagram. Notice that the area enclosed by the diagram is also equal to the net work done.

Figure 20-9

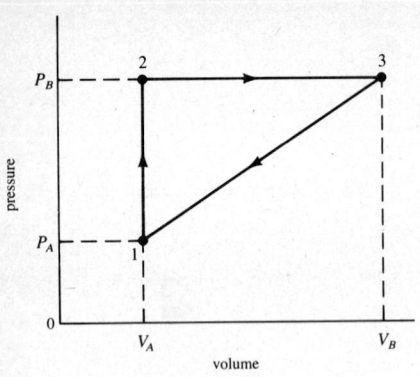

Figure 20-10

Supplementary Exercises

EXERCISE 20-1:   A small heat engine undergoes the cycle shown in Figure 20-10 with $P_A = 1.2 \times 10^5$ Pa, $P_B = 7.2 \times 10^5$ Pa, $V_A = 2.5 \times 10^{-4}$ m^3, and $V_B = 2 \times 10^{-3}$ m^3. (a) How much work is done per cycle by this engine? (b) If the engine runs at 6 cycles per second, what is its output power?

EXERCISE 20-2:   A heat engine operating on the Otto cycle has an efficiency of 40%. Find the minimum compression ratio necessary when $\gamma = 1.4$.

EXERCISE 20-3:   A certain diesel engine has a compression ratio of 18 and an expansion ratio of 6. Determine the theoretical efficiency of this engine if $\gamma = 1.4$.

EXERCISE 20-4:   A certain refrigerator removes heat at the rate of 2.2 kW from a cold reservoir at $-20°$C. This refrigerator requires 550 W to operate. (a) At what rate is heat transferred to the hot reservoir? (b) What is the coefficient of performance of this refrigerator?

EXERCISE 20-5:   If a Carnot refrigerator (a Carnot engine run in reverse) had a coefficient of performance of 4 and the cold reservoir had a temperature of $-20°$C, what would be the temperature of the hot reservoir?

EXERCISE 20-6:   In one cycle, a Carnot engine extracts 350 J of heat from a reservoir at 580 K and transfers the heat to a reservoir at 300 K. (a) What is the efficiency of this engine? (b) How much work is done per cycle?

EXERCISE 20-7:   Derive an expression for the efficiency of an engine operating on the Otto cycle in terms of the absolute temperatures at the points 1, 2, 3, and 4 in Figure 20-2.

EXERCISE 20-8:   Determine the relationship between the efficiency of a Carnot engine and the coefficient of performance of that engine operated in reverse as a refrigerator when operated between the same two temperatures.

EXERCISE 20-9:   One kilogram of water at $80°$C is mixed with one kilogram of water at $20°$C in an insulated container. The final temperature of the mixture is $50°$C. Determine the total entropy change during this process.

EXERCISE 20-10:   Determine the change in entropy of 1 kg of water when it freezes at $0°$C. The latent heat of fusion of water is 335 kJ kg^{-1}.

Answers to Supplementary Exercises

20-1:   (a) 525 J   (b) 3.14 kW

20-2:   3.59

20-3:   58.9%

20-4:   (a) 2.75 kW   (b) 4

20-5:   316 K $= 43°$C

20-6:   (a) 48.3%   (b) 169 J

20-7: $e_O = \dfrac{T_1 + T_3 - T_2 - T_4}{T_3 - T_2}$

20-8: $\eta_C = \dfrac{1 - e_C}{e_C}$

20-9: 3.63×10^{-2} J K^{-1}

20-10: 1.23×10^3 J K^{-1}

21 *MECHANICAL WAVES*

THIS CHAPTER IS ABOUT

- ☑ **Types of Waves**
- ☑ **Periodic Waves**
- ☑ **The Wave Equation**
- ☑ **Velocities of Waves**
- ☑ **Energy and Power in Waves**
- ☑ **Interference of Waves**

21-1. Types of Waves

Wave motion occurs when a disturbance or **pulse** travels (*propagates*) from one place to another. Waves transfer energy and momentum from one place to another with no net motion of matter. There are many types of waves but in this chapter we'll examine just mechanical waves.

- **Mechanical waves** require a **medium** in which to travel.

Examples: water is the medium for water waves, air is the medium for sound waves, and a guitar string is the medium for elastic waves in a string. The particles that constitute the medium oscillate as a wave moves through the medium. We can describe the motion of a wave by specifying the *position*, *velocity*, and *acceleration* of the particles of the medium as functions of time. We'll start with a simple case, one-dimensional wave motion along the x axis, then generalize our results to wave motion in three-dimensional space.

- The propagation of a **transverse wave** causes the particles of the medium to oscillate back and forth along a line *perpendicular* to the direction of motion of the wave.

Ordinary water waves and waves in a plucked or bowed string are examples of transverse mechanical waves.

- The propagation of a **longitudinal wave** causes the particles of the medium to oscillate back and forth along a line *parallel* to the direction of motion of the wave.

Pressure waves in a fluid, including sound waves in air, are examples of longitudinal mechanical waves.

21-2. Periodic Waves

- **Periodic waves** result from the displacement of a medium at regular intervals. Periodic waves can be either transverse or longitudinal.

Because of its mathematical convenience, we'll use the sine (or cosine) function to describe periodic wave motion. A wave that can be described with just one sine or cosine function is a **harmonic wave**. All other periodic waveforms (triangle, sawtooth, rectangle, etc.) can be described in terms of sine and cosine waves.

note: In our discussion of waves we will use many of the same concepts (amplitude, period, frequency, and angular frequency) introduced in our discussion of simple harmonic motion (Chapter 7). There is a fundamental connection between wave motion and simple harmonic motion: When periodic waves travel through a medium, each particle in the medium executes simple harmonic motion around an equilibrium point.

In Figure 21-1 we see five different views of a sine wave moving to the right along the positive x axis. In the first view, at time $t = 0$, we see that the wave has a characteristic **wavelength** λ, and in succeeding views we see that the wave moves to the right a distance $\frac{1}{12}\lambda$ in a time interval $\frac{1}{12}T$, where T is the **period**, the time the wave takes to move one full wavelength at the constant wave velocity v.

WAVE FORMULA
$$v = \frac{\lambda}{T}$$
(21-1a)

This relationship, called the **wave formula**, is more commonly written in terms of the **frequency** $v = T^{-1}$:

$$v = \lambda v$$
(21-1b)

In Figure 21-1 y represents the displacement of the particles of the medium from their equilibrium positions. A, the **amplitude**, is the maximum displacement. If we

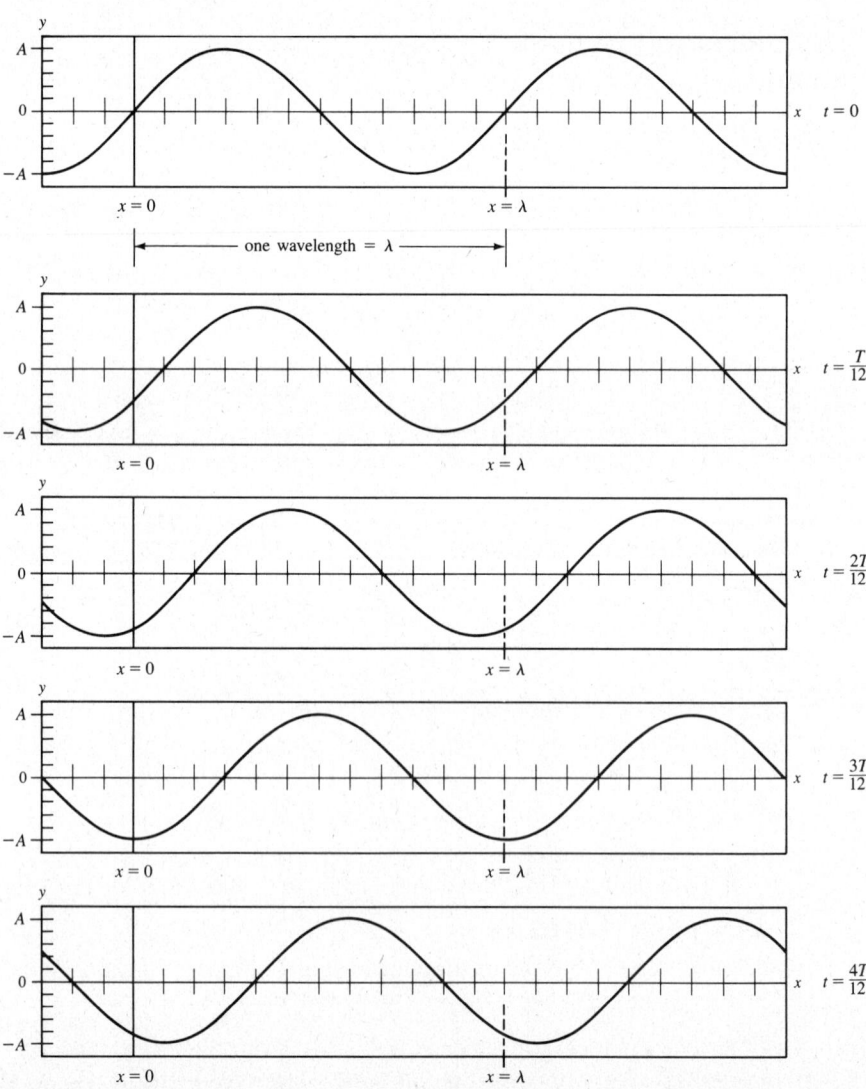

Figure 21-1. Five views of a sine wave propagating to the right along the x axis at velocity $v = \lambda v$. Each succeeding view shows the wave at a time $\frac{1}{12}T$ later.

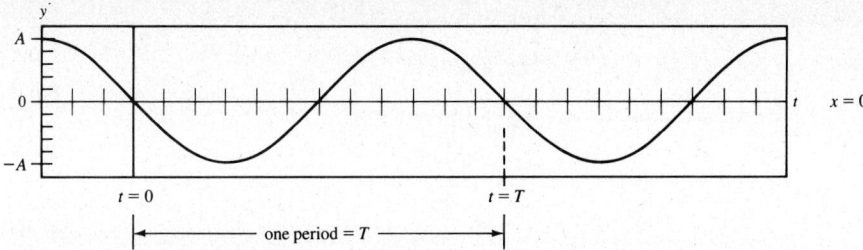

Figure 21-2. Displacement of the medium at the point $x = 0$ as a function of time. This function is represented by Eq. (21-1b).

define the **wave number** k as $2\pi/\lambda$, we can describe the position of any point on the wave at time $t = 0$ as

$$y(x) = A \sin(kx) \qquad \textbf{(21-2a)}$$

where kx is in radians. Likewise, if we define the **angular frequency** ω as $2\pi/T = 2\pi\nu$, we can describe the position of the medium at the location $x = 0$ at any time as

$$y(t) = A \sin(-\omega t) \qquad \textbf{(21-2b)}$$

where ωt is in radians. Figure 21-2 shows the position of the medium at $x = 0$ as a function of time. The complete description of the wave, the **wave function**, is the combination of Eqs. (21-2a) and (21-2b):

WAVE FUNCTION
FOR A HARMONIC WAVE $\qquad y(x, t) = A \sin(kx - \omega t) \qquad \textbf{(21-2c)}$

for propagation in the positive x direction. The description of a wave moving in the negative x direction is

$$y(x, t) = A \sin(kx + \omega t) \qquad \textbf{(21-2d)}$$

EXAMPLE 21-1: A sine wave of amplitude 15 cm, wavelength 1.2 m, and frequency 3 Hz travels along the positive x axis. (**a**) Find the wave velocity. (**b**) What is the displacement of the medium at the point $x = 0.4$ m at the instant $t = 0.1$ s?

Solution:

(**a**) Use Eq. (21-1b).

$$v = \lambda\nu = (1.2 \text{ m})(3 \text{ Hz}) = \boxed{3.6 \text{ m s}^{-1}}$$

(**b**) First compute k and ω,

$$k = \frac{2\pi}{\lambda} = \frac{2\pi}{1.2 \text{ m}} = 5.236 \text{ m}^{-1}$$

$$\omega = 2\pi\nu = 2\pi(3 \text{ Hz}) = 18.85 \text{ s}^{-1}$$

then use Eq. (21-2c), since the wave is moving in the positive x direction, with $A = 15$ cm.

$$y = A \sin(kx - \omega t)$$
$$= (15 \text{ cm}) \sin[(5.236 \text{ m}^{-1})(0.4 \text{ m}) - (18.85 \text{ s}^{-1})(0.1 \text{ s})]$$
$$= \boxed{3.12 \text{ cm}}$$

EXAMPLE 21-2: (**a**) Write Eq. (21-2c) in terms of λ and T instead of k and ω. (**b**) Write Eq. (21-2c) in terms of λ and v instead of k and ω.

Solution:

(a) Use the substitutions $k = 2\pi/\lambda$ and $\omega = 2\pi/T$.

$$y = A \sin(kx - \omega t)$$

$$= A \sin\left[\left(\frac{2\pi}{\lambda}\right)x - \left(\frac{2\pi}{T}\right)t\right]$$

$$\boxed{= A \sin\left[2\pi\left(\frac{x}{\lambda} - \frac{t}{T}\right)\right]}$$

(b) Use the substitutions $k = 2\pi/\lambda$, $\omega = 2\pi v$, and $v = \lambda v$.

$$y = A \sin(kx - \omega t)$$

$$= A \sin\left[\left(\frac{2\pi}{\lambda}\right)x - (2\pi v)t\right] = A \sin\left[\left(\frac{2\pi}{\lambda}\right)(x - \lambda vt)\right]$$

$$\boxed{= A \sin\left[\left(\frac{2\pi}{\lambda}\right)(x - vt)\right]}$$

Remember always to express the "argument" of the sine function in the dimensionless units *radians* when you use any of these forms of Eq. (21-2c).

21-3. The Wave Equation

The **wave equation** is a *partial differential equation* of second order in the space and time variables of the function that describes the wave, the wave function $\phi(x, t)$. The wave function $\phi(x, t)$ can be a simple function like Eq. (21-2c) describing a harmonic wave or a complex function describing more complex wave patterns. The wave equation is a partial differential equation because it contains more than one variable but is differentiated with respect to each variable separately. In one dimension, the wave equation is

WAVE EQUATION IN ONE DIMENSION
$$\frac{\partial^2 \phi}{\partial x^2} = \frac{1}{v^2}\frac{\partial^2 \phi}{\partial t^2} \qquad \text{(21-3a)}$$

where v is the wave velocity. In three dimensions, $\phi = \phi(x, y, z, t)$ and the wave equation is

WAVE EQUATION IN THREE DIMENSIONS
$$\frac{\partial^2 \phi}{\partial x^2} + \frac{\partial^2 \phi}{\partial y^2} + \frac{\partial^2 \phi}{\partial z^2} = \frac{1}{v^2}\frac{\partial^2 \phi}{\partial t^2} \qquad \text{(21-3b)}$$

EXAMPLE 21-3: Show that the wave function for a sine wave, Eq. (21-2c), is a solution of the one-dimensional wave equation, Eq. (21-3a).

Solution: Differentiate Eq. (21-2c) twice with respect to x:

$$y(x, t) = A \sin(kx - \omega t)$$

$$\frac{\partial y(x, t)}{\partial x} = kA \cos(kx - \omega t)$$

$$\frac{\partial^2 y(x, t)}{\partial x^2} = -k^2 A \sin(kx - \omega t) = -k^2 y(x, t)$$

Differentiate the wave function twice with respect to t:

$$\frac{\partial y(x, t)}{\partial t} = -\omega A \cos(kx - \omega t)$$

$$\frac{\partial^2 y(x, t)}{\partial t^2} = -\omega^2 A \sin(kx - \omega t) = -\omega^2 y(x, t)$$

Since we can substitute $\dfrac{1}{-\omega^2} \times -\omega^2 y(x,t)$ for $y(x,t)$,

$$\frac{\partial^2 y(x,t)}{\partial x^2} = -k^2\left(\frac{1}{-\omega^2} \times \frac{\partial^2 y(x,t)}{\partial t^2}\right) = \left(\frac{k}{\omega}\right)^2 \frac{\partial^2 y(x,t)}{\partial t^2}$$

Express k/ω in terms of v.

$$\frac{k}{\omega} = \frac{2\pi/\lambda}{2\pi v} = \frac{1}{\lambda v} = \frac{1}{v}$$

Therefore

$$\frac{\partial^2 y(x,t)}{\partial x^2} = \frac{1}{v^2}\frac{\partial^2 y(x,t)}{\partial t^2}$$

We conclude that $y(x,t) = A\sin(kx - \omega t)$ is a solution of the wave equation, as is $A\sin(kx + \omega t)$.

21-4. Velocities of Waves

A string, stretched between two fixed points, provides a medium in which one-dimensional transverse waves can propagate (see Fig. 21-3). If the string is uniform throughout its length, then the wave velocity will be constant throughout. The mass per unit length (the *linear density*) of the string shown in Figure 21-3 is $\mu = m/L$, where m is the mass of the string between the points $x = 0$ and $x = L$. If the string is under tension S (measured in newtons), the speed of

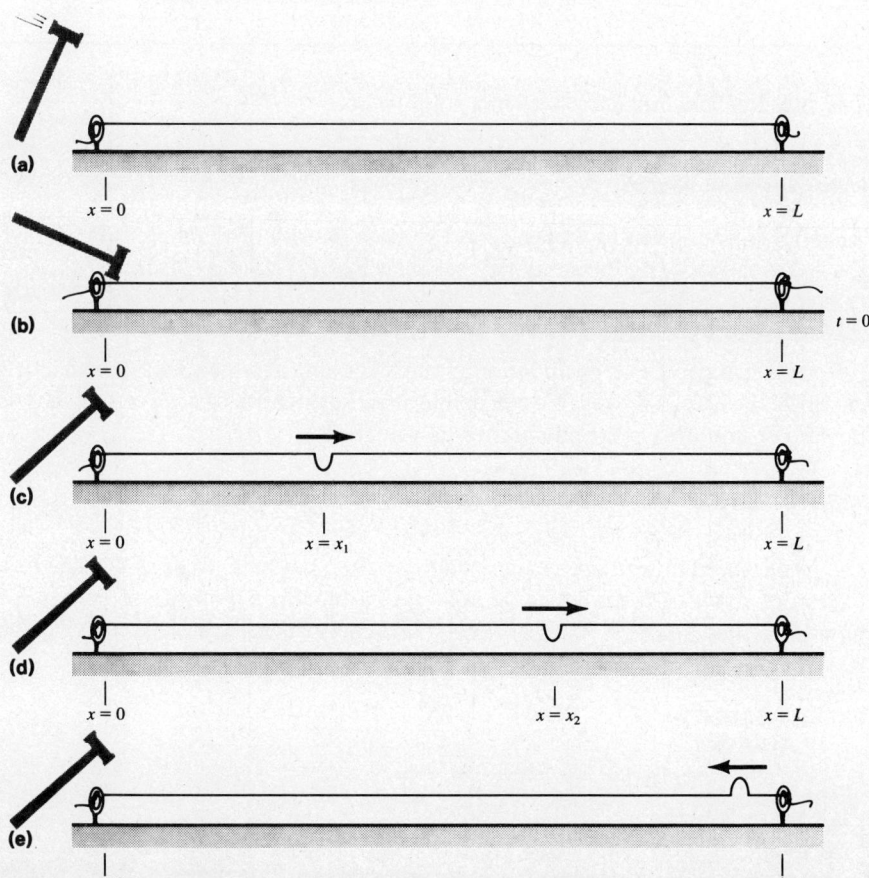

Figure 21-3. (a) The hammer is about to strike the string stretched between the two fixed points. **(b)** At $t = 0$ the impact of the hammer deforms the string at the point $x = 0$. **(c)** At $t = t_1$ the wave pulse has traveled a distance x_1. **(d)** At $t = t_2 = 2t_1$ the disturbance in the string has traveled twice as far: $x_2 = 2x_1$. **(e)** Upon reaching the fixed point at $x = L$, the wave pulse is reflected. The amplitude of a pulse reflected from a fixed point is *opposite* that of the incident pulse.

transverse waves traveling along the string is

**SPEED OF
TRANSVERSE WAVES**
$$v = \sqrt{\frac{S}{\mu}}$$
(21-4)

- When a wave traveling along a string reaches a *fixed* end point, it is *reflected* from the end point with an amplitude *opposite* its original amplitude (Fig. 21-3e).

EXAMPLE 21-4: A piece of string is stretched with a tension of 20 N between two fixed points 1.5 m apart. The mass of the string between these two points is 0.9 g. **(a)** With what velocity do transverse waves travel in this string? **(b)** Find the wave number for sinusoidal waves that have a frequency of 120 Hz traveling in this string.

Solution:

(a) First compute the linear density of the string, then use Eq. (21-4) to obtain the wave velocity.

$$\mu = \frac{m}{L} = \frac{0.9 \times 10^{-3} \text{ kg}}{1.5 \text{ m}} = 6 \times 10^{-4} \text{ kg m}^{-1}$$

$$v = \sqrt{\frac{S}{\mu}} = \sqrt{\frac{20 \text{ N}}{6 \times 10^{-4} \text{ kg m}^{-1}}} = \boxed{183 \text{ m s}^{-1}}$$

(b) Use the wave formula, $v = \lambda \nu$, to express λ in terms of the known quantities v and ν, then substitute this expression into the definition of wave number, $k = 2\pi/\lambda$.

$$\lambda = \frac{v}{\nu} \qquad k = \frac{2\pi}{\lambda} = \frac{2\pi\nu}{v} = \frac{(2\pi)(120 \text{ s}^{-1})}{183 \text{ m s}^{-1}} = \boxed{4.12 \text{ m}^{-1}}$$

The speed of *longitudinal* waves in a solid rod is

**SPEED OF
LONGITUDINAL WAVES**
$$v = \sqrt{\frac{Y}{\rho}}$$
(21-5)

where Y is Young's modulus and ρ is the mass density of the material of which the rod is made. Note the similarity between Eqs. (21-4) and (21-5).

21-5. Energy and Power in Waves

All waves transport energy and momentum. The *rate* at which a wave transports energy is the wave's power. We can define also the **intensity** of a wave, which is its power per unit area perpendicular to its velocity

WAVE INTENSITY
$$I = \frac{P}{a}$$
(21-6)

and is measured in watts per square meter in the SI system. Figure 21-4 shows a section of string of cross-sectional area a. In a time Δt a quantity of energy ΔE flows into the volume $V = a\Delta x = av\Delta t$. The **energy density** u of the wave is the wave energy per unit volume:

**ENERGY DENSITY
OF A WAVE**
$$u = \frac{\Delta E}{\Delta V} = \frac{\Delta E}{a\Delta x} = \frac{\Delta E}{av\Delta t} = \frac{P}{av}$$
(21-7)

where the power transported is $P = \Delta E/\Delta t$. Thus we can also express the intensity of a wave as the product of energy density and wave velocity:

$$I = uv$$
(21-8)

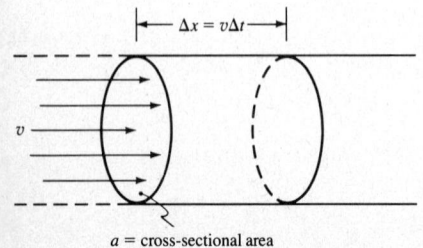

$\Delta x = v\Delta t$

v

a = cross-sectional area

Figure 21-4

For sine waves in a string that has a mass density ρ and is vibrating at an angular frequency ω, the energy density is

ENERGY DENSITY OF HARMONIC WAVES $u = \tfrac{1}{2}\rho\omega^2 A^2$ **(21-9)**

where A is the amplitude of the waves. Therefore the intensity of sine waves is

INTENSITY OF HARMONIC WAVES $I = \tfrac{1}{2}\rho\omega^2 A^2 v$ **(21-10)**

EXAMPLE 21-5: Derive Eq. (21-9).

Solution: Each particle or mass element of a string in which sine waves are traveling executes simple harmonic motion around its equilibrium location. The amplitude of this motion is A, identical to the amplitude of the wave. The total energy of a harmonic oscillator is, according to Eq. (7-11),

$$E = \tfrac{1}{2}kX^2 = \tfrac{1}{2}kA^2$$

where k is the constant that relates the angular frequency of oscillation and mass. According to Eq. (7-3), $\omega = \sqrt{k/m}$. Substitute for k above and let $m = \Delta m$, the mass of the oscillating particle or mass element.

$$\Delta E = \tfrac{1}{2}kA^2 = \tfrac{1}{2}\Delta m\omega^2 A^2$$

where ΔE is the total energy associated with the motion of the element Δm. Divide by the volume occupied by Δm to obtain the energy density.

$$u = \frac{\Delta E}{\Delta V} = \frac{1}{2}\left(\frac{\Delta m}{\Delta V}\right)\omega^2 A^2 = \boxed{\frac{1}{2}\rho\omega^2 A^2}$$

EXAMPLE 21-6: The string in Example 21-4 has a diameter of 0.5 mm and the waves traveling in it have an amplitude of 3 cm. (a) What is the energy density of these waves? (b) What is their intensity?

Solution:

(a) First compute the density of the string, where a is the string's cross-sectional area and d is its diameter.

$$\rho = \frac{m}{V} = \frac{m}{aL} = \frac{\mu}{a} = \frac{\mu}{\dfrac{\pi}{4}d^2} = \frac{4\mu}{\pi d^2}$$

$$= \frac{4(6 \times 10^{-4}\ \text{kg}\,\text{m}^{-1})}{\pi(0.5 \times 10^{-3}\ \text{m})^2} = 3.056 \times 10^3\ \text{kg}\,\text{m}^{-3}$$

Then use Eq. (21-9):

$$u = \tfrac{1}{2}\rho\omega^2 A^2$$

$$= \tfrac{1}{2}(3.056 \times 10^3\ \text{kg}\,\text{m}^{-3})[(2\pi)(120\ \text{Hz})]^2(3 \times 10^{-2}\ \text{m})^2$$

$$= \boxed{7.82 \times 10^5\ \text{J}\,\text{m}^{-3}}$$

(b) Use Eq. (21-8):

$$I = uv = (7.82 \times 10^5\ \text{J}\,\text{m}^{-3})(183\ \text{m}\,\text{s}^{-1}) = \boxed{1.43 \times 10^8\ \text{W}\,\text{m}^{-2}}$$

This last result may appear outrageously large for a string, but keep in mind that the cross-sectional area of the string is only $a = \dfrac{\pi}{4}d^2 = 1.96 \times 10^{-7}\ \text{m}^2$.

Therefore the power transmitted by this string in this example is only

$$P = Ia = (1.43 \times 10^8\ \text{W}\,\text{m}^{-2})(1.96 \times 10^{-7}\ \text{m}^2) = 28\ \text{W}$$

21-6. Interference of Waves

Wave motion obeys the **superposition principle**, which states

- The motion of the medium at a point where two or more waves are present is the algebraic sum of the motion due to the individual waves at that point.

When the motions of the waves reinforce one another, we say that **constructive interference** results. When the motions oppose one another, we say **destructive interference** results. Consider the superposition of two harmonic waves of the same amplitude and frequency but with a **phase difference** ϕ. From the superposition principle, we can express the displacement y_s of any point in the medium by $y_s = y_1 + y_2$. In terms of the wave function, Eq. (21-2c), this is

$$y_s = A \sin(kx - \omega t) + A \sin(kx - \omega t + \phi)$$

If we let $\alpha = kx - \omega t$ and use the trigonometric identities $\cos(-\theta) = \cos\theta$ and

$$\sin\alpha + \sin\beta = 2\sin\left(\frac{\alpha + \beta}{2}\right)\cos\left(\frac{\alpha - \beta}{2}\right)$$

we get

$$y_s = A[\sin\alpha + \sin(\alpha + \phi)]$$

$$= 2A\sin\left(\alpha + \frac{\phi}{2}\right)\cos\left(-\frac{\phi}{2}\right)$$

$$= 2A\cos\left(\frac{\phi}{2}\right)\sin\left(kx - \omega t + \frac{\phi}{2}\right) \qquad \textbf{(21-11)}$$

The result is a wave that has the same frequency as the original waves, a phase that differs from the phase of each original by $\phi/2$, and an amplitude that depends on ϕ (Fig. 21-5). The resultant wave repeats itself every π radians, so it is sufficient to know the amplitude dependence in the range $0 \le \phi \le \pi$. When the phase difference is zero or an even integer multiple of π the resultant wave has twice the amplitude of either original wave, but when the phase difference is π or an odd integer multiple of π,

$$\phi = (2n + 1)\pi, \quad n = 0, 1, 2, 3\ldots$$

and the resultant wave has an amplitude of zero.

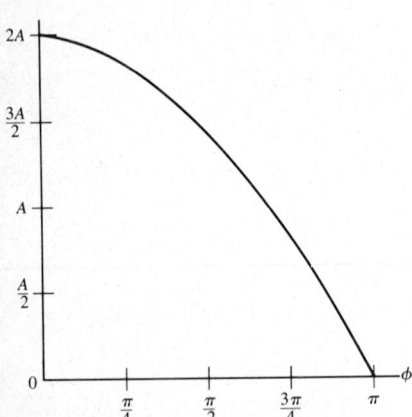

Figure 21-5. Amplitude of y_s as a function of phase difference ϕ.

EXAMPLE 21-7: Find the phase difference for which the resultant wave has an amplitude equal to that of each original.

Solution: Use Eq. (21-11), set the amplitude of the resultant wave equal to the amplitude of one of the original waves, and solve for ϕ.

$$2\cos\left(\frac{\phi}{2}\right)A = A$$

$$\cos\left(\frac{\phi}{2}\right) = \frac{1}{2} = 0.5$$

$$\phi = 2\cos^{-1}(0.5) = \boxed{2.094 \text{ rad} = 120°}$$

The superposition of two harmonic waves of equal amplitude may produce interference effects in three different ways.

A. Phase difference due to difference in path length

Think of two sources, each generating waves of identical amplitude and frequency. When waves from the two sources meet at a third point in space, they may differ in phase at that point because the distances between the sources and the point are unequal, i.e., because of a difference in *path length* Δx (see Fig. 21-6). Let the wave arriving at the point from source S_1 be $y_1 = A \sin(kx - \omega t)$ and the wave arriving from source S_2 be

$$y_2 = A \sin[k(x + \Delta x) - \omega t] = A \sin(kx - \omega t + k\Delta x)$$

where Δx is the difference in path length. The phase difference therefore is

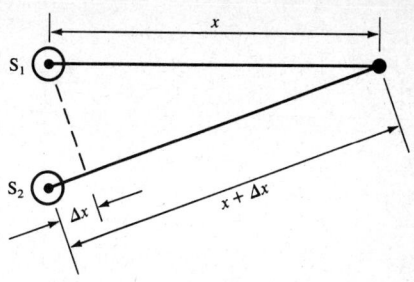

Figure 21-6

PHASE DIFFERENCE DUE TO PATH DIFFERENCE
$$\phi = k\Delta x = \frac{2\pi}{\lambda}\Delta x \qquad \text{(21-12)}$$

So when the path difference is zero or an integer multiple of λ,

$$\Delta x = n\lambda \qquad n = 0, 1, 2, 3 \ldots$$

$\phi = 2\pi n$ and $2\cos(\phi/2)A = 2A$ or $-2A$. The displacement of the medium has its maximum value and we say **complete constructive interference** results. When the path difference is an odd integer multiple of a half wavelength,

$$\Delta x = (n + \tfrac{1}{2})\lambda, \qquad n = 0, 1, 2, 3 \ldots$$

$\phi = \pi(2n + 1)$ and $2\cos(\phi/2)A = 0$. The two waves cancel each other completely and we say **complete destructive interference** results.

B. Phase difference due to difference in frequency

If the path lengths from two wave sources to a point are equal, the waves may differ in phase at the point if the two sources differ in frequency. For mathematical simplicity, let's assign the point the position $x = 0$. Then we can express the wave function due to the two waves at this point as

$$y_s = y_1 + y_2 = A \sin\omega_1 t + A \sin\omega_2 t$$

(The minus signs on ω_1 and ω_2 don't affect the result, so we can omit them.) If we express the difference between the two angular frequencies as $\omega_1 - \omega_2 = \Delta\omega$,

$$y_s = A \sin\omega_1 t + A \sin(\omega_1 t - \Delta\omega t)$$

So $\Delta\omega t$ represents the phase difference ϕ in this case. Notice that the phase difference is *time-dependent*. Figure 21-7c shows the function y_s that results from the superposition of y_1 and y_2 when $\omega_1 = 17\pi \text{ rad s}^{-1}$ (Fig. 21-7a), $\omega_2 = 15\pi \text{ rad s}^{-1}$ (Fig. 21-7b), and the time interval $\Delta t = t_3 - t_1$ is one second. At the instants t_1 and t_3, y_1 and y_2 are completely *out of phase*, and since they have the same amplitude, they cancel. At the instant t_2 they are *in phase* and their superposition results in an amplitude of $2A$. Applying again our same trigonometric identity for the sum of two angles gives us

$$y_s = 2A \cos\left(\frac{\Delta\omega}{2}t\right)\sin\left[\left(\frac{\omega_1 + \omega_2}{2}\right)t\right]$$

$$= 2A \cos\left(\frac{\Delta\omega}{2}t\right)\sin(\bar{\omega}t) \qquad \text{(21-13)}$$

where $\bar{\omega} = \dfrac{\omega_1 + \omega_2}{2}$ is the average angular frequency. Equation (21-13) tells us that y_s is a product of two functions, $\sin\bar{\omega}t$, which represents the *frequency* of y_s, and $2A \cos\left(\dfrac{\Delta\omega}{2}t\right)$, which represents the *amplitude* of y_s. This amplitude is time-dependent too, and if we plot it as a function of time, we get the dotted line in Figure 21-7c. The amplitude function $2A \cos\left(\dfrac{\Delta\omega}{2}t\right)$ is

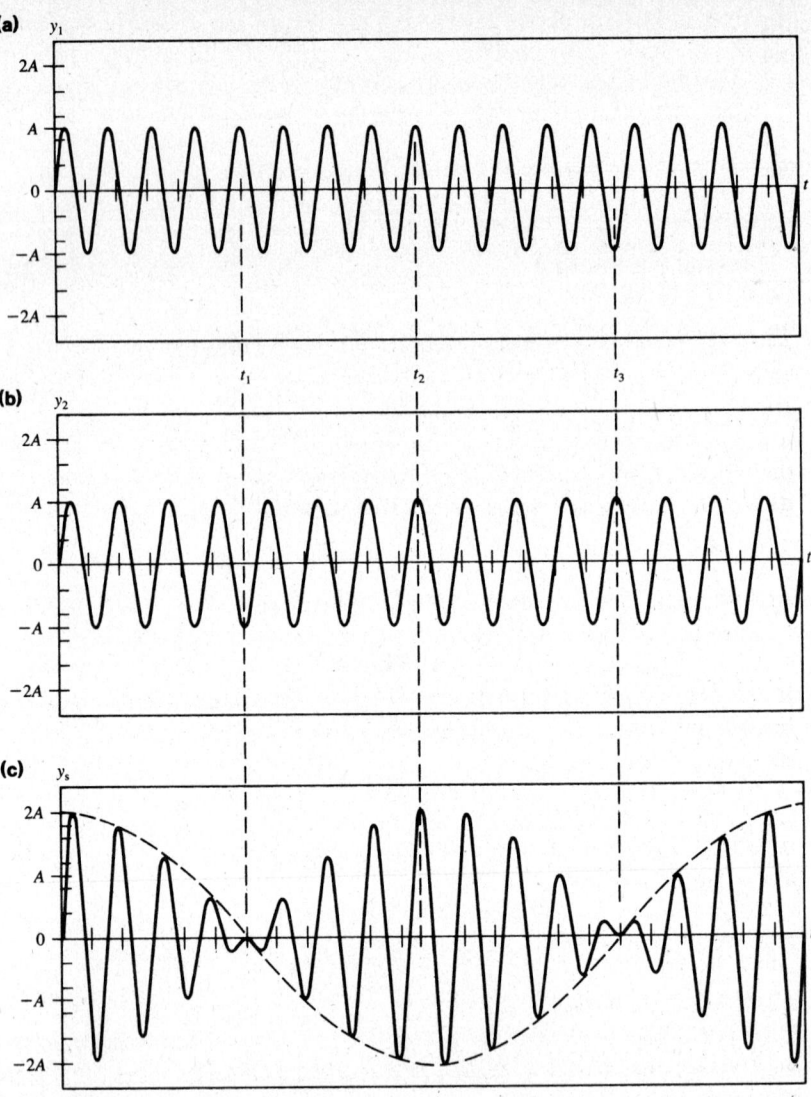

Figure 21-7. (a) $y_1 = A \sin \omega_1 t$ where $\omega_1 = 17\pi$ rad s^{-1}; $\omega_1 = 2\pi v_1$ so $v_1 =$ 8.5 Hz. **(b)** $y_2 = A \sin \omega_2 t$ where $\omega_2 = 15\pi$ rad s^{-1}; $\omega_2 = 2\pi v_2$ so $v_2 = 7.5$ Hz. **(c)** $y_s = y_1 + y_2$.

sometimes called the **modulation**. If y_1 and y_2 are sound waves, the sound we hear as the result of y_1 and y_2 interfering, y_s, has a frequency of $\bar{\omega}/2\pi = (v_1 + v_2)/2$ and an amplitude (loudness) that varies from 0 to $|2A|$ according to the modulation function. We hear a succession of discrete noises or **beats**, where each beat extends from one zero of the modulation function to the next. The reciprocal of the time between beats, the **beat frequency** v_b, is

BEAT FREQUENCY $v_b = v_1 - v_2$ **(21-14)**

where $2\pi v_1 = \omega_1$ and $2\pi v_2 = \omega_2$. We encounter beats most frequently in acoustics (the study of sound) but the same principle applies to all types of wave motion.

EXAMPLE 21-8: Start with Eq. (21-13) and derive Eq. (21-14).

Solution: The time between zeros of the modulation function $2A \cos\left(\dfrac{\Delta\omega}{2}t\right)$ is half the period T of the modulation function. Thus the beat frequency is the reciprocal of half the period: $v_b = 2/T$. Each period T corresponds to 2π radians, so

$$\frac{\Delta\omega}{2} T = 2\pi$$

Solving for T and substituting,

$$v_b = \frac{2}{T} = \frac{2\Delta\omega}{4\pi} = \frac{\Delta\omega}{2\pi} = \frac{2\pi v_1 - 2\pi v_2}{2\pi} = \boxed{v_1 - v_2}$$

C. Phase difference is zero but waves move in opposite directions

Interference effects also arise then two waves with the same frequency traveling in opposite directions are in phase. This situation is produced when a wave is reflected from a fixed end point, so that its amplitude is reversed (Sec. 21-4). We can express the sum of two waves traveling in opposite directions (see Eqs. 21-2c and 21-2d) as

$$y_s = y_1 + y_2 = A \sin(kx - \omega t) + A \sin(kx + \omega t)$$

Applying our handy trigonometric identity again gives us

$$y_s = 2A \sin(kx)\cos(-\omega t) \tag{21-15}$$

In this case the medium oscillates at the same frequency as the original waves but the amplitude, $2A \sin(kx)$, depends on position. The amplitude is zero at the points where

$$kx = n\pi \qquad \text{for} \qquad n = 0, 1, 2, 3, \ldots.$$

because $\sin(n\pi) = 0$ for $n = 0, 1, 2, 3 \ldots$. These locations of zero amplitude along the x axis are

$$x = \frac{n\pi}{k} = \frac{n\pi}{2\pi/\lambda} = n\left(\frac{\lambda}{2}\right) \qquad n = 0, 1, 2, 3 \ldots$$

and are called **nodes**. Note that the distance between nodes is a half wavelength. The locations halfway between nodes where the medium oscillates with its maximum amplitude of $2A$ are called **antinodes**. Figure 21-8 shows the medium at seven successive instants of time. The time span $t_7 - t_1$ is equal to half the period of the wave motion. Because the location of the nodes and antinodes remains stationary, this interference effect is called **standing wave** motion.

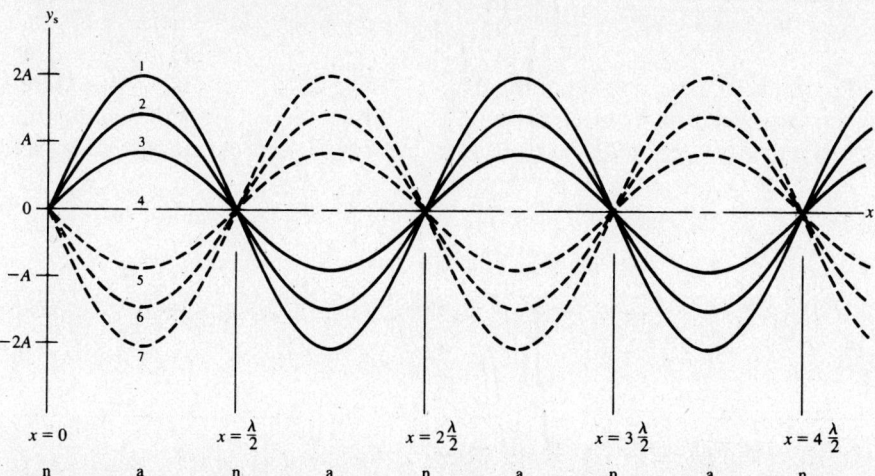

Figure 21-8. The standing wave pattern in a medium at seven successive instants of time, beginning at t_1. The letters n and a denote the locations of nodes and antinodes, respectively.

EXAMPLE 21-9: Waves of frequency 500 Hz are sent in opposite directions down a wire stretched with a tension of 6×10^3 N. A 1-m length of this wire has a mass of 6 g. What is the distance between the nodes of the standing wave pattern observed?

Solution: The distance between nodes is half a wavelength. Use the wave formula to calculate the wavelength, but first you must find the wave velocity with Eq. (21-4).

$$v = \sqrt{\frac{S}{\mu}} = \sqrt{\frac{6 \times 10^3 \text{ N}}{6 \times 10^{-3} \text{ kg m}^{-1}}} = 10^3 \text{ m s}^{-1}$$

From Eq. (21-1b), $v = \lambda v$,

$$\lambda = \frac{v}{v} = \frac{10^3 \text{ m s}^{-1}}{500 \text{ Hz}} = 2 \text{ m}$$

The distance between nodes is $\frac{1}{2}\lambda = \boxed{1 \text{ m}}$

SUMMARY

1. *Mechanical waves* are the propagation of a disturbance through a *medium*; they transport energy and momentum.
2. In *transverse waves* the medium oscillates in a direction *perpendicular* to the direction of wave propagation; in *longitudinal waves* the medium oscillates in a direction *parallel* to the direction of wave propagation.
3. The relationship between wavelength, frequency, and wave velocity is given by the *wave formula*, $v = \lambda v$.
4. When a harmonic wave moves through a medium in the positive direction along the x axis, the displacement from its equilibrium position ($y = 0$) of any point in the medium is given by

$$y(x, t) = A \sin(kx - \omega t)$$

where A is the amplitude of the motion (the maximum displacement of the medium), k is the wave number $k = 2\pi\lambda^{-1}$, and ω is the angular frequency $\omega = 2\pi v$. For waves moving in the negative x direction the equation becomes

$$y(x, t) = A \sin(kx + \omega t)$$

5. The *wave equation* for waves moving with velocity v in one dimension is

$$\frac{\partial^2 \phi}{\partial x^2} = \frac{1}{v^2}\frac{\partial^2 \phi}{\partial t^2}$$

where $\phi(x, t)$ is the *wave function*.
6. The speed of transverse waves traveling in a string that has a linear density μ, stretched with a tension S, is

$$v = \sqrt{\frac{S}{\mu}}$$

7. The speed of longitudinal waves traveling in a solid rod of mass density ρ that has a Young's modulus Y is

$$v = \sqrt{\frac{Y}{\rho}}$$

8. The *intensity* of a moving wave is its power per unit area:

$$I = \frac{P}{a} = uv$$

where u is the *energy density* of the wave, the energy per unit volume.

9. For a harmonic wave with amplitude A moving in a medium of mass density ρ

$$u = \tfrac{1}{2}\rho\omega^2 A^2$$

10. Two waves that have identical frequency and amplitude but a *phase difference* ϕ are

$$y_1 = A\sin(kx - \omega t) \quad \text{and} \quad y_2 = A\sin(kx - \omega t + \phi)$$

11. Interference effects are produced when two or more waves overlap because the medium obeys the *superposition principle.*

12. *Complete constructive interference* of two waves of identical frequency and amplitude occurs where the difference in the waves' path lengths $\Delta x = n\lambda$, $n = 0, 1, 2, 3\ldots$. *Complete destructive interference* of two waves of identical frequency and amplitude occurs where the difference in the waves' path lengths $\Delta x = (n + \tfrac{1}{2})\lambda$, $n = 0, 1, 2, 3\ldots$.

13. When two waves differ in frequency their interference produces *beats*. The *beat frequency* is $\nu_b = \nu_1 - \nu_2$.

14. Identical waves traveling in opposite directions interfere to produce *standing waves*. The distance between *nodes* of standing waves is $\tfrac{1}{2}\lambda$.

RAISE YOUR GRADES

Can you explain . . . ?

☑ why the velocity of transverse waves differs from the velocity of longitudinal waves in a given medium

☑ why the velocity of transverse waves is less than the velocity of longitudinal waves in a given medium

☑ why longitudinal waves (pressure waves) can propagate in a gas but transverse waves cannot

☑ why waves of different frequencies may have different velocities in a given medium

☑ why increasing the tension in a string increases the velocity of transverse waves

☑ why the intensity of a wave depends upon the square of its amplitude rather than on just the first power of the amplitude

☑ whether or not interference effects could result from the overlapping of two waves with somewhat different amplitudes

☑ how the phenomenon of beat frequency is used in tuning musical instruments

SOLVED PROBLEMS

Periodic Waves

PROBLEM 21-1: The speed of longitudinal waves in an aluminum rod is $5.06 \times 10^3 \text{ m s}^{-1}$. What is the wavelength of longitudinal waves in an aluminum rod if they have a frequency of 2.8×10^3 Hz?

Solution: Use the wave formula, Eq. (21-1b), and solve for λ.

$$v = \lambda \nu$$

$$\lambda = \frac{v}{\nu} = \frac{5.06 \times 10^3 \text{ m s}^{-1}}{2.8 \times 10^3 \text{ Hz}} = \boxed{1.81 \text{ m}}$$

PROBLEM 21-2: A wave that has an amplitude of 3 cm and a wavelength of 80 cm travels through a medium along the positive x axis with a speed of 22 m s^{-1}. **(a)** Find the displacement of the medium from equilibrium at $x = 7.1$ m and $t = 0.2$ s. **(b)** What is the wave's frequency?

Solution:

(a) Because the wave is moving along the positive x axis, use Eq. (21-1c). Use the wave formula to express k and ω in terms of the known quantities λ and v.

$$y(x,t) = A \sin(kx - \omega t) = A \sin\left(\frac{2\pi x}{\lambda} - 2\pi \nu t\right) = A \sin\left(\frac{2\pi x}{\lambda} - \frac{2\pi v t}{\lambda}\right) = A \sin\left[\frac{2\pi}{\lambda}(x - vt)\right]$$

$$= (0.03 \text{ m}) \sin\left\{\frac{2\pi}{0.8 \text{ m}}[7.1 \text{ m} - (22 \text{ m s}^{-1})(0.2 \text{ s})]\right\} = \boxed{+2.12 \times 10^{-2} \text{ m} = +2.12 \text{ cm}}$$

Remember that the units of the argument of the sine function are radians.

(b) Use the wave formula, Eq. (21-1b), and solve for v.

$$v = \lambda \nu$$

$$\nu = \frac{v}{\lambda} = \frac{22 \text{ m s}^{-1}}{0.8 \text{ m}} = \boxed{27.5 \text{ Hz}}$$

The Wave Equation

PROBLEM 21-3: Show that any continuous function of the form $\phi(x - vt)$ is a solution of the one-dimensional wave equation, Eq. (21-3a).

Solution: Let $z = (x - vt)$ so that $\phi(x - vt) = \phi(z)$. Differentiate twice with respect to x and use the chain rule:

$$\frac{\partial \phi}{\partial x} = \frac{\partial \phi}{\partial z} \times \frac{\partial z}{\partial x} = \frac{\partial \phi}{\partial z} \qquad \text{because} \qquad \frac{\partial z}{\partial x} = 1$$

$$\frac{\partial^2 \phi}{\partial x^2} = \frac{\partial}{\partial x}\left(\frac{\partial \phi}{\partial z}\right) = \frac{\partial^2 \phi}{\partial z^2} \times \frac{\partial z}{\partial x} = \frac{\partial^2 \phi}{\partial z^2}$$

Differentiate twice with respect to t and use the chain rule:

$$\frac{\partial \phi}{\partial t} = \frac{\partial \phi}{\partial z} \times \frac{\partial z}{\partial t} = \frac{\partial \phi}{\partial z}(-v) = -v\frac{\partial \phi}{\partial z} \qquad \text{because} \qquad \frac{\partial z}{\partial t} = -v$$

$$\frac{\partial^2 \phi}{\partial t^2} = \frac{\partial}{\partial t}\left(-v\frac{\partial \phi}{\partial z}\right) = -v\frac{\partial}{\partial t}\frac{\partial \phi}{\partial z} = -v\frac{\partial^2 \phi}{\partial z^2} \times \frac{\partial z}{\partial t} = -v\frac{\partial^2 \phi}{\partial z^2}(-v)$$

$$\frac{\partial^2 \phi}{\partial t^2} = v^2\frac{\partial^2 \phi}{\partial z^2} = v^2\frac{\partial^2 \phi}{\partial x^2} \qquad \text{because, from above,} \qquad \frac{\partial^2 \phi}{\partial z^2} = \frac{\partial^2 \phi}{\partial x^2}$$

Divide by v^2 and rearrange to obtain the wave equation.

$$\frac{\partial^2 \phi}{\partial x^2} = \frac{1}{v^2}\frac{\partial^2 \phi}{\partial t^2}$$

Notice that the function $\phi(x + vt)$ is also a solution of the wave equation. You can easily prove this statement by letting $z = x + vt$ and repeating the same procedure.

Velocities of Waves

PROBLEM 21-4: The distance between the bridge and the pin for the wire of a piano string is 1.4 m. This distance is also half a wavelength for the transverse waves that travel back and forth in the string when it vibrates at 440 Hz. The wire has a linear density of 4.5 g m^{-1}. What is the tension in the wire?

Solution: Use the wave formula to calculate the wave velocity and then solve Eq. (21-4) for the tension S. The wavelength is $\lambda = 2(1.4 \text{ m}) = 2.8$ m.

$$v = \lambda v = (2.8 \text{ m})(440 \text{ Hz}) = 1.232 \times 10^3 \text{ m s}^{-1}$$

$$v = \sqrt{\frac{S}{\mu}}$$

$$S = \mu v^2 = (4.5 \times 10^{-3} \text{ kg m}^{-1})(1.232 \times 10^3 \text{ m s}^{-1})^2 = \boxed{6.83 \times 10^3 \text{ N}}$$

There are three strings for most notes in a piano to increase the instrument's loudness.

PROBLEM 21-5: Compare the speed of longitudinal waves in a steel rod with those in an aluminum rod. Young's moduli for steel and aluminum are 22×10^{10} Pa and 6.9×10^{10} Pa, respectively. The metals' respective densities are $7.83 \times 10^3 \text{ kg m}^{-3}$ and $2.7 \times 10^3 \text{ kg m}^{-3}$ (see Tables 13-1 and 14-1).

Solution: Use Eq. (21-5) for each material.

$$v = \sqrt{\frac{Y}{\rho}}$$

$$\text{Steel:} \qquad v_S = \sqrt{\frac{22 \times 10^{10} \text{ Pa}}{7.83 \times 10^3 \text{ kg m}^{-3}}} = \boxed{5.30 \times 10^3 \text{ m s}^{-1}}$$

$$\text{Aluminum:} \quad v_A = \sqrt{\frac{6.9 \times 10^{10} \text{ Pa}}{2.7 \times 10^3 \text{ kg m}^{-3}}} = \boxed{5.06 \times 10^3 \text{ m s}^{-1}}$$

Longitudinal wave velocity in a steel rod is only about 5% larger than in an aluminum rod.

Energy and Power in Waves

PROBLEM 21-6: A small, spherically symmetric source emits energy in the form of waves uniformly in all directions. The rate of energy emission from this source is 6 W. **(a)** Derive a formula for the intensity of the waves as a function of the distance from the center of the source. **(b)** What is the intensity at $r = 2$ m, where r is the distance from center of the source to a point in space?

Solution:

(a) Define a three-dimensional coordinate system with the source at the origin and let r be the distance from the origin to a point in space (Fig. 21-9). All the energy radiated away from the source must pass through a spherical surface, the center of which is also located at the origin. The surface area of this sphere is $a = 4\pi r^2$. Because the source radiates uniformly in all directions, the intensity has the same value at all points on the spherical surface. So by Eq. (21-6), $I = P/a$,

$$I = \boxed{\frac{P}{4\pi r^2}}$$

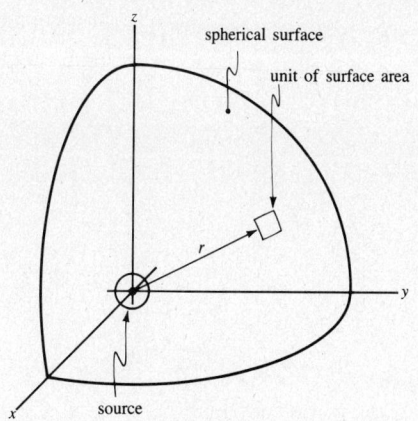

Figure 21-9

Notice that since the intensity is proportional to the amplitude squared for harmonic waves, Eq. (21-10), we can conclude that the amplitude of harmonic waves emitted uniformly from a small source is proportional to r^{-1}.

(b) At $r = 2$ m,

$$I = \frac{P}{4\pi r^2} = \frac{6 \text{ W}}{4\pi(2 \text{ m})^2} = \boxed{0.119 \text{ W m}^{-2}}$$

PROBLEM 21-7: Assume that the waves in Problem 21-6 have a velocity of 340 m s^{-1}. **(a)** Derive an equation for the energy density of the waves from a spherical source as a function of v and r. **(b)** What is the energy density of the waves in Problem 21-6 at $r = 2$ m?

Solution:

(a) Use Eq. (21-8) and the result of Problem 21-6a, and solve for u.

$$I = uv \qquad \text{and} \qquad I = \frac{P}{4\pi r^2}$$

$$u = \frac{I}{v} = \boxed{\frac{P}{4\pi r^2 v}}$$

(b) At $r = 2$ m with $v = 340$ m s^{-1},

$$u = \frac{6 \text{ W}}{4\pi(2 \text{ m})^2(340 \text{ m s}^{-1})} = \boxed{3.51 \times 10^{-4} \text{ J m}^{-3}}$$

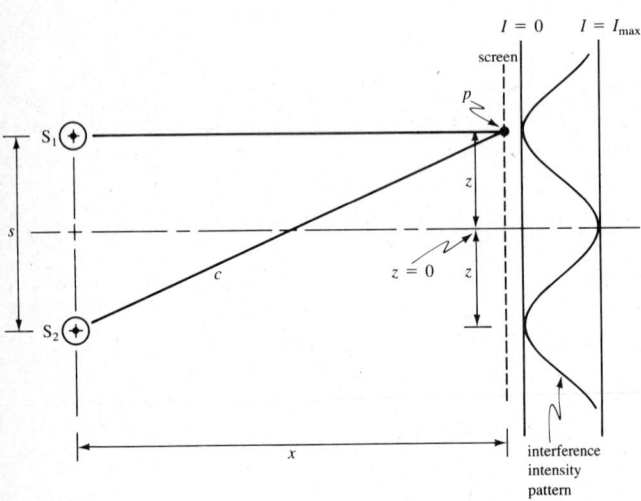

Figure 21-10

Interference of Waves

PROBLEM 21-8: Two sources generating identical waves are located a distance $s = 3$ cm from each other and a distance $x = 2$ m from a screen on which an interference pattern is visible (see Fig. 21-10). At $z = 0$ the intensity is a maximum; at the locations $z = 1.5$ cm, directly opposite from the sources, the intensity is a minimum. **(a)** Determine the wavelength of the waves. **(b)** Each source is moved 0.5 cm toward the other so that they are now 2 cm apart ($s = 2$ cm). Locate the points where the intensity first drops to a minimum; i.e., find the new value of z. [*Hint:* Use approximations when appropriate.]

Solution:

(a) Complete destructive interference results when the path difference between two identical waves is $\Delta x = (n + \frac{1}{2})\lambda$ for $n = 0, 1, 2, 3, \ldots$. Since complete destructive interference occurs at point p, you can find the wavelength by solving $\Delta x = (n + \frac{1}{2})\lambda$ for λ. First, however, calculate Δx. Let $c = \Delta x + x$, and apply the Pythagorean theorem.

$$s^2 + x^2 = c^2 \qquad c = \sqrt{s^2 + x^2}$$

$$\Delta x = c - x = \sqrt{s^2 + x^2} - x$$
$$= \sqrt{(0.03 \text{ m})^2 + (2 \text{ m})^2} - 2 \text{ m} = 2.25 \times 10^{-4} \text{ m}$$

Each value of n corresponds to a given distance from $z = 0$ to a point of destructive interference. Since we are calculating the distance to the *first* point of destructive interference, we substitute the *first* value of n: $n = 0$.

$$\lambda = \frac{\Delta x}{n + \frac{1}{2}} = 2\Delta x = \boxed{4.50 \times 10^{-4} \text{ m}}$$

(b) In this case the geometry is a bit more complicated but we can still express relationships between known quantities with the Pythagorean theorem (see Fig. 21-11). The difference in the paths must again be $\frac{1}{2}\lambda$, $c_2 - c_1 = \frac{1}{2}\lambda$. Now $s = 1$ cm, $\lambda = 4.5 \times 10^{-4}$ m, and $x = 2$ m.

$$c_2^2 = x^2 + \left(z + \frac{s}{2}\right)^2 \qquad \text{and} \qquad c_1^2 = x^2 + \left(z - \frac{s}{2}\right)^2$$

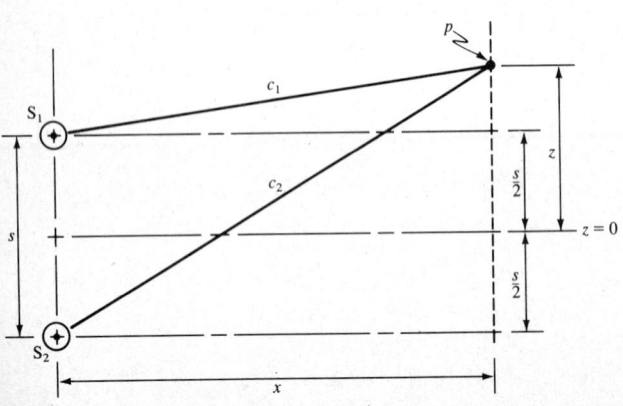

Figure 21-11

Subtract to eliminate x^2 and then expand.

$$c_2^2 - c_1^2 = \left(z + \frac{s}{2}\right)^2 - \left(z - \frac{s}{2}\right)^2$$

$$= z^2 + zs + \frac{s^2}{4} - z^2 + zs - \frac{s^2}{4} = 2zs$$

Substitute $c_2 = c_1 + \dfrac{\lambda}{2}$ and expand.

$$\left(c_1 + \frac{\lambda}{2}\right)^2 - c_1^2 = 2zs$$

$$c_1^2 + c_1\lambda + \frac{\lambda^2}{4} - c_1^2 = 2zs$$

$$\left(c_1 + \frac{\lambda}{4}\right)\lambda = 2zs$$

Use the approximations that $c_1 \simeq x$ and $\lambda/4$ is negligibly small compared to c_1 or x. Therefore to a good approximation:

$$x\lambda \cong 2zs$$

Solve for z.

$$z = \frac{x\lambda}{2s} = \frac{(2 \text{ m})(4.5 \times 10^{-4} \text{ m})}{2(2 \times 10^{-2} \text{ m})} = \boxed{2.25 \times 10^{-2} \text{ m}}$$

Note that in part (a), $z = 1.5$ cm; in part (b) its value increased to 2.25 cm. Moving the sources 1 cm closer together causes the intensity minima to move 1.5 cm farther apart.

PROBLEM 21-9: Two generators of slightly different frequency produce waves in a region of space. The resultant frequency detected near the generators is 1215 Hz and its amplitude changes (increases, decreases, and increases again) at a rate 30 times per second. The frequency of one of the generators is 1200 Hz. What is the frequency of the other generator?

Solution: When two generators of different individual frequencies interfere, the resulting frequency is their simple average. And the rate at which the resultant amplitude changes, the beat frequency, is $v_b = v_1 - v_2$ (Eq. 21-14). The beat frequency in this case is 30 Hz. If $v_1 = 1200$ Hz, then v_2 is 30 Hz either greater or less than v_1. Because their average $\frac{1}{2}(v_1 + v_2) = 1215$ Hz, we can conclude $v_2 = \boxed{1230 \text{ Hz}}$.

PROBLEM 21-10: Waves of the same frequency travel at 680 m s^{-1} through a medium in opposite directions. The distance between nodes in the resulting interference pattern is 24 cm. What is the frequency of the wave generators?

Solution: The distance between two nodes is one half of a wavelength, so use this relationship to find the wavelength, then use the wave formula to find the frequency.

$$x = \frac{\lambda}{2} \qquad \lambda = 2x$$

$$v = \lambda v \qquad v = \frac{v}{\lambda} = \frac{v}{2x} = \frac{680 \text{ m s}^{-1}}{2(0.24 \text{ m})} = \boxed{1.42 \times 10^3 \text{ Hz}}$$

PROBLEM 21-11: Two generators produce waves in a region of space. Generator 1 produces waves of 300 Hz; generator 2 produces waves of 200 Hz (Figure 21-12). (a) What is the beat frequency? (b) What is the modulation frequency? (c) What is the frequency of the resultant waves?

Solution:

(a)
$$v_b = v_1 - v_2 = (300 - 200) \text{ Hz} = \boxed{100 \text{ Hz}}$$

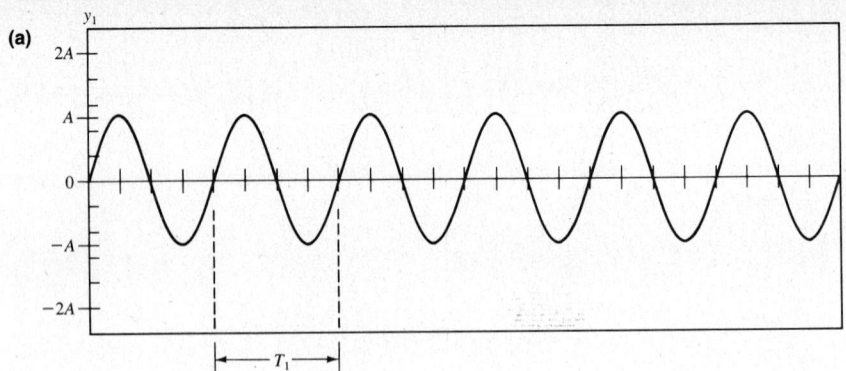

Figure 21-12. **(a)** Graph of $y_1 = A\sin(2\pi v_1 t)$ where $v_1 = 300$ Hz.
(b) Graph of $y_2 = A\sin(2\pi v_2 t)$ where $v_2 = 200$ Hz. **(c)** Graph of

$$y_s = y_1 + y_2 = 2A\cos\left(\frac{2\pi\Delta v}{2}t\right)\sin(2\pi\bar{v}t)$$

where $\Delta v = v_1 - v_2$ and $\bar{v} = (v_1 + v_2)/2$.
The dashed curve is a plot of the
modulation function, $2A\cos(\pi\Delta vt)$.
Notice that the complex resultant
waveform has a period of $2T_2$, which
corresponds to the period of the beat
frequency:

$$\frac{1}{2T_2} = \frac{v_2}{2} = \frac{200\text{ Hz}}{2} = 100\text{ Hz}$$

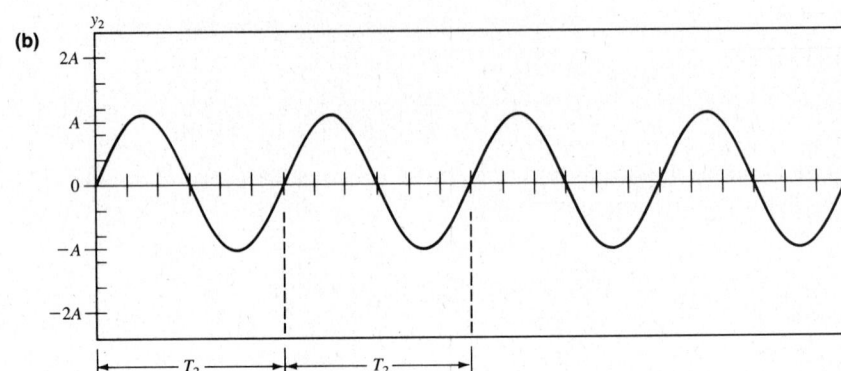

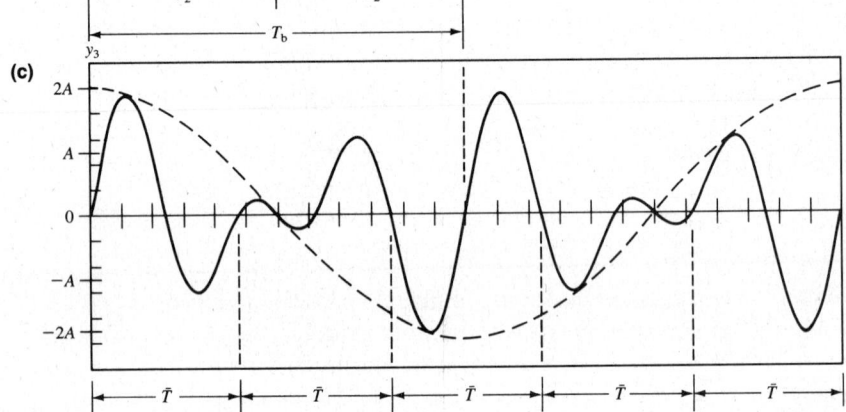

(b) The beat frequency is twice the modulation frequency.

$$\frac{v_b}{2} = \boxed{50\text{ Hz}}$$

(c) The resultant frequency is the average of the generators' frequencies.

$$\bar{v} = \frac{v_1 + v_2}{2} = \frac{(300 + 200)\text{ Hz}}{2} = \boxed{250\text{ Hz}}$$

Supplementary Exercises

EXERCISE 21-1: Waves of wavelength λ_1 travel in medium 1 at velocity v_1. When they pass into medium 2 they travel at velocity v_2. What is their wavelength in medium 2?

EXERCISE 21-2: A certain transverse wave can be described by the wave function

$$y = (4\text{ cm})\sin[\pi(3x - 4t)]$$

where the argument of the sine function is measured in radians. Determine **(a)** the wave amplitude,

(b) the wavelength, **(c)** the frequency, **(d)** the wave velocity, and **(e)** the direction of propagation of the wave.

EXERCISE 21-3: Which wave function is *not* a solution of the one-dimensional wave equation?

(a) $y = A\,e^{j(\omega t - kx)}$ where $j = \sqrt{-1}$ **(b)** $y = A\cos\left[2\pi\left(\dfrac{x}{\lambda}\right) - vt\right]$

(c) $y = A\sin\left(kx + \dfrac{2\pi t}{T}\right)$ **(d)** $y = A\cos\left(\dfrac{2\pi x}{v} - \dfrac{\omega}{v}t\right)$

EXERCISE 21-4: A string that has a mass per unit length of 4×10^{-4} kg m^{-1} is stretched with a tension of 16 N. Transverse waves of identical frequency and amplitude propagate in opposite directions along the string. The distance between nodes in the resulting standing wave pattern is 25 cm. What is the frequency of the wave generators?

EXERCISE 21-5: Calculate the velocity of longitudinal waves in a brass rod. The density and Young's modulus of brass are 8.6×10^3 kg m^{-3} and 9.1×10^{10} Pa, respectively.

EXERCISE 21-6: Transverse waves with an amplitude of 2 cm are sent down a 12 m length of string that has a mass per unit length of 4×10^{-4} kg m^{-1}. The source generating the waves vibrates at 440 Hz. The tension in the string is 16 N. How much kinetic energy resides in the vibrating string at any instant?

EXERCISE 21-7: How much power is transported by the string in Exercise 21-6?

EXERCISE 21-8: Figure 21-13 is a slight modification of Figure 21-11 to include c and θ. Use a procedure similar to that in Problem 21-8b to derive an expression for the location of the first n minima in the intensity interference pattern in terms of s, θ, n, and λ. Assume $x \gg s$, $c \simeq c_1$, and $c_2 - c_1 = \Delta$ where $\Delta = (n + \tfrac{1}{2})\lambda$, $n = 0, 1, 2, 3 \dots$ Ignore terms in Δ^2.

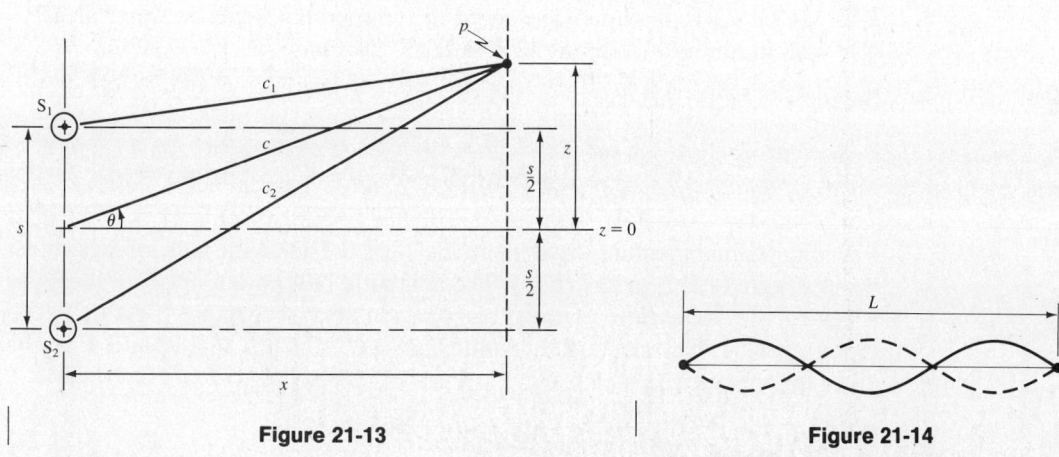

Figure 21-13 **Figure 21-14**

EXERCISE 21-9: A length of string ($L = 1.8$ m) stretched between two points has a mass per unit length of 6×10^{-4} kg m^{-1}. Waves with a frequency of 320 Hz are sent down the string and the standing wave pattern shown in Figure 21-14 results. What is the tension in the string?

Answers to Supplementary Exercises

21-1: $\lambda_2 = \lambda_1\left(\dfrac{v_2}{v_1}\right)$

21-2: **(a)** $A = 4$ cm **(b)** $\lambda = 0.667$ m
(c) $v = 2$ Hz **(d)** $v = 1.33$ m s^{-1}
(e) In the positive x direction.

21-3: **(d)**

21-4: 400 Hz

21-5: 3.25×10^3 m s^{-1}

21-6: $E = \tfrac{1}{2}\mu\ell\omega^2 A^2 = 7.34$ J

21-7: $P = \tfrac{1}{2}\omega^2 A^2\sqrt{S\mu} = 122$ W

21-8: $s\sin\theta = (n + \tfrac{1}{2})\lambda$

21-9: 88.5 N

22 SOUND

THIS CHAPTER IS ABOUT

- ☑ **The Speed of Sound**
- ☑ **Intensity Level**
- ☑ **Musical Intervals**
- ☑ **Resonance of Air Columns**
- ☑ **Beats**
- ☑ **The Doppler Effect**

22-1. The Speed of Sound

The speed of three-dimensional longitudinal pressure waves in a fluid is

SPEED OF PRESSURE WAVES IN A FLUID

$$v = \sqrt{\frac{B}{\rho}} \tag{22-1}$$

where B and ρ are, respectively, the bulk modulus and density of the fluid.

EXAMPLE 22-1: Compute the speed of longitudinal waves in water at 22°C. The bulk modulus of water at 22°C is 2.9×10^9 Pa.

Solution: Use Eq. (22-1).

$$v = \sqrt{\frac{B}{\rho}} = \sqrt{\frac{2.19 \times 10^9 \text{ Pa}}{10^3 \text{ kg m}^{-3}}} = \boxed{1.48 \text{ m s}^{-1}}$$

For longitudinal pressure waves in an ideal gas the **adiabatic bulk modulus** must be substituted for B in Eq. (22-1). The adiabatic bulk modulus of an ideal gas is γP where γ is the ratio of the molar specific heat at constant pressure to the molar specific heat at constant volume, $\gamma = C_p/C_v$ (Eq. 18-13), and P is the absolute pressure.

$$v = \sqrt{\frac{\gamma P}{\rho}} \tag{22-2a}$$

SPEED OF PRESSURE WAVES IN AN IDEAL GAS

or

$$v = \sqrt{\frac{\gamma RT}{M}} \tag{22-2b}$$

where R is the ideal gas constant, T the absolute temperature, and M the molecular weight of the gas. The human ear is sensitive to pressure waves in the air over a frequency range from about 20 Hz to 18×10^3 Hz. We call pressure waves in air **sound** and so Eq. (22-2b) gives us the speed of sound.

EXAMPLE 22-2: Compute the speed of sound in air at 22°C. Assume that air is a diatomic ideal gas with a molecular weight of 29 g mol^{-1}.

Solution: Use Eq. (22-2b) with $\gamma = 1.4$ for a diatomic ideal gas (Example 18-10) and $M = 29 \times 10^{-3}\,\text{kg mol}^{-1}$.

$$v = \sqrt{\frac{\gamma RT}{M}} = \sqrt{\frac{(1.4)(8.314\,\text{J mol}^{-1}\text{K}^{-1})(22 + 273)\text{K}}{29 \times 10^{-3}\,\text{kg mol}^{-1}}} = \boxed{344\,\text{m s}^{-1}}$$

22-2. Intensity Level

Sound pressure p is the slight variation above and below background atmospheric pressure, P_A, to which our ears respond. For sinusoidal sound waves

SOUND PRESSURE $\qquad p = p_0 \sin(kx - \omega t)$ $\qquad$ (22-3)

where p_0 is the **sound pressure amplitude**. The loudest sound that human ears can tolerate without pain, the *threshold of pain*, has a sound pressure amplitude of about 30 Pa. The faintest sound that can be heard, the *threshold of hearing*, has a sound pressure amplitude of about 3×10^{-5} Pa. Note that the ratio of these two thresholds is 10^6. The **sound intensity** is

SOUND INTENSITY $\qquad I = \dfrac{1}{2}\dfrac{p_0^2}{\rho v}$ $\qquad$ (22-4)

where ρ is the density of air and v is the speed of sound.

EXAMPLE 22-3: Calculate the sound intensity at the threshold of hearing. The density of air is $1.20\,\text{kg m}^{-3}$ and the speed of sound is $344\,\text{m s}^{-1}$ at 22°C.

Solution: Use Eq. (22-4).

$$I = \frac{1}{2}\frac{p_0^2}{\rho v} = \left(\frac{1}{2}\right)\frac{(3 \times 10^{-5}\,\text{Pa})^2}{(1.20\,\text{kg m}^{-3})(344\,\text{m s}^{-1})} = \boxed{1.09 \times 10^{-12}\,\text{W m}^{-2}}$$

The threshold of hearing is *defined* to be an intensity of exactly $10^{-12}\,\text{W m}^{-2}$ and denoted by I_0:

$$I_0 \equiv 10^{-12}\,\text{W m}^{-2} \qquad (22\text{-}5)$$

The **sound intensity level** is defined by

SOUND INTENSITY LEVEL $\qquad L = 10\log\left(\dfrac{I}{I_0}\right)$ $\qquad$ (22-6)

where the logarithm of the ratio of intensities is to base 10 and L is expressed in **decibels**, abbreviated db. Ordinary conversation has an intensity level of about 65 db.

EXAMPLE 22-4: What is the sound intensity of ordinary conversation, $L = 65$ db?

Solution: Solve Eq. (22-6) for I.

$$L = 10\log\left(\frac{I}{I_0}\right)$$

$$I = I_0\,\text{antilog}\left(\frac{L}{10}\right)$$

$$I = (10^{-12}\,\text{W m}^{-2})\,\text{antilog}\left(\frac{65}{10}\right) = \boxed{3.16 \times 10^{-6}\,\text{W m}^{-2}}$$

22-3. Musical Intervals

The physical quantity *frequency* is closely associated with the musical quantity **pitch**. On a standard musical instrument, such as a piano, there are seven musical intervals in an **octave**. On a piano, the difference between any two adjacent white keys is one interval, so the notes produced by the keys at the left and right ends of any group of eight consecutive white keys is an octave. The ratio of the frequency of one note to that of the note an octave below is 2. For example, the note A_5 has a frequency of 880 Hz and the note an octave below, A_4, has a frequency of 440 Hz.

$$\frac{\nu_{A_5}}{\nu_{A_4}} = \frac{880 \text{ Hz}}{440 \text{ Hz}} = 2$$

Note that the octaves are numbered from C to C. A musical scale that includes all the sharps and flats is called a **chromatic scale** and has 12 half intervals. The ratio of the frequency of any note to that of the note one half interval down in the *equally tempered* chromatic scale is the 12th root of 2.

$$\frac{\nu_{k+1/2}}{\nu_k} = 2^{1/12} = 1.05946309\ldots$$

The notes and frequencies of the equally tempered chromatic scale from A_4 to A_5 are shown in Figure 22-1. Fixed-note musical instruments are tuned to the equally tempered chromatic scale.

Figure 22-1. The equally tempered chromatic scale from A_4 at 440.0 Hz to A_5 at 880.0 Hz. Each step is half a musical interval.

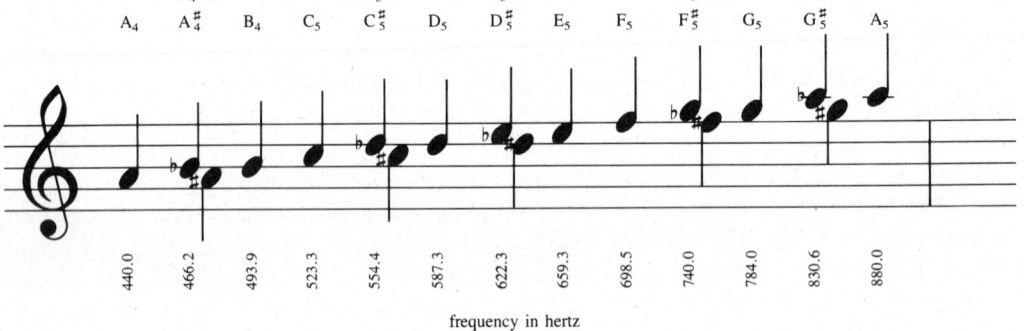

EXAMPLE 22-5: Determine the note that has a frequency of 1108.8 Hz.

Solution: Use the information in Figure 22-1. Because the given frequency is not in the range of those in the figure, divide by 2 to determine if the corresponding note one octave lower is in this range.

$$\frac{1108.8 \text{ Hz}}{2} = 554.4 \text{ Hz}$$

This frequency is $D_5^\flat$ or $C_5^\sharp$ on the equally tempered scale. Therefore 1108.8 Hz is $D_6^\flat$ or $C_6^\sharp$.

22-4. Resonance of Air Columns

When the air in a pipe or column is set into vibration, reflections of the sound waves at the ends of the pipe create standing wave patterns because the ends of the pipe reflect waves of equal amplitude and frequency in opposite directions. Figure 22-2 shows some of the possible patterns or **modes** for standing waves of the sound pressure in (a) a pipe open at both ends and (b) a pipe open at one end.

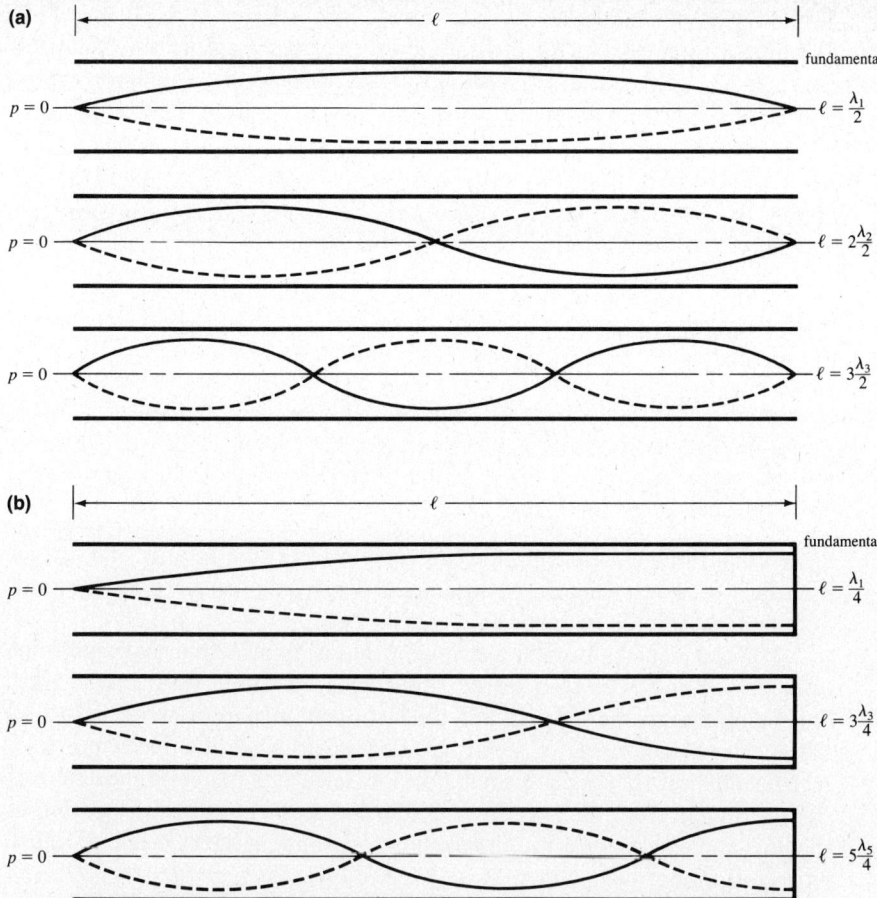

(a)

$p = 0$ fundamental $\ell = \frac{\lambda_1}{2}$

$p = 0$ $\ell = 2\frac{\lambda_2}{2}$

$p = 0$ $\ell = 3\frac{\lambda_3}{2}$

(b)

$p = 0$ fundamental $\ell = \frac{\lambda_1}{4}$

$p = 0$ $\ell = 3\frac{\lambda_3}{4}$

$p = 0$ $\ell = 5\frac{\lambda_5}{4}$

Figure 22-2. The first three resonant modes of air vibrating in **(a)** a pipe open at both ends, and **(b)** a pipe open at one end.

A. Pipe open at both ends

In a pipe with two open ends the sound pressure is always zero at both ends. Therefore in the mode in which the standing waves have their lowest possible frequency, the **fundamental mode**, the length of the pipe is half a wavelength. In all modes of vibration the length of the pipe is an integer multiple of half a wavelength. To find the frequencies (the **harmonic frequencies**) associated with these higher modes, use the wave formula. The first three, as shown in Figure 22-2a, are

$$\left.\begin{array}{c}\text{fundamental}\\ \text{or}\\ \text{first harmonic}\end{array}\right\} \qquad \ell = \frac{\lambda_1}{2} \qquad \lambda_1 = 2\ell \qquad v_1 = \frac{v}{\lambda_1} = \frac{v}{2\ell}$$

$$\text{second harmonic} \qquad \ell = 2\frac{\lambda_2}{2} \qquad \lambda_2 = \ell \qquad v_2 = \frac{v}{\lambda_2} = \frac{v}{\ell} = 2v_1$$

$$\text{third harmonic} \qquad \ell = 3\frac{\lambda_3}{2} \qquad \lambda_3 = \frac{2}{3}\ell \qquad v_3 = \frac{v}{\lambda_3} = \frac{3v}{2\ell} = 3v_1$$

Any higher harmonic frequency can be calculated in the same way. Note that the second harmonic frequency is one octave higher than the fundamental frequency ($v_2 = 2v_1$). The flute is a musical instrument that is, in principle, a straight pipe open at both ends. The mouthpiece approximately locates one open end and the last open hole approximately locates the other. By slightly "overblowing" the player can cause the air in a flute to vibrate in the second (instead of fundamental) mode and thereby produce a pitch an octave higher.

B. Pipe open at one end

In a pipe with one open end and one closed end the sound pressure is always zero at the open end and maximum at the closed end. Therefore, in the fundamental mode, the length of the pipe is one quarter wavelength for a closed-end pipe. In all modes of vibration the length of the pipe is an *odd* integer multiple of a quarter wavelength. The frequencies for the first three modes, as shown in Figure 22-2b, are

$$\left.\begin{matrix} \text{fundamental} \\ \text{or} \\ \text{first harmonic} \end{matrix}\right\} \quad \ell = \frac{\lambda_1}{4} \qquad \lambda_1 = 4\ell \qquad \nu_1 = \frac{v}{\lambda_1} = \frac{v}{4\ell}$$

$$\text{third harmonic.} \quad \ell = 3\frac{\lambda_3}{4} \qquad \lambda_3 = \frac{4\ell}{3} \qquad \nu_3 = \frac{3v}{4\ell} = 3\nu_1$$

$$\text{fifth harmonic,} \quad \ell = 5\frac{\lambda_5}{4} \qquad \lambda_5 = \frac{4\ell}{5} \qquad \nu_5 = \frac{5v}{4\ell} = 5\nu_1$$

Since the air in a closed-end pipe can vibrate according to only odd integer multiples of a wavelength, a closed-end pipe can produce only the odd-numbered harmonic frequencies.

The clarinet is a musical instrument that is, in principle, a straight pipe open at one end. The reed on the mouthpiece is closed most of the time, approximating the closed end, and the last open hole approximately locates the open end. The range of a clarinet is extended when the player presses a register shift key that opens a small hole located about one third the length of the instrument from the mouthpiece. This induces a node at the location of the hole and causes the air in the column to vibrate in the next higher (*third*) harmonic mode.

In actual practice the sound pressure drops to zero slightly beyond the open end of a tube so that the *effective length* of the tube is slightly greater than its actual length. To correct for this end effect, add $0.6r$ to the tube length for each open end, where r is the tube radius.

EXAMPLE 22-6: (a) A pipe 1.5 cm in diameter and 40 cm long is open at both ends. A student blows across the one end, causing the air to vibrate in the fundamental mode. What frequency is heard? (b) The student holds his hand over the far end of the pipe and again blows across the open end. What frequency is heard now, if the air still vibrates in the fundamental mode? Take the speed of sound to be 344 m s^{-1}.

Solution:

(a) When the air in a straight pipe open at both ends vibrates in the fundamental mode the pipe length is one half wavelength:

$$\ell = \frac{\lambda_1}{2} \qquad \text{so} \qquad \lambda_1 = 2\ell$$

where

$$\ell = \ell_{\text{actual}} + 2(0.6r) = \ell_{\text{actual}} + 0.6d = (40 + 0.6 \times 1.5) \text{ cm} = 40.9 \text{ cm}$$

Use the wave formula and solve for v.

$$v = \lambda \nu$$

$$\nu_1 = \frac{v}{\lambda_1} = \frac{v}{2\ell} = \frac{344 \text{ m s}^{-1}}{2(0.409) \text{ m}} = \boxed{420.5 \text{ Hz}}$$

(b) When one end is closed the length of the pipe is one quarter wavelength.

$$\ell = \frac{\lambda}{4} \qquad \text{so} \qquad \lambda = 4\ell$$

where

$$\ell = \ell_{\text{actual}} + 0.6r = (40 + 0.6 \times 0.75) \text{ cm} = 40.45 \text{ cm}$$

Use the wave formula and solve for v.

$$v = \lambda v$$

$$v = \frac{v}{\lambda} = \frac{v}{4\ell} = \frac{344 \text{ m s}^{-1}}{4(0.4045 \text{ m})} = \boxed{212.6 \text{ Hz}}$$

Notice that this frequency is approximately one octave below the frequency produced when both ends are open.

22-5. Beats

The phenomenon of beats (Section 21-5B) occurs in all types of wave motion, but its most familiar use is by musicians tuning their instruments. We can apply the equation for beat frequency, $v_b = v_1 - v_2$ (Eq. 21-14), directly to sound waves.

EXAMPLE 22-7: A musician plays A_4 on her flute. A second musician, who just came in from the cold, attempts to play A_4 but plays flat because her instrument has not yet warmed to room temperature. Six beats per second are heard. What is the frequency of the note played by the second musician?

Solution: A note played flat is below the desired pitch; a note played sharp is above the desired pitch. Since the second musician plays flat, the pitch produced is below A_4. Let $v_1 = 440$ Hz, the frequency of A_4, $v_b = 6$ Hz, use Eq. (21-14), and solve for v_2.

$$v_b = v_1 - v_2$$

$$v_2 = v_1 - v_b = 440 \text{ Hz} - 6 \text{ Hz} = \boxed{434 \text{ Hz}}$$

22-6. The Doppler Effect

The **Doppler effect** is a change in the apparent frequency of waves produced by *relative motion* between a source of waves and a detector. In this section we'll examine the Doppler effect for sound waves only, but the effect applies to all types of waves. In Chapter 35 we'll return to the Doppler effect as it applies to electromagnetic waves.

A. Fixed source, moving observer

Figure 22-3 shows the wave crests produced by a fixed source of sound. The distance between the crests is the wavelength of the sound in air. An observer moving directly toward the source at velocity u_o will detect a frequency v_o *greater* than that of the source:

$$v_o = v\left(1 + \frac{u_o}{v}\right) \qquad \text{(22-7a)}$$

where v is the frequency produced by the source and v is the speed of sound. When the observer moves directly away from the source, the frequency detected is *less* than that of the source:

$$v_o = v\left(1 - \frac{u_o}{v}\right) \qquad \text{(22-7b)}$$

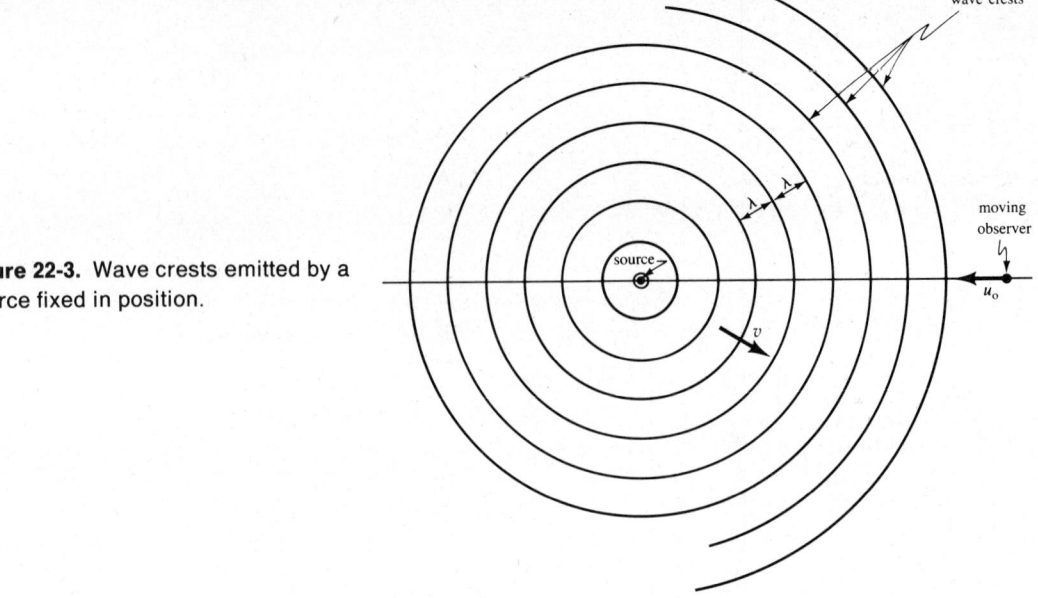

Figure 22-3. Wave crests emitted by a source fixed in position.

We can combine Eqs. (22-7a) and (22-7b) into one equation with the convention that + means the observer is moving toward the source and − means the observer is moving away.

APPARENT FREQUENCY (MOVING OBSERVER) $\qquad v_o = v\left(1 \pm \dfrac{u_o}{v}\right)$ $\qquad\qquad$ **(22-7c)**

B. Moving source, fixed observer

Figure 22-4 shows the wave crests produced by a source of sound moving along a straight line at speed u_s. Each numbered point along the line of motion is the point where the corresponding wave crest was generated. The distance between crests is the *effective* wavelength of the sound.

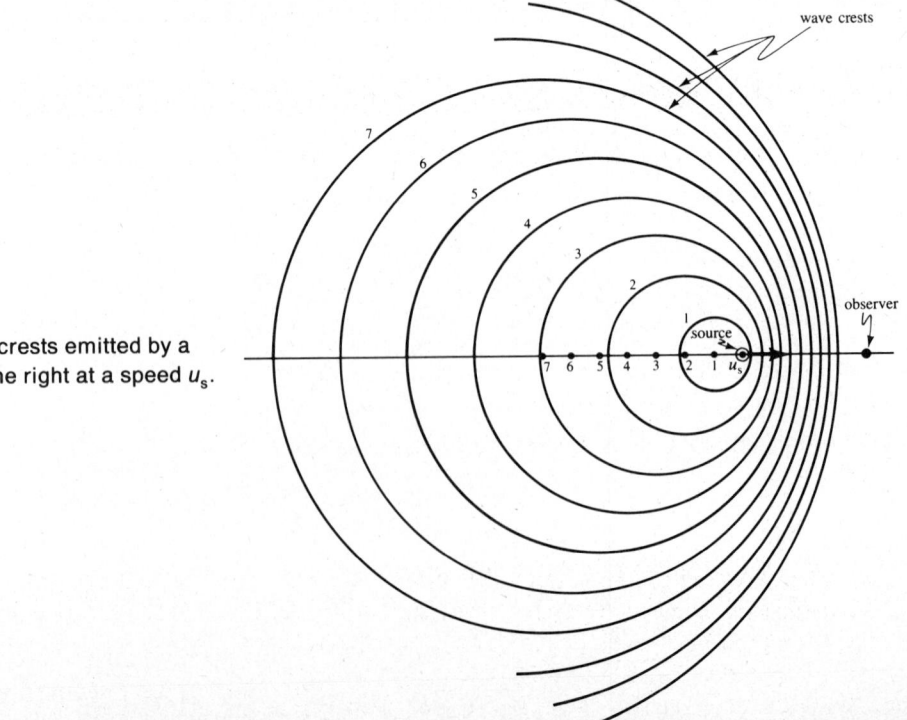

Figure 22-4. Wave crests emitted by a source moving to the right at a speed u_s.

An observer directly in line with the source moving toward him or her will detect a frequency *greater* than that of the source:

$$v_o = v\left(\frac{1}{1 - u_s/v}\right) \qquad \text{(22-8a)}$$

The frequency detected when the source is moving directly away from the observer is

$$v_o = v\left(\frac{1}{1 + u_s/v}\right) \qquad \text{(22-8b)}$$

We can combine these two equations into one equation with the convention that $+$ means the source is moving away from the observer and $-$ means the source is moving toward the observer.

APPARENT FREQUENCY $$v_o = v\left(\frac{1}{1 \pm u_s/v}\right) \qquad \text{(22-8c)}$$
(MOVING SOURCE)

Equations (22-7c) and (22-8c) can be combined into one equation that includes all four cases.

DOPPLER EFFECT $$v_o = v\left(\frac{1 \pm u_o/v}{1 \pm u_s/v}\right) \qquad \text{(22-9)}$$

EXAMPLE 22-8: A source of sound produces a steady frequency and moves to the right at half the speed of sound. An observer moves to the left, toward the source, at half the speed of sound. What is the frequency detected by the observer?

Solution: Use Eq. (22-9) with a plus sign in the numerator, because the observer is moving toward the source at $u_0 = 0.5v$, and a minus sign in the denominator, because the source is moving toward the observer at $u_s = 0.5v$.

$$v_0 = v\left(\frac{1 \pm u_0/v}{1 \pm u_s/v}\right) = v\left(\frac{1 + 0.5v/v}{1 - 0.5v/v}\right) = v\left(\frac{1.5}{0.5}\right) = \boxed{3v}$$

The frequency observed is three times the frequency generated.

SUMMARY

1. The speed of three-dimensional longitudinal waves in a fluid of density ρ is

$$v = \sqrt{\frac{B}{\rho}}$$

where B is the fluid's bulk modulus.

2. The speed of sound in an ideal gas is

$$v = \sqrt{\frac{B_{\text{adiabatic}}}{\rho}} = \sqrt{\frac{\gamma p}{\rho}}$$

3. The speed of sound in an ideal gas depends on temperature:

$$v = \sqrt{\frac{\gamma RT}{M}}$$

where M is the gas' molecular weight.

4. The *sound intensity I* in air of density ρ depends on the *sound pressure amplitude* p_0 according to

$$I = \frac{1}{2}\frac{p_0^2}{\rho v}$$

5. The *intensity level*, specified in decibels (db), of a sound is

$$L = 10 \log\left(\frac{I}{I_0}\right)$$

where I is the sound's intensity in $W\,m^{-2}$ and I_0 is the sound intensity at the threshold of hearing, defined to be $10^{-12}\,W\,m^{-2}$.

6. The frequency of each successive note in the *equally tempered chromatic scale* differs by a factor of $2^{1/12}$.

7. Air in a pipe of length ℓ open at both ends can vibrate only in modes in which the length of the pipe is an integer multiple of half a wavelength.

$$\ell = n\left(\frac{\lambda_n}{2}\right) \qquad n = 1,2,3,\dots$$

8. Air in a pipe of length ℓ open at one end can vibrate only in modes in which the length of the pipe is an odd multiple of one quarter wavelength.

$$\ell = (2n + 1)\frac{\lambda_n}{4} \qquad n = 0,1,2,3,\dots$$

9. The interference of two sounds of slightly different pitch produces a *beat frequency* equal to the difference in frequency between the two sounds: $v_b = v_1 - v_2$.

10. When there is relative motion between a source of sound and an observer, the apparent frequency differs from the source's frequency according to the *Doppler effect* formula

$$v_o = v\left(\frac{1 \pm u_o/v}{1 \pm u_s/v}\right)$$

where v_o, u_o, v, and u_s are, respectively, the observed frequency, the velocity of the observer, the frequency of the source, and velocity of the source. The velocity of sound is v. When the observer moves toward the source the $+$ sign is used; when the observer moves away the $-$ sign is used. When the source moves toward the observer the $-$ sign is used; when the source moves away the $+$ sign is used.

RAISE YOUR GRADES

Can you explain . . . ?

☑ why the *adiabatic* bulk modulus must be used for sound wave velocity

☑ why the velocity of sound increases rather than decreases as the temperature increases

☑ why the intensity of sound waves is inversely proportional to the wave velocity whereas the intensity of waves on a string is directly proportional to the wave velocity

☑ why a cold organ pipe plays flat

☑ why longitudinal waves travel faster in solids than in gases

☑ what happens to the wave crests when a source of sound travels at the speed of sound

☑ what happens to the wave crests when a source of sound travels faster than the speed of sound

☑ what happens to a person's speech when he or she breathes a mixture of helium and oxygen, as opposed to air, a mixture of nitrogen and oxygen

SOLVED PROBLEMS

The Speed of Sound

PROBLEM 22-1:   Show that the speed of sound in air, in meters per second, can be calculated from $v = 20.1\sqrt{T}$, where T is the absolute temperature in kelvins.

Solution: Use Eq. (22-2b).

$$v = \sqrt{\frac{\gamma RT}{M}} = \sqrt{\frac{\gamma R}{M}}\sqrt{T}$$

$$\sqrt{\frac{\gamma R}{M}} = \sqrt{\frac{(1.4)(8.314\ \text{J}\,\text{mol}^{-1}\,\text{K}^{-1})}{28.8 \times 10^{-3}\ \text{kg}\,\text{mol}^{-1}}} = \boxed{20.1\ \text{m}\,\text{s}^{-1}\,\text{K}^{-1/2}}$$

PROBLEM 22-2:   Calculate the speed of sound in helium gas at room temperature, 22°C.

Solution: Use Eq. (22-2b) with $\gamma = 1.67$ and $M = 4 \times 10^{-3}\ \text{kg}\,\text{mol}^{-1}$.

$$v = \sqrt{\frac{\gamma RT}{M}} = \sqrt{\frac{(1.67)(8.314\ \text{J}\,\text{mol}^{-1}\,\text{K}^{-1})(22 + 273)\text{K}}{4 \times 10^{-3}\ \text{kg}\,\text{mol}}} = \boxed{1.01 \times 10^3\ \text{m}\,\text{s}^{-1}}$$

Intensity Level

PROBLEM 22-3:   Derive an expression for the energy density of a sound wave as a function of the sound pressure amplitude, the density of air, and the speed of sound.

Solution: Recall that the intensity of a wave is the product of energy density times wave velocity, Eq. (21-8).

$$I = uv$$

Combine this with the equation for sound intensity, Eq. (22-4).

$$I = \frac{1}{2}\frac{p_0^2}{\rho v}$$

Thus

$$u = \boxed{\frac{1}{2}\left(\frac{p_0^2}{\rho v^2}\right)}$$

PROBLEM 22-4:   Two identical sources of sound each produce a sound intensity level of 70 db when operated separately. What is the sound intensity level when they are operated simultaneously?

Solution: Use Eq. (22-6) with I equal to the sound intensity due to one source operating by itself.

$$L = 10\log\left(\frac{I}{I_0}\right) = 70\ \text{db}$$

When both operate simultaneously the new intensity is $I_n = I + I = 2I$. Therefore the new level is

$$L_n = 10\log\left(\frac{2I}{I_0}\right) = 10\left[\log\left(\frac{I}{I_0}\right) + \log 2\right] = 10\log\left(\frac{I}{I_0}\right) + 10\log 2$$

$$L_n = 70 + 3 = \boxed{73\ \text{db}}$$

Note that regardless of the particular sound intensity level, doubling the sound intensity increases the sound intensity level by 3 db.

Musical Intervals

PROBLEM 22-5:   What is the frequency of the musical note D_2?

Solution: From Figure 22-1, $v_{D_5} = 587.3$ Hz. D_4 is one octave below D_5, D_3 is one octave below D_4, etc., so D_2 is three octaves below D_5. Therefore

$$v_{D_2} = \frac{v_{D_5}}{2^3} = \frac{587.3 \text{ Hz}}{8} = \boxed{73.4 \text{ Hz}}$$

Resonance of Air Columns

PROBLEM 22-6:   (a) Find the length of the organ pipe (open at one end) that produces the note E_5 when the pipe is played at room temperature. Assume that the pipe is operating in its fundamental mode. (b) By how many hertz is the note flat when the temperature of the pipe drops to 15°C?

Solution:

(a) When a pipe, open at one end, operates in its fundamental mode, the length of the pipe is one quarter wavelength.

$$\ell = \frac{\lambda}{4}$$

From Figure 22-1, you can obtain the frequency of E_5: 659.3 Hz. Let $v = 344$ m s^{-1}, use the wave formula, and solve for λ.

$$v = \lambda v \qquad \lambda = \frac{v}{v}$$

Thus

$$\ell = \frac{\lambda}{4} = \frac{v}{4v} = \frac{344 \text{ m s}^{-1}}{4(659.3 \text{ Hz})} = \boxed{0.13 \text{ m}}$$

(b) Use the result of Problem 22-1 to obtain the speed of sound at 15°C:

$$v = 20.1\sqrt{T} = 20.1\sqrt{15 + 273} = 341 \text{ m s}^{-1}$$

Use the result from part (a), $\ell = v/4v$, and solve for v:

$$v = \frac{v}{4\ell} = \frac{341 \text{ m s}^{-1}}{4(0.13 \text{ m})} = 655.8 \text{ Hz}$$

The drop in pitch is

$$\Delta v = v_{E_5} - v = (659.3 - 655.8)\text{Hz} = \boxed{3.5 \text{ Hz}}$$

PROBLEM 22-7:   To a good approximation, a flute can be considered a pipe open at both ends. The lowest note that can be played on a standard flute is C_4. Determine the length of a flute. Assume $v = 344$ m s^{-1}.

Solution: The length of a pipe open at both ends resonating in its fundamental mode is half a wavelength:

$$\ell = \frac{\lambda}{2}$$

From Figure 22-1, $v_{C_5} = 523.3$, so the frequency of C_4, one octave below C_5, is

$$v_{C_4} = \frac{v_{C_5}}{2} = \frac{523.3 \text{ Hz}}{2} = 261.6 \text{ Hz}$$

Use the wave formula to express λ in terms of v and v, then solve for ℓ.

$$v = \lambda v \qquad \lambda = v/v$$

$$\ell = \frac{\lambda}{2} = \frac{v}{2v} = \frac{344 \text{ m s}^{-1}}{2(261.6 \text{ Hz})} = \boxed{0.658 \text{ m} = 65.8 \text{ cm}}$$

Beats

PROBLEM 22-8: A violinist attempts to tune the A string on his instrument to A_4 ($v_{A_4} = 440$ Hz). While bowing the open string he listens simultaneously to a 440-Hz tuning fork and hears three beats per second. He tightens the string slightly and then hears four beats per second. Is he playing sharp or flat?

Solution: Tightening the string increases the tension and so increases the wave velocity (Eq. 21-4). Hence the frequency of vibration also increases. Because the beat frequency increased, the difference between the two frequencies also increased, and we can conclude that the frequency of the violin string is above that of the tuning fork—the violinist is playing sharp.

The Doppler Effect

PROBLEM 22-9: A person riding in a car has a tuning fork that vibrates at 523 Hz. As she approaches a source of sound she knows to emit 523 Hz, she hears 6 beats per second when she strikes her fork. How fast is the car moving?

Solution: Since the observer is approaching the source, she detects a frequency higher than the source's frequency. So, according to Eq. (21-14), $v_o = v + v_b = (523 + 6)$Hz $= 529$ Hz. To find the observer's velocity, use Eq. (22-7a) and solve for u_o. Assume v is 344 m s^{-1}.

$$v_o = v\left(1 + \frac{u_o}{v}\right)$$

$$u_o = v\left(\frac{v_o}{v} - 1\right) = (344 \text{ m s}^{-1})\left(\frac{529 \text{ Hz}}{523 \text{ Hz}} - 1\right) = 3.95 \text{ m s}^{-1}$$

PROBLEM 22-10: The upper limit in frequency that a certain person can perceive is 16 000 Hz. How fast is a source of sound approaching this person if it emits 1000 Hz and he is unable to hear it coming? Assume that the speed of sound is 344 m s^{-1}.

Solution: The speed of the source must be enough to increase the apparent frequency of the sound to at least 16 000 Hz. Use Eq. (22-8a) and solve for u_s.

$$v_o = v\left(\frac{1}{1 - u_s/v}\right)$$

$$u_s = v\left(1 - \frac{v}{v_o}\right) = (344 \text{ m s}^{-1})\left(1 - \frac{1000 \text{ Hz}}{16\,000 \text{ Hz}}\right) = \boxed{322 \text{ m s}^{-1}}$$

Supplementary Exercises

EXERCISE 22-1: A person sees a flash of lightning then hears the thunder clap 5 seconds later. How far is the observer from the lightning?

EXERCISE 22-2: Begin with

$$v = 20.1\sqrt{T} = 20.1(273 + T_C)^{1/2} = 332\left(1 + \frac{T_C}{273}\right)^{1/2}$$

and the approximation that $(1 + \alpha)^{1/2} \simeq 1 + \frac{\alpha}{2}$ for $\alpha \ll 1$ and derive an equation for the speed of sound in meters per second as a function of the Celsius temperature, T_C.

EXERCISE 22-3: A loudspeaker radiates sound uniformly into the hemisphere directly in front of it. The sound intensity 3 m from the speaker is 0.2 W m^{-2}. How much power is this speaker radiating?

EXERCISE 22-4:   Calculate the sound intensity level 3 m from the loudspeaker in Exercise 22-3.

EXERCISE 22-5:   What note, on the equally tempered chromatic scale, has a frequency of 98 Hz?

EXERCISE 22-6:   Two pipes, open at one end, both resonate at 523 Hz. One is 0.1525 m in length; the other is 0.4776 m. Use this information to determine the speed of sound.

EXERCISE 22-7:   A toy whistle (Fig. 22-5) has a piston so that the pitch the whistle emits when blown can be changed. In one position of the piston it emits 830 Hz. By how much must the piston be pulled out to decrease the pitch to 740 Hz? Take the speed of sound to be 344 m s^{-1}.

EXERCISE 22-8:   A pipe 32 cm in length and 2 cm in diameter is open at one end. Calculate the pipe's fundamental frequency of vibration when the speed of sound is 342 m s^{-1}.

EXERCISE 22-9:   A person rides with a source of sound that emits a frequency of 10^3 Hz. She is heading at a speed of 4 m s^{-1} toward a wall that reflects the sound. What is the beat frequency she hears? Assume the speed of sound is 344 m s^{-1}.

EXERCISE 22-10:   The angle between the *shock wave front*, the crests of all the waves (see Fig. 22-6) and the direction of motion of a source of sound is 35° on a day when the speed of sound is 344 m s^{-1}. How fast is the source traveling?

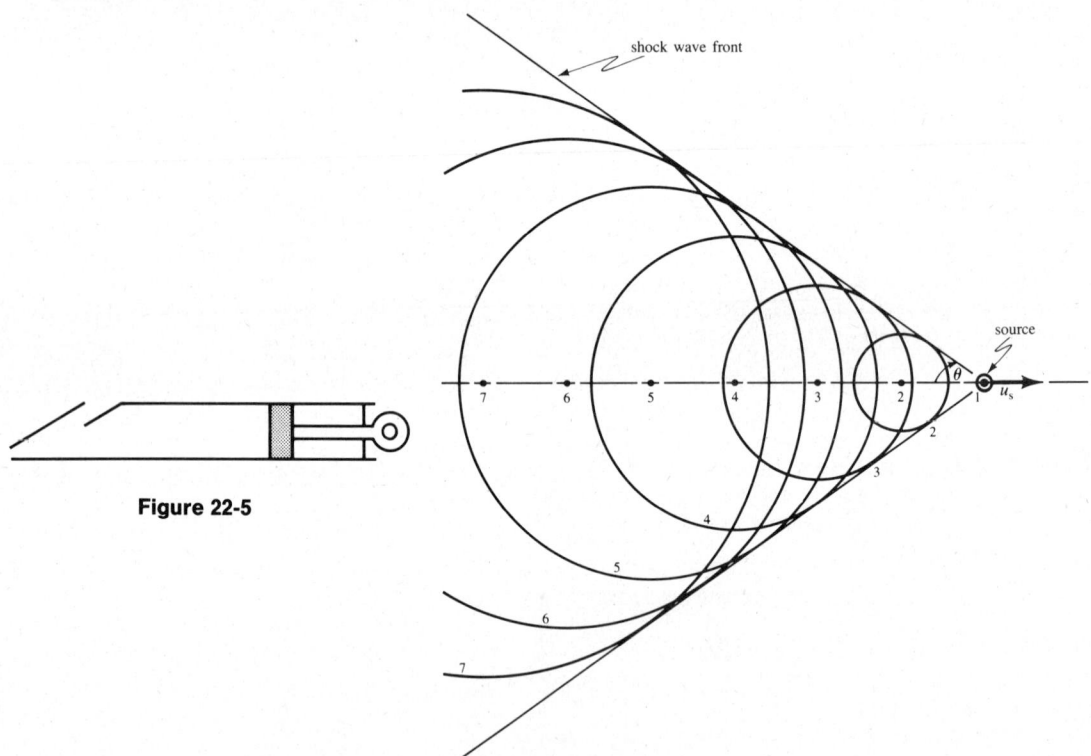

Figure 22-5

Figure 22-6. Wave crests emitted by a source moving at speed u_s greater than the speed of sound.

Answers to Supplementary Exercises

22-1: 1.72 km

22-2: $v \simeq (332 + 0.61\ T_C)$ m s^{-1}

22-3: 11.3 W

22-4: 113 db

22-5: G$_2$

22-6: 340 m s^{-1}

22-7: 1.26 cm

22-8: 262.3 Hz

22-9: 23.5 Hz

22-10: 600 m s^{-1}

FINAL EXAM

1. The cubic displacement of a certain engine is 350 in.^{3}. What is its displacement in liters? (One liter = 10^3 cm^3) [Ch. 1]

2. (a) Evaluate $\mathbf{V} = \mathbf{D} \times \mathbf{G}$ where $\mathbf{D} = 3\hat{\mathbf{i}} + 6\hat{\mathbf{j}}$ and $\mathbf{G} = -4\hat{\mathbf{i}} + 3\hat{\mathbf{j}}$. [Ch. 1]
   (b) Find the angle between $\mathbf{D}$ and $\mathbf{G}$. [Ch. 1]

3. A train travels 14 m while accelerating at 3 m s^{-2} to a speed of 10 m s^{-1}.
   (a) How fast was it going at the start of the 14-m stretch? [Ch. 2]
   (b) How long did the acceleration last? [Ch. 2]

4. A clay pigeon is launched directly upward with a speed of 30 m s^{-1}.
   (a) How much time passes until this object is traveling 15 m s^{-1} downward? [Ch. 2]
   (b) How far above the launch point does it rise? [Ch. 2]

5. A bomb is released from a plane flying at 150 m s^{-1} along a horizontal path 1.2×10^3 m above the ground. Determine the vertical and horizontal components of the velocity of the bomb just before it strikes level ground. [Ch. 3]

6. What is the centripetal acceleration of a point on the rim of a disk, 16 cm in diameter, spinning on a shaft at 1700 rev min^{-1}? [Ch. 3]

7. An 8-kg mass hangs from a cord. What is the tension in a horizontal cord that pulls the mass to the right so that the upper cord makes an angle of 30° with the vertical? [Ch. 4]

8. How fast would an object have to travel at the surface of the earth so that the centripetal force on it would be equal to its weight? The radius of the earth is 6.4×10^6 m. [Ch. 4]

9. A repulsive force, centered at the origin, can be expressed by $F = K/x$, where K is a constant (units: N m) and x is the distance from the origin to an object the force acts on. How much work is done by this force in moving an object from x_1 to x_2, where $x_2 > x_1$? [Ch. 5]

10. The engine of a small car (mass 1200 kg) can deliver 6×10^4 W of power. How fast is the car moving 3 s after starting from rest on level ground? [Ch. 5]

11. An 8-kg box slides down a straight ramp that reaches from a platform 3 m above the ground. The box is moving at 1.2 m s^{-1} when it reaches the ground. How much energy is transformed into heat during the slide? [Ch. 6]

12. Prove that the central force $\mathbf{F} = K(xy^2\hat{\mathbf{x}} + yx^2\hat{\mathbf{y}})$ is a conservative force. K is a constant that has units of N m^{-2}. [Ch. 6]

13. The position of a mass vibrating on the end of a spring is given by $x = (0.3 \text{ m}) \times \cos(8\pi t + 0.2\pi)$, where t is measured in seconds.
   (a) How fast is the mass moving as it passes the equilibrium position, $x = 0$? [Ch. 7]
   (b) What is the magnitude of the maximum acceleration with which the mass moves? [Ch. 7]

14. Use $T_\Theta \cong T[1 + (\frac{1}{2})^2 \sin^2(\Theta/2)]$ to calculate the error in using the small-angle approximation to determine the period of a pendulum swinging with an angular amplitude of 20°. [Ch. 7]

15. A 2-kg mass sliding along a frictionless horizontal track at 2.6 m s^{-1} strikes a 3-kg mass at rest. After the collision the pair, stuck together, slides on the track.
   (a) With what speed do they slide? [Ch. 8]
   (b) How much kinetic energy is lost in the collision? [Ch. 8]

16. When empty, a small rocket-powered car has a mass of 800 kg and can hold 200 kg of fuel. During a burn, the exhaust gases have a velocity of 400 m s^{-1} relative to the rocket. Determine the speed of the car after it starts from rest and burns a full load of fuel. Assume the car travels on a frictionless track and that air friction can be neglected. [Ch. 8]

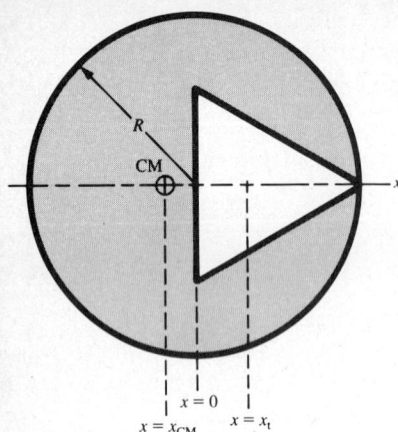

Figure E-10

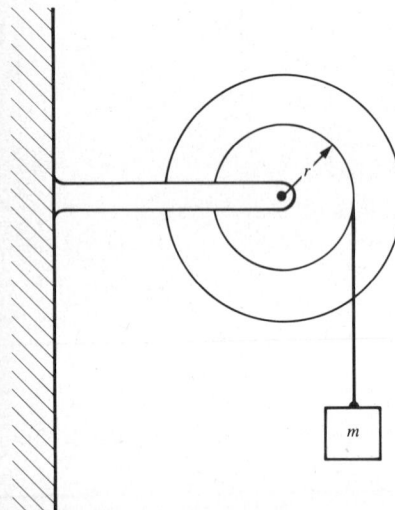

Figure E-11

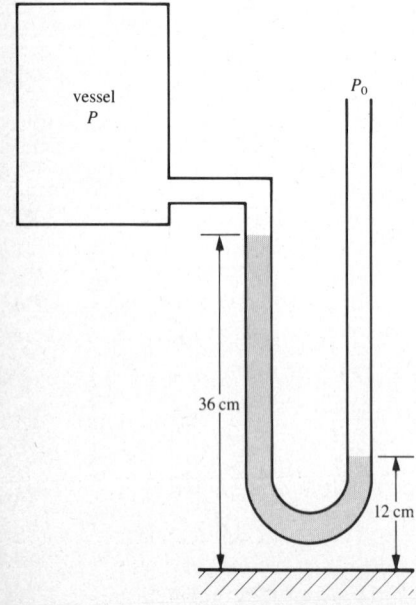

Figure E-12

17. An equilateral triangle of height R is cut out from a circular uniform sheet of aluminum. The radius of the circle is R (see Fig. E-10). Locate the circle's center of mass after the triangle is cut out. The area of an equilateral triangle of height R is $A_t = R^2 \tan 30°$ and its center is $x_t = R \tan^2 30°$ above its base. [Ch. 8]

18. A uniform meterstick is supported by a cord attached to the point on the stick at 80 cm. How much mass must be attached at the 90-cm point to balance the stick? [Ch. 9]

Questions 19 and 20 refer to the wheel shown in Fig. E-11. The wheel has a total moment of inertia I, is set in frictionless bearings, and the string attached to m is wrapped around a shaft of radius r.

19. Derive an expression for the angular acceleration of the wheel in Fig. E-11 in terms of m, I, r, and g. [Ch. 10]

20. Derive an expression for the angular velocity of the wheel in Fig. E-11 in terms of m, I, r, g, and the distance d that m descends from an initial position at rest. [Ch. 10]

21. Find the force of gravitational attraction between the two lead spheres of a Cavendish balance if the larger sphere has a mass of 12 kg, the smaller sphere has a mass of 50 g, and the centers of the spheres are separated by 8 cm.

22. If the moon had the same mass as the Earth $(6 \times 10^{24}$ kg) but the same separation as now $(3.8 \times 10^8$ m), with what period would these objects revolve about a central point? [Ch. 11]

23. A light in your laboratory blinks for 2 s and off for 2 s. What are the intervals reported by an observer passing in a rocket moving at $0.65c$? [Ch. 12]

24. Determine the momentum, in MeV/c, of a proton moving at $0.45c$. The rest energy of a proton is 938 Mev. [Ch. 12]

25. Young's modulus and Poisson's ratio for a material out of which a wire 2 mm in diameter is made are 9×10^{10} Pa and 0.3, respectively. Determine the decrease in diameter of this wire when it is subjected to a tensile stress of 4.2×10^7 Pa. [Ch. 13]

26. Derive an expression for the Poisson ratio of a material in terms of its shear and bulk moduli. [Ch. 13]

27. The liquid in the U-tube manometer shown in Figure E-12 is mercury, which has a specific gravity of 13.6. Take atmospheric pressure as $P_0 = 1.013 \times 10^5$ Pa.
    (a) What is the gauge pressure in the vessel? [Ch. 14]
    (b) What is the absolute pressure in the vessel? [Ch. 14]

28. A small gemstone of mass 124 g "weighs" 86 g when immersed in a light oil that has a specific gravity of 0.923. What is the specific gravity of the gemstone? [Ch. 14]

29. A light oil with viscosity of 2 poise (0.2 Pa s) flows through a pipe 1 cm in diameter at the rate of 0.1 L s^{-1} $(10^{-4}$ m^3 s$^{-1})$. What is the pressure gradient along the pipe? [Ch. 15]

30. To determine the viscosity of an oil that has a density of 0.9×10^3 kg m^{-3}, a student measures the terminal velocity of a sphere 1.2 cm in diameter as it falls through the oil. His result is 0.3 m s^{-1}. The density of the sphere is 3×10^3 kg m^{-3}. What is the oil's viscosity? [Ch. 15]

31. In reading an old book, a student discovers a forgotten temperature scale that she decides to call the "A scale" (A for ancient). She learns that the boiling and freezing points of water are 300°A and 80°A, respectively.

    (a) Derive a formula that converts temperature in degrees A to degrees Celsius. [Ch. 16]
    (b) What is the Celsius temperature of 0°A? [Ch. 16]

32. What is the Fahrenheit temperature of 100 K? [Ch. 16]

33. Aluminum has a coefficient of linear expansion of 2.4×10^{-5} $(\text{C}°)^{-1}$. Determine the rise in temperature that will increase the volume of a block of aluminum by 0.1%. [Ch. 16]

34. What pressure must be exerted on both ends of a steel rod to prevent the rod from increasing in length when its temperature is raised from 20°C to 70°C? Young's modulus and the coefficient of linear expansion of steel are 22×10^{10} Pa and 1.2×10^{-5} $(\text{C}°)^{-1}$, respectively. [Ch. 16]

35. A cup (whose specific heat capacity is negligible) contains 500 g of water at 22°C. A 650-g block of metal at 95°C is dropped into the water, causing the temperature of the water to rise to 25.5°C. What is the metal's specific heat capacity? The specific heat capacity of water is 4.186 $\text{kJ}\,\text{kg}^{-1}\,(\text{C}°)^{-1}$. [Ch. 17]

36. A 500-g piece of copper at 22°C is placed in a small furnace, also at 22°C. When the furnace is turned on, 3 kW of electrical power is supplied to its heating element. Assume that on the average 1 kW of this power is lost to the surroundings, 1 kW is needed to raise the temperature of the furnace, and 1 kW is absorbed by the copper. How long will it take to melt the copper? The specific heat capacity, melting temperature, and latent heat of fusion of copper are 0.385 $\text{kJ}\,\text{kg}^{-1}\,(\text{C}°)^{-1}$, 1083°C, and 134 $\text{kJ}\,\text{kg}^{-1}$, respectively. [Ch. 17]

37. The thermal conductivity of concrete is approximately 1 $\text{W}\,\text{m}^{-1}\,(\text{C}°)^{-1}$. Estimate the heat loss per square meter from a concrete building with walls 20 cm thick on a day when the outdoor temperature is 2°C and the indoor temperature is maintained at 18°C. [Ch. 17]

38. A 100-W electric heater is put inside a blackened copper sphere that has a surface area of 0.03 m² and emissivity of 0.9. Determine the temperature of the sphere when the heater is operating and the ambient temperature is 20°C. Assume that the sphere loses heat only by radiation. The Stefan–Boltzmann constant is $\sigma = 5.67 \times 10^{-8}$ $\text{W}\,\text{m}^2\,\text{K}^{-4}$. [Ch. 17]

39. A 0.1-m³ tank is filled with 2 kg of helium. What is the gauge pressure of the gas at 22°C? Assume that atmospheric pressure is 10^5 Pa and that helium is an ideal gas. Helium has a molecular mass of 4 $\text{g}\,\text{mol}^{-1}$ and $R = 8.314$ $\text{J}\,\text{mol}^{-1}\,\text{K}^{-1}$. [Ch. 18]

40. The speed of sound at 22°C is 345 $\text{m}\,\text{s}^{-1}$. What is the rms speed of a nitrogen molecule at this temperature? Take the mass of a nitrogen molecule as 4.65×10^{-26} kg and Boltzmann's constant as $k = 1.38 \times 10^{-23}$ $\text{J}\,\text{K}^{-1}$. [Ch. 18]

41. How much heat (in joules) must be transferred to 3 mol of an ideal diatomic gas confined to a tank to increase the gas' temperature from 22°C to 35°C? $R = 8.314$ $\text{J}\,\text{mol}^{-1}\,\text{K}^{-1}$. [Ch. 18]

42. To a first approximation, atmospheric pressure decreases with increasing height above the earth's surface according to

$$\frac{dP}{dh} = -\rho g \qquad \text{where} \qquad \rho = \frac{PM}{RT}$$

is the density of air. Find an expression for atmospheric pressure at a height h above the earth's surface. Let P_0 be atmospheric pressure at height $h = 0$. [Ch. 18]

For Problems 43 through 47 refer to Figure E-13, which is the *PV* diagram of a monatomic ideal gas confined to a cylinder equipped with a movable piston. The absolute pressures P_1 and P_2 are 2×10^5 Pa and 5×10^5 Pa, the volumes V_1 and V_2 are 0.1 m³ and 0.25 m³, and process C is isothermal at temperature 300 K.

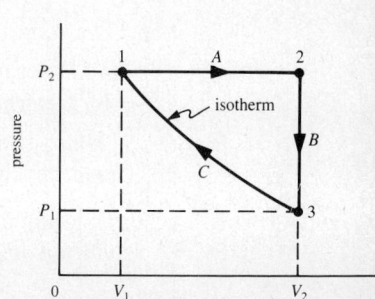

Figure E-13

43. (a) Identify process A. [Ch. 19]
    (b) Is work done on or by the gas during process A? [Ch. 19]

(c) Compute the amount of work done during process *A*.   [Ch. 19]

(d) What is the temperature at point 2?   [Ch. 18]

(e) How much heat is transferred during process *A*?   [Ch. 18]

44. (a) Identify process *B*.   [Ch. 19]

(b) How much work is done on or by the gas during process *B*?   [Ch. 19]

(c) Compute the amount of heat transferred during process *B*.   [Ch. 18]

45. (a) How much work is done on or by the gas during process *C*?   [Ch. 19]

(b) Compute the amount of heat transferred during process *C*.   [Ch. 18]

46. (a) How much net work is done per cycle?   [Ch. 19]

(b) What is the net heat transferred per cycle?   [Ch. 19]

47. Suppose the cycle shown in Figure E-13 represents a heat engine.

(a) What is the engine's efficiency?   [Ch. 20]

(b) What is the efficiency of a Carnot engine operating between these two temperatures?   [Ch. 20]

48. A small air conditioner that fits in a window is able to transfer 5000 BTU of heat per hour out of a room. It requires a power of 500 W to operate. Determine its coefficient of performance. (Recall that 1 BTU = 1055 J.)

49. An old steam engine has an efficiency of 25%. The input steam has a temperature 250°C, the exhaust temperature is 60°C, and the engine delivers 300 kW of power.

(a) At what rate does this engine exhaust heat?   [Ch. 20]

(b) What is the efficiency of a Carnot engine operating between these temperatures?   [Ch. 20]

50. Derive a formula for the increase in entropy of *m* kilograms of a substance heated reversibly from T_1 to T_2. The specific heat capacity of this substance depends on its absolute temperature according to $c = C_0 T^2$, where C_0 is a constant that has units of $J \, kg^{-1} \, K^{-3}$.   [Ch. 20]

51. A steel wire 2 m long has a mass of 12 g and is stretched with a tension of 5×10^3 N.

(a) What is the velocity of transverse waves in this wire?   [Ch. 21]

(b) At what frequency will it vibrate if plucked?   [Ch. 21]

52. A transverse sine wave moving with velocity 360 m s^{-1} in the positive *x* direction has a frequency of 220 Hz and an amplitude of 0.3 cm. Write an expression for the wave function $y(x, t)$ that describes this wave.

53. Light is an electromagnetic sine wave that travels at 3×10^8 m s^{-1} in air. A small laser emits a beam of light that has an intensity of 7.4×10^4 W m^{-2} distributed uniformly over a beam diameter of 3 mm.

(a) Calculate the energy density of the light in the beam.   [Ch. 21]

(b) What is the power output of this laser?   [Ch. 21]

54. Two identical sine waves, each of amplitude *A*, arrive a point with a phase difference of 90°. What is the amplitude of the sum of the two waves at this point?   [Ch. 21]

55. The speed of sound in carbon dioxide gas at 0°C is 258 m s^{-1} and the molecular mass of CO_2 is $M = 44$ g mol^{-1}. From this information determine $\gamma = C_p / C_v$.   [Ch. 22]

56. The sound pressure amplitude of a certain sound is 0.02 Pa.

(a) Determine the sound's intensity.   [Ch. 22]

(b) Determine the sound's intensity level.   [Ch. 22]

[*Hint*: The density of air and the speed of sound at 22°C are 1.2 kg m^{-3} and 344 m s^{-1}, respectively.]

57. Determine the length of a pipe, 1.9 cm in diameter and closed at one end, in which air can vibrate at a frequency of 440 Hz. Take the speed of sound to be $v = 345$ m s^{-1}.   [Ch. 22]

58. Derive an expression for the increase in observed frquency, $\Delta v = v_o - v$, of a source of sound moving toward a stationary observer at a speed u_s. Let *v* represent the speed of sound.   [Ch. 22]

Solutions to Final Exam

1. Cube the conversion relationship between inches and centimeters, 1 in. = 2.54 cm, to get 1 in.^{3} = 16.387 cm^3.

$$(350 \text{ in.}^3)\left(\frac{16.387 \text{ cm}^3}{\text{in.}^3}\right)\left(\frac{1 \text{ L}}{10^3 \text{ cm}^3}\right) = \boxed{5.74 \text{ L}}$$

2. (a)
$$\mathbf{V} = \mathbf{D} \times \mathbf{G} = \begin{vmatrix} \hat{\mathbf{i}} & \hat{\mathbf{j}} & \hat{\mathbf{k}} \\ 3 & 6 & 0 \\ -4 & 3 & 0 \end{vmatrix} = [(3)(3) - (6)(-4)]\hat{\mathbf{k}} = \boxed{33\hat{\mathbf{k}}}$$

(b) Let θ be the angle between $\mathbf{D}$ and $\mathbf{G}$. So $|\mathbf{D} \times \mathbf{G}| = DG \sin\theta$ where

$$D = \sqrt{D_i^2 + D_j^2} = \sqrt{3^2 + 6^2} = 6.71 \qquad \text{and} \qquad G = \sqrt{G_i^2 + G_j^2} = \sqrt{(-4)^2 + 3^2} = 5$$

$$\theta = \arcsin\left(\frac{|\mathbf{D} \times \mathbf{G}|}{DG}\right) = \arcsin\left(\frac{33}{(6.71)(5)}\right) = \boxed{79.6^\circ}$$

3. (a) Solve $v^2 = v_0^2 + 2ax$ for v_0.

$$v_0 = \sqrt{v^2 - 2ax} = \sqrt{10^2 - (2)(3)(14)} \text{ m s}^{-1} = \boxed{4 \text{ m s}^{-1}}$$

(b) Solve $v = v_0 + at$ for t.

$$t = \frac{v - v_0}{a} = \frac{(10 - 4) \text{ m s}^{-1}}{3 \text{ m s}^{-2}} = \boxed{2 \text{ s}}$$

4. (a) Let upward be the positive direction. Use $v = v_0 + at$ with $a = -g$ and solve for t.

$$t = \frac{v - v_0}{-g} = \frac{(-15 - 30) \text{ m s}^{-1}}{-9.8 \text{ m s}^{-2}} = \boxed{4.6 \text{ s}}$$

(b) Use $v^2 = v_0^2 + 2ax$ with $v = 0$ and $a = -g$. Solve for x.

$$x = \frac{v^2 - v_0^2}{-2g} = \frac{0 - (30 \text{ m s}^{-1})^2}{-2(9.8 \text{ m s}^{-2})} = \boxed{45.9 \text{ m}}$$

5. Let downward be the positive direction. To find the vertical component, first solve $y = v_{0_y}t + \frac{1}{2}gt^2$ for t, where $v_{0_y} = 0$. Then use this value for t in $v_y = v_{0_y} + gt$, with $v_{0_y} = 0$.

$$t = \sqrt{\frac{2y}{g}} = \sqrt{\frac{2(1.2 \times 10^3 \text{ m})}{9.8 \text{ m s}^{-2}}} = 15.65 \text{ s}$$

$$v_y = gt = (9.8 \text{ m s}^{-2})(15.65 \text{ s}) = \boxed{153 \text{ m s}^{-1}}$$

The horizontal component of the bomb's velocity is the same as that of the plane:

$$v_x = \boxed{150 \text{ m s}^{-1}}$$

6. Use $a_c = \omega^2 r$ with $r = 8 \times 10^{-2}$ m and

$$\omega = \left(\frac{1700 \text{ rev}}{\text{min}}\right)\left(\frac{2\pi \text{ rad}}{\text{rev}}\right)\left(\frac{\text{min}}{60 \text{ s}}\right) = 178 \text{ rad s}^{-1}$$

$$a_c = \omega^2 r = (178 \text{ rad s}^{-1})^2(8 \times 10^{-2} \text{ m}) = \boxed{2.535 \times 10^3 \text{ m s}^{-2}}$$

7. Figure E-14 is a free-body diagram showing the tension in the upper cord resolved into x and y components. Let T_1 be the tension in the upper cord, T_2 be the tension in the horizontal cord, and solve for T_2. The mass is in equilibrium so $T_2 = T_1 \sin\phi$ and $T_1 \cos\phi = mg$.

$$T_2 = mg\left(\frac{\sin\phi}{\cos\phi}\right) = mg \tan\phi = (8 \text{ kg})(9.8 \text{ m s}^{-2})(\tan 30^\circ) = \boxed{45.3 \text{ N}}$$

8. Set the weight of the object equal to the centripetal force and solve for v.

$$mg = m\frac{v^2}{R} \qquad v = \sqrt{gR} = \sqrt{(9.8 \text{ m s}^{-2})(6.4 \times 10^6 \text{ m})} = \boxed{7.92 \times 10^3 \text{ m s}^{-1}}$$

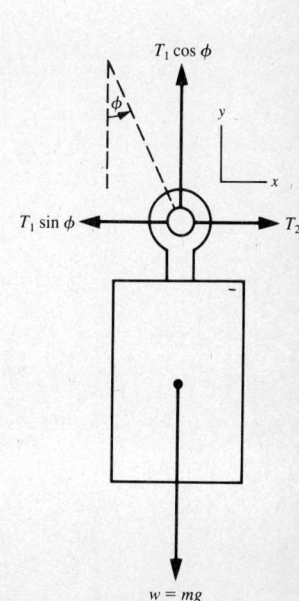

Figure E-14

9.
$$W = \int_1^2 F \cdot dx = \int_{x_1}^{x_2} \frac{K}{x} dx = K \int_{x_1}^{x_2} \frac{dx}{x} = \boxed{K \ln\left(\frac{x_2}{x_1}\right)}$$

10. The work $W = Pt$ done by the engine increases the kinetic energy of the car from zero to $\frac{1}{2}mv^2$. Solve for v.

$$W = Pt = \frac{1}{2}mv^2 \qquad v = \sqrt{\frac{2Pt}{m}} = \sqrt{\frac{2(6 \times 10^4 \text{ W})(3 \text{ s})}{1200 \text{ kg}}} = \boxed{17.3 \text{ m s}^{-1}}$$

11. Use the work–energy principle: The box's initial energy equals its final energy minus the work it does against friction, which is transformed into heat.

$$mgh = \frac{1}{2}mv^2 + W_f$$

$$W_f = m\left(gh - \frac{v^2}{2}\right) = (8 \text{ kg})\left[(9.8 \text{ m s}^{-2})(3 \text{ m}) - \frac{1.2 \text{ m s}^{-1}}{2}\right] = \boxed{229 \text{ J}}$$

12. A force is conservative if its curl is zero. Because the force is two-dimensional, F_z is zero and

$$\frac{\partial F_z}{\partial y} - \frac{\partial F_y}{\partial z} = 0 \qquad \text{and} \qquad \frac{\partial F_x}{\partial z} - \frac{\partial F_z}{\partial x} = 0$$

$$\frac{\partial F_x}{\partial y} = \frac{\partial(Kxy^2)}{\partial y} = 2Kxy \qquad \frac{\partial F_y}{\partial x} = \frac{\partial(Kyx^2)}{\partial x} = 2Kyx$$

So $\dfrac{\partial F_y}{\partial x} - \dfrac{\partial F_x}{\partial y} = 2Kyx - 2Kxy = 0$ and all three conditions for a force to be conservative are met.

13. (a) Use the equations relating the kinetic and potential energies of an oscillator and solve for v.

$$E = \frac{1}{2}kX^2 = \frac{1}{2}mv^2 + \frac{1}{2}kx^2$$

$$v = \pm\sqrt{\frac{k(X^2 - x^2)}{m}} = \pm\omega\sqrt{(X^2 - x^2)}$$

$$= (8\pi \text{ s}^{-1})\sqrt{(0.3 \text{ m})^2 - 0^2} = (8\pi \text{ s}^{-1})(0.3 \text{ m}) = \boxed{7.54 \text{ m s}^{-1}}$$

(b) The acceleration is $a = -\omega^2 x$ so

$$|a_{max}| = \omega^2 X = (8\pi \text{ s}^{-1})^2(0.3 \text{ m}) = \boxed{189 \text{ m s}^{-2}}$$

14.
$$\frac{T(\Theta)}{T} = \frac{T\left[1 + \left(\frac{1}{2}\right)^2 \sin^2(\Theta/2)\right]}{T} = 1 + \left(\frac{1}{2}\right)^2 \sin^2\left(\frac{\Theta}{2}\right)$$

$$= 1 + \frac{1}{4}\sin^2\left(\frac{20°}{2}\right) = \boxed{1 + 7.54 \times 10^{-3}}$$

The more exact equation yields a period 0.754% larger than that predicted by the small-angle approximation.

15. (a) Use the principle of conservation of linear momentum.

$$m_A v_{A1} + m_B v_{B1} = (m_A + m_B)v_2$$

Let $v_{B1} = 0$ and solve for v_2.

$$v_2 = \frac{m_A v_{A1}}{m_A + m_B} = \frac{(2 \text{ kg})(2.6 \text{ m s}^{-1})}{(2 + 3) \text{ kg}} = \boxed{1.04 \text{ m s}^{-1}}$$

(b)
$$\Delta E_k = E_{k\,initial} - E_{k\,final} = \frac{1}{2}m_A v_{A1}^2 - \frac{1}{2}(m_A + m_B)v_2^2$$
$$= \frac{1}{2}\{(2 \text{ kg})(2.6 \text{ m s}^{-1})^2 - [(2 + 3) \text{ kg}](1.04 \text{ m s}^{-1})^2\} = \boxed{4.056 \text{ J}}$$

16. Because all external forces on the car in the direction of its motion are to be ignored you can use the integrated form of the rocket equation, $v_2 - v_1 = u \ln(m_1/m_2)$. Let $v_1 = 0$ and $m_1 = m_2 + m_f$, where m_f is the mass of the fuel burned.

$$v_2 = u \ln\left(\frac{m_2 + m_f}{m_2}\right) = (400 \text{ m s}^{-1}) \ln\left(\frac{(800 + 200) \text{ kg}}{800 \text{ kg}}\right) = \boxed{89.3 \text{ m s}^{-1}}$$

17. Think of the cut-out triangle as a "negative" mass. The masses of both the circle and triangle are proportional to their area, so if K is a proportionality constant with units kg m^{-2}, the mass M_c of the complete circular sheet is

---

$K\pi R^2$ and the mass of the triangle M_t is $KR^2 \tan 30°$. Define the x axis to run through the centers of both the circle and the triangle, set the origin at the center of the circle, and use the equation for the coordinate of the center of mass,

$$x_{CM} = \frac{m_A x_A + m_B x_B + \cdots}{m_A + m_B + \cdots}$$

$$x_{CM} = \frac{m_c x_c - m_t x_t}{m_c + m_t} = \frac{(K\pi R^2)(0) - (KR^2 \tan 30°)(R \tan^2 30°)}{K\pi R^2 - KR^2 \tan 30°}$$

$$= \frac{-KR^2 \tan^3 30°}{KR^2(\pi - \tan 30°)} = \boxed{\frac{-R \tan^3 30°}{\pi - \tan 30°}}$$

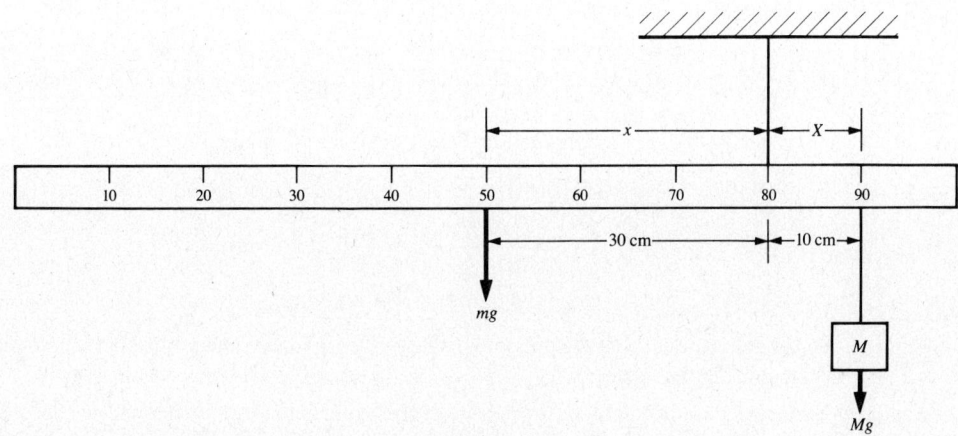

Figure E-15

18. Figure E-15 shows the forces on the stick. Let m equal the mass of the stick, set the sum of the torques about the point at 80 cm equal to zero, and solve for M, the mass that must be attached to the point at 90 cm.

$$\Sigma\tau = xmg - XMg = 0$$

$$M = m\left(\frac{x}{X}\right) = m\left(\frac{30\text{ cm}}{10\text{ cm}}\right) = \boxed{3\ m}$$

19. The torque acting on the wheel is $\tau = I\alpha = rT$ where T is the tension in the string. To find T, apply Newton's second law to m:

$$\Sigma F = ma = mg - T \qquad T = mg - ma$$

Because $a = \alpha r$, $T = mg - m\alpha r$, so

$$I\alpha = r(mg - m\alpha r) \qquad \text{and} \qquad \alpha = \boxed{\frac{mgr}{I + mr^2}}$$

20. The decrease in the potential energy is equal to the increase in the kinetic energy of rotation of the wheel plus the kinetic energy of translation of m.

$$mgd = \tfrac{1}{2}I\omega^2 + \tfrac{1}{2}mv^2$$

Because $v = \omega r$,

$$mgd = \frac{1}{2}\omega^2(I + mr^2) \qquad \text{and} \qquad \omega = \boxed{\sqrt{\frac{2mgd}{I + mr^2}}}$$

21. $$F = G\frac{m_A m_B}{r^2} = (6.672 \times 10^{-11}\ \text{N m}^2\,\text{kg}^{-2})\frac{(12\ \text{kg})(50 \times 10^{-3}\ \text{kg})}{(8 \times 10^{-2}\ \text{cm})^2} = \boxed{6.26 \times 10^{-9}\ \text{N}}$$

22. The radius of their orbit would be half their separation: $a = \tfrac{1}{2}(3.8 \times 10^8\ \text{m}) = 1.9 \times 10^8\ \text{m}$. Use Kepler's third law,

$$T^2 = \left(\frac{4\pi^2}{GM}\right)a^3 \qquad T = \sqrt{\frac{4\pi^2 a^3}{GM}} = \sqrt{\frac{4\pi^2(1.9 \times 10^8\ \text{m})^3}{(6.672 \times 10^{-11}\ \text{N m}^2\,\text{kg}^{-2})(6 \times 10^{24}\ \text{kg})}} = \boxed{8.22 \times 10^5\ \text{s}}$$

or about 9.5 days.

23. The proper time interval T_0 is 2 s.

$$T = \frac{T_0}{\sqrt{1 - (u/c)^2}} = \frac{2 \text{ s}}{\sqrt{1 - (0.65c/c)^2}} = \boxed{2.63 \text{ s}}$$

24. Multiply the equation for relativistic momentum by c^2/c^2.

$$p = \frac{mu}{\sqrt{1 - (u/c)^2}} = \left(\frac{mc^2}{c}\right)\left(\frac{u/c}{\sqrt{1 - (u/c)^2}}\right) = \left(\frac{938 \text{ MeV}}{c}\right)\left(\frac{0.45c/c}{\sqrt{1 - (0.45c/c)^2}}\right) = \boxed{473 \text{ MeV}/c}$$

25. First solve the equation relating tensile stress to tensile strain, $F/A = Y(\Delta\ell/\ell)$, for the strain $\Delta\ell/\ell$. Then use this result in the equation relating tensile and transverse strain, and solve for Δb.

$$\frac{\Delta\ell}{\ell} = \frac{F/A}{Y} = \frac{4.2 \times 10^7 \text{ Pa}}{9 \times 10^{10} \text{ Pa}} = 4.667 \times 10^{-4}$$

$$\frac{\Delta b}{b} = -\sigma\frac{\Delta\ell}{\ell} \qquad \Delta b = -b\sigma\frac{\Delta\ell}{\ell} = -(2 \text{ mm})(0.3)(4.667 \times 10^{-4}) = \boxed{-2.8 \times 10^{-4} \text{ mm}}$$

26. Use $S = Y/2(1 + \sigma)$ and $B = Y/3(1 - 2\sigma)$. Solve these two equations simultaneously for σ.

$$\frac{B}{S} = \frac{Y/3(1 - 2\sigma)}{Y/2(1 + \sigma)} = \frac{2(1 + \sigma)}{3(1 - 2\sigma)} \qquad 3B(1 - 2\sigma) = 2S(1 + \sigma)$$

$$\sigma = \boxed{\frac{3B - 2S}{2S + 6B} = \frac{1.5B - S}{S + 3B}}$$

27. From Figure E-12 it should be clear that the pressure inside the vessel is less than atmospheric pressure; that is, there is a "partial vacuum" in the vessel.

(a) The gauge pressure is $P_g = \rho g h$ where ρ is the density of the liquid in the manometer and h is the difference in heights of the two columns. In this case, you should report gauge pressure as a negative quantity to show that it is less than atmospheric pressure.

$$P_g = -\rho g h = -(13.6 \times 10^3 \text{ kg m}^{-3})(9.8 \text{ m s}^{-2})[(36 - 12) \times 10^{-2} \text{ m}] = \boxed{-3.20 \times 10^4 \text{ Pa}}$$

(b)
$$P = P_0 + \rho g h = 1.013 \times 10^5 \text{ Pa} + (-0.320 \times 10^5 \text{ Pa}) = \boxed{6.93 \times 10^4 \text{ Pa}}$$

28. Use the equation derived from Archimedes' principle, $w_i/w = 1 - (\rho_\ell/\rho)$. Note that $w_i/w = m_i/m$, where m_i is the "apparent mass" of the stone when it is immersed in the oil. Note also that $\rho_\ell/\rho_s = SG_\ell/SG_s$.

$$\frac{w_i}{w} = \frac{m_i}{m} = 1 - \frac{\rho_\ell}{\rho_s} = 1 - \frac{SG_\ell}{SG_s}$$

$$SG_s = \frac{SG_\ell}{1 - \dfrac{m_i}{m}} = \frac{0.923}{1 - \dfrac{86 \text{ g}}{124 \text{ g}}} = \boxed{3.01}$$

29. Use Poiseuille's equation and solve for the pressure gradient.

$$\frac{\Delta V}{\Delta t} = \frac{\pi r^4}{8\eta}\left(\frac{P_1 - P_2}{x}\right)$$

$$\frac{P_1 - P_2}{x} = \left(\frac{8\eta}{\pi r^4}\right)\left(\frac{\Delta V}{\Delta t}\right) = \left(\frac{8(0.2 \text{ Pa s})}{\pi(5 \times 10^{-3} \text{ m})^4}\right)(10^{-4} \text{ m}^3 \text{ s}^{-1}) = \boxed{8.15 \times 10^4 \text{ Pa m}^{-1}}$$

30. Use Stokes' low and solve for η.

$$v = \frac{2r^2 g(\rho - \rho_\ell)}{9\eta}$$

$$\eta = \frac{2r^2 g(\rho - \rho_\ell)}{9v} = \frac{2(6 \times 10^{-3} \text{ m})^2(9.8 \text{ m s}^{-2})(3 - 0.9) \times 10^3 \text{ kg m}^{-3}}{9(0.3 \text{ m s}^{-1})} = \boxed{0.55 \text{ Pa s} = 5.5 \text{ poise}}$$

31. (a) The difference between the boiling and freezing points on the A scale, $300 - 80 = 222$ A°, must equal the difference on the Celsius scale, 100 C°.

$$200 \text{ A}° = 100 \text{ C}° \qquad 1 \text{ A}° = \frac{100}{220}(1 \text{ C}°) = \frac{10}{22}(1 \text{ C}°)$$

At 0°C, $T_C = T_A - 80$, so $\boxed{T_C = \frac{10}{22}(T_A - 80)}$

(b)
$$T_C = \frac{10}{22}(T_A - 80) = \frac{10}{22}(0 - 80) = \boxed{-36.4°C}$$

32. Perhaps it is easiest to convert 100 K to Celsius degrees and then convert the result to Fahrenheit degrees.

$$T_C = T_K - 273 = 100 - 273 = -173°C$$

$$T_F = \frac{9}{5}T_C + 32 = \frac{9}{5}(-173) + 32 = \boxed{-279°F}$$

33. Use the equation for thermal volume expansion and solve for ΔT.

$$\Delta V = \beta V \Delta T = 3\alpha V \Delta T$$

$$\Delta T = \left(\frac{1}{3\alpha}\right)\left(\frac{\Delta V}{V}\right) = \frac{1}{3(2.4 \times 10^{-5})(C°)^{-1}}(10^{-3}) = \boxed{13.9 \ C°}$$

34. The pressure needed on both ends of the rod is equal to the thermal compressional stress on the rod when it can't expand.

$$\frac{F}{A} = Y\alpha \Delta T = (22 \times 10^{10} \text{ Pa})[(1.2 \times 10^{-5})(C°)^{-1}][(70 - 20)°C] = \boxed{1.32 \times 10^8 \text{ Pa}}$$

35. Use the method of mixtures and solve for c_m.

$$m_m c_m \Delta T_m = m_w c_w \Delta T_w$$

$$c_m = c_w \frac{m_w \Delta T_w}{m_m \Delta T_m} = [4.186 \text{ kJ kg}^{-1}(C°)^{-1}]\left[\frac{(500 \text{ g})(25.5 - 22)°C}{(650 \text{ g})(95 - 25.5)°C}\right] = \boxed{0.162 \text{ kJ kg}^{-1}(C°)^{-1}}$$

36. The quantity of heat needed to melt the copper is

$$Q = mc \Delta T + mL_f = m(c \Delta T + L_f)$$

$$- (0.5 \text{ kg})[(0.385 \text{ kJ kg}^{-1}(C°)^{-1})(1083 - 22)°C + 134 \text{ kJ kg}^{-1}] = 271 \text{ kJ}$$

The rate at which the copper absorbs heat is $P = Q/t = 1 \text{ kW} = 10^{-3} \text{ J s}^{-1}$. Solve for the time t.

$$t = \frac{Q}{P} = \frac{271 \times 10^3 \text{ J}}{10^3 \text{ J s}^{-1}} = \boxed{271 \text{ s} = 4.5 \text{ min}}$$

37. Heat is transferred through concrete by conduction, so

$$\frac{\Delta Q}{\Delta t} = KA\frac{T_H - T_C}{\Delta x} = [1 \text{ W m}^{-1}(C°)^{-1}](1 \text{ m}^2)\left[\frac{(18 - 2)°C}{0.2 \text{ m}}\right] = \boxed{80 \text{ W}}$$

38. Solve $\Delta Q/\Delta t = Ae\sigma(T_H^4 - T_C^4)$ for T_H. Remember to convert degrees Celsius to kelvins.

$$T_H = \left(\frac{\Delta Q/\Delta t}{Ae\sigma} + T_C^4\right)^{1/4} = \left[\frac{10^2 \text{ W}}{(0.03 \text{ m}^2)(0.9)(5.67 \times 10^{-8} \text{ W m}^{-2}\text{K}^{-4})} + (293 \text{ K})^4\right]^{1/4} = \boxed{519 \text{ K} = 246°C}$$

39. First express the quantity of helium in moles, then solve the ideal gas law for P. The ideal gas law yields the absolute pressure, so subtract atmospheric pressure to find the gauge pressure.

$$n = \frac{m}{M} = \frac{2 \times 10^3 \text{ g}}{4 \text{ g mol}^{-1}} = 500 \text{ mol}$$

$$PV = nRT \qquad P = \frac{nRT}{V} = \frac{(500 \text{ mol})(8.314 \text{ J mol}^{-1}\text{K}^{-1})(22 + 273) \text{ K}}{0.1 \text{ m}^3} = 1.23 \times 10^7 \text{ Pa}$$

$$P_{gauge} = P - P_{atm} = (1.23 \times 10^{-7} - 10^5) \text{ Pa} = (1.23 - 0.01) \times 10^7 \text{ Pa} = \boxed{1.22 \times 10^7 \text{ Pa}}$$

40.
$$v_{rms} = \sqrt{\frac{3kT}{m_m}} = \sqrt{\frac{3(1.38 \times 10^{-23} \text{ J K}^{-1})(22 + 273) \text{ K}}{4.65 \times 10^{-26} \text{ kg}}} = \boxed{512 \text{ m s}^{-1}}$$

41. The volume is constant, so use $Q = nC_v \Delta T$. For an ideal diatomic gas, $C_v = \frac{5}{2}R$.

$$Q = \frac{5}{2}nR \Delta T = \frac{5}{2}(3 \text{ mol})(8.314 \text{ J mol}^{-1}\text{K}^{-1})(35 - 22) \ C° = \boxed{811 \text{ J}}$$

42. For convenience of manipulation let $Mg/RT = B$. Then

$$\frac{dP}{dh} = -\rho g = -\left(\frac{PM}{RT}\right)g = -PB$$

Separate variables and integrate.

$$\frac{dP}{P} = -hB\,dh \qquad \int_{P_0}^{P} \frac{dP}{P} = -B\int_0^h dh$$

$$\ln P\Big|_{P_0}^{P} = \ln\left(\frac{P}{P_0}\right) = -B\left(h\Big|_0^h\right) = -B(h-0) = -Bh$$

$$\frac{P}{P_0} = e^{-Bh} \qquad P = P_0 e^{-Bh} = \boxed{P_0 e^{-Mgh/RT}}$$

43. (a) Process A is an isobaric (constant-pressure) expansion.

   (b) Because the gas expands, work is done *by* the gas.

   (c) In an isobaric process, $W = P\,\Delta V$.

$$W_A = P_2(V_2 - V_1) = (5 \times 10^5 \text{ Pa})(0.25 - 0.1)\text{ m}^3 = \boxed{7.5 \times 10^4 \text{ J}}$$

   (d) Use the ideal gas law and solve for T_2.

$$\frac{P_2 V_1}{T_1} = \frac{P_2 V_2}{T_2} \qquad T_2 = T_1\left(\frac{V_2}{V_1}\right) = (300 \text{ K})\left(\frac{0.25 \text{ m}^3}{0.1 \text{ m}^3}\right) = \boxed{750 \text{ K}}$$

   (e) Heat is transferred *to* the gas. At constant pressure, $Q = nC_p \Delta T$. For an ideal monatomic gas, $C_p = \frac{5}{2}R$. To find n, rearrange the ideal gas law:

$$n = \frac{P_2 V_1}{R T_1} = \frac{P_2 V_2}{R T_2}$$

The change in temperature $\Delta T = T_2 - T_1$, so

$$Q_A = nC_p \Delta T = \left(\frac{P_2 V_1}{R T_1}\right)\left(\frac{5}{2}R\right)(T_2 - T_1) = \frac{5}{2}(P_2 V_1)\left(\frac{T_2}{T_1} - 1\right)$$

$$= \frac{5}{2}(5 \times 10^5 \text{ Pa})(0.1 \text{ m}^3)\left(\frac{750 \text{ K}}{300 \text{ K}} - 1\right) = \boxed{1.875 \times 10^5 \text{ J}}$$

44. (a) Process B is an isovolumic decrease in temperature and pressure.

   (b) No work is done during process B because $\Delta V = 0$.

   (c) Heat is transferred out of the gas. At constant volume, $Q = nC_v \Delta T$. For an ideal monatomic gas, $C_v = \frac{3}{2}R$. As in Problem 43,

$$n = P_2 V_1/R T_1 \qquad \text{and} \qquad \Delta T = T_3 - T_2 \qquad \text{where} \qquad T_3 = T_1 = 300 \text{ K}$$

$$Q_B = nC_v \Delta T = \left(\frac{P_2 V_1}{R T_1}\right)\left(\frac{3}{2}R\right)(T_1 - T_2) = \frac{3}{2}(P_2 V_1)\left(1 - \frac{T_2}{T_1}\right)$$

$$= \frac{3}{2}(5 \times 10^5 \text{ Pa})(0.1 \text{ m}^3)\left(1 - \frac{750 \text{ K}}{300 \text{ K}}\right) = \boxed{-1.125 \times 10^5 \text{ J}}$$

The negative sign indicates that heat is transferred out of the gas.

45. (a) Integrate the equation for the work done by an ideal gas.

$$W = \int_{V_i}^{V_f} P\,dV = \int_{V_i}^{V_f} \frac{nRT}{V}\,dV = nRT \int_{V_i}^{V_f} \frac{dV}{V} = nRT \ln\left(\frac{V_f}{V_i}\right)$$

$$W_C = nRT_1 \ln\left(\frac{V_1}{V_2}\right) = \left(\frac{P_2 V_1}{R T_1}\right)(RT_1)\ln\left(\frac{V_1}{V_2}\right) = P_2 V_1 \ln\left(\frac{V_1}{V_2}\right)$$

$$= (5 \times 10^5 \text{ Pa})(0.1 \text{ m}^3)\ln\left(\frac{0.1 \text{ m}^3}{0.25 \text{ m}^3}\right) = \boxed{-4.58 \times 10^4 \text{ J}}$$

   (b) In an isothermal process, the amount of heat transferred is equal to the amount of work done, so $Q_C = W_C = -4.58 \times 10^4 \text{ J}$

46. (a) $$W_{net} = W_A + W_B + W_C = 7.5 \times 10^4 \text{ J} + 0 - 4.58 \times 10^4 \text{ J} = \boxed{2.92 \times 10^4 \text{ J}}$$

   (b) $$Q_{net} = Q_A + Q_B + Q_C = 1.875 \times 10^5 \text{ J} - 1.125 \times 10^5 \text{ J} - 4.58 \times 10^4 \text{ J} = \boxed{2.92 \times 10^4 \text{ J}}$$

47. (a) The efficiency of an engine is the ratio of output work to input heat.

$$e = \frac{W}{Q_\mathrm{h}} = \frac{W_\mathrm{net}}{Q_A} = \frac{2.92 \times 10^4 \text{ J}}{1.875 \times 10^5 \text{ J}} = \boxed{0.156 = 15.6\%}$$

(b)
$$e_C = 1 - \frac{T_\mathrm{c}}{T_\mathrm{h}} = 1 - \frac{300 \text{ K}}{750 \text{ K}} = \boxed{0.6 = 60\%}$$

48. The coefficient of performance is the ratio of output heat to input work — or the ratio of output power to input power.

$$\eta = \frac{Q_\mathrm{c}}{W} = \frac{dQ_\mathrm{c}/dt}{P}$$

$$\frac{dQ_\mathrm{c}}{dt} = \left(\frac{5 \times 10^3 \text{ BTU}}{\mathrm{h}}\right)\left(\frac{1055 \text{ J}}{\mathrm{BTU}}\right)\left(\frac{\mathrm{h}}{3600 \text{ s}}\right) = 1.465 \times 10^3 \text{ W}$$

$$\eta = \frac{dQ_\mathrm{c}/dt}{P} = \frac{1.465 \times 10^3 \text{ W}}{500 \text{ W}} = \boxed{2.9}$$

49. (a) Express the output heat Q_c in terms of the known quantities W and e, and differentiate.

$$e = \frac{W}{Q_\mathrm{h}} \qquad Q_\mathrm{h} = \frac{W}{e}$$

$$e = \frac{Q_\mathrm{h} - Q_\mathrm{c}}{Q_\mathrm{h}} = \frac{W/e - Q_\mathrm{c}}{W/e} \qquad W = \frac{W}{e} - Q_\mathrm{c} \qquad Q_\mathrm{c} = W\left(\frac{1}{e} - 1\right)$$

$$\frac{dQ_\mathrm{c}}{dt} = P\left(\frac{1}{e} - 1\right) = 300 \text{ kW}\left(\frac{1}{0.25} - 1\right) = \boxed{900 \text{ kW}}$$

(b)
$$e_C = 1 - \frac{T_\mathrm{c}}{T_\mathrm{h}} = 1 - \frac{(60 + 273) \text{ K}}{(250 + 273) \text{ K}} = \boxed{0.36 = 36\%}$$

50. First express the definition of heat capacity, $Q = mc\,\Delta T$, in differential form: $dQ = mc\,dT$. In this case $c = C_0 T^2$, so $dQ = mC_0 T^2\,dT$. Now substitute this into the definition of entropy change.

$$\Delta S = \int_{T_1}^{T_2} \frac{dQ}{T} = \int_{T_1}^{T_2} \frac{mC_0 T^2\,dT}{T} = mC_0 \int_{T_1}^{T_2} T\,dT = mC_0\left(\frac{1}{2}T^2\bigg|_{T_1}^{T_2}\right) = \boxed{\frac{mC_0}{2}(T_2^2 - T_1^2)}$$

51. (a) The speed of a transverse wave in a wire is $v = \sqrt{s/\mu}$ where $\mu = m/L = 12 \times 10^{-3} \text{ kg}/2 \text{ m} = 6 \times 10^{-3} \text{ kg m}^{-1}$.

$$v = \sqrt{\frac{5 \times 10^3 \text{ N}}{6 \times 10^{-3} \text{ kg m}^{-1}}} = \boxed{913 \text{ m s}^{-1}}$$

(b) Use the wave formula, $v = \lambda v$, where $\lambda = 2L = 2(2 \text{ m}) = 4 \text{ m}$.

$$v = \frac{v}{\lambda} = \frac{913 \text{ m s}^{-1}}{4 \text{ m}} = \boxed{228 \text{ Hz}}$$

52. For a wave traveling in the positive x direction the wave function is

$$y(x,t) = A \sin(kx - \omega t)$$

where the amplitude $A = 0.3 \text{ cm} = 3 \times 10^{-3} \text{ m}$,

$$k = \frac{2\pi}{\lambda} = \frac{2\pi v}{v} = \frac{2\pi(220 \text{ Hz})}{360 \text{ m s}^{-1}} = 3.84 \text{ m}^{-1}, \qquad \text{and}$$

$$\omega = 2\pi v = 2\pi(220 \text{ Hz}) = 1.382 \times 10^3 \text{ s}^{-1}$$

53. (a)
$$I = uv \qquad u = \frac{I}{v} = \frac{7.4 \times 10^4 \text{ W m}^{-2}}{3 \times 10^8 \text{ m s}^{-1}} = \boxed{2.47 \text{ J m}^{-3}}$$

(b)
$$I = \frac{P}{a} = \frac{P}{\pi d^2/4} \qquad P = \frac{\pi d^2 I}{4} = \frac{\pi(3 \times 10^{-3} \text{ m})^2(7.4 \times 10^4 \text{ W m}^{-2})}{4} = \boxed{0.52 \text{ W}}$$

54. The sum of two identical waves that have a phase difference of ϕ is

$$y_s = 2A \cos\left(\frac{\phi}{2}\right) \sin\left(kx - \omega t + \frac{\phi}{2}\right)$$

The amplitude of the resulting wave y_s is $2A \cos\left(\frac{\phi}{2}\right) = 2A \cos\left(\frac{\phi}{2}\right) = 2A \cos\left(\frac{90°}{2}\right) = \boxed{1.41A}$

55. Solve the equation for the speed of pressure waves in a gas for γ.

$$v = \sqrt{\frac{\gamma R T}{M}} \qquad \gamma = \frac{M v^2}{RT} = \frac{(44 \times 10^{-3} \text{ kg mol}^{-1})(258 \text{ m s}^{-1})^2}{(8.314 \text{ J mol}^{-1} \text{ K}^{-1})(0 + 273) \text{ K}} = \boxed{1.29}$$

56. **(a)**
$$I = \frac{1}{2} \frac{p_0^2}{\rho v} = \frac{(2 \times 10^{-2} \text{ Pa})^2}{2(1.2 \text{ kg m}^{-3})(344 \text{ m s}^{-1})} = \boxed{4.845 \times 10^{-7} \text{ W m}^{-2}}$$

(b)
$$L = 10 \log\left(\frac{I}{I_0}\right) = 10 \log\left(\frac{4.845 \times 10^{-7} \text{ W m}^{-2}}{10^{-12} \text{ W m}^{-2}}\right) = \boxed{56.9 \text{ dB}}$$

57. The acoustical length of the pipe is $\ell = \ell_{\text{actual}} + 0.6r$ where r is the pipe's radius. The acoustical length of a closed-end pipe vibrating in the fundamental mode is one-quarter wavelength. So substitute $\lambda = 4\ell$ into the wave formula $\lambda v = v$ and solve for ℓ.

$$\ell = \frac{\lambda}{4} = \frac{v}{4v} = \frac{345 \text{ m s}^{-1}}{4(440 \text{ Hz})} = 0.196 \text{ m} = 19.6 \text{ cm}$$

$$\ell_{\text{actual}} = \ell - 0.6r = 19.6 \text{ cm} - (0.6)\left(\frac{1.9 \text{ cm}}{2}\right) = \boxed{19.0 \text{ cm}}$$

58. The frequency a detector registers when a source emitting frequency v travels toward it at speed u_s is

$$v_o = v\left(\frac{1}{1 - u_s/v}\right)$$

$$\Delta v = v_o - v = v\left(\frac{1}{1 - u_s/v}\right) - v = v\left(\frac{1}{1 - u_s/v} - 1\right) = \boxed{v\left(\frac{u_s/v}{1 - u_s/v}\right)}$$

APPENDIX A: The International System of Units

The Système International d'Unités, abbreviated SI, was adopted by the Conférence Générale des Poids et Mesures (the General Conference on Weights and Measures) in 1960. It has three types of units: *base*, *supplementary*, and *derived*.

Quantity	Name	Symbol	Unit of measure
Base SI Units			
Length	metre (meter)	m	
Mass	kilogram	kg	
Time	second	s	
Electric current	ampere	A	
Thermodynamic temperature	kelvin	K	
Luminous intensity	candela	cd	
Amount of substance	mole	mol	
Supplementary SI Units			
Plane angle	radian	rad	
Solid angle	steradian	sr	
Derived SI Units			
Frequency	hertz	Hz	$1\ Hz = 1\ s^{-1}$
Force	newton	N	$1\ N = 1\ kg\,m\,s^{-2}$
Pressure, mechanical stress	pascal	Pa	$1\ Pa = 1\ N\,m^{-2}$
Work, energy, quantity of heat	joule	J	$1\ J = 1\ N\,m$
Power	watt	W	$1\ W = 1\ J\,s^{-1}$
Electric charge	coulomb	C	$1\ C = 1\ A\,s$
Electric potential, potential difference, electromotive force	volt	V	$1\ V = 1\ J\,C^{-1}$
Electrical capacitance	farad	F	$1\ F = 1\ C\,V^{-1}$
Electrical resistance	ohm	Ω	$1\ \Omega = 1\ V\,A^{-1}$
Electrical conductance	siemens	S	$1\ S = 1\ \Omega^{-1}$
Magnetic flux	weber	Wb	$1\ Wb = 1\ V\,s$
Magnetic flux density, magnetic induction	tesla	T	$1\ T = 1\ Wb\,m^{-2}$
Inductance	henry	H	$1\ H = 1\ Wb\,A^{-1}$
Luminous flux	lumen	lm	$1\ lm = 1\ cd\,sr$
Illuminance	lux	lx	$1\ lx = 1\ lm\,m^{-2}$

APPENDIX B:
Commonly Used SI Prefixes

Prefix	Abbreviation	Factor
Giga	G	10^9
Mega	M	10^6
Kilo	k	10^3
Deci	d	10^{-1}
Centi	c	10^{-2}
Milli	m	10^{-3}
Micro	μ	10^{-6}
Nano	n	10^{-9}
Pico	p	10^{-12}

APPENDIX C: Constants

Quantity	Symbol	Approximate value
acceleration due to gravity at sea level	g	$9.8 \ \text{m s}^{-1}$
radian	rad	$180°/\pi \approx 57.3°$
gravitational constant	G	$6.672 \times 10^{-11} \ \text{N m}^2 \text{kg}^{-2}$
atmospheric pressure at sea level	P_A	$1.103 \times 10^5 \ \text{Pa}$
speed of sound in air at 20°C		$344 \ \text{m s}^{-1}$
absolute zero		$0 \ \text{K} = -273.15°\text{C}$
Avogadro's number	N_A	$6.022 \times 10^{23} \ \text{molecules mol}^{-1}$
universal gas constant	R	$8.314 \ \text{J mol}^{-1} \text{K}^{-1}$
Boltzmann's constant	k	$1.381 \times 10^{-23} \ \text{J K}^{-1}$
calorie	cal	$4.186 \ \text{J}$
British thermal unit	BTU	$1055 \ \text{J}$
specific heat of water		$4.186 \ \text{kJ kg}^{-1} \text{K}^{-1}$
Stefan–Boltzmann constant	σ	$5.67 \times 10^{-8} \ \text{W m}^{-2} \text{K}^{-4}$
electron charge magnitude	e	$1.602 \times 10^{-19} \ \text{C}$
Coulomb's constant	k	$8.99 \times 10^{-9} \ \text{N m}^2 \text{C}^{-2}$
permittivity of free space	ε_0	$8.854 \times 10^{-12} \ \text{C}^2 \text{N}^{-1} \text{m}^{-2}$
permeability of free space	μ_0	$4\pi \times 10^{-7} \ \text{N A}^{-2}$
speed of light in a vacuum	c	$2.998 \times 10^8 \ \text{m s}^{-1}$
mass of an electron	m_e	$9.11 \times 10^{-31} \ \text{kg}$
mass of a proton	m_p	$1.673 \times 10^{-27} \ \text{kg}$
mass of a neutron	m_n	$1.675 \times 10^{-27} \ \text{kg}$
Planck's constant	h	$6.626 \times 10^{-34} \ \text{J s}$
Rydberg constant	R	$1.097 \times 10^7 \ \text{m}^{-1}$
electron volt	eV	$1.602 \times 10^{-19} \ \text{J}$
unified atomic mass unit	u	$1.661 \times 10^{-27} \ \text{kg} = 931.5 \ \text{MeV} \, c^{-2}$

APPENDIX D: Greek Alphabet

A	α	Alpha	I	ι	Iota	P	ρ	Rho
B	β	Beta	K	κ	Kappa	Σ	σ	Sigma
Γ	γ	Gamma	Λ	λ	Lambda	T	τ	Tau
Δ	δ	Delta	M	μ	Mu	Y	υ	Upsilon
E	ε	Epsilon	N	ν	Nu	Φ	ϕ	Phi
Z	ζ	Zeta	Ξ	ξ	Xi	X	χ	Chi
H	η	Eta	O	o	Omicron	Ψ	ψ	Psi
Θ	θ	Theta	Π	π	Pi	Ω	ω	Omega

INDEX

Acceleration, 15, 21, 26, 28, 35
  angular, 33–34, 135–136, 142
  centripetal, 31
  instantaneous, 15, 28
  total, 32
Adiabatic process, 239
Amplitude, 80, 260
  angular, 84
  velocity, 80
Angular acceleration, 33–34, 135–136, 142
Angular amplitude, 84
Angular displacement, 33
Angular frequency, 80, 87, 261
Angular frequency resonance, 87
Angular momentum, 134, 138, 142
  conservation of, 138, 143
Antinodes, 269
Archimedes' principle, 181, 185
Avogadro's number, 222

Base quantity, 1
Base unit, 1
Beat frequency, 268, 283, 289
Bernoulli's equation, 188, 193
Blackbody, 215
Boltzmann's constant, 222
Boyle's law, 222
British engineering system of units, 1
British thermal unit (BTU), 209
Bulk modulus, 170, 172, 176

Calorie, 209
Candela, 1
Capillary, 182
Carnot cycle, 249
Carnot efficiency, 249
Celsius temperature, 202, 206
Center of mass, 98–99, 107, 115, 121
Charles' law, 222
Circular motion, 31, 38
Collisions, elastic and inelastic, 97, 99, 104, 107
Compressibility, 172
Compression ratio, 247–248
Continuity, equation of, 188, 192
Conversion factor, unity, 2
Conversion relationship, 2
Conversion of units, 2
Cross product, 8, 116
Curl (of a vector field), 67

Deceleration, 15, 17, 21
Density, mass and weight, 178–179, 184
  of ideal gas, 226, 233
Derived quantities, 2
Dewar, 218
Diesel cycle, 246
Displacement, 14, 26, 35
  angular, 33
Doppler effect, 283, 289
Dot product, 8, 58, 62
Dynamics, 44, 51

Eccentricity, 151
Efficiency, 61, 247
  Carnot, 249

of heat engine, 254
Einstein's postulates, 158, 164
Elastic moduli (selected materials), 170
Electron-volt, mega (MeV), 161
Emissivity, radiative, 215
Energy
  conservation of, 70, 74
  density of waves, 264–265
  elastic potential, 69, 74
  equipartition of, 229–234
  gravitational (potential), 68–69, 73, 150, 156
  kinetic, 59, 62, 132, 141
  potential, 67
  mechanical, 70
  relativistic, 161
  rotational kinetic, 132, 141
Engine, heat, 246–247, 254
Entropy, 251, 257
Equilibrium
  condition for rigid bodies, 117, 123
  first condition of, 43, 51
  rotational and translational, 117
Equipartition of energy, 229, 234
Equivalence, principle of, 147
Escape velocity, 151
Expansion ratio (Diesel engines), 248

Fahrenheit temperature, 202, 206
Flow, laminar and turbulent, 191
Force
  centrifugal, 45, 48–49
  centripetal, 45
  conservative, 67, 70, 72
  Coriolis, 48–49
  frictional, 49, 70–71
  gravitational, 148, 154
  impulsive, 94–95
  inertial, 47
  moment of a, 116
  normal, 42
  velocity-dependent, 46, 54
Frame of reference
  inertial, 47
  accelerated, 55
Free-body diagram, 43
Freedom, degree of, 229
Free fall, 18, 20
Frequency, 80, 260
  angular, 80, 87, 261
  beat, 268, 283, 289
Friction
  coefficients, 42–43
  kinetic, 42
  static, 43

Galilean relativity, 158
Gas constant, universal, 222
Gravitation, Newton's law of universal, 147, 153
Gravitational constant, 147, 152
Gravitational energy, 150, 156
Gravitational field, 149, 155
Gravitational force, 148, 154
Gravitational mass, 147
Gravity, 18
Gyration, radius of, 129

Harmonic motion
  damped, 85, 91
  forced, 87, 92
  simple, 79, 89
Heat
  capacity, 209
  conduction, 213
  convection, 215
  engine, 246–247, 254
  of fusion, 211–212
  of vaporization, 211–212
  quantity of, 209–217
Hectare, 10
Hertz, 80
Hooke's law, 79, 89
Horsepower, 13, 60, 65

Ideal gas
  equation of state of, 222, 231
  density of, 226
  internal energy of, 225, 236
  law, 222
  pressure of, 224, 232
Impulse, 94, 104
Impulsive force, 94, 95
Inertia, moment of, 129, 131, 140
Inertial mass, 147
Insulation, thermal, 213
Intensity level (sound), 279, 287
Intensity, luminous, 1
Interference of waves, 266, 274
Isentropic process, 252
Isobaric process, 239
Isothermal process, 237–238
Isovolumic process, 238

Joule, 2, 58

Kelvin temperature scale, 203, 207
Kepler's laws of planetary motion, 151
Kinematics, 16, 19, 24, 29

Latent heat, 211–212
Length contraction, 159, 165
Lever arm, 117
Line integral, 62
Lorentz transformation, 159
Luminous intensity, 1

Manometer, 180
Mass, 41
  gravitational and inertial, 147
  relativistic, 161
  rest, 161
Mechanical system, 70
Mega-electron-volt (MeV), 161
Mixtures, method of, 210
Mode, 280–281
Modulation, 268
Molar specific heat, 227–228
Mole, 222
Moment
  arm, 117
  of a force, 116
  of inertia, 129, 131, 140

Momentum
  angular, 134, 138, 141
  conservation of, 95, 104
  linear, 94–95
  relativistic, 161
Musical interval, 280, 288

Newton, 2, 41
Newtonian fluid, 189–190
Newton's law of cooling, 215
Newton's law of universal gravitation, 147, 153
Newton's laws of motion, 41
Newton's rule, 97
Nodes, 269

Oscillator
  mass–spring, 79–80
  simple harmonic, 79
Otto cycle, 246–247

Parallel axis theorem, 129
Pascal, 2, 169
Pascal's principle, 180, 184
Pendulum, simple, 83, 91
Period, 80, 260
Phase
  angle, 80
  constant, 80
  difference, 266–267
Phase change, 211, 217
Planetary motion, Kepler's laws of, 151
Poise, 189
Poiseuille's equation, 189–190, 194
Poisson's ratio, 170–171, 175
Power, 60, 65
Pressure
  absolute, 180
  gauge, 180
  hydrostatic, 172, 180
  ideal gas, 224, 232
  static, 180, 182, 184
Product (vector multiplication)
  cross, 8, 116
  dot, 8, 58, 62
  scalar, 8
  vector, 8, 116
Projectile motion, 29

Quality factor, 87

R value (thermal insulation), 213
Radian, 33
Rankine temperature scale, 203, 207
Refrigerator, 248, 255

Relativistic energy, 161
Relativistic mass, 161
Relativistic momentum, 161, 166
Relativity
  Galilean, 158
  special, 158–163
Resonance, 280, 288
  angular frequency, 87
Restitution, coefficient of, 97
Rest mass, 161
Reversible cycle, 251
Reynolds number, 191, 195
Right-hand rule, 8
Rocket
  equation, 102
  propulsion, 102, 107
Root-mean-square (rms) speed, 225
Rotating coordinate system, 48
Rotational motion, 33, 39

Satellite motion, 151, 156
Scalar
  components, 4
  product, 8
  quantity, 3
Shear modulus, 170–171, 175
Shear stress and strain, 171–172
Shock wave, 290
Sound
  intensity level, 278–279, 287
  pressure amplitude, 279
Specific gravity (SG), 178, 184
Specific heat
  capacity, 209–210, 227, 233
  molar, 227–228
Speed, 14, 225
Spring constant, 79
Standing waves, 269
Stefan–Boltzmann law, 215
Stokes' laws, 190, 195
Stress, 169, 174, 205, 208
Strain, 169, 174
Surface tension, 182
Système International, 1

Temperature scales, Celsius, Fahrenheit, Kelvin
  and Rankine, 202–203, 206–207
Thermal conductivity, 213–214
Thermal expansion, 203–204
Thermal insulation, 213
Thermal stress, 205, 208
Thermodynamics
  first law of, 236–237, 239
  second law of, 250

Thrust, 102
Time dilation, 159, 165
Torque, 116, 122, 135, 142
Triple point of water, 202

Unit
  base, 1
  vector, 5
Units
  British engineering system, 1
  conversion of, 2
  derived, 2
  Système International (SI), 1
Unity conversion factor, 2

Vector
  addition, 3
  components, 4–5
  multiplication, 7, 11
  negative of a, 3
  product, 8
  quantity, 3
  unit, 7
Velocity
  amplitude, 80
  angular, 33–34
  average, 15, 27
  escape, 151
  instantaneous, 15, 27
  relative, 33, 39
  relativistic addition of, 160, 166
  terminal, 196
  transformation, 100
Viscosity, 189, 194

Wave
  equation, 262, 272
  formula, 260
  intensity, 264–265
  interference, 266, 274
  number, 261
  velocity, 263–264, 272, 278
Wavelength, 260
Waves
  energy density of, 264–265
  harmonic, 259, 265
  longitudinal, 259, 264
  shock, 290
  standing, 269
  transverse, 259, 264
Weight, 41
Work, 58, 62, 64, 236–239
Work–energy principle, 59–60, 64

Young's modulus, 170, 174

s Orbitals being filled

d Orbitals being filled

p Orbitals being filled

f Orbitals being filled

Noble gases

Transition elements

Period number = n, the highest occupied electron level

Group	IA ns^1	IIA ns^2	IIIB $(n-1)d^1ns^2$	IVB $(n-1)d^2ns^2$	VB $(n-1)d^3ns^2$	VIB $(n-1)d^4ns^2$	VIIB $(n-1)d^5ns^2$	VIIIB $(n-1)d^6ns^2$	VIIIB $(n-1)d^7ns^2$	VIIIB $(n-1)d^8ns^2$	IB $(n-1)d^{10}ns^1$	IIB $(n-1)d^{10}ns^2$	IIIA ns^2np^1	IVA ns^2np^2	VA ns^2np^3	VIA ns^2np^4	VIIA ns^2np^5	VIIIA ns^2np^6
1	**H** 1 $1s^1$ 1.0079																	**He** 2 $1s^2$ 4.0026
2	**Li** 3 $2s^1$ 6.941	**Be** 4 $2s^2$ 9.01218											**B** 5 $2s^22p^1$ 10.81	**C** 6 $2s^22p^2$ 12.011	**N** 7 $2s^22p^3$ 14.0067	**O** 8 $2s^22p^4$ 15.9994	**F** 9 $2s^22p^5$ 18.9984	**Ne** 10 $2s^22p^6$ 20.179
3	**Na** 11 $3s^1$ 22.9898	**Mg** 12 $3s^2$ 24.305											**Al** 13 $3s^23p^1$ 26.9815	**Si** 14 $3s^23p^2$ 28.086	**P** 15 $3s^23p^3$ 30.9738	**S** 16 $3s^23p^4$ 32.06	**Cl** 17 $3s^23p^5$ 35.453	**Ar** 18 $3s^23p^6$ 39.948
4	**K** 19 $4s^1$ 39.098	**Ca** 20 $4s^2$ 40.08	**Sc** 21 $3d^14s^2$ 44.959	**Ti** 22 $3d^24s^2$ 47.90	**V** 23 $3d^34s^2$ 50.9414	**Cr** 24 $3d^54s^1$ 51.996	**Mn** 25 $3d^54s^2$ 54.938	**Fe** 26 $3d^64s^2$ 55.847	**Co** 27 $3d^74s^2$ 58.9332	**Ni** 28 $3d^84s^2$ 58.70	**Cu** 29 $3d^{10}4s^1$ 63.546	**Zn** 30 $3d^{10}4s^2$ 65.38	**Ga** 31 $4s^24p^1$ 69.72	**Ge** 32 $4s^24p^2$ 72.59	**As** 33 $4s^24p^3$ 74.9216	**Se** 34 $4s^24p^4$ 78.96	**Br** 35 $4s^24p^5$ 79.904	**Kr** 36 $4s^24p^6$ 83.80
5	**Rb** 37 $5s^1$ 85.4678	**Sr** 38 $5s^2$ 87.62	**Y** 39 $4d^15s^2$ 88.9059	**Zr** 40 $4d^25s^2$ 91.22	**Nb** 41 $4d^45s^1$ 92.9064	**Mo** 42 $4d^55s^1$ 95.94	**Tc** 43 $4d^55s^2$ (97)	**Ru** 44 $4d^75s^1$ 101.07	**Rh** 45 $4d^85s^1$ 102.905	**Pd** 46 $4d^{10}$ 106.4	**Ag** 47 $4d^{10}5s^1$ 107.868	**Cd** 48 $4d^{10}5s^2$ 112.40	**In** 49 $5s^25p^1$ 114.82	**Sn** 50 $5s^25p^2$ 118.69	**Sb** 51 $5s^25p^3$ 121.75	**Te** 52 $5s^25p^4$ 127.60	**I** 53 $5s^25p^5$ 126.904	**Xe** 54 $5s^25p^6$ 131.30
6	**Cs** 55 $6s^1$ 132.905	**Ba** 56 $6s^2$ 137.33	**La*** 57 $5d^16s^2$ 138.905	**Hf** 72 $4f^{14}5d^26s^2$ 178.49	**Ta** 73 $5d^36s^2$ 180.948	**W** 74 $5d^46s^2$ 183.85	**Re** 75 $5d^56s^2$ 186.207	**Os** 76 $5d^66s^2$ 190.2	**Ir** 77 $5d^76s^2$ 192.22	**Pt** 78 $5d^96s^1$ 195.09	**Au** 79 $5d^{10}6s^1$ 196.967	**Hg** 80 $5d^{10}6s^2$ 200.59	**Tl** 81 $6s^26p^1$ 204.37	**Pb** 82 $6s^26p^2$ 207.19	**Bi** 83 $6s^26p^3$ 208.980	**Po** 84 $6s^26p^4$ (209)	**At** 85 $6s^26p^5$ (210)	**Rn** 86 $6s^26p^6$ (222)
7	**Fr** 87 $7s^1$ (223)	**Ra** 88 $7s^2$ (226)	**Ac†** 89 $6d^17s^2$ (227)	**Rf** 104 (260)	**Ha** 105 (260)													

Group numbers

* Lanthanides ~ $4f^x5d^{0-1}6s^2$

Ce 58 $4f^15d^16s^2$ 140.12	**Pr** 59 $4f^35d^06s^2$ 140.907	**Nd** 60 $4f^45d^06s^2$ 144.24	**Pm** 61 $4f^55d^06s^2$ (145)	**Sm** 62 $4f^65d^06s^2$ 150.35	**Eu** 63 $4f^75d^06s^2$ 151.96	**Gd** 64 $4f^75d^16s^2$ 157.25	**Tb** 65 $4f^95d^06s^2$ 158.925	**Dy** 66 $4f^{10}5d^06s^2$ 162.50	**Ho** 67 $4f^{11}5d^06s^2$ 164.930	**Er** 68 $4f^{12}5d^06s^2$ 167.26	**Tm** 69 $4f^{13}5d^06s^2$ 168.934	**Yb** 70 $4f^{14}5d^06s^2$ 173.04	**Lu** 71 $4f^{14}5d^16s^2$ 174.97

† Actinides ~ $5f^x6d^{0-1}7s^2$

Th 90 $5f^06d^27s^2$ 232.038	**Pa** 91 $5f^26d^17s^2$ (231)	**U** 92 $5f^36d^17s^2$ 238.03	**Np** 93 $5f^46d^17s^2$ (237)	**Pu** 94 $5f^66d^07s^2$ (244)	**Am** 95 $5f^76d^07s^2$ (243)	**Cm** 96 $5f^76d^17s^2$ (247)	**Bk** 97 $5f^96d^07s^2$ (247)	**Cf** 98 $5f^{10}6d^07s^2$ (251)	**Es** 99 $5f^{11}6d^07s^2$ (254)	**Fm** 100 $5f^{12}6d^07s^2$ (257)	**Md** 101 $5f^{13}6d^07s^2$ (258)	**No** 102 $5f^{14}6d^07s^2$ (255)	**Lr** 103 $5f^{14}6d^17s^2$ (260)